FOREST STAND DYNAMICS

UPDATE EDITION

**Blue Cliff Monastery
3 Mindfulness Road
Pine Bush, NY 12566**

FOREST STAND DYNAMICS

UPDATE EDITION

Chadwick D. Oliver

Professor of Silviculture
College of Forest Resources
University of Washington

Bruce C. Larson

Associate Professor of Forestry
School of Forestry and Environmental Studies
Yale University

John Wiley & Sons, Inc.

ISBN 0-471-13833-9

10 9 8 7

To my father, Hilary Rhodes Oliver;
my mother, Priscilla Trantham Oliver;
and my wife, Fatma Arf Oliver,
from Chad.

To my family, from Bruce.

TABLE OF CONTENTS

PREFACE

This book describes the various growth patterns of forests from a mechanistic point of view. Its purpose is to help silviculturists and forest managers understand and anticipate how forests grow and respond to intentional manipulations and natural disturbances. Demands on the forest have been increasing for timber production, wildlife habitat, water protection, recreation, and protection from fires, insects, and diseases. These demands have created an emphasis on prescribing site-specific treatments for individual stands of trees, rather than treating broad areas uniformly.

The book is a result of discussions with silviculturists in many parts of North America and elsewhere. Foresters in all regions have wanted to know how forests will grow with and without various manipulations so they can prescribe the most appropriate treatments. Informal discussions and seminars have led to short courses for midcareer silviculturists about stand dynamics. Travels to forests in many regions of the continent and world have led us to realize that similar patterns and processes of development are occurring in many places. As we synthesized material on forest development, we were further encouraged to make it available to a wide audience by publishing it as a book.

Strong interest in the 1990 edition of the book has led us to publish this updated version. Knowledge has advanced and expanded in many fields, and we have tried to update this knowledge and expand the references, although they are by no means exhaustive. Interest from abroad in the book and continued evidence that similar patterns and processes are occurring in forests in many places led us to expand the book to include more global references and examples. Recent advances in understanding climate change, multicohort stand development, mixed species stands, pruning, and stand edges has led us to update information in these areas. We are increasingly realizing that many of the values from the forest can not be provided by managing individual stands without regard to the surrounding landscape. Consequently, we have expanded our discussion on landscapes in many places in the book. Because concerns about protecting all species and "sustainability" directly relate to stand dynamics, we have included discussions of the relation of stand and landscape development to population changes of animals and plants. This discussion is different from many, since it regards the species from the perspective of the changing forest environment, rather than viewing the environment from the perspective of the species. We have expanded the discussion of applications to management include such topics as "sustainable development," "ecosystem management," and "adaptive management."

This book serves several purposes:

- It gives the state of knowledge of how forest stands develop, so future silvicultural practices can be done with a higher level of expertise.

- It synthesizes knowledge on the subject, to allow future research efforts to build on this knowledge.

- It shows that similar development patterns are occurring in many forests in North America and beyond.

- It puts a scientific perspective on the various ways society is considering relating to forests.

Recent appreciation of the role of disturbances and the continued responses of stands to natural disturbances has brought silvicultural and ecological research closer together. Many individual pieces of knowledge reported in this book are not new; however, no book has previously synthesized the information from physiology, ecology, and silviculture and compared patterns in different regions as is done here. We have intentionally given citations from a broad range of new and old applied forestry and basic ecology literature. The references show that many ideas have existed in forestry literature for many decades but have not been synthesized. Some of these ideas are more recently being reported in ecological literature.

The book emphasizes the constant change of all forests and shows that similar patterns of change are occurring within different regions. This book first gives a philosophical perspective from which to view forests—competition, the "niche," and "growing space." It then describes individual tree growth patterns. Disturbances are then described, not as black boxes, but as specific events with different characteristics and degrees of predictability The rest of the book builds on individual tree growth and disturbance patterns to show how forest physiognomies change in time both qualitatively and quantitatively. "Even-aged" development of stands beginning after stand-replacing disturbances are first described. Then, "uneven-aged" stands with new "cohorts" beginning after minor disturbances are described. Finally, the broad perspective of forest changes over landscapes is discussed.

The intent of the book is to describe patterns and to include and reconcile as many other perspectives of forest development as possible. In several places we suggest new terminology and/or refrain from using old terminology. The intent is not to establish new laws or to proliferate jargon, but to be more accurate in describing patterns. Time will determine if the new words are useful enough to last. Some statements in the book have not been referenced. These statements are generally based on personal observations or those of reliable colleagues. We felt it was important to raise certain issues to be scrutinized rather than err in omission.

The book is meant to give a biological basis for more applied books such as *The Practice of Silviculture* (Smith, 1986) and *Principles of Silviculture* (Daniel et al., 1979). On the other hand, it covers more of the patterns of stand development but delves less into basic physiology and ecology than do *Forest Ecology* (Spurr and Barnes, 1980), *Forest Ecology* (Kimmins, 1987), and *Forest Ecosystems: Concepts and Management* (Waring and Schlesinger, 1985).

Many people have contributed to this book. We are grateful to our students (graduate and undergraduate) who have provided ideas and sounding boards at Harvard University, University of Washington, Duke University, Yale University, and Middle East Technical University (Turkey). Many midcareer silviculturists around the country have given helpful input in midcareer short courses on the subject. Our scientific colleagues in North America and abroad have also been very helpful. We are especially grateful to Mr. E. C. Burkhardt of Vicksburg, Mississippi, for his enthusiasm and encouragement, which made us realize the importance of this subject; without his encouragement this book would not have been written. David M. Smith gets our special thanks for his encouragement and emphasis on studying stand dynamics. The encouragement of Walter Knapp of the U.S. Forest Service, Portland, Oregon, was also instrumental in our writing this book. Ernest Gould, Walter Lyford, and Hugh Raup will always be remembered by us for the education and encouragement we received from them at the Harvard Forest; many of their ideas are in this book. Kenneth J. Mitchell provided many of the ideas on forest growth through his careful research, for which we are grateful. We are also thankful for the careful reviews and many suggestions of Graeme Berlyn, Melih Boydak, Mary Ann Fajvan, Douglas Fredericks, Glenn Galloway, Judy Greenleaf, John Helms, John Kershaw, Karen Kuers, David Larsen, Jim McCarter, David Marquis, Xiandong Meng, Michael G. Messina, Melinda Moeur, Kevin O'Hara, Akira Osawa, Reza Pezeshki, Gerardo Segura, Fiorenzo Ugolini, Robert van Pelt, Daniel Vogt, and Kristiina Vogt. We appreciate the help and thoughtfulness of Mrs. Carol Green and Ms. Carol Bruun of the University of Washington Forest Resources Library, and of Mrs. Dorothea Kewley,

Ms. Sherry Stout, and Ms. Kaija Berleman of the University of Washington College of Forest Resources.

We would like to express our appreciation to Kevin O'Hara, Linda Brubaker, Scot Zens, and Ann Camp for specific inputs to updating this book. Many undergraduate, graduate, and mid-career stuents as well as scientific and professional colleagues have also given us ideas and input, which we appreciate. We also would like to thank Eric Stano of John Wiley & Sons for his enthusiasm and help during this updating, Bernice Heller for preparing the index, and Jack Hilton Cunningham for his chapter opener computer-generated art, page design and desktop publishing expertise.

The senior author would like to express his gratitude to his father, Hilary Rhodes Oliver, for giving him the silvicultural curiosity and perspective of this book; his mother, Priscilla Trantham Oliver, for her emphasis on creativity; his wife, Fatma Arf Oliver, for her strong encouragement, scientific objectivity, and moral support; and his children, Elif, Doa, Chadwick, and Renin, for their patience and understanding while the book was being written. He also appreciates the strong encouragement of Professor Cahit Arf and Mrs. Halide Arf. He wishes to thank Professors Charles Cheston, Henry Smith, and Charles Baird of the University of the South, Sewanee, Tennessee for their early concern, encouragement, and teaching.

The second author wishes to express his gratitude to his wife, Julie Larson, for her patience and understanding and to his children, Katherine and Heather, for their sharing their father with this manuscript. He also wishes to express his appreciation to Prof. Martin H. Zimmerman, formerly of the Harvard Forest, for his early inspiration which led to the study of silviculture.

Chadwick D. Oliver
Bruce C. Larson

CHAPTER 1

INTRODUCTION

Forest stand dynamics is the study of changes in forest stand structure with time, including stand behavior during and after disturbances. A *stand* is a spatially continuous group of trees and associated vegetation having similar structures and growing under similar soil and climatic conditions. *Stand structure* is the physical and temporal distribution of trees and other plants in a stand. The distribution can be described by species; by vertical or horizontal spatial patterns; by size of living and/or dead plants or their parts, including the crown volume, leaf area, stem, stem cross section, and others; by plants ages; or by combinations of the above. *Stand development* is the part of stand dynamics concerned with changes in stand structure over time. Stand *growth* or *productivity* refers to changes in total volumes of all plants or parts of plants. *Yield* refers to the standing volume of all or parts of the plants at a given time. *Site* refers to an area's potential for tree growth; site usually incorporates an area's soil and climate conditions. (These definitions are modified from Ford-Robertson, 1971.)

Disturbances are any relatively discrete events that disrupt the stand structure and/or change resource availability or the physical environment (Pickett and White, 1985). Both non-human and human-caused disturbances are considered, since tree response does not distinguish between non-human and human activities.

The study of stand dynamics uses as a foundation the power of observation and the knowledge of autecology, plant physiology, morphology, anatomy, and environmental influences. From these foundations it explores interactions within a stand. These interactions are the basis for both succession perspectives of many trophic levels (Whittaker, 1953; Odum, 1969; Waring and Schlesinger, 1985; and Kimmins, 1987) and silvicultural manipulations to produce specific stand growth patterns for specific objectives.

Studies of stand structure changes have been done with forests (Busgen and Munch, 1929; Assmann, 1970; Oliver and Larson, 1990) and other plant communities (Harper, 1977). Most attention to stand structure and development has been given to forests for several reasons:

1. Trees are large compared with grasses, herbs, and shrubs; therefore, differences in structure are more easily noticed.
2. Trees are perennial; tree structures build on themselves for many years, causing extensive and long-lasting variations in stand structures.
3. The values of forests for consumptive and nonconsumptive uses depend largely on their structures and species compositions; there is an economic incentive to study the changes in forest structure and species composition.

Patterns of forest stand development have been studied for many years. German silviculturists referred to the forest development patterns in their early writings (Busgen and Munch, 1929; Assmann, 1970). Clements (1916) and Gleason (1926) described changes following various disturbances. Forest mensuration studies have often assumed certain patterns of forest development when estimating growth (Meyer et al., 1952). Comparatively recently, the importance of disturbances on forest stand dynamics has been emphasized (Raup, 1964; Heinselman, 1970a,b; Loucks, 1970). The impacts of disturbances on forest communities are now being intensively addressed (White, 1979; Bormann and Likens, 1981; Oliver, 1981; Pickett and White, 1985).

The "Niche" of Stand Dynamics

Forest stand dynamics has a niche in the general hierarchy of scientific disciplines. If one imagines a very crude, abstract hierarchy of scientific disciplines from the most fundamental—mathematics—to disciplines derived from a combination of the fundamental ones and empiricism, a small portion of the hierarchy would appear as in Fig. 1.1. The observations of each discipline can be explained by (i.e., are caused by) processes on the immediately more fundamental level. Stand development patterns are based on tree physiology, soils, and micrometeorology. Similarly, physiological patterns can be explained at the biochemical level, which in turn can be explained at the chemical level, which can be explained at the subatomic physics level, which can be explained at the mathematics level. Appreciating degrees of causality is important in defining a discipline.

Attempts to explain successional patterns or ecosystem behaviors based on physiological causes have been generally cumbersome and unsuccessful. The physiology and ecosystem studies are separated by two hierarchical levels (Fig. 1.1); stand dynamics studies lie between them. Physiological processes act on individual trees. The trees, in turn, interact with other trees and create particular stand development patterns. Stand

development proceeds along variable paths which lead to limited numbers of successional patterns, ecosystem behaviors, and biomass productivity pathways.

DERIVED DISCIPLINES
(INTEGRATION)

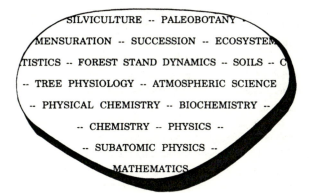

SILVICULTURE -- PALEOBOTANY --

MENSURATION -- SUCCESSION -- ECOSYSTEM

TISTICS -- FOREST STAND DYNAMICS -- SOILS -- C

-- TREE PHYSIOLOGY -- ATMOSPHERIC SCIENCE

-- PHYSICAL CHEMISTRY -- BIOCHEMISTRY --

-- CHEMISTRY -- PHYSICS --

-- SUBATOMIC PHYSICS --

MATHEMATICS

FUNDAMENTAL DISCIPLINES

(CAUSALITY)

Figure 1.1 The relative position of forest stand dynamics in a portion of an abstract hierarchy of scientific and technical disciplines; the hierarchy ranges from more fundamental, at the bottom, to more derived, at the top (*Oliver, 1982*). (See "source notes.")

Understanding stand dynamics has helped forest managers and scientists appreciate the role of natural disturbances in forest development, understand the biological basis of concerns about clearcutting (Marquis, 1973; Minckler, 1973), and separate the effects of natural forest processes from human influences such as air pollution (Johnson and Siccama, 1983; Plochman et al., 1983; Fritts and Swetnam, 1986; McLaughlin et al., 1986; Cook et al., 1987; Knight, 1987; Sheffield and Cost, 1987).

Stand dynamics is useful to more fundamental disciplines such as tree physiology and soil science (Fig. 1.1), where stand development patterns act as tests to these disciplines. For example, hypotheses about the physiological behavior of a tree species can be tested by observing the trees, interactions during stand development.

Understanding stand dynamics helps focus management objectives by predicting future stand structures and development patterns; reducing silvicultural costs; increasing stand productivity; obtaining management information rapidly and inexpensively; obtaining desirable species compositions or physiognomies; developing diagnostic criteria for determining, prescribing, and making decisions for silvicultural operations; and

enabling stands to be used as a general environmental "bioassay" for determining when stands are growing abnormally—perhaps as a result of pollution or other factors. Understanding stand dynamics and the relation of stands to the landscape also helps management provide values which can be best achieved at the landsape scale (Oliver. 1992a,b,c; Boyce and McNab, 1994; Boyce, 1995).

Historical Perspective

Many insights into forest development originated from forest ecology. Consequently, many of the helpful perspectives—as well as the misleading ones—have this origin. Like any scientific discipline, ecology is more than simply a collection of facts. It is also a way of assembling these facts into a logical construct. This way of thinking has changed with time as new facts have appeared but has built on the old ways of thinking. The perspective has prejudiced what research has been done, what information has been accepted, and what terminology has been used.

Anyone researching or applying a scientific discipline must be careful that the way of thinking is not erroneous, thereby causing incorrect scientific or management interpretations. Understanding the influences surrounding the development of ecological thought helps one be alert for possible errors in thought. Some errors are still latent in the discipline, while others have been identified and corrected. Silviculturists often have no opportunity to remain current on ecological thinking, and they may rely on the knowledge gained several decades ago. Since then, however, the way of thinking about ecology has changed. An understanding of the evolution of ecological thinking may help put the present concepts in better perspective.

Ecology emerged in the late 1800s when scientific inquiry was shifting from describing objects to examining processes and relationships. Many fields—geology, genetics, sociology, political science, physics, mathematics, psychology, and ecology—were influenced by this shift in thinking. The many nuances of this shift have been described elsewhere (A. N. Whitehead, 1925; von Bertalanffy, 1972) and will not be detailed here. Many ecological studies concentrated on patterns and processes of interactions and focused on *systems*. Other disciplines also adopted the systems approach. Systems theory, although excellent in itself, carries with it pitfalls which can lead scientists to regard their system—the ecosystem in the case of ecology (Tansley, 1935)—as being goal-directed, teleological, "a whole greater than the sum of its parts" or an interdependent whole—a "superorganism." Barring human activities, forests were assumed to grow to a "steady state" condition and remain there, with the only disturbances being deaths of individual trees, which were replaced by others.   Even where such perspectives are denied, they still subconsciously pervade parts of ecological thought (Whitehead, 1925; Odum, 1969, 1971; von Bertalanffy, 1972; Wilson, 1976, 1980; McIntosh, 1981).

A single approach to ecology has never been universally accepted. In fact, many ecologists have cautioned against the pitfalls described above (McIntosh, 1981). Nevertheless, sufficient concentration has been directed (perhaps misapplied) toward the systems approach to forest ecosystems as goal-directed, teleological, and interdependent that the "forest-as-a-superorganism" perspective has colored much of ecological theory.

Using elements of the systems approach, researchers confused patterns emerging from interactions among trees with obligatory, directional laws which forest communities must follow. Analogies between observed patterns in plant and animal communities further confused the emerging patterns by comparing animal community behavior with plant community behavior. In some cases, concomitance was considered a substitute for causality (Raup, 1942). Deductive reasoning then led to assumptions that patterns observed in some (often animal) communities apply to other (often plant) communities. Even empirical studies with good data were interpreted in the perspective of directional development toward a universal pattern (Hough and Forbes, 1943).

For several decades, scientific evidence began to question this "stable," "steady-state" perspective in favor of a more "dynamic" one. At the annual meeting of the Ecological Society of America in 1990, scientists acknowledged that enough evidence had accumulated that "the center of mass thinking among ecologists has shifted." (Stevens 1990).

> "...many scientists are forsaking one of the most deeply embedded concepts of ecology: the balance of nature.
>
> "Ecologist have traditionally operated on the assumption that the normal condition of nature is a state of equilibrium, in which organisms compete and coexist in an ecological system whose workings are essentially stable....A forest grows to a beautiful, mature climax stage that becomes its naturally permanent condition.
>
> "The concept of natural equilibrium long ruled ecological research and governed the management of such natural resources as forests and fisheries. It led to the doctrine, popular among conservationists, that nature knows best and that human intervention in it is bad by definition.
>
> "Now an accumulation of evidence has gradually led many ecologists to abandon the concept or declare it irrelevant, and others to alter it drastically...As a consequence...textbooks will have to be rewritten and strategies of conservation and resource management will have to be rethought."
>
> --Stevens, 1990

The interpretation of forest communities as being goal-directed and interdependent has pervaded terminology. "Climax" implies a specific structure, a process, and a "goal." "Succession," "early successional species," "pioneer species," and "climax species" all imply a teleological process as well as describe structures. Referring to plants as having "functions" or "strategies is also "Panglossian" (Gould and Lewontin, 1979; Harper, 1982). Many of these terms have been redefined by various ecologists to fit more current ways of thinking and their personal interpretations of processes. With reinterptetation, the terms no longer serve the scientific use of identifying a unique property or pattern; then the terms hinder the understanding of forest development.

Quite different from examinations of broad ecological laws, mechanistic research on individual tree and forest growth has also been done by forest scientists (e.g., M. Busgen, E. Munch, J. W. Toumey, C. F. Korstian, F. S. Baker, H. J. Lutz, D. M. Smith, and others). This research described behaviors of individual trees and groups of trees rather than trying to develop a more universal framework of plant interaction. It also examined both natural and managed stands and contributed strongly to the basic understanding of how forests develop.

The present understanding of how forests grow comes from the various influences described above and knowledge gained from fields related to forest stand dynamics (Fig.

1.1). The ultimate tests of this knowledge are provided by field foresters whose everyday practices determine if stands respond to manipulations as expected.

Approach of This Book

This book takes a mechanistic approach to forest stand development and response. It will discuss how individual trees behave and how they form stand patterns, using an inductive approach. This approach is reinforced by examination of physiology and other more "fundamental" sciences to determine if the expected response of individual trees is in agreement with how trees are known to grow. At times, extrapolations and interpretations of growth are made without direct evidence; when this is done, they are generally based on personal observations by the authors or upon discussions with other authorities. The mechanistic approach of this book is not necessarily more valid than the systems approaches; however, it offers an alternative which, from the authors' experiences, has been very helpful to those concerned with silviculture and forest ecology.

The above criticisms of past ecological research are not intended to denigrate the researchers' contributions to stand development. The early ecologists made a profound contribution when they defined *processes* instead of *objects* as the focus of study. The present book also must be taken in the perspective of the era in which it is written. Unfortunately, it is much easier to identify the prejudices of another era than those of one's own. It may be accurate to state that about half of what is written in this book is incorrect. Science progresses by later finding errors in previous doctrine. Every attempt has been made to avoid errors in this book; however, with sympathy for the manager who has to make a decision based on whatever knowledge is available, the authors have preferred to add material they feel is correct—err by commission—rather than to omit material. Most errors in the book, we hope, are errors of interpretation and not of fact. As much as possible, this book will avoid terms which either appear ambiguous or imply processes as well as describe patterns. In a few places, new terms will be used to avoid ambiguities of older ones.

Recently, research by forest scientists has become statistically very rigorous, while more fundamental ecological research has searched closely for causality. Many recent ecological and applied forestry studies are verifying observations previously made by practicing foresters and forest scientists. Much can be learned about forest development patterns by combining information from older writings of forest researchers and practitioners and modern studies by applied forest scientists and ecological scientists. The large amount of literature cited in this book is from a variety of sources and together shows a uniformity of processes in the patterns of forest development.

General Applications to Management

Silviculture is the manipulation of forest stands. As such, it is the ultimate application of stand dynamics knowledge, and also its ultimate test. Much knowledge of stand development has been gained by learning which manipulations have and which have not achieved the expected results in practice.

Silviculturists operate on a time and planning scale somewhere between scales of agronomists and geologic engineers. Agronomists plan for and invest heavily in their future resources—the plants. On the other hand, the geologic time scale is so long that geologic engineers primarily concern themselves with extraction of minerals. They do not invest in the minerals' renewal.

Silviculturists plan and invest in future resources, but not as intensively as agronomists because the longer time makes future uses less predictable. In addition, a shortage of forests or timber in North America is unlikely for the next few decades, although certain species and qualities of wood may be scarce (Oliver, 1986b) and catastrophic fires or curtailing of harvesting may make wood scarce (Peres-Garcia, 1993; Sampson and Adams, 1994; Sampson et al. 1994). Silviculturists predict both where and when forests will be suitable for certain uses. Within limits, they can also predict when and where they are at risk of being destroyed by insects, diseases, winds, and fires.   Forest uses, knowledge, management techniques, tools, and policies generally change between the time a tree begins growth and when it is harvested or otherwise dies.

Management intensity of American forests reflects the long time frame, uncertainty, and lack of wood shortage. Of the 21 percent of the United States classified as "commercial forestland" [over 195 million hectares (480 million acres)], only about 8 percent was seeded or planted between 1950 and 1978. The rest of the forests have regenerated naturally after human operations or natural disturbances. Most plantations are in the southeastern or Pacific northwestern United States. Less than 0.6 percent of the total commercial forestland is regenerated as plantations each year even in these most intensively managed regions. More stands have been harvested, but they have regenerated naturally either through planning or by accident. Even with a doubled rate of plantation establishment, the area of plantations established annually would become only about 1 percent of the total commercial forestland in the southeastern and Pacific coastal United States.

Extensive silvicultural practices—such as leaving seed trees and protecting from wildfires, grazing, and erosion—have been used successfully over large areas. Such practices often ensure that forests will grow in the future, but with little preplanning of the species, sizes, or other aspects of stand structure. Silviculturists play a predictive role when they try to estimate future development patterns of such stands.

The objectives of management often change over shorter time intervals than the harvest ages of trees. Wood became a source of pulp for papermaking only about 130 years ago; and now pulp is a dominant use of wood in many regions, while many stands originally managed for firewood and railroad ties have been used for other purposes. Earlier forest protection and management in North America to avoid an "impending timber shortage" has become less of an issue; increasing timber volumes have been grown and harvested in North America over each of the past five decades (MacCleery, 1992; Oliver 1986b)—and more importing of timber and use of substitute products (aluminum, steel, concrete, fossil fuels, etc) is done. Other concerns such as maintaining species' habitats, forest health (resistance from fires and epidemic levels of insects and diseases), aesthetic considerations, and other noncommodity values as well as production of high quality timber are becoming more of an issue(Frankin et al., 1986; Oliver, 1986; Kuehne, 1993; O'Hara et al., 1994). Silviculturists are continually expected to provide for the changed values in existing stands.

The silviculturist's task is often to manipulate *currently* existing stands, which either intentionally or unintentionally became established in a given species composition and spacing because of *historical* reasons, in order to meet some uncertain, *future* objective. Even when an objective is clearly defined, the stands often do not contain the most desirable species or spacings. It is rare, however, that the silviculturist has the luxury of destroying the stand and starting over with an ideal one—even if it were possible to predict the "ideal" stand for the future objective.

Because silviculturists generally deal with far less than perfect stands, it is usually their task to manipulate the stands as efficiently and effectively as possible. These stands exist in varied conditions, and a broad knowledge of stand dynamics is needed rather than simple prescriptions. Imperfect stands can be improved, but not to ideal conditions even with the best of knowledge. With erroneous knowledge, however, it is quite easy to ruin the ability of a stand to meet the landowner's objectives.

Knowledge of stand dynamics is useful when regenerating stands either naturally or artificially. Artificially regenerating a stand is the most expensive of silvicultural operations—especially when the cost is projected through compound interest to the end of the rotation. Such intensive operations are appropriate and efficient from both an economic and an ecological perspective in some places; however, alternative ways to obtain new stands are preferred in other places. Most silvicultural manipulations are mimics of natural disturbances and other processes. A species can be favored in a new stand by artificially creating appropriate disturbances to favor the preferred species' regeneration mechanisms.

It is now appreciated that all stands change in structure with growth, management, and disturbances. Each structure is useful for some values but not others; and different species depend on different structures or combinations of structures across a landscape for their survival. Consequently, no single stand provides all values simultaneously—even if a stand could be prevented from changing through growth and disturbances. For a forest to provide all values, a variety of structures need to be maintained by coordinating the changes among stands. Historically, this coordination of activities among stands has been part of the discipline of Forest Management; however, silviculturists have begun to make advances in this field (Boyce, 1985, 1995; Oliver, 1992c; Boyce and McNab, 1994; O'Hara et al., 1994).

In North America, expansive forests containing stands with particular distributions of structures are important for timber production, recreation, watershed, and extensive wildlife uses. A small stand containing a unique structure may be locally valuable for tourist and aesthetic purposes or for preservation of rare plants and animals; however, these small areas are primarily valuable because of their rarity. Most forests support relatively few recreationists, animals, acre-feet of water, or board feet of timber. Consequently, the value of each hectare (acre) used as forestland is relatively low compared with its value when used, for example, for urban development or agriculture. To be valuable for most uses, forest stands must have structures and development patterns similar to those of many other stands in the vicinity which can also have similar uses. Silviculturists can anticipate future patterns of stand structures over large areas and thus estimate their potential for various uses and manipulations by understanding forest development patterns, land use and disturbance histories, and soil conditions and by knowing which species are present.

CHAPTER 2

PLANT INTERACTIONS AND LIMITATIONS OF GROWTH

Introduction

Tree species are not randomly distributed across a landscape. Each one is commonly found on certain soils, in specific climatic zones, in a broad geographic zone, at predictable times following a disturbance, in particular layers in a forest canopy, and in close proximity to certain other species. While the patterns seem to be predictable, they are not universal. The same area can be occupied at different times by totally different groups of tree species. Alternatively, the same species can be commonly found in two different environments. Species which characteristically occupy a certain site or canopy position can occur elsewhere. Similarly, a species can sometimes grow quite vigorously with other species not usually associated with it.

The patterns are predictable enough that identifying where a species is found has become a basic part of its description. A species may also be found growing in concert with particular other species with enough regularity that the species are often classified together and associated with certain soils, in certain climates, and at certain times following disturbances. These groups of species have usually been named for one or a few representative species, depending on the objective of the classification or the behavior of the group. Characteristic groups of species are used to identify the soil or climatic conditions of an area [e.g., habitat types (Daubenmire and Daubenmire, 1968); life zones (Piper, 1906; Scott, 1962); climax associations (Braun, 1950); climax types (Scott, 1962); biogeoclimatic zones (Krajina, 1969); vegetation zones (Franklin and Dyrness, 1973); and indicator species (Hodgkins, 1970). See also Orr (1965), Daniel et al. (1979), Daubenmire (1952), Korstian (1917), Lemieux (1965), Gleason (1926, 1936), Cajander (1926), Cain (1936), Spurr and Barnes (1980), and Braun (1950).]

The earth can be divided into six geographically distinct groups of plants—"floristic realms" (Walter, 1973). The Holarctic realm occupies the northern hemisphere's temperate and boreal forests. The Neotropic realm occupies central and South America; the Paleotropic realm, Africa and southern Asia; the Australia realm, Australia; the Capensis realm, extreme southwest Africa; and the Antarctic realm, coastal southwest South America and the subantarctic islands. The realms generally contain different angiosperm genera because they evolved in relative isolation (Gentry, 1982; Tallis, 1991), while the gymnosperm families can be found in several realms. For example, the *Cupressaceae* family is found on all continents; the *Podocarpaceae* family and the *Araucaria* genus occurs naturally only in the southern hemisphere; and the *Pinaceae* family and nearly all *Taxodiaceae* are found naturally in only the northern hemisphere.

Some genera are common in certain environments within each realm (Bruenig and Y-w Huang, 1989; Gentry, 1989; Junk, 1989; Ashton, 1989; Geesink and Kornet, 1989). Spruces and firs are generally found in cooler sites and pine and oak species often dominate droughty sites throughout North America, Europe, and Asia—the Holarctic realm. Stand structures can be very similar in geographically isolated places within a realm where similar climates and soils exist because the related species often have similar physiological and morphological processes. Similar stand structures can also be found in different realms where genera are not evolutionarily related or in the same realm among unrelated groups of species (Oliver, 1992).

Mutualism and Competition among Species
Similar patterns of forest development are found in distant forests because similar patterns of interaction occur among trees in most forests. Understanding the interaction among tree species is fundamental to understanding, predicting, and manipulating forest development. Two differing processes of interaction—mutualism and competition—have been used to explain forest development patterns.

Mutualism
Historically, trees within a stand were implicitly or explicitly assumed to be mutually interdependent for their survival and growth (Clements, 1916; McIntosh, 1981; Vandermeer, 1982; Finegan, 1984). Just as a lichen is a mutualistic association of an alga and a fungus which have coevolved so that they cannot live independently, species in a forest were perceived as having coevolved. Observations that species were found in predictable groups, as well as post-Darwinian advances in concepts of evolution and coevolution among animals, the systems approach to the sciences, and assumptions

about correlation between sizes and ages of trees, have reinforced the mutualistic concept (Vandermeer, 1982).

At its extreme, a community of plants was considered a "superorganism." Each plant lived in delicate balance with the others. Community integrity continued only if unaffected by external forces. Interpretations of how stands behave under different outside forces were predicated on the interdependence of the various plants (Clements, 1916; Odum, 1971).

The superorganism idea has influenced many philosophical, political, and ecological ideas of forest development and management. It is important to be aware that many attributes, perspectives, and descriptions of forest development may subconsciously be rooted in the mutualistic concept of forest communities. Such ideas as steady state, constant final yield, habitat types, climax, plant functions and strategies, plant community directions, and ecological concerns about some forest management practices may have this bias. [For more information, see discussions by Braun (1950), Mueller-Dombois and Ellenberg (1974), Richardson (1980), McIntosh (1981), and Kimmins (1987).]

Competition

Most forest scientists now feel that the interaction between tree species usually results in one individual's having an advantage and dominating or killing another. This interaction is termed *competition*. Unlike mutualism, competition implies that trees have not evolved a dependence on each other but actually can impede each other's growth. Competition, not mutualism, is now considered the primary pattern of interaction among holarctic tree species, and probably among trees in other floristic realms.. Trees are not highly specialized plants; they are versatile organisms with many adaptations which allow them to survive in a variety of biological and physical environments. Many of their adaptations may not be exhibited under all growing conditions.

Some evidence that tree interactions are more competitive than mutualistic includes the following:

1. *Behavior of managed and logged stands.* Some forestry and logging practices affect plant communities in far different ways than occur naturally. The plants have responded to these treatments resiliently by regrowing rather than behaving like organisms which were too specialized and interdependent on each other to respond to a completely different environment than they had previously experienced. Forests on fragile soils even regrew vigorously after being clearcut and the regrowth suppressed for 3 years by herbicides (Bormann and Likens, 1981).

2. *Species migrations and longevity of forest communities.* Recent evidence suggests that many apparently stable forest communities are in fact relatively recent assemblages of tree species which did not necessarily live together as communities in the past (Davis, 1981; Brubaker, 1991; Sprugel, 1991). In fact, changes in climates cause interactions among species to differ (Davis, 1986). Consequently, plant interactions observed at any time are unique to the climate as well as the species.)

Evolution occurs by genetic changes in progeny. The degree of coevolution is closely related to the number of generations during which coevolving species have been together. For forest species, a new generation is generally produced when a significant distur-

bance occurs to the forest. Even then, regeneration sometimes occurs by asexual mechanisms. Forest trees therefore produce successful genetic recombinations at very infrequent intervals—much less frequent than the time when a species first becomes fecund.
Few presently identified tree communities have been together long enough for coevolution toward mutualism to have occurred to an appreciable extent. Mixed stands of
Douglas-fir, western hemlock, western redcedar dominate the lower elevation forests of
western Washington, U.S.A.; however, they have existed as a community for only about
6,000 years, and "No compelling evidence...suggests that this forest type existed elsewhere before..." (Brubaker 1991). If a disturbance large enough to initiate a new stand
occurred approximately every 400 years (Agee, 1981; Fahnestock and Agee, 1983),
only 15 generations of these species have lived together in these mountains. These are
about as many generations as have occurred in humans since Columbus came to
America. Similarly, western hemlock has been present in the forests of Idaho and
Montana only about 2,000 years (Mack et al., 1978a,b; Karsian, 1995), but is a major
component of some stands.

Other evidence (Spaulding and Martin, 1979; Van Devender et al., 1979; Van
Devender and Spaulding, 1979) has shown that plant species do not necessarily migrate
together as communities in response to climatic shifts (which are still occurring today;
Lamb, 1977). Apparently stable plant communities often consist of assemblages of
species which migrated from different locations and will probably continue to migrate
quite independently of each other.

3. *Recent introductions and eliminations of species.* Elimination of one species in a
stand should negatively affect other components of the stands if strong mutualistic interdependencies exist among tree species. Species which have been or are being eliminated
by pathogens—such as the American chestnut (Korstian and Stickel, 1927; Braun, 1942;
Woods and Shanks, 1959; Hepting, 1971; Jaynes et al., 1976), the American elm (Parker
and Leopold, 1983), and the western white pine (Miller et al., 1959; Hepting, 1971)—
have been replaced by other species. Often, the remaining species grow better without
the eliminated one. The tree species in each community had not evolved such an interde-
pendency that they could not live without these other species.

The ability of introduced plant species to live within a native plant community—and
not be excluded by the original species—also indicates that mutual interdependencies are
not a strong force in tree community interactions. Japanese honeysuckle is now a common wild species in many communities in the southeastern United States. Douglas-fir
has been growing for many decades in mixed stands with native species in central
Europe. Incursion of nonholarctic genera such as *Eucalyptus* into parts of North
America, Europe, and Asia suggests that plants are readily able to grow with new
species, and therefore do not have a strong dependency on a specific, surrounding set of
plants.

4. *Recent evidence of age distributions and disturbances in forests.* Part of the evidence for mutualism between forest trees has been the assumption that different tree
species became established at progressively later times after a disturbance when the preceding species had created the appropriate environment. This younger tree then grew
and established the appropriate environment for even another species to invade. The vertical stratification of tree species within a forest and the small sizes of the understory

trees was interpreted as the smaller trees' being younger (Meyer and Stevenson, 1943; Phillips, 1959; Mueller-Dombois and Ellenberg, 1974). This interpretation reinforced the concept of a mutualistic, all-aged forest development pattern. More detailed evidence (Fig. 2.1; and Wilson, 1953; Blum, 1961; Gibbs, 1963; Marquis, 1967; Roach, 1977; Oliver, 1980) shows that the understory stems are not necessarily younger; rather, they have been outcompeted by the dominating species and are now growing very slowly.

Mutualism and Competition

Actually, both mutualism and competition probably occur to some extent among tree species, since the two are not mutually exclusive. Species which have not been in close proximity would probably behave more competitively, since no opportunity for coevolution would exist. Plants which have lived together in a stable environment for many generations would probably tend toward coevolution and mutualistic behavior. In this book, forest stand development patterns can be best understood if competition is assumed to be the primary interaction.

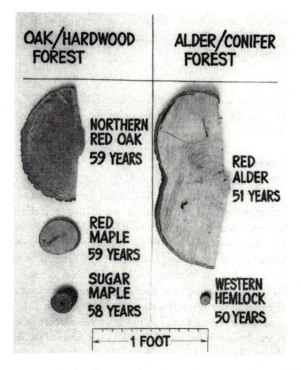

Figure 2.1 When many species invade at one time after a disturbance, some outcompete others and grow larger. Relative tree size does not indicate age, and trees of greatly different sizes can be the same age, as shown by these tree sections taken at 1.4 m (4.5 ft) in mixed-species stands (*Oliver, 1980*). Shown are oak/hardwoods from northern Connecticut and alder/conifers from western Washington. (*See "source notes."*)

Where Plants Are Found

Tree species are not predestined or obligated to grow only on certain sites or with certain other species. They are found where their seeds (or stumps or roots, in the case of sprouting species) are present and where they can survive and compete successfully with the surrounding vegetation. Local conditions or geographic barriers can prevent the presence of a plant. For example, an area may not contain a species, even if all other conditions are appropriate, if the area is upwind from seed sources and the species regenerates from seeds dispersed by wind. Similarly, distant regions may contain appropriate climates, soils, and other conditions for a species to grow; but the species may not have migrated to this area because of oceans, mountain ranges, or other obstacles.

Whether a tree can survive a given physical environment depends on the species' physiological characteristics and the characteristics of the environment. Whether a tree can compete successfully depends on the attributes of the species, its competitors, and the environment.

Physiological characteristics are continually being researched; however, studies of forest development generally do not wait for a resolution of all physiological processes to define important relations among trees. Empirical evidence of where each species can and cannot survive is accepted, and physiological evidence is used to refine the knowledge as it becomes available.

The Niche

Describing where a species can survive and compete under the many influences of the environment can be conceptually systematized under the framework of environmental gradients and the niche.

The gradient of soils

The environment varies in many ways. For simplicity, only one component—the soils—will be treated at first. Soil variations and their influence on forest growth have been widely studied. Consequently, soil influences will be alluded to in this book but rarely described in detail. Soil influences can often be inferred from the geomorphologic and climatic conditions. The geomorphology, of course, determines the potential minerals, particle sizes, chemical composition, and landscape of an area. The climate strongly governs the rate of weathering and leaching, and hence nutrient availability, soil texture, and organic material. Understanding how these interact with the vegetation allows one to anticipate the responses of the different trees to the soil condition.

Two contrasting soils will first be used to illustrate the concept of competition and shed light on where species are found and why. Two large areas of the different soils are shown schematically in Fig. 2.2. All other environmental factors are equal. One area contains a droughty, sandy soil of poor productivity. The other area contains a highly productive, moist, well-drained, sandy clay loam soil. If a species (tulip poplar, for example) is planted in both areas, it will grow best in the moist soil. If another species (Virginia pine, for example) is planted in both areas instead, it will also grow best in the moist soil. Both species grow best on the productive soil; however, pines do not grow as well as tulip poplars on the productive soil, but they grow better than tulip poplars on the less productive soil (Doolittle, 1958; Broadfoot, 1970).

If both species are now planted together in each area and the most vigorous species in each area outcompetes and suppresses the other, two outcomes will occur. Tulip poplars will be found on the best site, and pines on the poorest. Pines will be found on the poorest site *not* because they grow best there, but because they can outcompete tulip poplars there.

As a general rule tree species can grow best in moist, well-drained, sandy clay loam soils; however, species tend to grow where they can compete successfully, *not* where they can grow best.

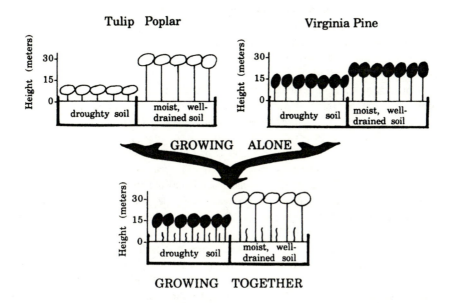

Figure 2.2 Most species are not found naturally where they grow best; rather, they are found where they can compete successfully. A schematic relation of growth and competition of two species is shown. Both species grow best on moist, well-drained soil; however, Virginia pines are found on "droughty soil" when both species grow together because they outgrow tulip poplars there (*Doolittle, 1958*). (Here, height growth is assumed to represent ability to compete.)

A species which grows rapidly on productive sites but poorly on unproductive sites is termed *site-sensitive*. Site sensitivity is relative. A species may be considered site-sensitive compared with one species and insensitive compared with another. When tulip poplars and Virginia pines are compared, tulip poplars are considered site-sensitive and Virginia pines are considered site-insensitive. As a rule, hardwoods—with tulip poplars, cottonwoods, and sweetgums as extremes—are quite site-sensitive and conifers are generally less site-sensitive.

The above example implied that the species which attained the greatest height would be the most competitive and dominate the site. Other factors of spacing, regeneration mechanism, and disturbance type are also important in determining dominance. Competitiveness of a tree on a site does not necessarily mean greatest biological productivity [greatest volume per area (hectare or acre) per year]. The most competitive trees on a site sometimes produce less biomass than less competitive ones would have, had they been allowed to grow. Pine and other conifer stands often produce much more biomass (especially when averaged over several years) than hardwood stands on productive sites as well as on poor ones (Frederick and Coffman, 1978). Conifers, however, are often outcompeted by hardwoods on good sites and so are more often found on poor sites.

Silvicultural activities of site preparation and weed control in conifer plantations are intended to give the more productive conifers an initial competitive advantage on sites where hardwoods would otherwise outcompete them. Less silvicultural weed control is necessary to establish conifer stands on poor sites than on more productive sites. On very productive sites, expensive efforts are required to reduce the hardwood competition; and silviculturists often allow hardwood stands to grow even though more volume could be produced from conifers.

A gradient of soil environments exists, instead of just two discrete types. In fact, there is an almost infinite series of soil gradients such as soil moisture, pH, cation exchange capacity, and others (Coile, 1952); however, only a single, generalized gradient of soil moisture will be considered first.

Figure 2.3 shows four schematic gradients of soil moisture ranging from droughty, sandy soils (xeric) through moist, well-drained soils (mesic) to overly moist (hydric) soils (similair to Whittaker, 1965). Available soild moisture often limits growth on xeric sites, while soil oxygen limits growth on hydric sites. Within each general forest type, tree species are usually found at specific places along this gradient. (Simplifications are made here for illustrative purposes.)

In all four examples (Fig. 2.3), all species grow best on the mesic soils—the moist, well-drained ones. Some species are very site-sensitive—e.g., yellow poplar in the eastern upland forest, sweetgum in the southeastern Coastal Plain forests, grand fir in the northern Rocky Mountains, and red alder on the Pacific Coast. They decline dramatically in vigor as soil conditions become worse. Some species cannot even survive under some soil conditions.

A few species can tolerate both very dry, and very wet soils—Atlantic white cedar in New England (not shown in Fig. 2.3) and lodgepole pine in the Pacific northwestern United States, as examples. These trees often live under physiological drought conditions on both sites, since there is no water on the xeric site and the roots cannot live in the anaerobic soil on the wet site to obtain the water.

When many species grow together, each species predominates along that portion of the gradient where it can compete best. The result is a general pattern to the vegetation. Each species could grow on a much wider range of sites; however, each is found where it can compete successfully with the other species present. If one species were eliminated—such as western white pine from the Pacific northwestern region—the next mostly competitive species would predominate—lodgepole pine on the drier sites and Douglas-fir on the moister sites.

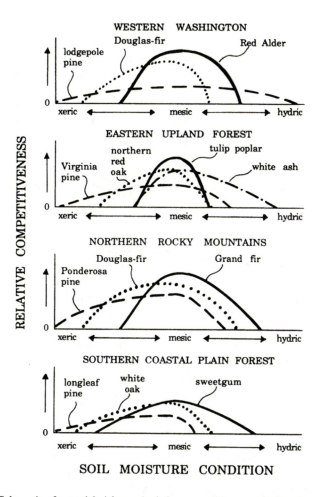

Figure 2.3 Schematic of potential niches and relative competitiveness of selected species growing together in four forest types. All species grow most vigorously on mesic sites, but each is found in its "realized niche" where it outcompetes other species—or where other species cannot grow at all.

In some cases (eastern uplands, northern Rocky Mountain, and southeastern Coastal Plain forests, Fig. 2.3), each species has the competitive advantage in one continuous segment along the soil moisture gradient. Consequently, the number of situations which allow a different species to dominate equals the number of species. Where a species has a competitive advantage in more than one segment of an environmental gradient (e.g., lodgepole pine in western Washington), the number of segments in which one species dominates exceeds the number of species.

Very few species can survive on the extreme sites, and sometimes a species is found on these sites simply because no other species can live there. Presence of a species on these site extremes can generally be correlated with soil productivity above the minimum level necessary for the plant to exist. Where many species compete on mesic sites, the dominant one may have gained the competitive advantage from gradients other than soil conditions. Identifying sites using plant indicators—*vegetation typing*—is generally more advanced and more accurate in environmental extremes—such as in the droughty Rocky Mountains or at upper elevations. There, the few species indicate the limits of different plants' abilities to survive on certain soils.

Where each species can grow along the soil gradient will be defined in this book as its soil *fundamental niche* (Hutchinson, 1957; Kimmins, 1987). In fact, this fundamental niche would be defined by an almost infinitely large number of subtle gradients of pH, texture, nutrient availability, and other soil conditions; however, for conceptual purposes, it will be regarded as a single gradient.

A plant's competitive vigor varies within its niche; therefore, it may rarely be able to compete within parts of its fundamental niche. Where a species is actually found is its *realized niche* (Kimmins, 1987). The realized niche of each species is flexible and depends on the species with which it grows on each site.

Two or more species can have nearly identical vigor in many places along a gradient. [For example, Virginia pine and red oak have nearly identical vigor on moderately xeric sites (Fig. 2.3).] Environmental conditions which allow two or more species to compete equally well at places along a gradient will be referred to as *transition conditions*. Other factors—or gradients—may give different species a competitive advantage at transition conditions.

Other gradients besides soil exist in the environment, and tree species vary in their abilities to survive and compete along them as well. They are as complex as the soil gradient, but will also be simplified for this discussion. Climatic variations will be considered next.

The gradient of climate

Climate variations and their influences on forest growth have also been widely studied, and so will not be described in detail in this book. Climatic patterns are somewhat predictable based on the physics of the circulating earth (MacArthur, 1972), and similar climates can be found in many different parts of the earth.

Temperature and moisture regimes exist in a gradient within each climatic zone. Like the soil gradient, climate contains an almost infinite number of gradients in temperature, humidity, wind speed, and other variations throughout the day and year; however, it is considered a single gradient for this discussion. The vigor and competitiveness of each tree species vary with temperature and moisture conditions, just as they do with soil conditions (Woodward, 1987). Consequently, species' niches are generally found along a climatic gradient (Daubenmire, 1966).

The climate generally becomes progressively cooler and moister as one moves either toward the poles or to higher elevations. Consequently, the same species are often found at high elevations in low latitudes and at low elevations in high latitudes. An elevation change of 300 m (1000 ft) is roughly equivalent to 160 km (100 mi) of latitude in similar climatic zones (*Hopkins Bioclimatic Law*; MacArthur, 1972).

Unlike the situation with soil conditions, there does not seem to be an optimum climatic condition for all species. This lack of an optimum climate is because "climate" integrates several environmental influences on plants. Most plants do have a similar optimum temperature range for photosynthesis (Kozlowski et al., 1991); however, different climates require plants to withstand different extremes and durations of heat and cold, as well as different triggering mechanisms for growth (such as day length and degree-days). In fact, the climatic niches of some species do not overlap at all, and it is difficult even to keep some species alive in the same climate under the very artificial conditions of an arboretum.

Three-dimensional niche

The interrelation of two gradients such as climate and soil can be conceptualized as a three-dimensional niche. Figure 2.4 shows the conceptual three-dimensional niches and the relative vigors of Douglas-firs, red alders, and lodgepole pines in western Washington within each niche. Lodgepole pines can grow on many sites in western Washington but are usually found in very moist areas, on very droughty glacial outwash soils, or at very high elevations—where other species do not survive. The more mesic sites are occupied by the more competitive Douglas-firs and alders. As this example illustrates, there is no single set of environmental variables which defines where a species can be found; it can be found under several environmental conditions depending on its behavior and that of its competitors.

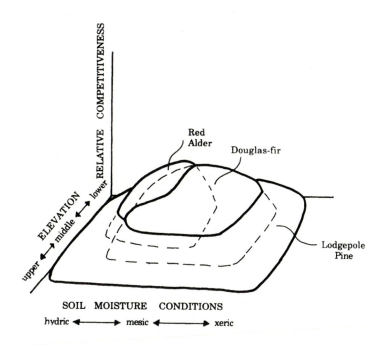

Figure 2.4 When several species' growth potentials are plotted on many gradients, each species' potential and realized niches are described. Niches and relative competitiveness of Douglas-fir, red alder, and lodgepole pine in western Washington distributed along the gradients of soil moisture and elevation are shown.

If three species exist along two environmental gradients and each species is most vigorous along one segment of each gradient, a total of 3 x 3 = 9 different combinations of
plant interactions can exist. Each plant would be dominant in an average of three interactions. In the above example, lodgepole pine is most vigorous along two segments of
the soils gradient; therefore, a total of 3 (climate) x 4 (soil) = 12 different combinations
of situations can exist where a different species can gain a competitive advantage. It can
be seen that the distribution of trees cannot be explained by a single set of environmental
conditions. The same species can exist under quite different conditions for quite different reasons.

While forest patterns are quite complicated, they are not random because given interactions do have predictable outcomes *provided all environmental factors are known.*
Often, groups of species are found together under similar soil and climate conditions
although the species may have different niches in other gradients. These groups of
species have been described as *communities or forest zones* and described according to
various environmental gradients (Franklin and Dyrness,1973).

Multidimensional niche

Other factors besides soil and climate influence where and how vigorously a species can
grow. These other factors also exist as gradients. An increasingly complex multidimen-
sional niche can be found by combining all gradients into an abstract "hypervolume"
containing more dimensions than the two shown in Fig. 2.4 (Hutchinson, 1957). The
survival of a plant will depend on whether its vigor at a given place along all combined
gradients allows it to compete successfully with other plants which happen to be present.

This concept of the niche is useful for estimating which factors influence plant growth
significantly, for systematizing the influences, and for understanding the complexity of
interactions among factors.

The complexity of forest patterns increases dramatically both with the number of
species and with the complexity of the environment as one goes from relatively simple
stands of few species, such as in artificial plantations or at polar latitudes, to stands of
many species at lower latitudes. For example, with 2 species and 4 gradients in which
each species had a competitive advantage along one segment of each gradient, there
would be a total of $2^4 = 16$ competitive situations in which one species could have an
advantage. Each species would dominate an average of 8 competitive situations. If the
number of species were doubled, a total of $4^4 = 256$ different situations, each species
would dominate in an average of 64 different situations.

Other gradients

Although many gradients exist, two other factors in addition to soils and climate will be
discussed as gradients in this book. These factors are disturbances and time since disturbance. Like soil and climate, each could also be subdivided into an almost infinite number of subunits. Disturbances are covered in detail in Chap. 4. Time since disturbance is
covered in Chaps. 5 through 14.

Limitations to Growth

Trees growing in suitable climates and soils will increase in size and number until one or more factors necessary for growth are no longer available. Plant growth is unlike growth of many animal populations in which behavior patterns control the size and number of individuals before the physical resources are depleted.

Which factor becomes limiting to plants varies. The degree to which each species slows in growth as each factor becomes limiting varies also, although the growth variation is usually not great. Oversimplified generalizations are often made about what is limiting growth. Some sites are described as "droughty," or "moisture limiting"; light is considered limiting in shaded conditions; nutrients are considered limiting on some sites. Such oversimplifications are sometimes useful if one appreciates that growth limitations are in fact more complicated.

Factors necessary for tree growth, and consequently those which can become limiting, can be inferred from the oversimplified photosynthetic and respiration equations:

$$\text{Photosynthesis: } 6CO_2 + 6H_2O + \text{sunlight energy} \rightarrow C_6H_{12}O_6 + 6O_2$$

$$\text{Respiration: } 6C_6H_{12}O_6 + 6O_2 \rightarrow \text{energy} + 6CO_2 + 6H_2O$$

These two equations express opposite chemical reactions but are not "mass balance" equations per se which exist in equilibrium. The photosynthetic reaction produces organic sugars and releases oxygen by using carbon dioxide from the atmosphere, water from the soil, and energy from sunlight. Photosynthesis occurs within a certain range of temperatures and requires mineral elements as catalysts. The sugars then either are converted to other organic compounds and used as structural or functional components of trees or are used as an energy source for plant survival and growth, resistance to insect and disease attacks, transport of substances, or conversion of organic molecules from one form to another.

Energy is obtained from the compounds produced by photosynthesis through the respiration reaction. Here, oxygen from the atmosphere (often the soil atmosphere for root respiration) is necessary, as are minimum temperatures, certain nutrients, and the sugars produced by photosynthesis. Unlike the situation in photosynthesis, in respiration the release of energy from organic compounds is done by plants which do not photosynthesize (fungi and bacteria) and by animals, as well as by photosynthesizing plants. The release of energy can also occur abiotically when a fire burns and follows the same reaction. Without oxygen some bacteria carry out anaerobic respiration with some organic molecules through a different, two-step reaction and produce methane (CH_4) and water:

$$\text{Step 1: } CH_3COOH + 2H_2O \rightarrow 2CO_2 + 8H^+$$

$$\text{Step 2: } CO_2 + 8H^+ \rightarrow CH_4 + 2H_2O$$

The factors necessary for tree growth, therefore, are sunlight, water, certain mineral nutrients, suitable temperatures, oxygen, and carbon dioxide. Each factor must be available to the tree in the appropriate form and location. Their physiological relation to tree growth has been discussed in detail elsewhere (e.g., Salisbury and Ross, 1969; Leopold and Kriedemann, 1975; Kramer and Kozlowski, 1979; Larcher, 1983; Chabot and Mooney, 1985; Kozlowski et al., 1991; Milner and Coble, 1995) and will be discussed only briefly here.

Sunlight

Trees generally photosynthesize increasingly faster in bright light up to full sunlight (Fig. 2.5; Boysen-Jensen, 1932; Kimmins, 1987) if other factors do not become limiting. Some species can survive in more shaded conditions than others, since their leaves photosynthesize enough at lower light intensities to stay alive. Such species are described as *shade-tolerant.* The lowest light intensity at which a leaf can sustain itself (the *compensation point*) is not very different among species. For example, the compensation point of individual leaves of woody plants ranges from 0.3 to 1.5 percent of full sunlight—a difference of only 1.2 percent (Larcher, 1983). More light is needed for whole trees to survive, since the leaves must photosynthesize enough to keep stems, branches, and roots alive. The extra sunlight utilized by the shade-tolerant species contributes little to growth, although it may enable the species to survive in understory, shaded conditions. For survival of the entire tree, the architecture, spatial arrangement of leaves, and respiration rates of stems, branches, and roots are important in allowing some species to survive at lower light intensities than others. Seedlings of very shade-intolerant species such as ponderosa pine need a minimum of 30 percent full sunlight to survive in the understory. Very shade-tolerant species such as sugar maple can stay alive at levels below 5 percent (Burns, 1923; Bates and Roeser, 1928).

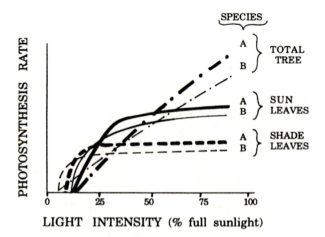

Figure 2.5 Schematic photosynthesis rate for sun and shade leaves of two species (A and B). The magnitudes of the curves vary for different species. Light intensity at which zero net photosynthesis occurs—the "compensation point"—varies slightly, but the general curve shape stays similar. Total tree photosynthesis continues above about 25 percent full sunlight because the leaf angles alter (*Leopold and Kriedemann, 1975*).

A few forest shrub species cannot withstand direct sunlight; they grow best at intermediate sunlight intensities but grow very poorly in very bright or very dim light. These plants, such as salal, are usually found beneath the forest overstory shade. Some tree species may behave similarly. The inability to grow well in full sunlight is probably related to heat stress in the leaves or water stress caused by rapid evapotranspiration.

Water

Water is available to plants from the soil primarily as capillary water in small pores and surrounding individual soil particles. The remaining water adheres to soil particles at progressively stronger tensions as soil water is depleted. As the plant exerts tension (through evapotranspiration) greater than the attraction tension of the soil, water moves from the soil into the plant. Trees vary in their abilities to extract water from the soil and endure conditions of high soil water tensions without injury. Most tree growth occurs at soil moisture tensions of less than —6 bars. Little growth seems to occur even in site-insensitive species at higher moisture tensions. Site-insensitive species may have a competitive advantage on the drier sites both through the ability to survive periods of high soil moisture tension (dry periods) and from the slight growth advantage gained from soil water extracted at very high tensions (Levitt, 1980).

Leaves of different species vary in ability to reduce evapotranspiration to minimal rates while living under high temperatures without the cooling effects of evapotranspiration. Under droughty conditions, the stomata of some species (*drought avoiders*) close, shutting off transpiration and maintaining a high internal moisture level, while others (*drought endurers*) continue transpiring until their internal moisture is quite low (Hinckley et al., 1981; Pezeshki and Hinckley, 1982).

Nutrients

Elements are taken up by trees in a few inorganic molecular forms. Each element is absorbed optimally in a single ionic form for most tree species, but other ionic forms are usable by some species. Carbon (in carbon dioxide), hydrogen (in water), and oxygen are discussed separately in this chapter. Species are found along a gradient of availability of different nutrients (Voigt, 1968; McColl, 1969). Once inside the tree, most nutrients are concentrated in the leaves, phloem (inside the bark), buds, root tips, and reproductive organs. To varying extents, tree species can internally recycle certain nutrients before their foliage is abscised. This recycling varies greatly between nutrients and species.

Soil pH can also be considered a gradient which gives different species an advantage. Most tree species grow optimally on slightly acid soils, and trees generally live between pHs of 4.0 (very acidic) to 7.5 (slightly basic; Spurr and Barnes, 1980). Certain species have the ability to grow at more extreme pHs than others. Some species are found on extremely acidic or basic soils *not* because they grow best there, but because they are uniquely able to survive there. Plants of the Ericaceae family (Rhododendron, Kalmia, and Vaccinium, for example) grow well in acidic soils. Other species seem to grow best on basic soils (Brooks, 1987).

Species vary in their ability to extract elements from either organic or inorganic media. The large size of conifer fine roots may give them an advantage in extracting nutrients from primary minerals, perhaps giving conifers a growth advantage on less well developed soils (Voigt, 1968). Alternatively, the smaller, more numerous fine roots of hardwoods may provide more surface area for nutrient absorption from organic matter in well-developed soils.

Mutualistic associations between microorganisms and tree roots increase the ability of the tree to absorb essential elements. Mycorrhizal associations between fungi and roots expand the tree root systems' surface area for absorption because of the large surface area of fungal mycelia. Mycorrhizal fungi also extend the life of fine, absorbing roots; they break down complex organic molecules and pass the extracted nutrient elements to the roots. It is unclear if different fungal species absorb nutrients from different sources and in different forms. Under high soil nutrient conditions (such as in a green house or nursery), the mycorrhizal association develops less. Seedlings without mycorrhizae which are planted in soils without the fungal inoculum present often do not survive, unless inoculum is incorporated into the soil (Hatch, 1936; Lutz and Chandler, 1946). The mycorrhizal association also does not develop well if the seedling is very weak.

Nitrogen is unique among essential elements in the way it becomes available. It is absorbed from the soil; however, most soil nitrogen must be fixed from the atmosphere by bacteria, actinomycetes, and possibly other microorganisms. Certain tree species form close associations with nitrogen-fixing microorganisms. These lower plants grow in root nodules—abnormal growths of the tree roots. Alders and leguminous species such as black locust, mesquite, and acacia are especially known for their nitrogen-fixing associations. Nitrogen-fixing abilities may give these plants temporary advantages on poor soils. Alders and leguminous trees have rapid early growth on poor soils, but their best growth is on more mesic, well-developed sites. On poor sites, even where nitrogen appears limiting, alders and legumes can be eventually outcompeted by more site-insensitive conifers. Other tree species may develop a soil environment around their roots which promotes nitrogen incorporation (*fixation*; Richards and Voigt, 1963).

Free-growing microorganisms and nutrient availability

Microorganisms require the same factors for survival and growth as photosynthesizing plants do, except that nonphotosynthesizing microorganisms obtain their energy from organic compounds in litter from dead roots, abscised leaves, fallen branches or trees, and dead animals (or living plants and animals in the case of parisites). In the process of obtaining energy, microorganisms decompose (respire) the organic material into small organic compounds and release carbon dioxide, water, heat, and inorganic nutrients. Microbes reabsorb these nutrients for further growth until a shortage of nutrients is no longer the limiting factor in the microbes' growth. This change occurs when some other factor, usually available organic matter for energy, becomes limiting. Microbes can extract nutrients from both mineral sources and decomposing organic matter. Microorganisms take up nutrients more efficiently than higher plants do; therefore, higher plants are usually able to get nutrients only through their mycorrhizal association, or after the microorganisms no longer require them for growth.

Nitrogen is the most commonly limiting nutrient for tree and microorganism growth, in part because it is used in higher concentrations than the other nutrients. Even where nitrogen is present in the forest litter, it is not available where chemically bound in organic matter until microorganisms decompose the litter and release it (Switzer and Nelson, 1972; Gosz et al., 1976; Turner, 1977; Nadelhoffer et al., 1983). The rate of nitrogen release depends on the digestibility of the litter and the need of nitrogen by the microbes. Where the litter has a high lignin content, it is not very rapidly digested (Fogel and Cromack, 1977; Melillo et al., 1982).

The carbon/nitrogen ratio is an index of nitrogen availability to higher plants. In fresh litter, carbon/nitrogen ratios over 100 are common, while decomposed material near the bottom of the soil organic layers has carbon/nitrogen ratios of 10 or less (Lutz and Chandler, 1946; Buckman and Brady, 1964; Staff and Berg, 1982; Waring and Schlesinger, 1985). Energy eventually becomes limiting to microorganisms with the depletion of organic matter. Inorganic nitrogen is then less needed and so may be released by the microorganisms. Nutrient fertilization can have a "snowballing" effect on fertilizer availability to the trees. Some fertilizer is immediately taken up by the trees, while that taken up by microorganisms increases their growth and respiration, which leads to increased decomposition and nutrient release from previously undecomposed organic matter.

In warm, moist climates such as in river floodplains in the southeastern United States and in lowland tropical forests, microbial decomposition of forest floor organic matter proceeds rapidly, and little litter remains on the ground. Consequently, nutrient recycling through the litter and back to the plants occurs rapidly. In cooler or drier climates, microbial activity proceeds more slowly, and the forest floor organic matter decomposes more slowly than it accumulates from dying roots and falling leaves, branches, and trees (Olson, 1963). Litter builds up, and nutrients become unavailable to trees.

Deep litter fuels fires in dry climates. The burned organic matter releases nutrients to the soil, except for some nitrogen which is volatilized (Grier, 1975; Klemmedson, 1976; Chandler et al., 1983a; Covington and Sackett, 1986). Other disturbances such as windthrows or clearcutting accelerate decomposition and nutrient release to the soil by overturning the horizons and putting the organic matter in contact with more oxygen above and nutrient-rich soil horizons beneath. Such disturbances also allow the warmth of direct sunlight to reach the forest floor, thus increasing microbial activity, litter decomposition, and nutrient release (Ford, 1984; Binkley, 1986).

Overstory tree growth and vigor may decline as nutrients become tied up in the organic matter in the absence of disturbances in cool, moist areas (Heilman, 1966, 1968; Ugolini and Mann, 1979; Bowers, 1987). Extreme examples of the reduction in site productivity caused by nutrient tie-up in forest floor organic matter are seen in sphagnum moss layers, wetland bogs, and thick litter layers in Alaskan spruce-hemlock forests. The declining overstory permits more light and heat to reach the forest floor, which may lead to higher decomposition rates and nutrient availability. Often, however, sphagnum moss grows under declining overstories, ties up nutrients, and insulates the soil, thereby further reducing litter decomposition. Sphagnum moss obtains nutrients from rainwater and so continues to grow under these nutrient-poor conditions. This condition has created upland bogs in Ireland, Scotland, Canada, Finland, Russia, and Alaska which have lasted for thousands of years.

Bogs also develop when microbial activity removes the oxygen from stagnant water and prevents decomposition of organic matter (or limits decomposition to anaerobic respiration). The wet bog fills with undecomposed organic matter, or a floating mat of vegetation may grow on its surface. This mat consists of sphagnum, ericaceous shrubs, and other shrub and tree species which grow under very nutrient-poor conditions.

Temperature

Life processes for most animals and plants are restricted at slightly below $0°C$ ($32°F$) by the freezing of water and above $55°C$ ($130°F$) by the death of cells. Respiration of photosynthesizing and nonphotosynthesizing plants increases steadily through this range, although the rate of change varies with species. Gross photosynthesis, on the other hand, increases rapidly from the freezing point to a relatively constant "plateau" of photosynthesis within a narrow range of 8 to $18°C$ (55 to $75°F$; Tranquillini, 1955; Nobel, 1976). The plateau level is species- and genotype-specific (Fryer and Ledig, 1972). Temperatures above the plateau create a decrease in net photosynthesis because of the increased respiration. The extremely high growth rates of forests in the cool, rainy Pacific northwestern United States have been partly attributed to the lack of summer heat, which may cause trees in warmer regions to respire away some of their potential growth (Franklin and Dyrness, 1973).

Tree species vary greatly in their ability to tolerate temperature extremes or fluctuations even though they vary little in their optimum temperature ranges for growth. At all higher elevations and at all but the most tropical latitudes, temperatures occasionally drop below freezing. Summer temperatures just above the soil surface can commonly exceed $65°C$ ($150°F$) in direct sunlight in most parts of North America even though ambient temperatures are much lower. Plants which cannot survive these extremes are found only in nonfreezing climates or where soil surface conditions are less extreme. Newly germinating seedlings are particularly susceptible to extreme temperatures because protective mechanisms and structures have not yet formed. Some plants avoid the extremes by completing their life or annual cycle in the spring after the frosts no longer occur and before the extreme temperatures of summer begin. Other plants endure the temperature extremes by becoming "frost hardened" (not to be confused with dormant; see Kramer and Kozlowski, 1979; Larcher, 1983) to low temperatures and by developing thick, insulating bark and other structures before summer temperatures reach extremes at ground level.

Different species and genotypes respond to changes in photoperiod and other environmental stimuli to different extents. They respond to changing seasons (Kramer and Kozlowski, 1979; Larcher, 1983; Kimmins, 1987). Different generations seem quite able to adapt to changing stimuli even though the responses are apparently inherited (Vaartaja, 1959).

Oxygen

Oxygen is absorbed through bark lenticels, leaves, buds, and roots. Oxygen becomes limiting when not available to roots in the soil. As with other growth factors, most tree species grow best when soil oxygen is optimum; however, some species have mechanisms which allow them to live in anaerobic soil conditions. Species such as tupelo and bald cypress can survive several years of standing water which kill most species within weeks by "pumping" air to the roots from the stem (Fowells, 1965). Other species, such as coastal redwood, cottonwoods, aspens, willows, and black spruce, avoid death when their root systems are buried in silt or organic bogs by developing adventitious roots from their stems. Still other species, such as northern white cedar, are able to live with a few roots near the surface of anaerobic waters in a condition of physiological drought.

Carbon dioxide

Carbon dioxide is diffused from the atmosphere through stomata on leaves. Except when leaf stomata are closed to reduce water loss, absence of carbon dioxide does not limit growth. About 0.033 percent (by volume) of the atmosphere consists of carbon dioxide. Photosynthesis increases with increased concentrations of carbon dioxide, although species vary in their rate of response to an increase (Strain, 1978; Larcher, 1983). Carbon dioxide concentrations in the atmosphere vary slightly with time of day (Huber, 1958), weather conditions (Wilson, 1948; Selm, 1952), and elevation (Decker, 1947). On calm nights and mornings concentrations may reach levels 25 to 50 percent higher than normal close to the forest floor where respiring microorganisms are emitting carbon dioxide (Assmann, 1970; Woodwell and Botkin, 1970). These levels may allow understory herbs, shrubs, and trees to photosynthesize efficiently and survive at lower light intensities than they otherwise could.

Other factors

Other environmental factors, such as atmospheric pollution and nuclear radiation, may affect growth. Species vary in their susceptibility to atmospheric pollution (Larcher, 1983).

The requirements for different growth factors change as plants grow larger. Some species can tolerate low light intensities during their first growing season, probably because they can live from stored food reserves in their seeds; however, many seedlings cannot tolerate water-saturated soils as well as older trees of the same species can. As trees age or otherwise decline in vigor, they become less able to adjust to irregular fluctuations in the environment or to tolerate adverse conditions such as atmospheric pollution or flooding.

The Concept of Growing Space

Each tree in a forest utilizes the growth factors discussed above until its growth becomes limited by unavailability of one or more factors. Any one factor may limit growth because it is not present in the area, it has been taken up by another plant, or it is available at such a slow rate that it is essentially absent. When the limiting factor becomes available, growth will proceed until another factor becomes limiting, in a fashion similar to the *Law of the Minimum* concept of Liebig (Taylor, 1934; Odum, 1971). The availability of one factor can also influence the efficiency with which a tree can use another factor—the *Law of Compensation* (Assmann, 1970).

The limiting factor varies between areas and between times of the day, year, or longer in the same area. Lack of one factor may prevent a plant from acquiring another factor. For example, a lack of sunlight in an understory may reduce an understory plant's growth so that it cannot grow enough roots; water then becomes limiting, and the plant dies. Research has long been concerned with which factor is limiting (Tourney, 1929b; Korstian and Coile, 1938; Kozlowski, 1943). While resolutions of these ecophysiological questions are important, the study of stand dynamics can proceed by utilizing a simplistic generalization—provided the limitations of this generalization are recognized.

It is often convenient to consider that a site contains a certain amount of intangible *growing space,* or capacity for plants to grow until a factor necessary for growth becomes limiting. Growing space is similar to the *root capacity* of Coile (1937) and the *biological space* of Ross and Harper (1972) and Hutchings and Budd (1981). The term "growing space" avoids the tantalizing but distracting debate about which factor is limiting growth of an individual plant at a given site and time. Growing space is dimensional when growth is most limited by the volume of space available for shoot or root penetration or by the surface areas available for nutrients, water, or light accumulations. Growing space may also describe more abstract situations, such as when nutrient conditions limit the site's capacity to support growth.

Factors which limit growing space

The amount of growing space varies spatially and temporally. Some species are capable of utilizing growth factors in a form unavailable to other species. Such differences among species are not great but can give different species competitive advantages on different sites.

Different species survive and compete successfully under slightly different environmental conditions. It might be assumed that optimum conditions for growth are quite different between species and that the growing space is different for each species, even on the same area. In fact, the opposite seems true. Species vary little in their ability to grow under different availabilities of growth factors, so that what is growing space on a site is approximately the same for nearly all species.

Species' various competitive advantages under different conditions seem to lie largely in two characteristics:

1. Species vary in their allocation of photosynthate to shoot extension, root extension, height growth, limb spread, insect and disease resistance, and other uses. Consequently, species vary in their rate of obtaining growth factors, their stability, and their ability to resist adverse conditions.

2. Species vary in ability to endure low levels of certain growth factors, even though the amount of each growth factor which will produce optimum growth varies little among species.

The specific factors which can limit growing space are sunlight, water, mineral nutrients, temperature, oxygen, and carbon dioxide and have been described above.

Variations in growing space between areas

The growing space of a given area can be considered general for all tree species rather than species-specific, since tree species vary relatively little in the required concentrations and forms of growth factors. The combinations of availability and unavailability of different factors give one or another species a competitive advantage at a given time and on a given site (Grubb, 1977). Different places vary dramatically in their amount of growing space. For example, dry areas with little growth potential may be adjacent to moist areas with much higher ability to support growth.

Variations in growing space within a site

The growing space and limiting factors fluctuate on a site in response to daily and seasonal cycles and to unpredictable or long-term events such as storms or climate changes. Daily and seasonal cycles are so predictable that plants have physiologically adapted to cope with periods of good and poor conditions. Plants have adapted to less regular variations, such as fluctuations in the normal weather cycle, long-term climatic changes, disturbances, and forest development following disturbances, by asexual or sexual regeneration mechanisms. Changes in available growing space caused by plant interactions are referred to as "autogenic" changes (Tansley, 1935; Spurr and Barnes, 1980), while changes caused by events external to the plants are "allogenic" processes. Allogenic processes include weather and climatic changes.

Daily changes. Daily fluctuations in growing space follow a general pattern (Helms, 1965; Hodges, 1967) which varies somewhat depending on the severity of the climate and soils. The lack of light at night limits growth. Temperature may limit growth on cool sites during the early morning. High evapotranspiration demand (at midday) can cause stomatal closure and temporarily reduce photosynthesis. Midmornings can be optimum for growth when temperatures are moderate and soil moisture is available from dew or soil water migration during the night. Late afternoons can be favorable if water is not limiting.

Seasonal changes. Seasonal fluctuations in the growing space follow more varied trends than daily fluctuations. Seasonal fluctuations are regular enough, however, that seasonal physiological patterns of growth and organ development occur in many parts of the temperate zone. Cold temperatures generally limit growth in winter. Shoots are generally dormant, and root growth is very slow. All growth factors are generally available in early spring. Rapid root development occurs then, followed by budbreak, shoot and leaf growth, and cambial growth. Shoot growth can continue into summer for a short or long time, depending on the environment and the species, as will be discussed later. Cambial growth can continue longer than shoot growth if growing space is adequate. Buds develop early or late in the season depending on the species. In the autumn, aboveground growth slows, and deciduous leaves develop abscission layers and fall; root growth increases for a while, after which trees lapse into a condition of deep dormancy and resistance to winter cold.

Figure 2.6 shows the variations of rainfall, temperature, and potential evapotranspiration for selected areas of the United States (from Eagleman, 1976). In each area, temperature is the limiting factor for much of the winter. Where temperatures become high enough in winter, evergreen plants may photosynthesize, thus utilizing a part of the growing space unavailable to deciduous trees. Winter photosynthesis may explain the ability of some evergreen species such as eastern hemlocks, western hemlocks, and western redcedars to survive and grow quite large beneath deciduous overstories.

In early spring, most regions contain enough soil moisture (from snow, rain, or low winter evapotranspiration demand) and available nutrients to permit rapid growth when temperatures are high enough. Overstory leaves do not expand for several weeks after the temperatures rise in deciduous forests, so growing space at the forest floor is not limited by shade. Vernal herbaceous understory plants complete most of their life cycle during this short interval, tree seedlings germinate, and understory shrubs such as mountain laurel photosynthesize enough to survive in the understory.

Species vary in the time of budbreak and subsequent leaf emergence both between years on a given site and between sites, depending on weather conditions. Generally, however, species follow a certain order of budbreak. In the Pacific northwestern United States, conifers generally break buds slightly later than red alders do, while oaks generally delay breaking buds until 1 or 2 weeks after other hardwoods in the eastern United States.

Late spring offers optimal growing conditions for overstory trees in most regions. Growth is generally curtailed by a summer drought, increased evapotranspiration, or both. Summer rains or extended soil moisture may renew or extend the growing season in some regions, such as on good sites in the Pacific northwestern, southeastern, coastal Alaskan, and southern Rocky Mountain regions of the United States. A combination of moderate temperatures and available soil moisture may occur again in autumn, and this permits some understory herbs and tree species such as the white oaks to germinate. Active growth occurs in midsummer in cool areas where moisture is adequate (Zimmerman and Brown, 1971; Kramer and Kozlowski, 1979).

Weather fluctuations. Extremely wide fluctuations in seasonal weather patterns can be of such magnitude that trees are abnormally affected. Warm periods in early spring can melt snow or cause early growth and make plant roots or shoots vulnerable to later freezing. On the other hand, late spring frosts, snows, or lingering snowpacks can decrease the growing space available for spring herbs. Floods or standing water late in the spring kills less flood-resistant species. Late warm periods in the autumn or unexpected cold periods may also cause extra growth or death to some trees.

Droughts have a significant effect on both tree mortality and fire hazard. Greatest fire danger in hardwood forests is in spring and autumn when no leaves are on deciduous trees. Extended droughts are common in many forests in the western United States during midsummer, but unusually long ones make the stands susceptible to fire. Droughts also cause trees to die, especially when weak anyway. New shoot tips of species with "zigzag" growth forms (Chap. 3) die during droughts. Unusual summer rains increase the growing space previously limited by a lack of moisture and cause some species to add second, *lammas* flushes of growth.

Unusual weather patterns act as a disturbance in a stand (Chap. 4). They often cause a recognizable deformity in plant growth which can be used to reconstruct the stand's his-

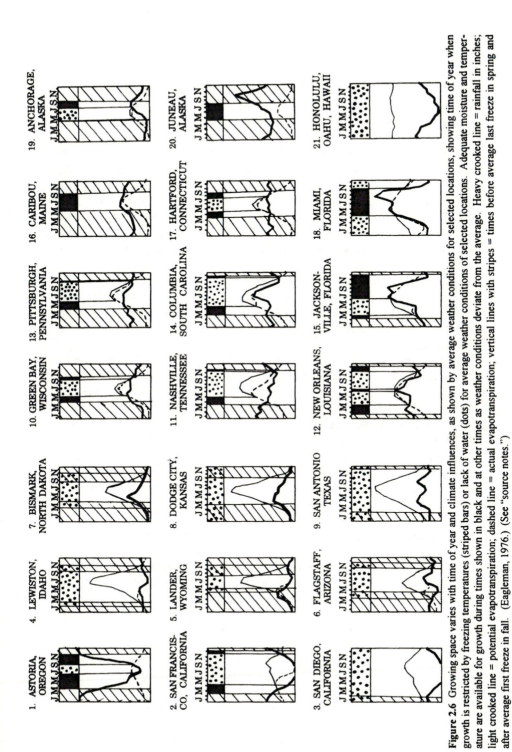

Figure 2.6 Growing space varies with time of year and climate influences, as shown by average weather conditions for selected locations, showing time of year when growth is restricted by freezing temperatures (striped bars) or lack of water (dots) for average weather conditions of selected locations. Adequate moisture and temperature are available for growth during times shown in black and at other times as weather conditions deviate from the average. Heavy crooked line = rainfall in inches; light crooked line = potential evapotranspiration; dashed line = actual evapotranspiration; vertical lines with stripes = times before average last freeze in spring and after average first freeze in fall. (Eagleman, 1976.) (See "source notes.")

tory. Abnormal weather can also cause different species to dominate the stand for many years or even many centuries afterward.

Long-term climatic fluctuations. The earth's climates fluctuate over decades and centuries and affect where species are and have historically grown and how vigorous they are. Climate fluctuations occur at shorter intervals than the life span of most trees. The fluctuations are attributed to changes in sea surface temperatures; changes in the extent of ice sheets (glaciers); changing patterns of solar radiation caused by the earth's orbital characteristics and sunspot activity; changes in atmospheric gases—both greenhouse gases and aerosols that cool—from natural causes such as volcanic eruptions and, recently, from increasing carbon dioxide and other human-caused pollutants in the earth's atmosphere (Heikkinen and Tikkanen, 1981; Heikkinen, 1982; Gilliland, 1982; Skinner and Porter, 1987; Gates, 1993; Wright et al., 1993). Longer-term fluctuations are associated with moving of continents and changes in the earth's tilt and orbit (Skinner and Porter, 1987).

Climates do not fluctuate uniformly throughout the earth, so the entire earth does not necessarily become warmer or cooler at the same time. Similar climatic fluctuations can occur in different regions, although sometimes at slightly different times. Parts of the earth reached a peak of warmth in the mid-1930's (A.D.), which had not been exceeded for thousands of years.

The period from early 1400's (A.D.) to the mid-1800's A.D. is known as the "little ice age" and may have been a global event (Gates, 1993). Glaciers in many parts of Europe, North America, and Asia began to expand. Villages at high elevations in Europe were abandoned, vineyards were moved to lower elevations, and people were considering abandoning settlements in Iceland because wheat could not be grown there (Lamb, 1963, 1977, 1988). "It was as if the countries of Europe had moved 300 miles to the north." (Gates, 1993).

Trends of mean annual temperatures for northwestern Europe are shown in Fig. 2.7. Differences of a few degrees become more important when one realizes the differences in mean annual temperatures are only 5 to 6°C (10°F) between Atlanta and Washington, D.C.; between Washington, D.C., and Brunswick, Maine; between Sacramento, California, and Spokane, Washington; and between St. Louis, Missouri, and Denver, Colorado (Eagleman, 1976).

In the short term, climatic trends are difficult to detect because weather fluctuations within these long-term trends are often large and mask any longer trend. Several cool and warm periods have been identified for northwestern Europe (Lamb, 1963, 1977; Hansen et al., 1981; Henderson and Brubaker, 1986; Skinner and Porter, 1987). These patterns are similar to some parts of North America (Gates, 1993):

1. *Postglacial climatic optimum warm epoch.* This period reached a maximum about 6000 to 3000 B.C., with generally the warmest conditions since the glaciation.

2. *Postglacial cool periods.* Early iron age cold epoch about 1000 to 500 B.C. and another cool period about 500 to 1000 A.D.

3. *Little climatic optimum (Medieval optimum).* This period reached its maximum about 900 to 1200 A.D., also with conditions nearly as warm as those of the Postglacial climatic optimum.

4. *Little ice age.* This cooler period reached its maximum about 1430 to 1850 A.D.

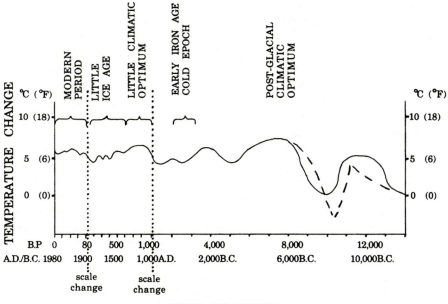

TIME IN YEARS

Figure 2.7 Average temperature variation in northwestern Europe since the last glaciation (*from Lamb, 1977*). Global climatic conditions have not been stable and have altered growing space and the species which are most competitive on a site. The patterns may change with different regions of the world. Dashed lines on right show a recent possible interpretation.

5. *Modern period.* The earth's climate generally warmed from about 1850 until a peak in the 1930s. Between about 1945 and 1970 the earth's climates appeared to be cooling again. Because of the daily and annual fluctuations, discerning a trend since 1970 is difficult. Much of the concern that atmospheric pollution will cause global temperature increases apparently has not been borne out to date.

Both temperature and precipitation patterns shift as the climate fluctuates. During a cool period the moist subtropical region may contract while the upper latitudes may become cooler and drier, although the pattern varies depending on location on the continental masses.

The effect of climate variations on forests can be dramatic but not immediately obvious (Franklin et al., 1971; Henderson and Brubaker, 1986; Brubaker, 1988, 1991; Gates, 1993). Most trees initiate in a stand shortly after disturbances. The climate during the first few years is a large factor in determining which species will become dominant; but once in a dominant position, a tree can endure quite large climate changes. Species which live hundreds of years may be surviving in a stand because they had a competitive advantage when the climate was quite different. Foresters prescribing species to plant in a harvested area sometimes mistakenly assume that the species dominating the previous stand would best survive if reestablished. Actually, the species growing on an area may be a relic of a past climate, rather than a key to present climatic conditions. Relic stands may be especially common in mountainous terrain since a given temperature and precipitation regime rapidly moves up and down a mountainside as the global climate fluctuates.

Plant species are continually migrating. Since tree species generally invade an area after a disturbance, the migration is often extremely slow. Stand-initiating disturbances occurred only every 100 years or more in North America before European colonization. Most tree seeds are disseminated less than a few miles. Tree migration onto and across deglaciating areas does not appear to have been more rapid than contemporary migration across nonglaciated areas in response to the changing climate.

The climatic fluctuations have also changed the physical landscape dramatically. During the glacial maximum—about 20,000 years ago—most of Canada and some of the United States were under a sheet of ice, the sea level was about 110 meters (m) [360 feet (ft)] lower, and the Atlantic coast was as much as 145 kilometers (km) [90 miles (mi)] further eastward (Shepard, 1963; Emery, 1967; Skinner and Porter, 1987). Suitable locations for plant growth changed dramatically as ice advanced and retreated, climates cooled and warmed, and seas fell and rose (Wright, 1968; Whitehead, 1972). Species which did not move to suitable areas became excluded or extinct during the changes. The paucity of species in central Europe compared with either North America or Asia Minor has been attributed to the inability of species to migrate over the Alps during the various weather fluctuations and glacial advances and retreats. According to one hypothesis, east-west pattern of the Alps formed a barrier to migration during climatic changes in Europe. The north-south directions of mountain ranges in North America allowed much more southerly movement of species as the glacier advanced, and the southerly latitudes of Asia Minor and its high plateau surrounded by water created ice-free environments for many different species. In central Europe, there are three or four native species of oak trees and a total of about 35 commercial tree species. There are about 60 native species of oak trees in North America and over 120 commercial tree species altogether (Harlow et al., 1979). In Asia Minor there are 18 native tree species of oak (Yaltirik, 1984) and over 60 commercial tree species in all (Davis, 1965).

Variations in growing space caused by disturbances

A disturbance causes changes in growing space in two ways: it can eliminate the existing plants on an area, making growing space available to other plants, and it can alter the total amount of growing space available to all plants by altering the availability of water, nutrients, or oxygen.

Disturbances which lead to soil erosion or soil compacting reduce the total growing space within the soil by reducing its moisture-holding capacity, volume where roots can penetrate, and oxygen and nutrient availability. Alternatively, disturbances such as windthrows overturn the soil and incorporate organic matter into the soil, break up potential hardpans, and thus increase the availability of nutrients, moisture, and oxygen to roots. Similarly, soil deposition can increase the nutrients available to the site although initially the existing root systems may suffocate. The heat from direct sunlight on the exposed forest floor may increase microbial activity and thus nutrient availability following a disturbance. Fire may increase the availability of certain nutrients by burning the organic matter in which they were previously incorporated. Fires may also reduce growing space in soils if most of the moisture retention and cation exchange capacity is in the organic matter.

Changes in growing space during forest development after a disturbance

When a disturbance kills a plant, it makes available the growing space which was previously occupied by this plant. After a disturbance, plants expand to reoccupy the newly available growing space. As the plants develop, they extend roots and branches and produce leaves to utilize the available light, moisture, nutrients, warmth, and other growth factors. Consequently, the amount of unoccupied growing space diminishes even if the total amount of growing space remains unchanged.

Both the occupied and unoccupied growing space can change as a forest develops. A growing stand can actually expand the amount of moisture available as tree roots penetrate the mineral soil and expand pore space for moisture retention, as occurred when forests regrew on abandoned farmlands in the Piedmont of the southeastern United States (Kittredge, 1952; Gaiser, 1952). Stands in coastal fog areas may increase their available moisture as they grow by providing branches and leaves on which the fog can condense and drip to the ground. Clearcutting Douglas-fir stands in the fog belt of the Pacific northwestern United States can reduce the moisture reaching the forest floor.

Soil oxygen and available nutrients similarly increase as the depth of optimum soil pore space increases. Nutrient availability increases as roots, microorganisms, and the soil solution take up nutrients from the rock lattices and circulate them within the tree and soil.

Total nitrogen in the plant-soil system increases as a stand grows, since more nitrogen is fixed from the atmosphere. Plants such as legumes and alders with nitrogen-fixing capabilities increase the rate at which nitrogen accumulates in the system.

Especially in cooler or drier climates, total growing space may actually decrease if a forest stays too long without a disturbance and the nutrients become tied up in the very slowly decomposing organic matter.

Temperature regimes also change as a forest develops. The rooting zone cools in very northerly climates with the resulting tie-up of nutrients, decline of forest growth, and eventual development of bog conditions. Immediately following a disturbance, sunlight directly on the soil may create warm temperatures and allow heat-tolerant species to initiate. As the stand grows, shade cast on the ground creates cooler conditions again and individuals less tolerant to the heat may grow.

The decrease in evapotranspiration immediately after a disturbance can cause saturation and anaerobic soil conditions. More water is transpired as the new stand develops, and the depth of suitable soil for rooting can again increase.

Growing space available to each plant

It is now possible to describe a scenario of tree growth within a stand. When growing space (i.e., the sum of factors necessary for growth) is present but not occupied by other plants, the growing space can be considered "available." This occurs when a flowerpot is filled with soil, seeds are added and watered, and the pot is placed in the sun; when a moist field is newly plowed; or when a disturbance kills trees or parts of trees.

Plants which have germinated, been planted, or survived the disturbance expand into the available growing space. The rate of this expansion is limited by the availability of growth factors and by the genetically predetermined growth rates of the species.

The expanding plants come literally or figuratively into contact with others also expanding into the available growing space. After contact, the rate of growth of each plant is limited by the growing space it occupies. The amount of growing space each plant occupies is defined by the surrounding plants.

Because of their unique anatomies, plants must expand in size to live. A plant first allocates the energy obtained through using its available growing space to maintain its presently living cells (*maintenance respiration*; Assmann, 1970). After respiration demand is fulfilled, any extra energy is used for growth. A tree occupying a fixed growing space increases in size at progressively slower rates, because it obtains a fixed amount of energy through photosynthesis, while increasing amounts of energy are allocated to maintenance respiration demands of the increasingly larger living system. Size eventually reaches a maximum in a fixed growing space when all photosynthesis is used for respiration. The tree cannot grow larger unless its growing space is increased.

The above scenarios are shown graphically in Fig. 2.8. Together, the figure shows the fundamental considerations for tree growth: tree size, growing space per tree, and time. [To be compatible with later discussions (Chap. 15), the growing space axis has been reversed to read from large to small values.] When a tree begins growth (e.g., from a seed), it expands in amount of growing space along XY, beginning at X. At some point (e.g., *c*), the growing space becomes limited, and the plant expands in size at this growing space (along *cC*). If it occupied more growing space at the time growing space became limited (e.g., *b*), it would grow larger. Eventually, tree size reaches an upper limit for each amount of growing space. Empirical studies (Yoda et al., 1963; White and Harper, 1970; Drew and Flewelling, 1977, 1979) have shown the upper limit of growing space to follow a relation of approximately

$$\text{Tree size} = K \text{ (growing space)}^L$$

so that the slope of the upper limit is L when both axes are logarithmic. K is a species-specific constant. In trees growing in full sunlight, L has a value of approximately $-3/2$ (White and Harper, 1970; Ford, 1975; Lonsdale and Watkinson, 1983; Westoby, 1984; Weller, 1987; Osawa and Sugita, 1989). Reasons for this slope and the relation of growth to time, volume per tree, and growing space will be discussed in Chap. 15.

When plants have filled all available growing space, they begin competing with other plants to obtain and to maintain growing space. The varied ways the different species survive and grow allow a species to have an advantage and hence expand into another's growing space under certain conditions but to give it up under others.

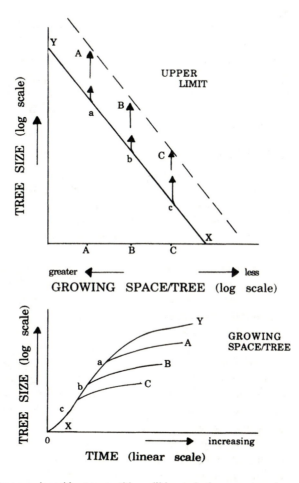

Figure 2.8 A tree growing without competition will increasingly occupy growing space and grow in size along line XY. Trees encountering competition at points a, b, or c increase in size along *aA*, *bB*, or *cC*, respectively, with limited growing space until an upper size limit is reached for that amount of growing space (this concept will be discussed in more detail in later chapters).

If one plant has a competitive advantage, it expands at the expense of another. The plant whose growing space is reduced may be able to survive if it can utilize some growing space which the more aggressive plant cannot, or it may die if there are no such differences in growing space utilization. Plants of the same species utilize the same growing space, and so a plant soon dies if outcompeted by another of the same species.

Different species sometimes have slight differences in growing space requirements, and an individual can survive even if outcompeted by a plant of another species. For example, red alders often outcompete western hemlocks and Douglas-firs for sunlight by growing above them, expanding their crowns, and shading the other species. The hemlocks survive beneath the alders because they can live with lower light intensities than the lowest foliage of the alders can; however, the Douglas-firs die beneath the alders because the alders and Douglas-firs utilize sunlight to the same minimum levels of intensity (Stubblefield and Oliver, 1978).

The different growth behaviors which give a species an advantage are sometimes referred to as "strategies." This term implies anthropomorphic preplanning which is inappropriate for trees (Gould and Lewontin, 1979; Harper, 1982). Much has been written about the theoretical aspects of competition and strategies which is beyond the scope of this book (Slobotkin, 1961; Harper, 1977; Grime, 1979).

Diversity and Stability in Plant Communities

The growth factors required by trees are inorganic; consequently, trees are not dependent on other living organisms for immediate survival. Tree species may form mutualistic associations with certain lower plants—fungi or bacteria—in which a coevolved interdependence is probable. Less immediate mutualistic interdependencies also seem to form between tree species and animals, especially involving pollination, seed dissemination, and sometimes germination.

Although increased species diversity is sometimes assumed to imply stability (Whittaker, 1965; Brookhaven National Laboratory, 1969; May, 1973), the lack of organic dependencies of trees makes interactions between trees much different from interactions between animals. Animals exist in a food web, where each animal feeds on other animals or plants. These animals are dependent on others for their energy source and certain organic vitamins. Species diversity in animals is critical for survival and community stability, since the elimination of one animal or plant may jeopardize the food source (and hence survival) of many predator animals unless other species are present as substitutes (Hutchinson, 1959). Conversely, all trees live at the "bottom" of the food hierarchy. The lack of other trees or animals does not destabilize a tree's food supply. Other plant and animal species do probably add stability by not providing the continuity of food of a single species to a predatory pathogen or animal, by aiding in seed dispersal and pollination, and perhaps in other ways; however, the relationship between diversity and stability is not so close as in animal communities.

Applications to Management

Concerns that human manipulations cause unalterable changes to the natural processes of forests do not seem well founded. Human activities appear less drastic when people realize that forests are casual, temporary assemblages of plants which are not closely coevolved and which compete vigorously with each other and are constantly subjected to disturbances. Human activities basically mimic different aspects of natural disturbances.

Use of "indicator species" to identify growth potential of a site needs to be adjusted for disturbance history, since different species can dominate depending on which gains the competitive advantage following a disturbance (O'Hara, 1995). Indicator species are most useful at extreme site conditions. Here, a species' presence indicates that conditions are not too extreme for it to survive, although absence does not indicate that the species could not grow there.

Moving a species to more severe sites than where it is naturally found can make it susceptible to insects, diseases, and weather disturbances. Where no barriers to migration exist but a species is not found in an area, it is usually because climate or soil conditions prevent its successful growth. Alternatively, a species may be able to grow successfully in another part of the world but has been unable to migrate there.

Trees react to silvicultural manipulations according to the physiological and morphological characteristics of the species. To maintain conifers and other site-insensitive species on good sites requires intensive management activities and large costs. Likewise, a site-sensitive species can often be grown on a poor site, but the reduction in growth and susceptibility to insects and diseases may not allow them to survive.

Much of silviculture is either making the stand's growing space available to desirable species and individuals or putting these individuals in a competitively advantageous position. Elimination and continuous exclusion of all plants except crop trees, although giving most rapid growth of crop trees, is rarely done in North American forestry because of the high cost compared with the value of wood. Various thinning and regeneration practices give the competitive advantage to desirable species. Partial shade of shelterwoods often gives a desired species the advantage over competitive weeds until the desired species grows large, after which it can be released to full sunlight.

Management activities also increase or decrease the total growing space of an area. Timber harvesting operations and site preparations can decrease total growing space when they damage the soil structure or reduce nutrients by causing them to be leached, volatized in the case of fire, or taken from the site during harvest.

Management activities can increase the total growing space by fertilization or site preparation which increases soil rooting depth or reduces unfavorable microsites.

Changing the growth factors on a site sometimes creates management problems. Nitrogen fertilization increases needle biomass and length of needle retention of Douglas-firs. If the fertilized trees have small diameters and are not stable anyway, the added weight on the crown increases the stress on the bole and can cause breakage especially in areas of heavy, wet snows.

Management activities should be timed to utilize periodic fluctuations in growing space. Planting when roots are dormant in late autumn, winter, or early spring not only prevents damage to rapidly growing root tips, but establishes the seedlings when water is plentiful and the temperature is low. Growing space is less likely to be limited by a lack of moisture. Where advance regeneration is to be saved, the overstory is best not removed in late spring, after the newly developed shoots have formed shade needles which will be subjected to full sunlight during the droughty, hot summer.

Mortality from climatic changes is most likely when the trees are in the seedling stage. Later in life the trees can also die, especially if they are otherwise under stress. Recognition of climatic change trends allows conservative managers not to establish species in stands near historically cooler or warmer extremes of their ranges, since these sites may not be suitable for the duration of the trees' lives. Reestablishing the same species on a site where it formerly dominated also may not be successful, since the previous stand may have become established under a much different climatic regime.

Increased competitiveness of a species at an extreme of its range can suggest the species is responding to a change in climate—a warming climate where the increase is at the cooler and higher elevations, a cooling climate where the increase is at the cooler and lower elevations, a drier climate where the increase is at the wetter extreme of it range, and a wetter climate where the increase is at the drier extreme of its range.

In general, younger stands are more adapted to the current climate, since climatic shifts occur gradually. A decline in vigor of a species in many old stands at an extreme of its range often indicates a significant climate change. This decline in vigor can be accompanied by insect attacks and mortality.

Simply moving all species to higher and lower elevations and expecting them to behave as they did before may not be an effective response to changing climate, since different species will become more or less vigorous under the changed climate. New or accentuated variations in the patterns of interaction among species can be expected (Davis, 1986).

CHAPTER 3

TREE ARCHITECTURE AND GROWTH

Introduction

Stands of given species, spacings, and sites develop only in certain patterns. Stand development patterns emerge because individual trees with innate growth habits respond with characteristic growth habits both to the physical surroundings—such as site, disturbances, and weather—and to other plants competing for the same growing space. While it is not possible to predict the exact development pattern of a stand, the possible patterns can be reduced to a few. The probability of each pattern can be estimated by understanding each species' growth habits and responses to different conditions.

Similarly, the past history of a stand can be inferred from the shapes of its trees. Each tree shape—be it crooked or straight; limbed, knotted, or cylindrical—provides evidence of its past history because trees constantly add new growth onto their old structures developed under different environments. Forests are often said to "remember" their past because these shapes are preserved as trees grow.

Tree shape is largely determined by its *primary growth* (Esau, 1965), which reflects the environment of the time it grows. Primarily growth refers to growth from a bud, root tip, or other apical meristem. *Secondary growth* is growth from a cambium and is responsible for the thickening of tree stems, branches, and roots. These terms should not be confused with the ecosystem terms of "primary production" and "secondary production" (Whittaker, 1975; Waring and Schlesinger, 1985).

Detailed physiological, morphological, or anatomical descriptions of tree growth can be found in Busgen and Munch (1929), Romberger (1963), Esau (1965), Larson (1969), Wilson (1970b), Kozlowski (1971), Zimmerman and Brown (1971), Kramer and Kozlowski (1979), and Funakoshi (1985).

General Growth Patterns
Primary vegetative growth of the stem

Each year a tree's *terminal* (the top of the main stem) and its *lateral* branches increase in length by adding new primary growth (Fig. 3.1). The year's growth from the terminal is referred to as a *leader*, or *terminal shoot*. The primary growths from lateral branches are referred to as a *lateral shoots*.

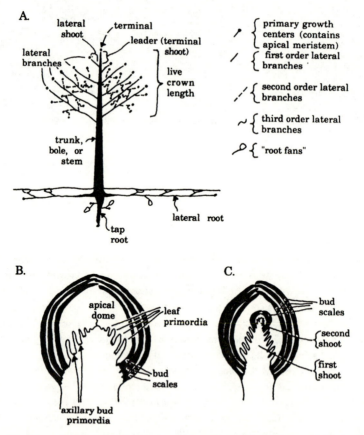

Figure 3.1 Schematic tree and buds showing various tree architecture terms. A. Parts of tree. B. Detailed parts of bud. Sustained growth bud is shown, without inner bud scales. C. Preformed bud, showing inner bud scales.

Shoot growth entails both *primordia* development (Fig. 3.1) from the apical meristem and active shoot expansion. Primordia are small precursors of stems or leaves and are formed by the apical dome both before and during active elongation. A primordium can develop during most seasons when it is within the protected *bud* or bud-like folded leaves. Primordia development often proceeds during more adverse seasons than does active shoot elongation. Most woody plants have protective buds covering the shoot apical meristems (unlike the case for root apical meristems) and developing primordia. Budbreak signals the beginning of active shoot elongation.

Active shoot elongation occurs as the primordia expand rapidly and develop new stems, leaves, and accompanying lateral primary meristems (usually shoot or flower buds). Most elongation occurs when growing space is at its maximum and the possibility of injury from frost, desiccation, or heat damage is low. During active elongation, most newly exposed shoots are tender, unsuberized, and susceptible to frost, heat desiccation, and browsing. Elongating shoots can be resistant to frost or browsing in some cases.

Time of active elongation varies with climate, species, genotypes within species, and local weather conditions. Initiation and cessation of elongation are triggered by environmental stimuli such as day length, accumulated warmth (heating degree days), moisture availability, or a combination of factors (Zimmerman and Brown, 1971). The beginning of shoot elongation can vary by over a week during the same year within a species at a single location. Greater but more predictable variation occurs between species. Certain understory vernal herbs elongate before the overstory in springtime, when available sunlight and other growing space are greater. Deciduous species usually elongate slightly before conifers. Many ring porous oak species characteristically elongate vegetative shoots much later than other species.

Cessation of shoot elongation is also triggered by various environmental stimuli—especially drought and photoperiod. Time and stimulus vary between sites, species, climates, and weather conditions. Pattern of cessation differs depending on the shoot growth pattern, as will be discussed. Shoot elongation may reinitiate in late summer or autumn if weather and soil conditions are favorable.

Many tropical species also have periods of active shoot elongation and dormancy. These periods do not necessarily occur in all species at the same time or on an annual basis (Zimmerman and Brown, 1971) and may be linked to times of available moisture.

Floral shoot growth
Floral shoot growth will not be discussed in detail here. Floral shoots usually begin development and elongation in spring or fall and mature in 1 or 2 years (Kozlowski, 1971; Allen and Owens, 1972; Edwards, 1976).

Root growth
Root growth has been studied less intensively than shoot growth. Most studies have been with small trees or seedlings. Seedling root growth of most species occurs throughout the year in favorable soil moisture and temperature conditions (Webb, 1977). Generally, peaks of root growth occur in early spring before active shoot elongation and

in late autumn after cessation of shoot elongation and cambial activity (Zahner, 1958; Kramer and Kozlowski, 1960; Kozlowski, 1971; Zimmerman and Brown, 1971).

Mycorrhizal activity also seems to be periodic, occurring mostly during periods of favorable root growing space (Vogt et al., 1981a,b).

Secondary growth

Secondary growth occurs as xylem and phloem thickness increases through growth of the cambium and as bark thickness increases through growth of the cork cambium.

Shoot elongation activates hormones which stimulate the secondary growth in a "wave" from the top of the tree toward the base (Larson, 1969). Consequently, secondary growth begins slightly later in spring than primary growth but can continue later into the summer. The extent of secondary growth of each tree is very sensitive to the amount of growing space available.

The annual rings in temperate trees are formed by xylem cells produced at different times of the year. These rings allow estimations of tree ages and height and diameter growth rates. Many species first add xylem cells of large, thin-walled rings referred to as *early wood* or *spring wood*. Later in the season they add smaller, thick-walled cells known as *late wood* or *summer wood*. Hardwoods with this difference in early and late wood appearance are referred to as *ring porous*, and those with a more uniform distribution of large and small cells are referred to as *diffuse porous*. To a lesser extent, conifer species exhibit these different xylem cell types, with some species showing more distinction than others between early and late wood. Conifers and hardwoods with relatively distinct early and late wood (and relatively distinct annual rings) have preformed shoot growth patterns (see below), while species without distinct early and late wood generally have sustained shoot growth patterns.

Cell size and cell wall thickness are influenced by many environmental factors (Larson, 1969; Kozlowski, 1971). Summer cycles of drought and rain, periodic flooding, or elongation of new shoots late in the growing season can change the xylem cell shapes and sizes, creating what mistakenly appear as annual rings (Kozlowski, 1971). "False rings" in cypress and other species produce mistaken estimates of tree ages or growth rates. On the other hand, if growing space is very limited, a tree may not produce enough photosynthate to create xylem cells in lower parts of the tree, leading to "disappearing," or "missing," rings.

Rings in tropical species are often not linked to annual cycles. Trees of the palm family (a monocot) do not have annual rings, since their xylem and phloem are arranged in paired bundles similar to a cable of many fibers (Zimmerman and Brown, 1971). In tropical monocots and dicots, the trees are difficult to age accurately.

Secondary growth on roots does not seem to be as regular as on stems. Growth rings do occur but are not necessarily produced annually (Wilson, 1964). Secondary thickening and rings occur near the tree base, on the upper sides of roots, and near the soil surface where the root is exposed to light or heat (Zimmerman and Brown, 1971).

Shoot Development Patterns

Growth of the terminal shoot dictates a tree's architecture by controlling branch growth, crown shape, and stem growth. Temperate and tropical trees grow in many patterns which vary within and between species. Some trees–palms, for example—have no lateral primary meristems or branches; if the terminal meristem is killed, the tree dies. Other trees such as cascara, eastern redcedar, western redcedar, Alaska yellowcedar, and some tropical species do not form buds during resting phases. Their leaves simply enclose the apical meristem. Halle and Oldeman (1970) and Halle, Oldeman, and Tomlinson (1978) identified 23 generalized growth models which describe the growth patterns of most woody plants.

Five growth patterns most common to temperate forests of the northern hemisphere will be described: preformed (fixed growth), sustained, recurrent, terminal florescence, and aborted tip (sympodial, or zigzag pattern). [These patterns do not directly match the models of Halle and Oldeman (1970).] First, the general development of the apical meristem and its primordia will be described.

Apical meristem

The tip of the apical meristem consists of a rounded dome (Fig. 3.1; Romberger, 1963; Larson, 1969; Zimmerman and Brown, 1971; Funakoshi, 1985). Bulges form on the flanks of the dome and expand by cell division to form leaf primordia. As these leaf primordia develop, the dome expands upward and adds younger leaf primordia within and above the older ones. The older leaf primordia expand upon rapid shoot elongation to form leaves or needles. In some cases the leaf primordia become modified. If the apical meristem becomes a flower, the leaf primordia develop into bracts or petals. In most plants, a series of leaf primordia will develop at the appropriate time into bud scales and enclose the apical dome and any younger primordia. In mature pines, the leaf primordia may develop as paperlike brown scales, and the "needles" arise from short shoots of buds developing in their axils.

Species which have lateral branches develop apical meristems in the inner axils of the leaf primordia as they become older. Sometimes these *axillary meristems* (or *axillary buds* in species in which bud scales develop) will form branches after the parent shoot stem has expanded. Arrangement of the leaf primordia, and hence axillary meristem primordia, strongly determines the branch pattern and resulting tree form in hardwoods. Leaf primordia appear as pairs of opposite shoulders of the apical dome in species such as maples and ashes. Consequently, the trees have an opposite, paired branching pattern. In other species, the leaf primordia appear sequentially in spirals around the stem, producing other branching patterns.

The axillary meristems often do not form distinct branch patterns in conifers. In single-needle plants such as true firs, spruces, Douglas-firs, hemlocks, and juvenile pines, most axillary meristems remain quiescent; only a few develop into branches. In mature pines, most axillary meristems form *short shoots* (extremely small branches) which produce needles at their terminals. The entire shoot is a *fascicle*, and the number of needles in the fascicle is characteristic of the species.

Short shoots also develop from nonterminal shoots in other genera such as cherry and larch. These short shoots break buds and produce foliage each year but elongate only enough to allow the leaves to extend. A terminal short-shoot condition occurs in longleaf pine seedlings for several years (Wahlenberg, 1946).

Preformed growth

Preformed (or fixed) growth describes a pattern of bud development, dormancy, and activation found in many temperate trees which survive periods of environmental extremes through dormancy (Fig. 3.2A). New bud scales are formed from leaf primordia at the tip of the new shoot before or during active shoot elongation in the spring and early summer (Funakoshi, 1985). The apical meristem within the new bud scales then develops new leaf primordia before active shoot elongation occurs again the next year. All the new primordia are formed by (or shortly after) the time of rapid shoot elongation (budbreak), and the new shoot expands simply by telescoping out its already preformed primordia. Any extra primordia development occurs within the bud to be expanded the following spring. Most species of oaks, true firs, Douglas-fir, hickories, spruces, ashes, some pines, and other genera have this growth habit.

Dissynchronies can occur between primordia development and expansion, since the two processes are separated in time. Many leaf or needle primordia develop under vigorous conditions. If subsequent conditions during shoot expansion are less favorable, expansion is limited and the new shoot contains many leaves close together. This results in the "bottle brush" pattern seen in conifer seedlings. It can occur in seedlings which grew vigorously in the nursery (where many primordia developed) but which were damaged during outplanting before budbreak, reducing shoot elongation and confining the many needles to a small shoot.

Sometimes moisture and other conditions continue to be favorable for shoot expansion after the preformed primordia are fully expanded. In these cases, the bud containing the shoot which would ordinarily expand the following year may instead open and elongate precociously, creating a second flush (Fig. 3.2A; also known as *lammas growth, Johannistriebe,* or *St. John's growth*—after mid-summer's day). Lateral buds as well as terminal buds can develop as lammas growth. Which shoots elongate influences subsequent tree form. The propensity for terminal or lateral buds to form lammas growth appears to be genetically controlled (Weber, 1983). Premature breaking of the bud by this second flushing does not seem to reduce height growth during the following year; in fact, total height growth may be increased (Rudolph, 1964).

During rapid elongation to a preformed shoot, the amount of stem expansion between leaves varies. (*Internodes* are the stem lengths between leaves.) Internodal elongation is generally greater near the shoot base, so that leaves and axillary buds are closer together progressively higher on the stem and form a cluster (*whorl*) near the top. Progressively higher buds on each annual shoot usually produce more vigorous shoots the next season.

In trees with preformed growth, neither the lateral (axillary) buds nor the new terminal bud breaks and elongates during the year they are formed (except in the case of lammas growth). Terminal and lateral buds break during the spring, rapid shoot expansion takes place, and the shoots develop as the main stem and branches. The terminal and upper lateral shoots develop most vigorously, while progressively lower shoots grow less so. Lower branches may form short shoots in some species. Some lower buds may not even break, and the enclosed apical meristem may stay quiescent, growing outward with the cambium but hidden within the bark. Shoots from these buds can emerge later as *water sprouts* (*epicormic sprouts*). Lower branches on each year's shoot often grow little in subsequent years and may soon disappear. While alive, lower branches add greatly to the leaf area of the upper crown (Jensen and Long, 1983). The upper cluster of branches grows largest, increasing the whorled appearance of many preformed trees.

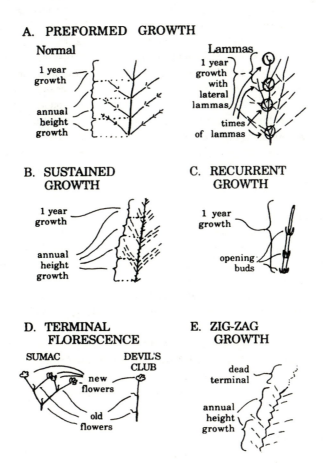

A. PREFORMED GROWTH

Normal

Lammas

B. SUSTAINED GROWTH

C. RECURRENT GROWTH

D. TERMINAL FLORESCENCE

E. ZIG-ZAG GROWTH

Figure 3.2 Growth patterns commonly found in North American trees. A. Preformed growth without lammas growth (*left*). Strong apical dominance can lead to weaker apical control, creating strong lateral branches (compared with B). Preformed growth with one, repeating pattern of lammas growth (*right*). Lammas growth of a lateral bud in the terminal cluster occurred four times (*circled*) in the tree. Each time, the lammas shoot "competed" with the normal terminal for dominance, creating a fork. If one shoot eventually asserts dominance, the other becomes a steeply angled lateral branch. B. Sustained growth. Annual height growth can be identified from the suppressed buds at the base of each annual shoot. Note that weak apical dominance leads to strong apical control, creating weak lateral branches (compare with A). C. Recurrent flush growth pattern, in which several "telescoped" buds break during the same year, often at overlapping times. D. Terminal florescence. Flowers are produced in some trees and shrubs from terminal primary meristems instead of lateral primary meristems. Lateral buds assert dominance when these flowers die, creating either dichotomous branching as in sumac (*left*), where two lateral branches "compete" for control, or a straight stem as in devil's club (*right*), where a subterminal branch assumes control. E. Aborted tip growth pattern, in which the terminal often dies back at the end of the growing season. This pattern is usually associated with a zig-zag shoot pattern (*Zimmerman and Brown, 1971*).

The tendency for current lateral (or terminal) buds to remain dormant as the shoot expands (except through lammas growth) is termed "apical dominance" and occurs in trees with preformed growth. Trees with strong apical dominance exhibit vigorous growth of previous years' lateral shoots (Brown et al., 1967), which develop into large, strong lateral branches able to endure intense physical abrasion during stand development. The large lateral branches suggest that the terminal has not "controlled" these branches; consequently, trees with strong apical dominance have weak "apical control."

"Control" of lateral shoots is so weak in some preformed trees (e.g., oaks) that lateral branches grow upward and act as terminals. A central stem cannot be distinguished in trees with these multiple terminals and the result is a *decurrent* tree form (Fig. 3.3). When such trees are shaded from the sides, the terminal with the greatest access to sunlight becomes dominant, creating a single, although crooked, stem. This opportunistic determination of the central stem probably gives oak trunks their sweeping shape. Reaction wood added to the stem provides necessary support (Sorenson and Wilson, 1964).

In other preformed trees such as Douglas-firs or true firs, the terminal controls the angle and length of lateral shoots (Mitchell, 1969; 1975), and an *excurrent* growth form with a distinct central stem or trunk is maintained. These laterals grow outward and do not compete with the terminal for dominance, although the lateral branches remain quite strong.

In preformed pines, such as ponderosa and the white pines, shoots from both the terminal and the lateral buds first grow vertically after budbreak, creating an appearance of green "candles" on the tree in early spring. Eventually, all except one shoot reverts to a horizontal position, leaving a single shoot as the central stem.

Sometimes the terminal bud in oaks will not form the following year's main stem. Rather, the terminal shoot develops into a lateral branch, and a lower bud grows upward as the main shoot. In most other genera the terminal bud forms the central stem more consistently.

Sustained growth

In the sustained growth pattern, not all primordia which form the shoots and leaves develop before active shoot elongation in the spring (Fig. 3.1B). Instead, a bud is set near the end of a growing season, and a small number of primordia for next spring's growth form before budbreak. As existing primordia elongate during spring budbreak, new primordia continue to form which elongate shortly afterward. Development of new primordia and subsequent shoot elongation continue into the summer so long as environmental conditions permit; this can be viewed as "ad lib" growth. Sweetgums, hemlocks, red alders, yellow poplars, and red maples have sustained growth patterns. Sustained growth also occurs in trees which do not form buds. Sustained growth is sometimes referred to as "free growth" (Kramer and Kozlowski, 1979; Daniel et al., 1979); however, "free growth" will be used in this book to describe crown expansion without restriction from adjacent trees (after Hummel, 1951; Jobling and Pearce, 1977).

Shoots with the sustained growth pattern generally do not elongate as rapidly as trees on the same site with preformed growth, possibly because only a small part of the shoot is preformed and ready for expansion at budbreak (Zimmerman and Brown, 1971). On the other hand, trees with the sustained growth pattern continue to grow longer into the growing season and hence become taller if environmental conditions are adequate, since they do not have a preformed inner bud which stops shoot development.

Figure 3.3 Weak apical and epinastic control of open-grown oaks. A central stem is lost as several stems compete for the dominant position. Where the species grows with more side shade, the tree keeps a central stem.

Many species with sustained growth are considered more site-sensitive than associated species with preformed growth. Sustained growth allows continued growth late in the growing season on sites where moisture continues to be adequate. Preformed growth allows rapid shoot expansion on poor sites during the spring before soil moisture becomes limiting. Western hemlocks and red alders, for example, have sustained growth and are considered more site-sensitive than the preformed Douglas-firs and western white pines. Similarly, sweetgums and yellow poplars have sustained growth and are considered more site-sensitive than most ashes, oaks, or hickories. Of course, other physiological factors besides growth form influence site sensitivity of a species.

Sustained growth shoots appear much differently from preformed shoots. Lateral, axillary buds on sustained growth shoots expand during the same growing season in which they are formed, unlike those in preformed shoots which remain dormant for one year (except as lammas growth). A few axillary buds at the base of sustained shoots generally do not expand and grow during the same season, however. These buds may be part of the shoot which was performed before budbreak and so act as normal, preformed shoots (Zimmerman and Brown, 1971). During subsequent seasons these lower buds are suppressed by more vigorous shoots above them. Consequently, the lower buds rarely expand very much and create an easily recognizable stem section free of large branches which can be used to identify annual height increments of trees with sustained growth for many years (Fig. 3.2B).

Shoots exhibiting sustained growth have weak apical dominance since the terminal meristem does not "dominate" lateral buds on its shoot by preventing their breaking. Sustained growth trees generally have weaker lateral branches than preformed trees have, but they maintain more of a central stem than preformed trees such as oaks do. This strong control of the lower lateral branches by the terminal is referred to as "strong apical control" (Brown et al., 1967).

The weak lateral branches of trees with the sustained growth form give these trees a competitive disadvantage against preformed trees. The strong lateral branches of pre-formed trees protect the terminal from abrasion during heavy winds while simultaneous-ly abrading or breaking the less protected terminal shoots of trees with sustained growth forms. This abrasion has been observed between oaks and maples, birches (Oliver, 1979a), and hemlocks (Kelty, 1986) and between Douglas-firs and western hemlocks (Wierman and Oliver, 1979).

Compared with preformed species, species with sustained growth forms produce more uniform, diffuse porous wood in hardwoods and less distinction between early and late wood in conifers (Zimmerman and Brown, 1971).

Variations occur within the sustained growth pattern. Species which do not form buds may not produce any preformed primordia, making annual height growth patterns very difficult to identify.

The sustained growth eastern, western, and mountain hemlocks as well as American beeches have leaders which tend to droop during the year they are formed. If grown in full sunlight, western hemlock leaders will straighten up shortly after shoot elongation the following year and the new stem will arise from the top of the old stem. When grown in shade, however, the drooping stem does not always straighten, in which cases the new terminal arises from a lateral bud farther back along the shoot.

Recurrent flush growth

Southern pines superficially appear to possess the preformed growth habit, with an inner bud established on top of an enclosed shoot before the outer bud opens. Telescoped inside this inner bud can be a preformed shoot with another bud at the end and still another preformed shoot and bud inside of this (Fig. 3.2C). After the outer bud breaks in spring, each bud and shoot will break and elongate in sequence, often before the previous shoots are fully elongated (Allen and Scarbrough, 1969; Boyer, 1970; Griffing and Elam, 1971). Vigorous trees produce more flushes of elongating shoots than weak trees do.

Recurrently flushing pines do not produce annual whorls of branches. Where whorls appear, they reflect the end of a single shoot, rather than the end of a whole year's growth. If flushes occur later in the growing season after a drought or other unfavorable condition has temporarily stopped expansion, a "false ring" may appear in the annual xylem.

Terminal florescence growth

The apical meristem of terminal shoots can develop into a flower (Fig. 3.2D) in sumacs, devil's club, and other woody plants. A terminal flower asserts epinastic control over the lower vegetative branches, forcing them to grow more laterally. In sumacs a pair of vegetative branches grow beneath the flower, creating a dichotomous fork in the stem when the flower dies and falls away. In devil's club, only one lateral branch develops subjacent to the terminal flower. This lateral branch gradually straightens after the flower dies, pushing the flower's remains to one side and reforming a central stem. Dichotomous branching also occurs if the terminal bud or shoot dies on maples or other oppositely branched trees following injury to a terminal, or after lammas growth in which two branches assume the terminal position.

Aborted tip growth

The terminal shoot tips of some species commonly abort, leaving lateral branches to resume growth. This growth pattern coincides with species with zigzag (sympodial) twig shapes (Fig. 3.2E; Zimmerman and Brown, 1971). Although this growth form may be derived from either preformed or sustained growth forms, the appearance and behavior is striking enough that it produces a distinctive appearance.

In species which grow in a zigzag pattern, each leaf acts as a straight continuation of the stem, and the distal stem grows at an angle. Tips of zigzag shoots lack vigor and often die during unfavorable summers (Busgen and Munch, 1929). Growth resumes the following year; the uppermost living axillary bud assumes the new terminal position, while lower axillary buds become branches. Sugarberry, other members of the elm family, birches, redbud, and other hardwoods exhibit this growth form.

Other growth forms

Many woody shrubs grow as rapidly in height each year as trees and can offer severe competition to small trees. Unlike trees, these fast-growing shrubs do not accumulate one year's height growth onto the tip of the previous year's, and consequently they remain short. Terminal shoots of shrubs sometimes bend over rather than remain upright, possibly because of a physiological dissynchrony between cell elongation and lignification (Zimmerman and Brown, 1971). The bowed appearances of salmonberries, vine maples, and winter honeysuckles arise from this dissynchrony. The terminal shoots of many shrub species do not develop from the tip of the previous year's growth, possibly because of some inherited physiological blockage to the plant's transport or hormone system.

Changes in shoot development patterns

Shoot development patterns can be different between species, between individuals within a species, within an individual at different ages, and under different environments. Very young plants of most species have sustained patterns, and the tendency to revert to preformed growth increases with age. Seedlings of many species which later have a preformed habit grow with sustained growth during the first 1 to 3 years in a nursery bed. This pattern may be maintained even longer under field conditions, where maturation from juvenile characteristics proceeds more slowly (Bormann, 1955, 1956).

Environments alter growth patterns. Extreme drought or shaded conditions do not allow much ad lib growth by trees normally exhibiting sustained growth. Trees which initiate or curtail shoot elongation according to temperature or day length stimuli often do not function normally when moved to different latitudes, and abnormal growth patterns can result. A *foxtail* develops when a terminal shoot grows without producing lateral branches and often occurs in many temperate pines when grown in tropical latitudes. Foxtails can be over 6 m (20 ft) tall. This condition has been partially attributed to the lack of stimuli for the terminal shoot to cease growth or produce lateral buds and branches (Lanner, 1966; Kramer and Kozlowski, 1979).

Crown Shapes

Crown shapes vary from very columnar to almost flat-topped. They are determined by the numbers, lengths, positions, and directions of growth of the central stem and lateral branches. Some palms have no lateral branches and so form a "tufted" crown shape of leaves around the primary meristem of the central stem (Dransfield, 1978). Crown shapes are the result of both the inherent growth form of the species and environmental influences. Variations in crown shape occur between and within species at the same or different latitudes and elevations. Genetic variations in crown shape even occur between trees of the same species in the same stand (Golomazova et al., 1978). Increased nutrients may shift a tree's crown to a more flat-topped shape (Goudie, 1983).

Columnar crowns are often found at high elevations and latitudes, while species and individuals with conical and flat-topped crowns are more frequently found at lower latitudes (Kuuluvainen and Pukkala, 1989). The gradient in crown shape and latitude has been attributed to efficiency of light absorption. Light rays are perpendicular to the tree stem for much of the day and year at high latitudes, and a columnar shape gives a crown maximum exposure to direct sunlight. At low latitudes the overhead sun gives flat-topped crowns maximum exposure (Fritschen et al., 1980). Columnar crowns also efficiently shed snow commonly found at higher latitudes and elevations.

Crown shape is greatly modified by inherent responses to shading or breakage. Each species behaves in a rather predictable way, leading to a predictable crown shape when subjected to given environmental conditions.

Crown shapes are determined by the relative extensions of terminal and lateral branches (Fujimori and Whitehead, 1986). Much of a crown's shape can be understood from the terminal growth form. Lateral branches generally follow the same growth form as the terminal shoot, except they are oriented at some angle from vertical and usually grow less vigorously. Lateral branches extending from the main stem—*first-order branches*—are controlled to varying degrees by the terminal. Similarly, *second-order branches*—those growing from first-order branches—are controlled by the terminal of this parent first-order branch, and so on. Theoretically, the number of orders of branches should be greatest on the lowest (oldest) branches and could be one less than the age of the tree in years. In fact, branches of higher orders fail to form or die in the interiors of tree crowns; rarely are more than five orders of branches found (Jensen and Long, 1983).

Epinastic control

Crown shape is also influenced to various degrees by internal, physiological controls. A tree's terminal bud controls the length and orientation of lateral branches to different extents. This control is referred to as *epinastic control*. The degree of epinastic control depends on the species, tree vigor, and position of the lateral branch. In a species with strong epinastic control, lateral shoots will grow to a predictable angle and length relative to the terminal soon after shoot elongation. Douglas-firs, true firs, spruces, and hemlocks have this strong epinastic control under most conditions of full sunlight, as do many hardwoods such as yellow poplars, sweetgums, black cottonwoods, and red alders when young (Fig. 3.4). This strong control results in an excurrent growth form.

If the terminal is killed in a tree with strong epinastic control, several uppermost laterals will turn upward and grow at steep angles. None will have complete enough epinastic control to grow vertically until one becomes taller than the others. When it is taller, the most vigorous branch will grow straight up, and the other branches will resume their horizontal orientations.

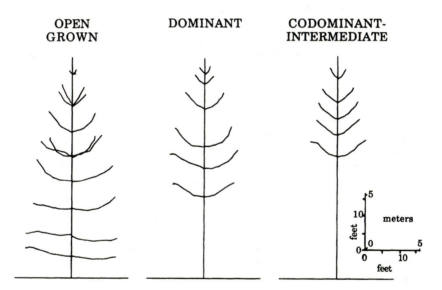

OPEN GROWN   DOMINANT   CODOMINANT-INTERMEDIATE

Figure 3.4 Change in branch length and angle in Douglas-firs. Strong epinastic control of the upper branches keeps them at similar angles and lengths. The lower branches become weighted down, and the tips of very low branches often turn up. (*See "source notes."*)

Often, however, rounded crowns result from weak epinastic control (Fig. 3.3). Weak epinastic control can allow several shoots to behave as the central stem, forming the decurrent growth form. Oaks and some other species have relatively weak epinastic control; their branches grow at many angles and lengths. Other species have intermediate degrees of epinastic control.

In open-grown trees with strong epinastic control, the upper laterals form an angle to the stem determined primarily by the terminal's influence (Fig. 3.4). This angle is similar within each species, with variations between genotypes. Lower on the stem, the branches bend more downward, pulled down by the weight of the branches as well as influenced by the terminal's epinastic control. Still lower in the crown, the inner portions of the branches are pulled downward by gravity, but the outer portions are so far from the terminal that the epinastic effect is depleted and the branches often turn upward at a steep angle. In some species—such as western redcedar and Norway spruce—lower branches can become free of the terminal's control when grown without side shade. These lower branches turn upward and grow with the appearance of a separate tree top.

Trees with more rounded crowns can have strong but less obvious epinastic control if the angle and length of the laterals is under the strong influence of the terminal. Sugar maples often have rounded crowns, but the lateral branches grow in a controlled fashion.

Changes in crown shapes

The innate crown shape can be modified in two ways: external influences can act directly on the limbs, such as through physical abrasion, and/or external or internal factors can influence the ability of the terminal to maintain epinastic control over the rest of the crown. In both cases, the resulting crown shape and size are characteristic of these influences and can be used to infer the history of the tree, or the stand when many crowns are observed. Crown shape is also important in determining a tree's ability to compete and photosynthesize efficiently.

Factors affecting the crown shape directly. Side shade, shade on the sides of a crown, directly affects growth of those branches shaded. Except for the terminal shoot, each branch must photosynthesize at least enough for its own growth and survival (Sprugel and Hinckley, 1988). Lateral shoot growth (branch extension) is reduced where shade limits photosynthesis on a side of a crown (Cochrane and Ford, 1978). Branch growth becomes significantly less, lateral buds become fewer and smaller, and increasingly more branches die in the lower parts of a shaded crown. Thus, side shade determines the crown length.

The effects of side shade on branch shape have not been well studied. Shaded branches lose their rigidity and bend downward under their own weight more than unshaded branches do. Even if exposed to full sunlight, previously shaded branches retain some of their former, droopy shape.

The diameter of a branch at death determines the size of knot on the tree trunk and, where visible, indicates the amount of sunlight that reaches the limb before it was shaded and killed. Formerly open grown or widely spaced trees have larger dead branches than trees grown at narrower spacings since stand initiation. Of course, species vary in how long the dead branches remain attached to the tree.

Sunlight from an overhead angle can cause trees of some species to appear to grow toward the sunlight, creating crooked, leaning stems. It is unclear whether each annual shoot actually leans phototropically toward the sun, or whether the heavier crown on the sunny side of the tree pulls the tree toward sunlight. Other species have gravimorphic tendencies instead and tend to grow vertically even if light is from an angle. Generally, conifers have gravimorphic tendencies, as do some hardwoods. Gravimorphically con-trolled trees appear to "lean" away from another tree and toward the sunlight, but it is often the weight of branches on the sunny side rather than a phototropic response causing the lean. A sweeping stem and vertical upper tree can also be caused by the tree's bend-ing when small. This bending can be the result of snow pressure, especially on a hill-side, or can be caused by the tree's germinating on an unstable medium—such as an old stump or log—which fell or decomposed when the tree was small.

Side shade in trees with weak epinastic control (e.g., oaks) may help maintain a single trunk by shading the steep lateral branches and keeping them too weak to grow as addi-tional trunks. Unequal shade from the sides or top allows one or another of these steep laterals to replace the central stem and leads to a sweeping or crooked stem (Sorensen and Wilson, 1964). This behavior differs from other phototropic behaviors in which a single terminal leans to the light; instead, the oak leader closest to the light gains domi-nance.

Branch abrasion occurs when normally quite distant branches or terminals rub togeth-er during windstorms. This abrasion contributes to the dominance of trees with strong lateral branches, such as those with preformed growth patterns. The abrasion can rub the leaves and bark from branches or can break branches (Oliver, 1978a; Wierman and Oliver, 1979). Such abrasion can give crowns sculpted appearances. Physical abrasion becomes more severe as trees grow taller. This abrasion and separation of crowns is sometimes referred to as "crown shyness."

Very severe abrasion occurs when a nearby tree falls and shears off part of another tree's crown. Even after the broken crown has regrown, locations of breaks can be noted by forks, absences of branches, or branches of unusual shapes, such as those created by water sprouts.

A sudden change in hormonal balance leads some species to develop *water sprouts* along the stem and branches (Fig. 3.5). Water sprouts appear as new shoots along old stems and arise from dormant buds *(epicormic sprouts)* or newly differentiating buds *(adventitious sprouts)*. Water sprouts below the previous live crown give the tree a dis-tinctive appearance, since they are generally shorter than other primary limbs. It is unclear how much water sprouts contribute to total photosynthesis of the tree, although they alter its appearance and produce knots in wood grown after emergence of the sprouts. Species vary in their tendency to produce water sprouts when disturbed.

Factors affecting epinastic control. Side shade can eventually reduce a tree's height growth if it restricts the size of the photosynthesizing crown too much (Mitchell, 1975). The tree becomes less vigorous, and eventually the terminal loses its control over crown shape.

Foresters have long recognized overhead shade cast on an overtopped tree as being of two types: *low shade* and *high shade* (Fig. 3.6; Chapman, 1944a). Low shade is created by taller trees within the same stratum immediately above a crown. High shade is cast by foliage far above the tree, generally in a different stratum. Trees behave differently in high shade than they do in low shade because the light has different properties.

Most shade cast by foliage is discontinuous, with "flecks" of sunlight passing through. A crown in full sunlight contains foliage in a zone several meters thick. Consequently, a plant living immediately beneath the upper leaves of this zone will not be completely shaded and can live in this partial "low shade." Immediately below a leaf or branch, distinct areas of shade and sunlight are found delineated by a sharp boundary. The shaded and sunny areas move with the passage of the sun, so much of the area is illuminated

Figure 3.5 Water sprouts (epicormic or adventitious sprouts) in yellow birch (*above*) and noble fir (*below*). Water sprouts grow on some species when a disturbance upsets the physiological balance within the tree.

with bright sunlight for at least part of the day. Foliage in low shade is either in very deep shade where it photosynthesizes very little, or in full sunlight where it actively photosynthesizes.

Until a terminal is overtopped by the canopy of taller trees, it receives enough direct sunlight in the flecks to maintain strong epinastic control, even though leader growth is reduced by the partial shade.

High shade, from far above a tree, has different properties because sunlight diffuses into the shaded area as it passes an object, and the shade appears to diffuse outward (Fig. 3.6). The boundary between lighted and shaded areas becomes less distinct, and a broad area of intermediate light intensity develops, growing larger with increasing distance of the shading object. Roughly, the area of full sunlight completely disappears at a distance of 70 times the diameter of the opening, and an area of diffuse sunlight is found that is larger than the original opening; shadows, on the other hand, become much less distinct only a few feet below an opening. As partial shade is projected farther from the foliage creating it, therefore, the environment contains fewer distinct "flecks" of intense sunlight and deep shade. Rather, it contains more areas of diffuse sunlight. Sunlight entering from an angle also travels farther from the overhead shade and further creatures diffuse light conditions. Wavelengths of this diffuse light may also be less favorable for photosynthesis, since the higher leaves absorb the more photosynthetically active radiation.

Under the weak light of high shade, epinastic control weakens, and crown shapes change dramatically (Fig. 3.7) in species which can survive at all. A *physiological shift* occurs which favors growth on horizontal shoots instead of vertical ones. Consequently, branches become more horizontal, and the ratio of lateral-to-terminal shoot length increases with high shade. Terminals of trees which maintain strong epinastic control in high shade become short compared with the laterals and give the tree a flat-topped, or "umbrella," appearance. This shape occurs in spruces, firs, and Douglas-firs (Kohyama, 1980, 1983; Meng, 1986; Tucker et al., 1987; for contrast, see Klinka et al., 1993). Terminals of species with weaker epinastic control under high shade grow sideways as lateral branches and no distinct, central leader remains (Busgen and Munch, 1929; Trimble, 1968a). Hemlocks, oaks, beeches, and many other species lose their terminal under high shade. The amount of high shade required for a shift to more flat-topped growth varies with species. More shade-tolerant species endure lower light levels before losing their epinastic control. Similarly, the degree of flat-topped suppression a tree can endure before dying is directly related to its shade tolerance.

Upon release from high shade, the leader reasserts control in trees which maintained their central leader (Fig. 3.7). The crown eventually returns to its open-grown form, although the shaded branch patterns are retained where the trees were suppressed (Tucker et al., 1987). In nearly all trees where height growth has been severely curtailed by overstory shade and then rapid growth resumed after release, characteristically large branches grow from where height growth had slowed, leaving a distinguishable pattern on the tree stem. In trees whose terminals reverted to laterals, several laterals turn upward and compete for the terminal position. Eventually a crook forms after one lateral assumes dominance and the others return to the horizontal, or a fork forms if more than one branch becomes dominant (Trimble, 1968a).

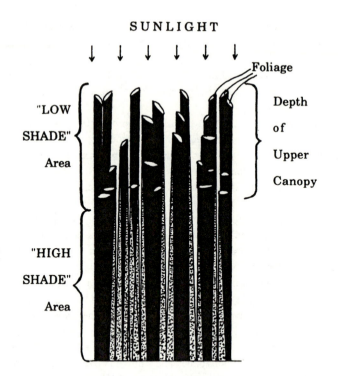

SUNLIGHT

"LOW SHADE" Area

"HIGH SHADE" Area

Foliage

Depth of Upper Canopy

Figure 3.6 Change in light properties as light passes into and beneath the upper canopy. Because of the response of the trees, zones of low shade and high shade can be delineated. *Low shade:* All except the top leaves in the upper canopy are either in bright sunlight (unshaded in figure) or in deep shade (black in figure) as the sun changes position during the day. While in direct sunlight, they photosynthesize vigorously. *High shade:* Well below the canopy, light from sun flecks diffuses with shaded areas and creates large areas of low-intensity sunlight (speckled area in figure). Tolerant species can live, but not grow rapidly, in this more diffuse light.

In western redcedars, the extensive horizontal growth of upper lateral branches continues even after the tree has been released from high shade. Control seems to be lost over lateral branch length. The result is a tree with long lateral branches, which may take away wood that would otherwise be added to the main stem.

Perturbations which injure or destroy the terminal bud release the laterals from its epinastic control. The uppermost lateral branches then grow upward and reassert epinastic control. If one branch is uppermost, it grows vertically and causes the other laterals to revert to horizontal growth, creating a bayonet-like appearance in the main stem. Sometimes two or more branches grow straight up, creating a fork where the old terminal existed.

Forks or bayonets can also be caused by other factors. When an upper lateral bud adds lammas growth but the terminal does not, the lammased shoot converts into a terminal and grows upward the following spring, as does the original terminal bud. The initial result is a forked tree until one of the stems gains dominance (Fig. 3.2A; Jump, 1938; Weber, 1983). Then, the nondominant stem grows as a lateral branch, albeit at an abnormally steep angle. If two or more lateral buds produce lammas growth while the terminal bud does not, the laterals often offset each others' influence and the original terminal maintains epinastic control. If the terminal bud produces lammas growth, it will retain its epinastic control. The pattern of lammas—the terminal, one or two laterals, or a combination—seems inherent within each tree (Weber, 1983; Carter et al., 1986*a*).

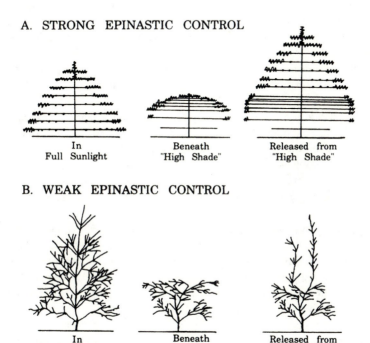

Figure 3.7 As trees become overtopped by high shade, their terminals lose "control" over lateral shoots; the terminals shoots become shorter relative to the laterals; and in some species, the terminal may lose epinastic control, droop, and form a lateral branch. More shade-intolerant species become flat-topped under less shade and die before becoming extremely flat-topped. A. Strong epinastic control. Upon release from high shade, these trees can regain their symmetrical shape and central stem (provided they can withstand the shock of intense sunlight), leaving branches close together at the past point of suppression. B. Weak epinastic control. These trees may respond to release by sending several lateral branches upward as terminals, leaving forks or crooks at the point of release.

Flowering and seed production utilize energy which would normally be used for height and diameter growth (Kozlowski, 1971). During good seed years some trees slow dramatically in height growth. Reduced vigor or death of the terminal may reduce its epinastic control and initiate the decurrent growth form.

Drought can reduce the epinastic control by causing dieback of the terminal of some species. It can also weaken the epinastic control without dieback and may cause the early shift of trees from excurrent to decurrent forms on poor sites.

Trees often shift from strong epinastic control and excurrent forms to decurrent forms as they grow larger and older. It is unclear if this shift is caused by a physiological maturing or is simply a reaction to accumulated years of drought, flowering, and other impacts on the terminal's vigor.

Height Growth
General patterns

Tree height generally follows a sigmoid curve with age when the tree is growing in full sunlight. Height growth is slow at first when the tree is young and too small to accumulate energy for rapid terminal growth. As the tree's size and foliage increase, more energy is available for the terminal shoot, causing a rapid increase in height until it reaches its highest growth rate—the "grand period of growth." Eventually, the growth rate slows as increased stress created by the extreme height, exposure, or crown size limit the extension of the terminal.

Height growth in full sunlight is roughly the same within a species under similar site conditions and across a broad range of crown sizes. Even here, the height growth patterns can vary both polymorphically and anamorphically (Fig. 3.8).

Different species have characteristically different height growth patterns (Fig. 3.9; Carmean, 1970a,b; Herman and Franklin, 1976). These differences influence crown expansion and strongly influence the ability of a species to compete under different situations.

Crown size is proportional to tree height, especially in trees with strong epinastic control, because lateral branch extension is proportional to terminal extension. Tree height, therefore, is closely correlated with foliage area and total tree volume increment in trees growing without competition. In stands where live crown length is determined by shade from surrounding trees, heights and spatial patterns of the component trees determine their crown sizes and growth rates, as will be discussed.

The height of dominant trees in even-aged stands is used as an index of a site's growth potential, since dominant trees of a given species grow in height at approximately the same predictable rate over a wide range of spacings. Height, expressed in site index, is generally considered a better measure of site quality than diameter or total volume growth for several reasons. Height growth of the terminal shoot is influenced by growing conditions both during the season the bud (or resting terminal) is formed and during elongation the subsequent year; consequently, height growth is the average of influences from more than 1 year and is not affected by weather fluctuations so much as diameter growth is. Photosynthate allocation to height growth is given priority over diameter growth. Consequently, a tree's height growth is relatively independent of its degree of crowding and amount of foliage except at extremely narrow spacings (Fig. 3.10; Eversole, 1955; Oliver 1967; Reukema 1970, 1979; Mitchell, 1975; Mitchell and Goudie, 1980; Lloyd and Jones, 1983; Hann and Ritchie, 1988). Some evidence suggests trees slow in height growth at very wide spacings (Allen and Marquis, 1970; Reukema, 1970; Hamilton and Christie, 1974), while other evidence does not (Baker, 1953; Mitchell, 1975). A reduction in height growth probably occurs more readily at very wide spacings in trees with weak epinastic control. Excessive artificial pruning of lower, living limbs can reduce a tree's height growth (Staebler, 1964).

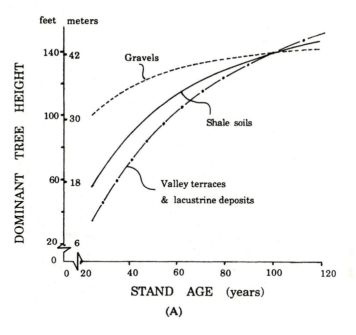

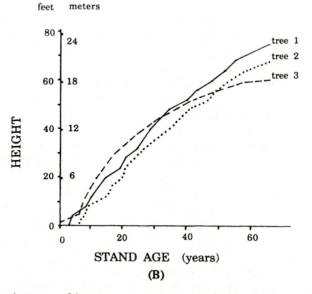

Figure 3.8 Dominant trees of the same species can grow with different height patterns on different site conditions (A) and even within the same stand (B). A. Douglas-firs of the same site index grow differently in different soils (*Carmean, 1956*). B. Dominant northern red oaks from the same stand in central New England show distinctly different height growth patterns (*Oliver, 1975*). (*See "source notes."*)

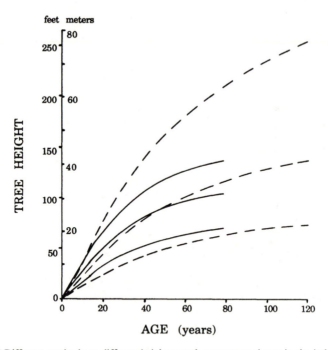

Figure 3.9 Different species have different height growth patterns, as shown in site index curves of Douglas-fir (*dashed lines; King, 1966*) and loblolly pine (*solid lines; USDA Forest Service, 1929*). (See "source notes.")

The time required for a species to reach a given height varies with site. For a given species, however, the stand structure and accumulated volume of individual trees is more closely correlated with tree height than with age. For example, stands at each spacing and height will appear as in Fig. 8.1 whether they took 15 or 35 years to achieve each height. Similarly, when Douglas-fir plantations regularly spaced 2.4 m (8 ft) apart reach about 21.3 m (70 ft) tall, diameter growth slows dramatically at about 20 cm (8 in) diameter [at 1.4 m (4.5 ft)], regardless of whether it took 20 years (on a good site) or 45 years (on a poor site) to reach this height (Fig. 3.17; Oliver et al., 1986a).

Height of dominant trees is often used instead of age when comparing stand structures. Dominant trees are used because heights of more suppressed trees are not as independent of stand history as dominants are. Stand growth models such as TASS (tree and stand simulator; Mitchell and Cameron, 1985) index mortality, tree size, and stand volumes to dominant heights instead of stand ages. Measures of stand crowding, such as the spacing/top height ratio (Wilson, 1946) or the Douglas-fir competition/density management diagram (Drew and Flewelling, 1979) also use dominant tree heights as the index of stand development.

Dominant heights cannot be considered a complete substitution for age in all aspects of stand development, since foliage age, annual ring width, and number of branch whorls are directly related to stand age. Some evidence also suggests diameter increase is slightly greater for a given height change and spacing on poor sites. Also, intermediate and suppressed trees tend to live longer at a given position on droughty sites, since more light reaches them through the thinner dominant and codominant foliage compared with moister sites. The longer life of intermediate and suppressed trees can also give poorer sites greater standing volumes than good sites at the same dominant heights and initial spacings.

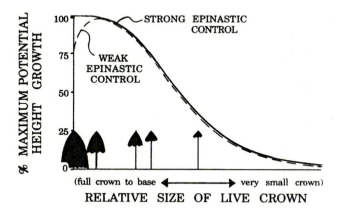

Figure 3.10 If the relative crown size becomes too small, height growth is curtailed. Species with relatively weak epinastic control (*dashed line*) may also curtail height growth when open-grown, unlike species with stronger epinastic control (*solid line; from Mitchell, 1975*). (See "source notes.")

Variation in height growth

Although more constant than diameter growth on a given site, a species' height growth patterns vary with shading, tree vigor, elevation, and climate change as well as between sites and individuals.

Both extreme side shade and shade from above can reduce height growth. Understory trees often exhibit periodic slowdowns and accelerations of height growth as they are released and later suppressed again by a disturbed and reclosing overstory. During times of unsuppressed growth, small trees which were formerly suppressed by high shade grow at a rate which parallels height growth of dominant trees which were never suppressed on the same site (Fig. 3.11; Oliver, 1976; Hoyer, 1980; Larson, 1982; Jaeck et al., 1984). Determining site growth potential through site index is difficult in these stands unless dominant trees are dissected and rates of unsuppressed growth are compared with normal site index curves. Obtaining a site index from "effective" or "economic age" (Chapman and Demeritt, 1932) by extrapolating diameter growth after release to the core has not proven accurate (Oliver, 1976).

Height growth of very young trees is apparently as strongly affected by the tree's vigor as by site productivity. As an extreme example, longleaf pine seedlings can remain in a *grass stage* with arrested height growth for up to 20 years, depending on the seedlings' vigor (Wahlenberg, 1946). When vigorous enough, they pass out of the grass stage and grow tall rapidly. During some years, Pacific silver fir saplings grow less than

5 cm (2 in) compared with 30 cm (12 in) in a normal year, apparently because of environmental influences (Crawford et al., 1982).

By utilizing stored energy of the preformed root system, stump and root sprouts initially grow more rapidly in height than individuals of the same species originating from seeds do. Sprouting vigor and subsequent height growth rates in many species decline with age, size, or degree of suppression of the parent tree (Solomon and Blum, 1967). Alternatively, lammas growth does not appear to reduce height growth of trees the following year. Height growths of many young Douglas-fir plantations in western Washington have exceeded expectations where weed competition has been controlled and vigorous seedlings planted. Where 50-year height of the previous stand was measured at 41.1 m (135 ft), growth of new, 20-year-old plantations is projected to reach 47.2 m (155 ft) at 50 years if continued.

Tree height at a given age is an imperfect indication of site productivity, since two stands of the same species may vary greatly in growth rate and accumulated volume even if the dominant trees are the same height and age. The variation may result from different stand structures, or different soil and climatic conditions governing the height growth. For example, excessively moist and excessively dry sites can both lead to slow tree height growth; however, moist sites allow more cambial growth later in the growing season, thus leading to more volume growth in the stand.

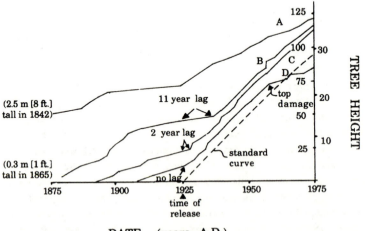

DATE   (years A.D.)

Figure 3.11 Height-age curves of four selected hemlocks released in approximately 1925. Upon release from high shade, shade-tolerant trees with large crowns often act as nonsuppressed trees of the same size, regardless of age. After an initial "lag period" proportional to their height, released trees grew in a height-age trajectory similar to seedlings of the same species which were never suppressed, except for very tall trees (*Oliver, 1976*). (See "source notes.")

Regeneration Mechanisms

Trees can regenerate by several sexual and asexual methods. These regeneration mechanisms will be discussed in Chap. 4.

Root Growth

Primary growth of roots creates an underground framework similar to the aboveground shoot growth. Unlike most shoots, roots do not have buds, although they do differentiate and elongate at varying rates during different seasons.

Like shoots, root patterns vary among species and with changes in the environment (Toumey, 1929a; Lyr and Hoffmann, 1967; Sutton, 1969). Root growth begins during germination with extension of the radicle into the soil. In monocots, this taproot soon aborts, and all subsequent roots are relatively weak, adventitious roots. In species such as hemlocks, this taproot grows relatively little. Other species such as longleaf pines have a very strong and woody taproot which dominates the lateral roots somewhat in the way the trunk dominates the branches. The deep taproot provides stability in heavy winds (Wahlenberg, 1946). The extent of this taproot, as with other roots, can be modified by the soil conditions. High water tables or impermeable layers can inhibit taproot development even in species which form large taproots in deep, well-drained soils. Wind can cause cracks and breaks in roots (Hintikka, 1972).

Lateral roots grow from four locations on the tree (Zobel, 1975):

1. The radicle, or primary root
2. The pericycle of other roots, giving rise to "lateral" roots
3. Plant parts other than roots, producing adventitious roots
4. The basal region—approximately the root-collar yielding outward growing "basal roots"

Basal and other lateral roots grow outward generally parallel to the soil surface. In uniform soils with easy penetration, these roots grow in a straight line until meeting an obstacle. They follow a path of least resistance around it, then generally resume growth in the original direction (Lyford and Wilson, 1966; Wilson, 1966). Growth requires pressure along the root to allow penetration of the tip through the soil. Where an air space does not provide pressure, the root may elongate backward, filling the space and creating a root ball rather than extending through the soil.

Basal and lateral roots produce secondary roots at distinct angles, similar to secondary branches along a stem. These roots also produce smaller branches and, eventually, very branched, fine, nonwoody roots which are the primary nutrient absorption sites on the roots (Lyford and Wilson, 1964; Lyford, 1980; Fig. 3.12A). These nonwoody, branched root masses are sometimes called *root fans* (Lyford and Wilson, 1964). The combination of woody and nonwoody roots on a tree is referred to as *heterorhizy*. It is unclear if all species have this heterorhizy condition.

In addition to the regular branching patterns, roots may form distinctive forks when the root tip is killed by the browsing of animals, self-abrasion against a rock, unfavorable weather, or other factors (Lyford and Wilson, 1964; Lyford, 1980). The nonwoody roots may become woody roots or act simply as temporary absorbing appendages parallel in function to leaves. Roots also develop various mycorrhizal associations with fungi which allow them to absorb more nutrients and/or water (Kramer and Kozlowski, 1979). As the lateral roots expand and proliferate outward, the root fans grow outward in "waves," die, and then regrow all along the woody roots from adventitious meristems, thus perpetuating the absorptive surfaces. Litter produced by shedding of fine roots often approximates litter produced by shedding of leaves but may be much higher on poor sites. Root shedding may also contribute more nitrogen to the forest soil (Kozlowski et al., 1991).

A. Northern red oak

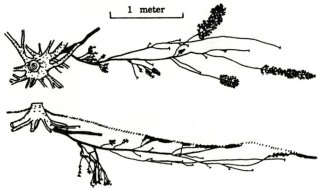

B. Chestnut oak
Red maple

"c", solid lines --10.0 meters (33 ft.) tall:
8.6 cm (3.4 in.) DBH
"r", dashed lines--10.4 meters (34 ft.) tall;
9.7 cm (3.8 in.) DBH

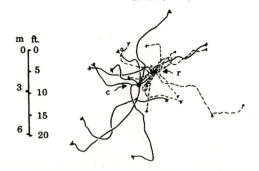

Figure 3.12 Root growth patterns. A. Much of the absorption from roots comes from the small-diameter, nonwoody roots in the forest floor. These roots grow from smaller woody roots returning to the soil surface from the major lateral roots. Top view (*above*), side view (Lyford, 1980). B. Unlike tree crowns, tree roots overlap and intertwine with neighbors' roots profusely, as shown by maps of the major lateral roots of a juxtaposed red maple and chestnut oak (*top view*). Roots often grow beyond neighboring trees and even under another's main stem (Stout, 1956). (See "source notes.")

The depth of lateral roots varies with species. Genera such as spruces and birches frequently have major lateral roots within 20 cm (8 in) of the surface, while many oaks have lateral roots over 50 cm (20 in) deep. In more aerated and easily penetrable soils, roots of all species are found deeper; however, they seem to maintain a predictable stratification by species (Stout, 1956). Taproots have been found to penetrate 4 m (12 ft) into dry sands, and lateral roots have been found at nearly this depth; however, most roots of temperate trees are found within the top 60 cm (2 ft) of soil (Hopkins and Donahue, 1939; Sculley, 1942; Pritchett, 1979; Vogt et al., 1981*a,b*). Roots extend more rapidly and with fewer branches through sands with little water or nutrient value, and proliferate in more favorable nutrient and moisture conditions (Heyward, 1933; Ruark et al., 1982). Trees in less competitive positions have less extensive, more shallow, less branched, and lower densities of roots compared to more competitive trees of the same species (Kozlowski et al., 1991).

Roots can extend laterally 40 m (125 ft) or more from the tree trunk. No territorial separation of roots among trees or concentration of roots beneath a tree's own crown occurs. Roots grow entwined with those of other trees, beneath the crowns of other trees, and often directly beneath the stem of a neighbor (Fig. 3.12B; Stout, 1956; McMinn, 1963). Roots of some tropical species in nutrient-poor soils grow beneath other trees and up their stems, absorbing nutrients in the rainwater flowing down the other trees' stems (Sanford, 1987).

Conifers generally have many fewer but larger roots than hardwoods do (Kozlowski and Scholtes, 1948; Kozlowski, 1971). The thicker roots of conifers may allow them to penetrate less well developed soils effectively and may partly account for their relative site insensitivity (Voigt, 1968).

Soon after a major disturbance the warmth of direct sunlight on the forest floor and possibly the migration of nutrients through the soil may stimulate fine absorbing roots to develop within the mineral soil. The roots may grow upward if litter accumulates with stand age. In many cases woody roots will grow upward from the major laterals toward the soil surface and form numerous absorptive roots. Lyford (1975) found 1,100 absorbing root tips in a cubic centimeter of soil near the soil's organic-mineral interface. Fine roots tend to reoccupy a site rapidly and parallel leaf cover (Vogt et al., 1981*b*), although the amount of large woody roots continues to increase throughout the life of the tree (Santantonio et al., 1977).

Roots vary among species in their ability to grow in different soil media. Hemlock roots often grow in rotting wood, many pines can endure extremely dry soils, and tupelos and baldcypresses can tolerate extended anaerobic periods. Many species' roots cannot live in anaerobic conditions. Where a water table fluctuates, the flooded roots of some species die, only to regrow downward as the water table recedes. Other species endure a temporarily high water table for different lengths of time. Buried soil horizons and organic matter caused by windthrows and other disturbances create diverse soil microenvironments in which roots develop. Consequently, absorption roots are not evenly distributed throughout the soil.

Trees often germinate on unstable media such as logs, stumps, and the upturned soil of windthrow mounds. The tree will remain stable if the roots can grow down to a stable substrate and enlarge enough to support the tree before the substrate rots or erodes. The ability of tree roots to form stable props depends both on the disintegration rate of the substrate and on the growth rate of the roots. This root growth is strongly related to the amount of sunlight the tree receives. As the tree grows, it becomes heavier and presses downward on its base, often creating a curved stem and cracks between its roots. Even trees with strong prop roots develop tension cracks between the roots. These cracks eventually spread upward through the stem. Like most wood, tree roots are much stronger in tension than in compression; consequently, the support roots keeping a tree from falling over are the tension ones on the upper side of a leaning tree (Mergen and Winer, 1952; Mergen, 1954).

Where roots of the same or different trees overlap and enlarge, a graft may form which transmits water, nutrients, chemicals, and pathogens from one tree to another (Lyford, 1980; Reynolds and Bloomberg, 1982). Root grafts usually form between trees of the same species (Bormann, 1966). These grafts may allow survival of trees which would otherwise die from suppression; however, only a limited quantity of substances can be transported over the small cross section of a graft; so these root grafts usually do not support vigorous growth of otherwise suppressed trees.

Crown and Tree Development

A tree of simple architecture, preformed growth, and strong epinastic control will be used first to describe development of the crown and resulting volume growth (Fig. 3.13). Variations of crown development in other growth forms will then be discussed.

TREE # 1. OPEN GROWN

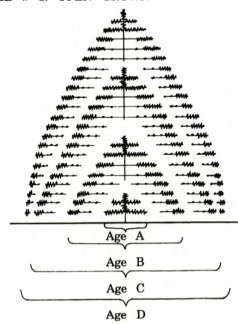

Figure 3.13 Schematic, open-grown tree with strong epinastic control. In all crowns, the leaves are found primarily in a shell around the periphery.

When a tree grows in the open, the crown expands upward and horizontally as its terminal and lateral branches grow outward. The tree alters the surrounding light, moisture, oxygen, and nutrient regimes and soil area within and beneath the canopy as it expands. The present discussion will concentrate on changes in light regimes, although other factors are also important.

Shade cast by leaves and twigs reduces the light intensities from 100 percent of full sunlight just above the canopy to between 3 and 30 percent below the lowest living leaves, depending on the species, moisture, and other conditions (Andersson, 1966, 1969; Kira et al., 1969; Swank and Schreuder, 1974; Honda and Fisher, 1978; Waring, 1983). Leaf temperatures and wind speeds also decrease and humidity generally increases within and beneath the canopy. These changes are quite dramatic over the short distance through the foliage layer of 1 to 3 m (several feet) or less.

The orientations, morphologies, and physiologies of leaves on the same tree differ at different levels within the canopy. Photosynthesis of leaves perpendicular to the sun does not increase markedly with increasing light intensity above about 25 percent of full sunlight (Fig. 2.5). Below a lower threshold of light for each species, respiration exceeds gross photosynthesis, net photosynthesis becomes negative, and leaves die.

During shoot elongation, leaves expanding into full sunlight develop thick cuticles, extra cell layers, and other mechanisms which make them photosynthetically efficient as well as resistant to heat desiccation and dehydration at high light intensities. These leaves are termed *sun leaves*. Leaves developing in shade, *shade leaves*, are much thinner and less resistant to heat and moisture stress but photosynthesize more efficiently at low light intensities.

Leaves (and needles) grow from all sides of a branch; however, the *pulvinus* (at the base of the petiole) orients leaves at the most efficient angle to sunlight. Leaves which first expand after budbreak into full sunlight are nearly parallel to the sun's rays. This orientation reduces overheating of the leaf but does not proportionately decrease photosynthesis. Leaves become increasingly perpendicular to the sun if they expand after budbreak into lower light intensities inside the crown or in other shaded conditions (Busgen and Munch, 1929). The flatter orientation allows them to absorb more light inside the shaded crown. Pulvini also move leaves in high shade during expansion so they will not overlap, but form single, flat layers (Wilson, 1966). These flat, monolayers of leaves at low light intensities have been noted in many forests (Horn, 1971; Oldeman, 1972). The entire crowns of trees in lower canopy strata often have this shaded, monolayer appearance, although the crown would contain more layers if grown in full sunlight. The physics of leaf placement deep within a crown has been described by Horn (1971). Deeply shaded leaves are usually found parallel to the ground, since most filtered light comes from directly above, through the path of least intervening foliage. A canopy opening or stand edge producing light from a more horizontal direction may lead to shade foliage oriented perpendicular to this direction, at an angle to the ground.

Leaf orientation in most species is fixed during initial shoot elongation and does not change even if sunlight conditions change. In trees with stationary, perennial leaves, the year of release of a branch from shade to full light can be dated. Annual shoots predating the release maintain leaves or needles oriented in planes perpendicular to the sun, while annual shoots which grow after the release contain leaves or needles pointing at steep angles to the sun as well. By contrast, the pulvinus of some oak species continues to change the leaf angle throughout the day in response to sunlight (Wells and Shunk, 1931).

Leaf angles change less regularly through an evergreen crown than a deciduous one. Leaf angles become increasingly flatter as one moves downward through a deciduous crown. On the other hand, evergreen foliage originally developed in the sun is outgrown by new shoots and relegated to the shade in subsequent years. These needles retain their sun-induced angle and anatomy even in deep shade. They are as photosynthetically efficient as shade-grown needles of younger twigs around them. A greater total leaf surface area is generally found in evergreens, since leaves of several years can persist. In deciduous trees with sustained growth, leaves which expand late in the growing season can have a different appearance (Critchfield, 1960; Clausen and Kozlowski, 1965; Zimmerman and Brown, 1971). They shade older, sun-adapted leaves of the same season and reduce their photosynthetic efficiency.

Light intensity eventually becomes so low within the crown that leaf respiration exceeds photosynthesis and no leaves of that species are found (Monsi et al., 1973). The minimum light intensity at which leaves are found within a crown varies by species and site. Certain species, such as hemlocks, beeches, and some true firs and hickories can survive under more shade than other species can, since their leaves can photosynthesize at lower light intensities. These are referred to as *relatively shade-tolerant species*. They generally keep dense canopies which cast dark shadows, since leaves are maintained at low light levels. Other, shade-intolerant species such as tulip poplars, cottonwoods, and many pines cannot survive under as much shade. Consequently, they have less dense canopies, which cast less shade, since their leaves are not retained at very low light intensities.

Total leaf surface area of a canopy is greater than the amount of ground area because of the overlap and steep angles of leaves. The values of the *leaf area index* (Watson, 1947)—the ratio of total leaf area to ground area covered—varies dramatically between species and sites. Shade-tolerant and evergreen species on mesic sites generally have the greatest leaf area index, with values of 12 for Norway spruce (Assmann, 1970) and higher estimates for Douglas-fir, although the accuracy of the bigger estimates is uncertain (Marshall and Waring, 1986). One-sided leaf area index values for deciduous species range from 3 for relatively intolerant oaks to over 6 for shade-tolerant beeches (Assmann, 1970).

Trees retain fewer shade leaves on dry sites; therefore, more light reaches the forest floor. Leaf surface area per hectare generally declines from moist sites to dry sites (Grier and Running, 1977). Even for the same species, more light is generally found beneath stands on dry sites, indicating that the trees have less leaf area. Suppressed trees on droughty sites often die after dry summers, and other trees loose their lower leaves. Growth in the stand would not be reduced in proportion to the reduction in leaves, since primarily lower, shaded leaves which photosynthesize very little are lost (Waring et al., 1981; Waring, 1983).

The dramatic change in light intensity from full sunlight to shade beneath a forest canopy is often unnoticed by the human eye, which adjusts quickly to changing light intensities. The light beneath the canopy can be as low as 3 percent of full sunlight, but may be over 30 percent for shade-intolerant species on dry sites (Reifsnyder and Lull, 1965; Geiger, 1965). Differences in minimum light intensities for survival of leaves of shade-tolerant and shade-intolerant species may not be more than 5 to 10 percent of full sunlight on any given site (Burns, 1923; Bates and Roeser, 1928). Heavily shaded foliage of shade-tolerant species probably contributes little to volume growth; nonetheless, this foliage enables tolerant trees to survive low light intensities in the understory of mixed stands and so affects the stand structures and responses to disturbances.

Growth of a tree in the open

As terminal and lateral branches expand in an open-grown tree, a shell of foliage expands with sun leaves on the outside and shade leaves and (in evergreens) older sun foliage on the inside. Farther inside the shell, living leaves are not found. The greatest depth of live shell is parallel to the most intense sunlight, which comes from high angles at midday, rather than perpendicular to the tree crown's surface. The thickness of the foliage shell perpendicular to the crown's surface is greater at the top of the tree than on the sides for this reason.

The crown surface area exposed to sunlight is in two dimensions (width and depth), while wood accumulates on a tree in three dimensions (width, depth, height). The relation of this area to volume may partly account for certain growth patterns which are described by squared and cubed mathematical powers (Fig. 2.8; Yoda et al., 1963; White and Harper, 1970; Drew and Flewelling, 1977, 1979; Westoby, 1977; Miyanishi et al., 1979; Lonsdale and Watkinson, 1983; Perry, 1984). Of course many other resources are also available in the roughly two-dimensional soil surface—soil moisture, nutrients, root oxygen, and even soil for the physical support of trees.

Trees occupy more light growing space as their crowns expand. Simultaneously, roots occupy more soil growing space. Full occupancy of available growing space is often assumed to be when crowns of adjacent plants touch, thus utilizing all available light. Especially on poor sites this assumption may be an oversimplification, since moisture or nutrients can become limiting before sunlight. On these sites, all available growing space may be occupied while the crowns of adjacent trees are still far apart. The association of crown closure with full site occupancy is a useful generalization if its shortcomings are recognized. It may also be realistic on seasonally droughty sites where light is the limiting factor during active growth in the spring. Where moisture is more limiting, growing space may become limiting at a time of "root closure." The intertwining of roots suggests this root closure is more complicated than crown closure.

Growth with side shade

A light-impermeable, circular wall surrounding a tree will first be used to illustrate tree growth in a forest (Fig. 3.14).

The walls of tree 2 resemble conditions in a close, evenly spaced, even-aged stand where full sunlight is cast from a large arc above the tree and its neighbors, which are about the same height. Shade cast by the walls, as by the neighbors, is *side shade*. The eccentric wall around tree 3 resembles an unevenly spaced stand. Both trees are assumed to be preformed and geomorphic and to exert strong epinastic control. They have the same potential height growth and crown expansion as tree 1 (Fig. 3.13A) when not modified by the environment.

Open growth. Trees within both the narrow (tree 2) and wide (tree 3) walls expand freely as if growing in the open until the lower limbs reach the walls. Before this, the trees expand into unoccupied growing space (or growing space occupied by relatively noncompetitive shrubs and herbs). Growth into unoccupied space is referred to as *free growth* (Hummel, 1951; Jobling and Pearce, 1977). Tree growth before crown closure is termed *open growth*. O'Conner's (1935) "free growing" is here described as "open growth." "Free growth" has also been used to describe the sustained growth form; this usage will not apply in this book. Open-grown trees always are free-growing; however,

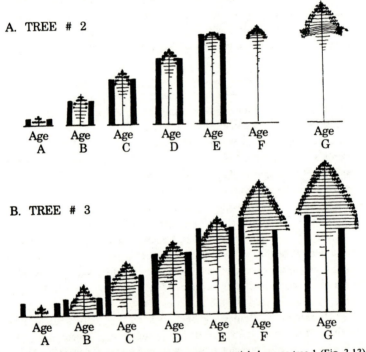

Figure 3.14 Growth of individual trees with the same potential shape as tree 1 (Fig. 3.13), except that a constantly rising wall surrounding the tree changes the tree growth. Lower branches abrade against the wall, and the wall and upper branches shade lower branches, causing the lower branches to slow their growth and then die, and the crown base to "rise." Dead branches may fall off or persist, depending on the species. A. Tree 2. At age D, there is so little photosynthetic surface area relative to the respiration area of the limbs, stem, and roots that height growth slows. The tree responds to removal of the wall (time F) by increase of the foliage area with height growth (and retention of the lower crown), buildup of nonfunctional crown, production of water sprouts, and densification of foliage. If the species had weaker epinastic control, the limbs would spread faster, creating a more flat-topped crown. Because tree 2 became repressed in height growth, it took very long to rebuild a crown and increase growth. B. Tree 3. The wall around tree 3 is at irregular distances from the tree, causing an asymmetric crown. At time F, the wall stops rising, and tree 3 grows above it. Crown size increases with tree height, since the base of the crown no longer recedes.

free growth reinitiates following a thinning or similar disturbance which removes adjacent trees and allows the crown to expand unimpeded, even if the trees are no longer open-grown.

Crown closure. Lower limbs of tree 2 reach the walls first, since both trees expand in height and crown width at the same rate. Limbs shaded by the walls grow slowly. They eventually touch the wall and, if there is wind, abrade against the wall. If an expanding tree touches adjacent trees instead of a wall, branches of the trees first overlap and then interlock, and concomitantly slow in lateral extension (Cochrane and Ford, 1978). The limbs above this grow over the restricted limbs and further shade the lower foliage. These branches in turn slow as they reach the wall and become abraded (Tarbox and Reed, 1924; Spurr, 1964), only to be shaded later by even higher limbs. The growth rate of the tree in a confined space is reduced compared with that of the open-grown tree because photosynthesis is reduced in the shaded lower branches. Eventually, the lower, confined branches become so shaded that they die and the base of the live crown retracts upward. The period when the branches first begin to slow down because of lateral confinement is referred to as the beginning of *crown closure*. Complete crown closure occurs when the lower, shaded branches die and the crown begins receding. Time of crown closure and size of the living crown are directly related to the distance of the walls, i.e., the spacing in single-species, even-aged stands (Gevorkiantz and Hosley, 1929; Eversole, 1955; Kramer, 1966; Mitchell, 1969, 1975; Curtis and Reukema, 1970; Cole and Jensen, 1982). The modification of tree shape has been referred to as a *plastic response* (Harper, 1977; Hutchings and Budd, 1981). Consequently, at crown closure the trees enter a *plastic phase*—also described as a *zone of suppression* by O'Conner (1935).

Crown closure occurs unevenly around a tree where distances to shade are unequal (Stiell, 1982). For example, the branches on tree 3 (Fig. 3.14) reach the closer part of the wall first, slow in growth, become shaded, and eventually die as in tree 2, but at a later time since the wall is farther away. Branches on the other side of tree 3 continue growing, creating an unbalanced crown. The crown of tree 3 is longer and wider after crown closure than that of tree 2, since the lower branches live longer in the larger space between the walls.

Functional live crown. Each branch must photosynthesize enough to provide its own stem and needle respiration requirements or it will die (Rangnekar et al., 1969; van der Wal, 1985). Before dying, shaded branches eventually produce so little photosynthate that no excess is available to contribute to growth of the main tree stem (Mar-Moller, 1960; Woodman, 1971). These branches are termed *nonfunctional*. Nonfunctional branches may take photosynthate away from the main tree stem for the branch to use for respiration, thus detracting slightly from total stem growth. Pruning of lower, shaded branches has been suggested to remove a photosynthate "sink" and thus increase tree growth. (Removing branches to produce knot-free wood is generally a stronger reason for pruning.) The live crown above these nonfunctional branches makes a net positive contribution to stem growth and is referred to as the *functional crown*.

Branch and knot size. The length of the branches upon death depends on distance to the confining wall or adjacent tree; (Bennett, 1960; Smith et al., 1965; Abetz, 1970; Drolet et al., 1972; van Laar, 1973; Carter et al., 1986b). The diameter of a branch is determined partly by genetics (Smith, 1961) and partly by the branch age and tree spacing (Eversole, 1955; Cromer and Pawsey, 1957; Grah, 1961; Stiell, 1964, 1966; Sjolte-Jorgensen, 1967; Mitchell, 1969, 1975; Jack, 1971; James and Revell, 1974; Brazier, 1977; Sonderman, 1985). Branch size varies considerably within a species under similar environmental conditions. Some individuals produce many small branches, and others produce a few large branches. A given size and shape of tree crown with many small branches produces smaller knots in the stem but contains more total branch stemwood and cambial surface area than do a few large branches supporting a crown of the same size, shape, and amount of foliage. Each small branch has less foliage, more respiration area, and less photosynthate left for stem growth. A tree genetically predisposed to produce many small branches may thus be less efficient if its crown expands widely than a tree with fewer, large branches.

Dead branches remain attached to the bole for varying times depending on the species and size of branches. Larger branches generally take longer to fall off. Southern pine, yellow poplar, northern red oak, and red alder branches fall off ("self-prune") soon after they die, while Douglas-firs, eastern white pines, Sitka spruces, and southern red oaks can retain their dead branches for many decades.

Branches are invaded by fungi and begin to decompose when they die. Small branches do not contain heartwood; upon death, the living sapwood of the main stem resists movement of the decay into the stem. Large branches of some species contain heartwood, which allows heart-rotting fungi to enter the main stem upon death of the branch. Wide spacings which encourage development of large limbs may ultimately lead to heartrot in susceptible species. These species may have an upper limit to the size of limbs which can be naturally or artificially pruned without the tree's becoming susceptible to heartrot. This upper limit would be reached at the limb size in which heartwood extended into the limbs.

Pathogens, such as Indian paint fungus (*Echinodontium tinctorium*) in grand fir, can enter the main stem of advance regeneration through very small dead branches. The fungus remains quite inactive even if the tree is released and grows larger. Much later, a disturbance which weakens the tree—especially if it opens a wound and exposes the xylem to air—will stimulate the fungus to grow vigorously and rot the center of the standing tree (Aho, 1982).

Growth and allocation of photosynthate

Tree volume increases with time in a sigmoid curve if there is no hindrance to its crown expansion [tree 1, Fig. 3.15; similar to Ovington (1957)]. It grows slowly at first until it produces enough foliage, roots, and other tissues to allow rapid assimilation. Then, it grows rapidly during the grand period of growth as the foliage and absorptive roots increase more rapidly than the respiring tissue.

A tree's photosynthates are allocated to different uses in an order of priorities (Boysen-Jensen, 1932; Bormann, 1965; Gordon and Larson, 1968; Assmann, 1970; Rangnekar and Forward, 1973; Kellomaki and Kanninen, 1980; Mitchell and Goudie,

1980; Waring and Pitman, 1983; Kozlowski et al., 1991). Although the details and mechanisms of these priorities have not been worked out, certain generalities hold. The hierarchy is not absolute, and allocations often overlap:

1. The first priority for photosynthates is maintenance respiration of the living tissue. The respiration is needed to keep the tree alive increases as it grows in height, diameter, and number of branches and roots. Respiration rates partly depend on temperatures; high temperatures cause plants to respire more rapidly (Kramer and Kozlowski, 1979).
2. Production of fine roots (Harris et al., 1978; Fogel and Hunt, 1979) and leaves follows respiration in order of priority. This production not only incorporates carbohydrates into the new roots and leaves, but also respires more photosynthates in the process. Root production may have slightly lower priority than production of new leaves (Kotar, 1972; Magnussen, 1980; Magnussen and Peschl, 1981).
3. Flower and seed production take next priority (Eis et al., 1965). Height and diameter growth are often slowed during active seed years (Kozlowski, 1971), and foliage production may even be reduced (Morris, 1951). Flower and seed production can be stimulated by release and fertilization (Croker, 1952; Allen, 1953; Wenger, 1954; Kozlowski, 1971; Yocum, 1971).
4. Lower priority goes to primary growth—terminal and lateral branch growth and root extension (Fig. 3.10). Reaction tissue—such as tension and compression wood and scar tissue—may take a higher priority than height growth. Renewal of phloem may take equal or higher priority.
5. If still more photosynthate is available, growth will go to adding xylem (diameter growth; Gordon and Larson, 1968; Rangnekar and Forward, 1973; Kellomaki and Kanninen, 1980; Waring and Schlesinger, 1985) and to developing resistance mechanisms to insects and diseases (Keen, 1936; Westveld, 1954; Blais, 1958; Sartwell, 1971; Coulson et al., 1974; Schoeneweiss, 1975; Lorio and Hodges, 1977; Amman, 1978; McLaughlin and Shriner, 1980; Waring and Pitman, 1980; Waring et al., 1980; Manion, 1981; Mitchell et al., 1983; Waring, 1983; Shigo, 1985). Resistance mechanisms include resins and other chemicals in many species. Some resistance mechanisms may take higher priority. Of course, added respiration is necessary for all growth (Chung and Barnes, 1977).

As an example, allocation of photosynthates among respiration, root and needle growth, and stem growth varied among three classes of European ash trees of different vigor (Boysen-Jensen, 1932; Assmann, 1970). More vigorous, dominant trees satisfied respiration demands with 26 percent of their gross photosynthesis, while their root and leaf growth demands used 32 percent. The remaining 42 percent was allocated to stemwood growth. The least vigorous, suppressed trees allocated 50 percent of gross photosynthesis to respiration, 42 percent to root and leaf growth, and only 8 percent to stemwood growth. The less vigorous trees with smaller crowns manufactured less gross photosynthate than more vigorous trees, and only a very small amount of this was available for stem growth.

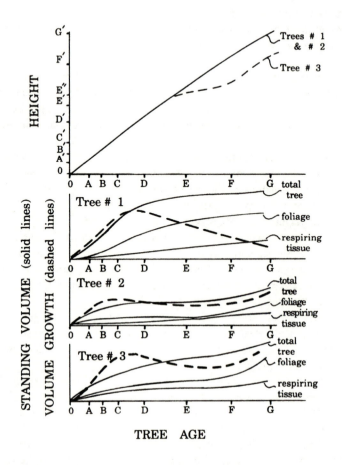

Figure 3.15 Schematic of the trees in Figs. 3.13 and 3.14 showing changes of different characteristics with time (the letters on the horizontal axis correspond to the times shown in Figs. 3.13 and 3.14). Potential height growth is assumed linear except when tree 2 becomes severely suppressed and then released. In open-grown tree 1 (Fig. 3.13), the foliage increases so much more rapidly than the respiring tissue that residual photosynthate for total tree growth (*dashed line*) increases, and then decreases. In tree 2, grown in a narrow space, the foliage soon stops increasing, respiring tissue continues increasing, and total tree growth (*dashed line*) slows until the tree is released and the crown again increases. In tree 3, grown in a wide space, the crown stops increasing when much larger than in tree 2, and total tree growth (*dashed line*) slows less dramatically before release.

Suppressed fir seedlings produce proportionately fewer roots than more vigorous seedlings, suggesting photosynthate allocation to root growth is curtailed (Kotar, 1972; Herring and Etheridge, 1976; Magnussen, 1980; Magnussen and Peschl, 1981). Suppressed trees of many species can live for many years without adding annual rings throughout their stems, but may create partial rings in parts of the stem (Larson, 1956; Bormann, 1965; Oliver, 1978*a*; Stubblefield, 1978).

In pure stands of western redcedar the dominant trees were found to grow vigorously, while less vigorous trees with smaller crowns and less direct sunlight were dying from *Armillaria mellea,* a fungus which attacks weakened trees (Nystrom, 1981). Apparently the less vigorous trees could not allocate resources to resistance mechanisms. Lodgepole and ponderosa pines growing slowly in narrow spacings cannot produce enough photosynthate to resist insect attacks, unlike more vigorous trees in wider spacings with larger crowns (Waring and Pitman, 1980; Larsson et al., 1983; Mitchell et al., 1983).

A tree's foliar canopy is sometimes referred to as a carbohydrate *source,* and all other living tissue as a *sink.* Individual tree growth is closely related to crown size (Holsoe, 1948; Krajicek and Brinkman, 1957; Berlyn, 1962; Hamilton, 1969; Mitchell, 1969, 1975; Jobling and Pearce, 1977; Oliver, 1978a), and survival and growth can be considered in terms of relative sizes of the source and sink. *Live crown ratio*—an index of tree vigor—is the ratio of crown length [an estimate of the potential photosynthetic surface area (the source)] to tree height [an estimate of the total respiration area (the sink)].

Inside the sapwood of large stems is the dead heartwood, which serves no function except to provide stability and as a place to dump metabolic wastes. In fact, hollow trees whose heartwood has decomposed can grow as well as trees of comparable size with sound centers. The volume of living, respiring tissue in a stem, therefore, may be more proportional to its surface area or to the leaf canopy area than to bole volume (Dixon, 1971; Grier and Waring, 1974; Waring et al., 1977; Whitehead, 1978; Gholz et al., 1979; Running, 1980; Kaufmann and Troendle, 1981). This relation is similar to the "pipe stem" model in which the functional xylem supplies water to the branches and heartwood xylem reflects the past history of water conductance (Shinozaki et al., 1964a,b; Oohata and Shinozaki, 1979). The dead, nonfunctional centers of large trees are often considered as living tissue in many estimates of standing volume, basal area, or biomass (Waring, 1983). The presumption that the entire tree is alive until it completely dies may help explain confusion in growth and yield estimates when total tree or stem volume or basal area is used as the dependent variable (Chap. 15).

Allocation of photosynthate varies greatly among sites. More photosynthate is allocated to fine root production on infertile or arid sites, so that less is available for stem growth, as has been found for Douglas-firs, yellow poplars, and Scotts pines (Agren et al., 1980; Aronsson and Elowson, 1980; Axelsson, 1981; Keyes and Grier, 1981). Greater allocation to fine roots may serve to increase root absorption area to compensate for the scarcity of soil moisture and nutrients. Total photosynthesis on poor sites may have been underestimated when only stem growths were compared between sites.

Sudden reductions in crown size. Sudden reductions in crown size by artificial pruning (Petruncio, 1994; Maguire and Petruncio, 1995), volcanic tephra deposition (Segura, 1991; Segura et al., 1994), or insect defoliation (Twery, 1987) affect tree growth in many ways. The sudden reduction reduces both total photosynthesis (Helms, 1964) and evapotranspiration (Oksbjerg, 1962), unless the stomata are artificially kept open (e.g., by volcanic tephra; Seymour et al., 1983; Hinckley et al., 1984). Small reductions in lower, relatively nonfunctional branches (e.g., one third of the crown length from the base) may not dramatically reduce growth (Maguire and Petruncio, 1995) and may even slightly stimulate growth (O'Hara, 1991). Greater reductions in living branches reduce diameter growth, then height growth (Maguire and Petruncio, 1995; Petruncio, 1994; Segura, 1991). If too much of the live crown is killed rapidly, such as by extreme pruning, the tree apparently can not reduce its respiring tissue rapidly enough to balance the lower photosynthesis and the tree will die. Sitka spruces and western hemlocks in Southeast Alaska sometimes died if more than 50 percent of their live crowns were pruned at a single time (Petruncio, 1994). Tephra depositions or insects killing large amounts of the living crown on large trees similarly caused the tree to die, probably because the remaining foliage could not maintain the respiration demands (Segura, 1991).

Regular mortality. As a tree grows, its respiring tissue increases (Rook and Carson, 1978; Waring and Schlesinger, 1985). Unless its photosynthetic surface area also increases sufficiently, a tree's respiratory requirements cannot be met and the tree eventually dies of starvation. Often, a weakened tree is attacked by insects or diseases or cannot support its own weight and falls over before this. When differential tree growth occurs in a stand, the crown sizes of slower growing trees become reduced, accelerating their weakening (Chap. 8). Death of weakened trees from insect or disease pests or simply by starvation is termed *regular mortality,* since it is an emergent (autogenic) property of the stand. By contrast, *irregular mortality* occurs when trees not normally becoming suppressed die by allogenic processes—such as fires, windstorms, freezing, or attack by pests.

Exactly when mortality occurs is difficult to ascertain. In general, the maximum number of living stems in a stand is inversely related to the stem sizes (Reineke, 1933; Yoda et al., 1963; White and Harper, 1970; Gorham, 1979; White, 1980; Westoby, 1981; Weller, 1987), and the maximum number of trees a stand can support decreases with age, since trees constantly grow larger. Regular mortality fluctuates with available growing space between seasons and years. Suppressed trees obtain enough moisture to survive during rainy years in normally droughty soils, but the lack of moisture in droughty years kills them. Death can occur in late summer droughts or in late spring when the suppressed trees do not have enough stored photosynthate to sustain shoot elongation and root growth. Intolerant trees or those with rapid height growth, broad crowns, or narrow spacing constraints most rapidly succumb to regular mortality, since all three factors rapidly decrease the proportion of photosynthesis to respiration tissues.

Individual tree growth. Growth of trees 2 and 3 diverges from that of the open-grown tree (tree 1) as their lower branch growths first slow and prevent their crowns from expanding (Figs. 3.13 and 3.14). Tree 2 slows in diameter growth first, and then tree 3, as the crowns become restricted, since less photosynthate is allocated to the low-priority diameter growth.

The confined trees continue normal height growth in the fixed growing space after crown closure (as the branches reach the walls; Fig. 3.14). As height growth continues, the live crown size and amount of foliage remain constant in each tree instead of expanding as in open-grown trees (Mar-Moller, 1947; Tadaki, 1966; Marks and Bormann, 1972; Long and Turner, 1975; Gholz and Fisher, 1982; Gholz, 1986). The gross photosynthate produced by each tree remains constant because of the constant crown size; however, the total living (respiring) tissue increases as the stem becomes longer and thicker. With growth, increasingly more photosynthate is utilized for maintenance respiration, and increasingly less is available for diameter (stemwood) growth (Fig. 3.16). Consequently, diameter growth closely reflects crown size and, hence, spacing (Fig. 3.17; Holsoe, 1948; Ralston, 1953; Bennett, 1960; Mackenzie, 1962; Hansbrough et al., 1964; Sjolte-Jorgensen, 1967; Hansbrough, 1968; Barrett, 1970, 1973, 1979, 1982; Curtis and Reukema, 1970; Jack, 1971; Mohn and Krinard, 1971; Balmer et al., 1975; Hatch et al., 1975; Krinard and Johnson, 1975; Oswald and Parde, 1976; Stiell, 1976; Sarigumba and Anderson, 1979; Johnstone, 1982; Lloyd and Jones, 1983; Omule, 1984; Oswald, 1984a,b; Liegel et al., 1985; Jones, 1986; Oliver et al., 1986a; J. H. G. Smith, 1986). Tropical tree plantations can dramatically reduce their diameter growth at the time of crown closure. This reduction is probably caused by a combination of the rapid reduction in crown size of the relatively flat-topped trees (reduction in photosynthesis) and the trees' continued high respiration demands in the hot climate (Galloway et al., 1995).

Early in a tree's life, proportionately more photosynthate is allocated to root growth (Ovington, 1957; Santantonio et al., 1977; Eis, 1986), mostly to nonwoody roots which die after several months or years and do not add permanent volume to the trees. Standing volume per tree, therefore, continues to increase after crown closure as the photosynthate allocation shifts from roots to stem. Insufficient cambial surface area to utilize the photosynthate may also limit growth in short (young), vigorous trees. The growth rate may increase until the cambial surface area becomes large enough to utilize all the photosynthate produced. Volume growth gradually slows as the respiratory tissue volume increases (Figs. 3.15 and 3.16). The sigmoid volume growth curve is common in tree growth, although its exact shape may vary dramatically (Hunt, 1982).

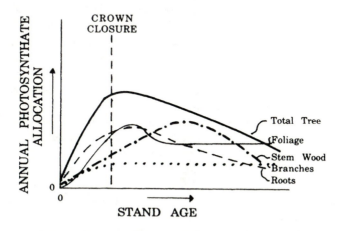

Figure 3.16 Schematic of annual photosynthate allocation to various tree parts before and after crown closure on sites where sunlight most limits growing space (*similar to Mar-Moller, 1954*). Note that the total tree volume growth per year probably peaks shortly after crown closure; stem volume growth (that which is not lost through root, limb, or foliage shedding) probably peaks even later (Fig. 3.15) because foliage (*Tadaki, 1966*) and fine root production (*Ovington, 1957; Eis, 1986*) do not decline until well after crown closure. On droughty sites, fine root production continues at a steady rate, rather than declining. Cumulative growth (Fig. 3.15) reflects an increase in stem volume.

Eventually, all trees slow in height growth and crown expansion (Fig. 3.9) even though photosynthesis and diameter and volume growth continue (Jensen and Long, 1983). Stand differentiation, with some trees dominating others, occurs more slowly; and the increase in diameter leads to more stable trees. Where height growth slows but diameter growth continues, dominant height may no longer be a suitable substitute for age in estimating tree and stand volumes.

Tree growth efficiency. Some parts of both conifer and hardwood crowns photosynthesize more efficiently than others (Labyak and Schumacher, 1954; Ford, 1985). The central part is most efficient, with the crown's top second, and the lower part least efficient (Assmann, 1970; Woodman, 1971).

Species with relatively flat crowns have less foliage at crown closure at a given spacing than species with conical or columnar crowns. Consequently, trees with flat crowns grow less rapidly unless greater photosynthetic efficiency compensates for their smaller crowns. Where the flat top is caused by weak epinastic control, any reduced growth caused by the flat crowns is accentuated, since the number, size, and vigor of respiring limbs is generally greater than in trees with strong epinastic control. Often trees with flat, inefficient crown shapes overtop and suppress trees with more efficient growth; there is not necessarily a close relation between efficiency of volume accumulation and competitive ability of trees. Trees with flat crowns may have an advantage at very low latitudes, where sunlight comes from directly above at midday, although sunlight still comes from a steep angle in the mornings and evenings.

Stands of trees with conical crowns have a greater canopy area per hectare than those with flat or rounded crowns. This greater exposed crown surface allows more photosynthesis—and hence greater volume growth per hectare. The greater volume growth of many conifer stands compared with that of broadleaf stands has been partly attributed to the ability of the conifers' cone-shaped crowns to photosynthesize more efficiently than the relatively flat crowns of broadleaf forests. Even within conifer forests, greater productivity of stands of certain species—hemlocks compared with Douglas-firs, for example—has been attributed to their (the hemlocks') narrower crowns. Hemlocks in a pure stand are generally shorter than Douglas-firs in a pure stand of the same age and site; but the hemlock stand often contains greater volume per hectare because of their narrower crowns (Atkinson and Zasoski, 1976). Other factors such as the greater shade tolerance of hemlocks may also contribute to its greater volume.

A columnar crown may photosynthesize most efficiently at very high latitudes (Fritschen et al., 1980). Trees with columnar crowns may have a smaller respiratory demand than trees with broad crowns of the same height, species, site, and foliage cover because their shorter limbs have less respiring tissue. Branches on trees with columnar crowns such as lodgepole pines do not extend outward very rapidly. The lower branches become more shaded and die as the wall (or surrounding trees) grows taller, even if these branches do not actually touch the wall.

Tree growth efficiency is measured by growth per tree, growth per unit of foliage, or growth per unit of ground surface area covered (O'Hara, 1988). Growth per tree is usually greatest for open-grown, or nearly open-grown, trees (Jobling and Pearce, 1977). Trees with slightly less than full live crowns grow more than open-grown trees if the lower branches of the open-grown trees are importing photosynthate from the stem to stay alive. The lower branches of open-grown trees can have relatively little foliage and cannot maintain the respiration requirements of the increasingly long branch stem. Eventually such branches die; however, they may survive temporarily by obtaining photosynthate from more vigorous branches via the main stem.

Growth per unit of ground surface area covered (crown projection area) is used to express tree growth efficiency. Where crown size is restricted by adjacent trees, this measure is an indication of growth per hectare at different spacings (Fig. 3.18; Boldt, 1971; O'Hara, 1988). Maximum growth per unit of foliage for individual trees (growth efficiency or growing space efficiency) is often in tall trees with crowns of medium size (Senda and Satoo, 1956; Satoo et al., 1959; Hamilton, 1969; Assmann, 1970; Ford, 1975; O'Hara, 1988).

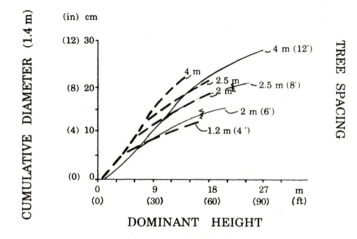

Figure 3.17 Cumulative diameter growth for Douglas-firs (solid lines; *Oliver et al., 1986a*; see "source notes") and loblolly pines (dashed lines; *Bennett, 1960; Hansbrough et al., 1964; Hansbrough, 1968; Balmer et al., 1975*) at different tree heights and regular spacings where little differentiation is occurring. The effect of different sites on age can be accounted for in some species by substituting plantation height for age.

Trees with strong epinastic control become more efficient at wider spacings as they become older until they become very widely spaced. The lower crowns of open-grown trees extend the area occupied by a tree's foliage far more than they increase its growth and so reduce growth efficiency (Fig. 3.18). In trees with weak epinastic control, several shoots compete for the terminal position in each widely spaced tree and create trees with decurrent growth habits (Fig. 3.3). Narrow spacings create side shade on shoots not centered within the opening, make these shoots less vigorous, and give more apical control to the center terminal in full sunlight. Consequently, trees with weak epinastic control develop a more excurrent growth habit at relatively narrow spacings and, unlike trees with strong epinastic control, grow more volume per hectare (Assmann, 1970; Fig. 3.18).

Tree stem stability. A tree with a restricted crown (e.g., tree 2, Fig. 3.14) remains proportionally thinner (Fig. 3.17) and less resistant to pests as it grows taller. Its reduced photosynthate also prevents the bracket-shaped root connections at the stem base from growing strong. If not first killed by pests, the tree eventually becomes so tall and thin that it buckles under its own weight in a slight wind or tips over from breaking roots near

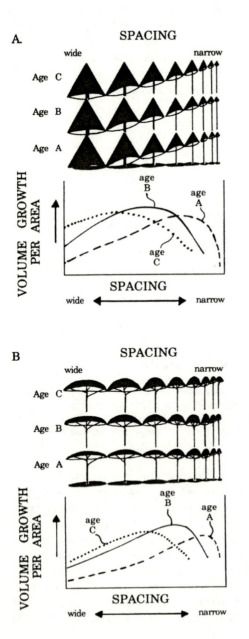

Figure 3.18 Growth efficiency (volume growth per unit of ground surface occupied) varies with spacing, age, crown shape, and species. A. Trees with strong epinastic control become more efficient at wider spacings with age, since their crown shape stays constant and foliage efficiency varies with location within the crown (*Assmann, 1970; Woodman, 1971*). B. Trees with weak epinastic control are more efficient at narrow spacings, where side shade creates more rounded crowns and more foliage surface area (*Assmann, 1970*).

its base (Fig. 8.3). Little work has been done on stem buckling in North America; however, studies elsewhere suggest that a stem diameter of less than 1 percent of total tree height is very unstable for most species (Abetz, 1976; Kramer, 1977; Halle et al., 1978). Other internal and environmental factors can also contribute to a tree's buckling or tipping (Abetz and Kunstle, 1982). Under extremely unstable conditions, trees lean on each other for mutual support. Removal of some trees can cause the others to fall over (Groome, 1988). The increase of crown size and greater drag force in winds at wide spacings are apparently more than compensated by the stronger stem and roots, so that trees at wider spacings are generally more stable (Blackburn and Petty, 1988).

Buckling or tipping (Fig. 8.3) from crowding should not be confused with uprooting, which creates large windthrow mounds (Fig. 6.5). Windthrows which create large mounds are more related to soil conditions and to the tree's exposure than to narrow spacings.

Height growth repression. If it does not succumb to pests, buckling, or tipping first, a tree with a constant crown size will eventually grow so tall that respiration and foliage and root growth consume all photosynthates, and normal height growth is not maintained (tree 2; Figs. 3.14 and 3.15; Horton, 1956; Curtis and Reukema, 1970; Mitchell, 1975; Mitchell and Goudie, 1980). The crown becomes flat-topped as terminal and branch growth decrease. This height growth repression occurs first in trees with small crowns (Figs. 3.10 and 3.14A). It can occur in individual trees as they become suppressed or in whole stands as they approach *stagnation* (Eversole, 1955; Maisenhelder and Heavrin, 1957; Oliver, 1967; Allen and Marquis, 1970; Curtis and Reukema, 1970; Reukema, 1970, 1979; Balmer et al., 1975; Johnson and Burkhardt, 1976; Johnstone, 1976; Mitchell and Goudie, 1980; Lloyd and Jones, 1983; Liegel et al., 1985; Jones, 1986). Since height growth does not entirely stop, even very dense lodgepole stands of over 100,000 stems/ha (40,000 stems/acre) do not truly stagnate and the more dominant trees very slowly outgrow the more suppressed ones (Mitchell and Goudie, 1980). Such stands are referred to as repressed rather than as truly stagnated. The smaller foliage areas of more flat-topped trees lead to suppression at shorter heights for a given spacing than for trees with more conical or columnar crowns. As the trees' height growths slow, they essentially suspend growth, if height is substituted for age in considering stand structures.

Increases in crown size

Growth above a confinement. Figure 3.14 shows the crown of tree 3 after it has grown above the confining walls. The branches first extending over the wall would continue to be longer than the ones above it, and would stay alive. The base of the live crown would not recede, and the crown size would thus increase. Volume growth would increase as the live crown increased (Fig. 3.15).

Release from side shade. Tree response to release from confinement—such as when the walls of tree 2 (Fig. 3.14) are suddenly removed—depends on the environment, the species, and the tree's physiological condition. The tree is susceptible to heat scald, windthrow, or stem breakage at first. If it survives, the live crown increases by the non-functional crown's becoming functional, by development of water sprouts beneath and within the live crown, and/or by branch and height growth.

If the environment is very hot or dry, direct sunlight and heat can kill newly exposed leaves and the phloem and cambium within a stem which have not developed resistance to direct sunlight. Scars develop on the sunny (generally southern in the northern hemisphere) sides of upper stems and branches. Sun leaves survive when released from side shade, but sometimes lower, shade leaves are killed or become necrotic and relatively inefficient.

A very tall and thin released tree may fall over or break if left without the surrounding support and protection it had in confinement (Groome, 1988). A tree's stem in confinement develops as a pencil, with uniform strength throughout; however, a tree grown in a windy situation develops as a fishing rod, with increasing flexibility toward its top, allowing it to bend rather than break in winds. As a tree sways in the wind, reaction tissue develops in its xylem and it becomes more wind-tolerant. A tree released from confinement eventually produces enough reaction tissue to become more wind-resistant.

When exposed to full sunlight, even weak buds of previously shaded branches produce shoots with sun leaves. Subsequent shoots, buds, and foliage are very vigorous and greatly increase the branch weight, causing it to sag. The increased foliage can also cause the entire tree to become top-heavy and susceptible to stem breakage from wind, snow, or freezing rain.

Lower branches of the released tree remain alive as the tree grows taller, since they are no longer shaded (Ausennac et al., 1982). They grow outward, the base of the live crown stays fixed, and the crown becomes longer as new foliage is added to the taller growing tree. Lateral shoots of released branches increase their number of buds, which increase the crown density when they form branches (Maguire, 1983). In trees with strong epinastic control, the crown shape remains fixed, and crown spread occurs only in proportion to height growth. Response to release is often more rapid in younger trees and those on good sites; in both cases, height growth is quite rapid. Crowns under less epinastic control spread outward rapidly even without the proportional height growth increase.

Many broadleaf species and some conifers develop water sprouts when released from suppression. These sprouts occur within the present live crown and increase the density of foliage. They can also grow along the bare stem beneath the live crown. It is unclear how much the lower sprouts contribute to total photosynthesis, but they certainly change the tree's appearance and wood quality.

Trees with weak epinastic control, but which grow excurrently because of side shade, may develop multiple terminals after release and revert to the decurrent habit.

In trees for which the confinement is not so severe that height growth has been repressed, normal height growth and crown buildup continue after release. Response to release is much slower where a tree has been very confined and unvigorous (Krueger, 1959; Allen and Marquis, 1970; Barrett and Roth, 1985; Roth and Barrett, 1985), especially where height growth became repressed (tree 2; Fig. 3.15; Mitchell and Goudie, 1980). Normal growth may not ever resume after release from severe crowding of very intolerant species such as eastern cottonwoods (Krinard and Johnson, 1975). After release from severe confinement, little photosynthate is available for increasing branch growth, leaf area, and height growth, since most is consumed by maintenance respiration. In fact, increased heat along the stem and lower crown may increase respiration and take photosynthate away from possible crown development. If the tree was not too suppressed, the little photosynthate not used in respiration and root development would be used to increase foliage and height growth after release. The increased height would increase crown size. This increase in crown size creates a snowballing effect of increasing total photosynthate and thus increasing photosynthate available for crown development. Severely suppressed trees may take years after release before beginning rapid growth. Alternatively, a physiological alteration may occur in some severely suppressed trees, and the trees may not grow rapidly even if photosynthates become available. This pattern may be similar to topophysis, the often irreversible changes of growth patterns as parts of a plant mature (Zimmerman and Brown, 1971).

Growth upon release from side shade. Although it may decrease initially, volume growth generally increases after a tree is released from side shade, providing the initial exposure does not kill the tree, the tree is not blown over before it becomes more stable, or more aggressive plants do not occupy the newly available growing space and then weaken the tree. How rapidly growth occurs and where it occurs depend on the tree's condition at the time of release (Krueger, 1959). Release may increase respiration by increasing heat on the previously shaded stem and branches and cause a temporary shock (Staebler, 1956a,b; Bradley, 1963; Harrington and Reukema, 1983; Maguire, 1983). Photosynthesis of the lower branches increases after release, since more intense sunlight reaches them, although they photosynthesize less efficiently than sun leaves. The photosynthate available for growth increases as crown size increases (Siemon et al., 1976).

A tree has a vigorous living crown when released if side shade had not been close or if the tree was still small (e.g., tree 3, Fig. 3.14). In either case the photosynthesis rate is not much less than that for an open-grown tree of the same species, height, and site. Only a slight increase in diameter and volume growth (Fig. 3.15), if any, occurs upon release, and the major effect is to maintain the tree's rapid growth.

A more confined tree (tree 2, Fig. 3.14) shows a notable increase in volume growth upon release, since the increase in crown size is in upper lateral branches where photosynthesis occurs quite efficiently (Fig. 3.15). Total growth after release may be less in the previously more confined tree than in less confined ones; however, the increase in growth is more dramatic because the tree had been growing far below its potential before release. Growth after release accelerates as crown size increases. Increased photosynthate first goes to producing needles and buds. If height growth had not been curtailed, photosynthate is then added to the main stem beginning within the live crown. Xylem is added basipetally along the trunk so long as photosynthate is available. When photosynthesis is limited, diameter growth increase is more noticeable near the base of the live crown than farther down the stem.

If a tree is released to a given spacing and allowed to grow to the same height and crown size as another tree which has always grown at the postrelease spacing, the released tree will always be smaller in diameter and total volume. Both trees will add volume at the same rate once crown sizes are similar; however, the initially more confined tree was once smaller and cannot add the missing volume to "make up" for its past history (Pienaar, 1965).

Release from side shade can have indirect effects as well. A new cohort of plants can begin growing around the larger tree where the forest floor receives more sunlight. These plants sometimes hinder growth of the larger tree (Johnson and Kovner, 1956; Zahner, 1958; Oliver, 1984). Increased sunlight on the forest floor also increases the soil temperature, which aids root growth and availability of nutrients through litter break-down in cold climates. If the release is associated with the death of other trees, more moisture becomes available (Douglass, 1960; McClurkin, 1961), which may or may not be beneficial to surviving trees. Increased moisture increases growing space in a droughty soil but saturates very moist soils and reduces soil oxygen.

Effects of high shade

Growth with high shade. High shade creates trees with weak terminals, little height growth, and often broad, flat crowns (Fig. 3.7; Busgen and Munch, 1929; Kohyama, 1980; Meng, 1986; Tucker et al., 1987). Trees grow slowly under high shade because the little photosynthate they produce at these light intensities is used primarily for mainte-nance respiration. Unlike trees in side shade, trees in high shade have crowns living completely beneath their shading competitors. Because their crowns are flatter, they require more lateral room free of side shade to maintain a large live crown than would trees of the same species and height growing without high shade.

Trees beneath high shade can have small or large living crowns. They have small crowns if they are very shaded and if lateral extension of upper branches shades and kills the foliage immediately beneath on lower branches of the same tree. The only living foliage on lower branches is that extending beyond the upper branches. When this foliage's photosynthesis is not enough to support the long respiring branch, the whole lower branch dies and the crown length is reduced. Trees beneath high shade can also have small crowns when nearby trees create side shade and/or abrade the lateral branch-es. The flatter crowns cause side shade and competition to occur at wider spacings for a given height among understory trees than among trees in full sunlight. Trees in high shade can have large living crowns where the trees are not extremely shaded by the over-story or crowded from the sides by other trees.

Growth upon release from high shade. Sometimes trees surviving under high shade receive enough light to maintain some height growth. They grow taller to where the light intensity is greater and thus grow even more rapidly. These trees can eventually emerge from the high shade. As they receive more sunlight, their growth and develop-ment shifts to forms more characteristic of side-shaded or open-grown trees (Fig. 3.7).

Some shaded trees receive so little light that they do not grow and so remain in the low light levels. The rooting medium of these suppressed trees is very important for sur-vival. A tree growing on an unstable substrate such as a rotting log, dead stump, or windthrow mound must grow enough for its roots and stem to hold the tree up as the wood or mineral substrate deteriorates. Otherwise, the suppressed tree falls over or becomes tilted. Upon release, extra photosynthate may be used to produce reaction wood to straighten the stem, so little is left for height growth.

If high shade is suddenly removed from trees growing on stable substrates, they respond by dying, by maintaining their shape without growing, or by growing slowly or rapidly. The released tree forms a stem and crown characteristic of the species' open growth form on top of the older shade-grown base (Fig. 3.7). The shade leaves and stems not thickened to resist direct sunlight respond as lower branches released from side shade do. Depending on the severity of the environment, the entire tree may die, leaves may die or become necrotic (Tucker and Emmingham, 1977), or patches of the phloem and cambium may die of sun scald. Some evergreen trees cannot add new foliage if most leaves are killed, and the tree dies. In other evergreen trees and most deciduous ones, a second flush of leaves develops during a growing season if most of the first set is killed; however, enough leaves have to be killed so that the remaining ones do not inhibit development of new leaves. A *false ring* can occur in the xylem where a new flush of leaves develops. In other cases the tree does not die but may not grow because all excess photosynthate produced in the full sunlight is consumed by the rapidly respiring stem in the increased heat. A physiological blockage to growth similar to topophysis or to continued height growth repression in released trees with small crowns may also develop.

Trees may also grow vigorously after release from high shade. Trees in high shade beneath an upper canopy are often classified as "suppressed' and are mistakenly expected to respond to release in the same manner as trees in the upper canopy which are suppressed by side shade; however, behavior upon release is much different when trees beneath high shade have large live crowns compared with their respiration surface areas. The large crown of such a tree under high shade does not photosynthesize much, but it does maintain many limbs with terminal and lateral buds. Although these buds are not vigorous, they rapidly increase in vigor within a few years after release. They build up rapidly to a large crown. Unlike trees with short crowns suppressed by side shade (tree 2, Fig. 3.14) or high shade, trees under high shade with large crowns do not have to wait for new height growth to produce more lateral branches before a large crown can develop.

Upon release, the formerly shaded trees with large crowns often exhibit a lag in height (and possibly diameter) growth (Oliver, 1976; Zens, 1995) and then grow in height at the same rates as trees of the same species and site which have never been suppressed (Fig. 3.11; Oliver, 1976; Hoyer, 1980; Larson, 1982; Jaeck et al., 1984).

Applications to Management

The forester has no control over some aspects of growth forms for each species; however, other aspects can be influenced. Two major factors alter inherent growth forms: the growing conditions of the tree and disturbances.

Before removing an overstory with the expectation that the understory will respond, the forester needs to examine the condition of the understory trees. If the trees will not develop straight growth forms because of loss of epinastic control, release of the understory may not create future timber value but may enhance some wildlife. Expensive corrective action will be needed to eliminate crooked trees if straight ones are desired.

Some disturbances affect the growth forms of trees. Insects can destroy the leaders of species such as eastern white pine and Sitka spruce, and serious forking will occur as lateral branches turn upward; Similar forking in other species occurs if wind or freezing weather kills the terminal. In a dense stand a single lateral will usually exert dominance, and it may be advisable to maintain relatively dense stands where such disturbances are common.

Species with sustained growth can take advantage of a few good growing seasons and gain a competitive advantage over more preformed species, shading the preformed species and causing them to become flattopped. Subsequent bad growing seasons can then cause the sustained growth species' upper crowns to die back, resulting in poor growth and form by both species. Where an area has a history of several good seasons followed by several bad ones, the forester may prefer to promote single-species stands during early operations to avoid such problems.

Since lammas growth patterns are genetically controlled, favoring those trees which maintain desirable forms can reduce adverse impacts of lammas growth.

Some species are more likely to develop water sprouts than others. If the manager is interested in growing lumber without "pin knots," heavy thinnings should be avoided in species which readily develop water sprouts. Water sprouts as well as persistent, normal branches on the lower stems can be avoided by keeping species which are subordinant in height as *trainers*. Beeches are used in parts of Europe to keep a central stem free of branches (and knots) in preformed species such as oaks. Red maples, some birches, and sweetgums will fill this role in some stands in the eastern United States (Chap. 9).

Silviculturists in the northern hemisphere have generally grown trees at narrow spacings and conservatively thinned stands, creating relatively crowded stands. A general problem in many regions is: How can very crowded stands containing trees of small diameters be treated? (Oliver, 1986). The benefits of wide spacing are beginning to be appreciated for individual tree growth, as well as for tree vigor, higher quality timber, and desirable stand structures for many wildlife and recreation uses. Tree vigor is directly related to the amount of sunlight received. Although wide spacings and heavy thinnings result in periods with unoccupied growing space, silvicultural regimes which make use of wide spacings may prove advantageous. The tradeoffs need to be analyzed between higher planting costs, narrow growth rings, small tree sizes, and small branch sizes at narrow spacings and lower planting costs, wide growth rings, large tree sizes, greater resistance to wind and (some) insects, and large branch sizes or pruning costs at wide spacings.

CHAPTER 4

DISTURBANCES AND STAND DEVELOPMENT

Introduction

Disturbances have a profound effect on forest development since they kill vegetation and thus release growing space, making it available for other plants to occupy. Individuals and species which first occupy growing space during forest development have an advantage in maintaining it and excluding other plants. Forest compositions of most areas are strongly influenced by disturbances, since natural and human disturbances occur in all forests and tree species generally have longer potential life spans than the intervals between large disturbances. Subtle variations in disturbances and other events during the initiation of a stand can determine which species first occupy the growing space.

The significance to development of a new stand is not just the disturbance itself, but the creation of available growing space by elimination of plants previously occupying the growing space. Disturbances can last for a short time in the case of fires or windstorms or for much longer periods, such as during a glacier's advance and retreat. Growing space is also made available by events which are often not thought of as disturbances. A "disturbance" may actually consist of several events. For example, a combination of overgrazing and the unusual rainfall pattern in 1919 provided growing space for Ponderosa pines to become established and therefore acted as a disturbance (Pearson, 1950; Daniel, 1980). In this book, events making growing space available will be referred to as "disturbances," as is done elsewhere (Neilson and Wullstein, 1983; White and Pickett, 1985).

After a disturbance, both surviving plants and new ones expand into the newly available growing space. Those able to expand most rapidly before the growing space becomes occupied become larger and have an advantage in taking growing space from other plants. If soil, climate, or other conditions do not give other plants an overriding advantage, these plants often dominate the stand for decades or even centuries. Especially during *transition conditions* in which two or more plants have equal advantages on the soil, climate, and time gradients, a plant gains the competitive advantage because of specific characteristics of the disturbance.

Awareness of Disturbances

The role of disturbances in determining forest structures and species compositions has been actively studied in the past two decades (Raup, 1964; Heinselman, 1970a,b, 1973; Loucks, 1970; Henry and Swan, 1974; Horn, 1974; Levin and Paine, 1974; Pickett, 1976; Sprugel, 1976; Lorimer, 1977b; Oliver and Stephens, 1977; Whittaker and Levin, 1977; Caswell, 1978; Connell, 1978; Veblen and Ashton, 1978; White, 1979; McIntosh, 1980; Oliver, 1981; Veblen et al., 1981; Hemstrom and Franklin, 1982; Harmon et al., 1983;Sousa, 1984; Pickett and White, 1985). Foresters and ecologists have always been aware of disturbances (Clements, 1905, 1916; Lutz, 1940; Goodlett, 1954; Mooney and Godron, 1983); however, their importance was minimized both because the pervading perspective of ecology—paralleling that of other fields of thought—regarded disturbances as relatively superfluous influences and because poor communication and transportation kept scientists from being aware of how common, frequent, large, and devastating disturbances can be.

Recent advances in communication and transportation have made scientists aware of the magnitude, frequency, and universality of disturbances. The first public satellite transmission of television was in 1962. The first regular television broadcasting service was in about 1935. The first commercial radio broadcast was in 1898. Before the automobile, travel and observations of landscapes were restricted to routes along railroads or to horses, which traveled slowly. Even though communication has always occurred, it has been slow and limited until very recently. Scientists throughout the world are already aware of a forest fire in China which burned over 350,000 ha (865,000 acres) in 1987 (Zhou, 1987), a fire of 3.2 million ha (8 million acres) in Borneo in 1983 (Davis, 1984), and the Yellowstone fires of 1988, which burned 570,000 ha (1,408,000 acres; Christensen et al., 1989). The eruption of Mount Saint Helens in 1980 was widely known, unlike similar ones occurring in past centuries (Yamaguchi, 1983). The 1938 hurricane in New England which destroyed 243,000 ha (600,000 acres) of forest (Baldwin, 1942; Smith, 1946; Spurr, 1956) was widely reported. A similar hurricane occurred there in 1815, but most within the area thought it was a local storm, and those living elsewhere were unaware of it. The 1962 Columbus Day storm in the Pacific northwestern United States is better documented than a similar storm in the same region in 1921 (Washington State Library, 1975). The dramatic valley damming and draining like that by the Hubbard Glacier in Alaska in 1986 has occurred many times before throughout the world (Meier, 1964), but without such widespread attention.

Disturbances which seem rare to people can be a major determinant of forest structures. Trees of different species can live from 70 to over 1,500 years. Human life spans are, of course, much shorter, and human memories are even shorter. Consequently, if a hurricane crosses inland New England about every century on the average (Perley, 1891), it can affect many forest stands, but will be remembered by relatively few people. Devastating fires in the Olympic Peninsula and Cascade Range of Washington State 400 to 750 years ago created many of the present forest stands—and hence determined their species compositions and structures (Agee, 1981; Hemstrom and Franklin, 1982). This time scale is much longer than even the recorded history of Washington State.

Early ecologists interpreted the effects of disturbances in different ways. Some recognized that disturbances shift the species composition temporarily. Others recognized that certain disturbances—especially fires—were natural and considered certain end points of succession to have been achieved when plants could endure the natural fires—e.g., "fire climax." Others felt disturbances were simply a temporary setback during the progression of plants toward an ultimate end point. Gleason (1926, 1939) felt that disturbances were a strong determiner of plant communities and that plant communities did not have an ultimate direction in themselves.

At times there has even been an attitude that disturbances—especially human disturbances—irrevocably damage a forest stand. This attitude was based on such concepts as a coevolved interdependence of species (Chaps. 1 and 2). Under such coevolution concepts, any extraneous event was thought to disrupt the genetically guided interactions that had evolved between individuals so that the forest could no longer grow. This attitude was partly responsible for the concern about clearcutting in the 1960s, 1970s, 1980s, and 1990s (Bormann et al., 1968; Likens et al., 1970; Duffield, 1972; Herman and Lavender, 1973; Minckler, 1973; Marquis, 1973).

Forest cutting, burning, land clearing, and other human disturbances partly mask and minimize the effects of natural disturbances. Human disturbances are generally more frequent than natural ones, although human fire prevention lengthens many natural fire cycles (Cooper, 1960; Houston, 1973; Cole, 1977). Human disturbances are remembered by more people and can minimize the effects of natural disturbances. For example, a stand which began after a clearcutting 40 years ago is shorter, more stable, and less prone to be blown over in a windstorm than a natural stand of 300 years on the same area; likewise, a stand which began after a recent human fire is less likely to burn following a lightning storm than a stand which had not burned for many years.

At other times, human interference exacerbates the effects of natural disturbances. Control of mild, natural surface fires allows fuels to accumulate in many parts of North America until intense natural fires, which cannot be controlled, burn severely. (O'Laughlin et al., 1993; Sampson et al., 1994; Sampson and Adams, 1994.)

Recent studies have brought attention to forest disturbances. Use of techniques similar to those found in archaeology to reconstruct forest history has only recently become acceptable to the forest science community. Detailed reconstruction of 0.4 ha on the Harvard Forest between 1950 and 1955 (Stephens, 1955b) inspired many similar studies.

In nearly all studies in which the history of a stand was reconstructed, evidence of natural disturbances strongly affecting the species composition and age distribution have been found (Hough and Forbes, 1943; Uhl and Murphy, 1982; Pickett, 1983).

A controlled study was conducted to determine the effects of disturbances on forest development in central New Hampshire (Bormann and Likens, 1979, 1981). Several areas were clearcut and vegetation regrowth was inhibited by herbicides for several years. Even after this drastic disturbance—far more devastating than windstorms, forest harvesting, or other silvicultural operations—the forest grew back rapidly as soon as the chemical inhibition was removed. This study and others (Vitousek and Reiners, 1975), along with evidence such as the rapid regrowth of vegetation following the 1980 eruption of Mount Saint Helens in Washington (Bilderback, 1987), indicate that forest communities in diverse regions recover quite rapidly even from extreme disturbances.

Realization of the frequency (on a geologic scale) and devastation of disturbances is increasing as more is understood about world weather patterns, continental drift, and meteoric impacts on the earth. In total, natural disturbances are so common and so devastating that there are probably few human disturbances for which a counterpart cannot be found in nature. Only the chemical effects of pesticides, air pollution, and dumping of chemical wastes do not seem to mimic natural disturbances, although this does not imply such effects are necessarily more harmful than others.

People impact forests as technology, economics, and population migrations change the frequency of forest clearing for agriculture, farmland abandonment, forest harvesting, fires, grazing, and other activities (Lutz, 1959; Cooper, 1960; Tomaselli, 1977; Pyne, 1982; Edwards, 1986; Abrams and Scott, 1989; Uhl and Vieira, 1989; Glitzenstein et al., 1990; Uhl et al., 1990; Galbraith and Anderson, 1991; Abrams, 1992; Baker, 1992; MacCleery, 1992; Denevan, 1992; Nowacki and Abrams, 1992; O'Laughlin et al., 1993; Covington et al., 1994; Lehmkulh et al., 1994; Oliver et al., 1994b; Irwin et al., 1994; Abrams and Ruffnen, 1995).

Human disturbances often occur at different frequencies and are of different types than would naturally be found in a given forest stand or region. Clearcutting and burning of slash on the Olympic Peninsula may occur every 50 to 80 years, whereas natural fires which created similar stands occurred about every 400 years or longer (Agee, 1981). Similarly, controlling natural fires can create longer intervals between fires in many regions of the southeastern and western United States (Cooper, 1960; Houston, 1973; Cole, 1977). Such changes affect the structures and species compositions of these stands. The changes may even threaten the survival of some species dependent on disturbances of certain types and frequencies for their existence. For example, the Franklinia bush in northern Georgia was driven to near extinction, apparently because of land clearing there in the late 1700s (Ayensu and DeFilipps, 1978). The Kirtland's warbler was threatened with extinction because control of fires in the Great Lakes region was eliminating its habitat—the young jack pine stands which grew after fires (Walkinshaw, 1983). The spotted owl is considered endangered as more of the very old stands of Douglas-firs, hemlocks, and true firs are cut in the Pacific northwestern United States.

Impact of Disturbances

Disturbances to forest stands have generally been considered allogenic phenomena—that is, phenomena which are not generated by changes within the stand. Actually, disturbances are partly allogenic and partly autogenic, since the impact of a disturbance is the result of both the magnitude of the disturbance and the predisposition of the stand to the particular disturbance type (Odum, 1971; White, 1979; Spurr and Barnes, 1980).

Magnitude of the disturbance agent

Disturbances of different strengths generally follow a somewhat random pattern, but the probability of an event greater than a given magnitude is inversely proportional to the magnitude. For example, windstorms and floods of increasing magnitudes occur at progressively less frequent intervals; and increasingly severe combinations of prolonged droughts, appropriate winds, and lightning storms for fires occur at increasingly longer intervals.

Predisposition of the stand to the disturbance

A stand often becomes more susceptible to disturbances as it grows, so that the magnitude of disturbance required to disrupt the stand becomes less with age (Fig. 4.1; Runkle, 1984). In areas of North America such as the southeastern pine forests or the western conifer forests, it is very difficult even to force a fire to burn an area for 2 consecutive years. As years pass since the last fire, dead needles, branches, stems, and other fuels accumulate and climatic requirements necessary to make another fire possible become less extreme. (Agee, 1993; O'Laughlin et al., 1993; Sampson et al., 1994; Sampson and Adams, 1993.) Similarly, the susceptibility of a stand to windthrow increases as trees grow taller and less windfirm and produce large crowns. At some time, the susceptibility of a forest is great enough to coincide with a disturbance event greater than the threshold magnitude necessary to disrupt the stand—and a disturbance occurs. Disturbances such as fires and wind throws occur in cycles characteristic of the climate, soils, and species (Van Wagner, 1978). Some insect outbreaks can be cyclic in cases in which trees of a certain age, size, and/or vigor are most susceptible to attack. Lodgepole pines are attacked by the mountain pine beetle (*Dendroctonus ponderosa*) on a 20- to 40-year cycle (Amman, 1977). The spruce budworm (*Choristoneura fumiferana* Clem.) periodically attacks balsam fir stands as the firs become weakened by age and dense stand structures (Ghent et al., 1957; R. S. Seymour et al., 1983). Other disturbances have less regular recurrence intervals (Swanson, 1979).

One disturbance can increase the predisposition of a forest to another disturbance. Very heavy grazing or fires can make some soils more susceptible to erosion (Hendricks and Johnson, 1944; Krammes, 1960; Rich, 1962). Dead, dry wood after a windstorm makes the area more susceptible to fires. Resins in a lightning scar may allow a tree to be easily ignited when struck again by lightning (Taylor, 1969). Insect or disease outbreaks or animals can kill trees and increase the fuel load and thus the stand's susceptibility to fires (Tinnin, 1981; Tinnin et al., 1982; Petersen, 1992; Agee, 1993; Haack and Byler, 1993; Johnson et al., 1993; Mealey, 1994; Sampson et al., 1994). Silvicultural thinnings at first can leave residual trees less able to stand up in heavy winds.

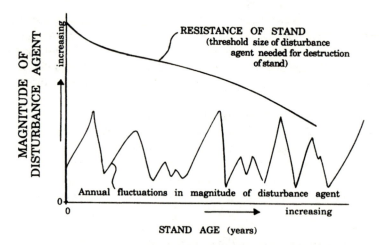

Figure 4.1 A disturbance occurs when a stand's resistance to a disturbance is less than the magnitude of the disturbance agent in a given year. The predisposition of a stand to different disturbances changes with the stand structure and type of disturbance but generally increases with age.

A disturbance killing some trees can also decrease a stand's susceptibility to other disturbances. The death of some trees caused by insect outbreaks or silvicultural thinnings will increase the photosynthate available for diameter growth and insect and disease resistance, decreasing a stand's vulnerability to winds and insects if the trees survive the first few years of exposure.

Classification of Disturbances

Disturbances can be described by their frequency, distribution, return interval, area covered, magnitude and other factors (Bazzaz, 1983; White and Pickett, 1985). For this book, disturbances are discussed in four ways: amount of existing forest removed above the forest floor; frequency of the disturbance; size and shape of the area disturbed; and amount of forest floor vegetation, forest floor, and soil removed. *Forest floor vegetation* is arbitrarily defined as that vegetation less than 2m (6 ft) tall. Two meters is an arbitrary height to be used when it helps elucidate a process and will not be treated as a strict guideline.

Many disturbances vary in the amounts of vegetation removed and the reoccurrence interval of the disturbance. Combinations of the amount of vegetation removed—whether major (stand replacing) or minor—and the recurrence interval could be considered as disturbance "regimes," similar to "fire regimes" described by Heinselman (1973). For example, a stand could have a low intensity, frequent wind disturbance regime if it is chronically subjected to winds which only blow over a few trees; on the other hand, a stand could have a high intensity, infrequent wind regime if it were infrequently subjected to catastrophic, stand-replacing windstorms.

Amount of overstory removed: major (stand replacing) and minor disturbances.

Disturbances can roughly be divided into two types based on the amount of overstory removed: those which remove or kill all the existing trees above the forest floor vegetation are referred to as *major disturbances,* or *"stand-replacing disturbances";* and those which leave some of the predisturbance trees alive are referred to as *minor disturbances* (Oliver, 1981). Following a major disturbance, newly regenerating trees compete only with other trees also regenerating after the disturbance. Following a minor disturbance, newly regenerating trees compete for growing space with previously established, large trees which survived the disturbance as well as with trees regenerating after the disturbance. Differences between major and minor disturbances are not clearly delineated if the forest floor vegetation is well developed and survives the disturbance or if only an occasional overstory tree remains.

Glaciers, severe fires, abandonment of farmland, or severe windstorms which kill all trees taller than the forest floor vegetation are major disturbances. Growth and competition is then among regenerating species. Resulting species compositions and stand structures are free of influence by larger, well-established trees which, if present, would have expanded into the existing growing space and would have altered the sunlight and moisture of the regenerating plants as they grew.

Windthrows, fires, insect or disease outbreaks, commercial clearcuts, thinnings, or other partial cuttings are minor disturbances if some trees (excluding forest floor vegetation) predating the disturbance remain alive.

Forest development after minor disturbances is more complicated than after a major disturbance. In fact, stands beginning after a major disturbance can be considered as only one extreme of a gradient of "overstory competition." The rest of the gradient would be increasing densities of overstory. Minor disturbances influence stand structure, species composition, and growth rates depending on how much of the existing forest is left alive and in what species and crown positions. Minor disturbances—such as "high-grade" logging—which kill vigorous dominant trees leave unvigorous residual trees which do not rapidly grow into the unoccupied growing space. On the other hand, a silvicultural thinning in the same area which removes the less vigorous trees would allow residual overstory to grow rapidly.

Frequency of disturbances

Disturbances vary from very infrequent and at irregular intervals to very frequent and at regular intervals. Hypothesized meteor collisions with the earth and large glacial advances and recessions occur at infrequent and irregular intervals of thousands or millions of years. Stand replacing disturbances reoccurred about every 1,000 years in parts of the north central and northeastern United States (Canham and Loucks, 1984; Lorimer and Frelich, 1994). Stand replacing fires and windstorms reoccurred at intervals of about 300 to 700 years, depending on the topographic and climatic features, on the Olympic Peninsula of Washington state, U.S.A. Severe hurricanes have impacted parts of New England about every one or two centuries (Smith, 1946); and stand-replacement fires have occurred in many parts of the Inland West (U.S.A.) every 150 years or less (Agee, 1993). Until fire suppression policies of the past 50 years, fires were very frequent in

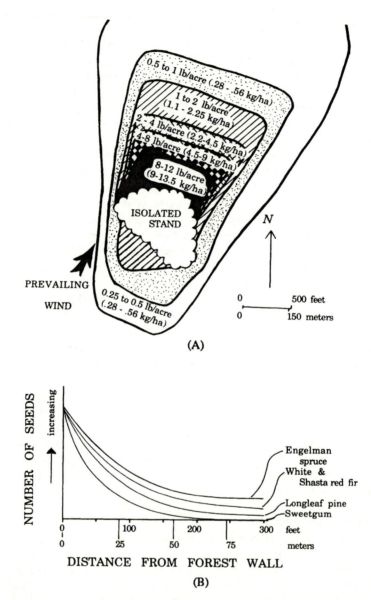

(A)

(B)

Figure 4.2 (A) Pattern of average annual Douglas-fir seedfall around an isolated stand surrounded by an open area in western Washington. The winds are primarily from the southwest during the season of dispersal (*after Dick, 1955*). (B) Dispersal of different seeds from seed trees along a forest wall. Heavy longleaf pine seeds (*Boyer, 1958*) and sweetgum seeds (*Guttenberg, 1952a,b*) with small wings did not disperse as far as light Engelman spruce seeds (*Roe, 1967*) or white and shasta red fir seeds (*Franklin and Smith, 1974*), which often grow in windy areas. (See "source notes.")

ponderosa pine stands in the southwestern United States (Covington and Moore, 1994) and many longleaf pine stands in the southeastern United States (Chapman, 1944b); however, they generally left some overstory trees intact. Fires were less frequent in Douglas-fir/true fir (Abies species) stands in western Washington and Oregon and in the Northern Rocky Mountains, U.S.A., but were more commonly stand-replacing (Wellner, 1970; Hemstrom and Franklin, 1982; Johnson et al., 1993).

Seasonal cycles of heat and cold or drought and rain and daily cycles of light and dark occur at such extremely regular and frequent intervals that plants have adapted to these cycles through physiological mechanisms. While trees similarly endure seasonal changes through physiological dormancy, some nonwoody plants have adapted to this annual regularity by completing their life cycle—or their aboveground life cycle—during a single growing season.

Plants have adapted to less regular cycles of fire and other disturbances to some extent. Raspberries, some violets (Viola fimbriatula), and other small plants grow during the few years immediately following a disturbance before the taller-growing trees shade and kill them. After raspberries die, the seeds stay dormant in the soil until the next disturbance, which then allows the seeds to germinate and the plants to grow for the short interval before the canopy again closes (Cook, 1978; Osawa, 1986). Such species spend more time as seeds than as growing plants but could produce seeds every year under artificial, constantly disturbed conditions; however, their actual generation time—the time for successful development of a genetically new individual—is naturally much longer and is determined by the time between disturbances.

Several tree species—pin cherry in the eastern United States (Marks, 1974) and bitter cherry in the Pacific Northwest—grow rapidly after a disturbance which removes the previous stand but are short-lived. They, too, produce seeds which stay dormant in the soil for many years and are stimulated to grow by a disturbance. Lodgepole and jack pines, similarly, have serotinous cones in many areas which stay closed until a fire (or other heat source) forces them open.

Many plants have less specialized mechanisms for responding to infrequent disturbances such as glacial retreat, devastating fires, and windstorms. They regrow in disturbed areas by regeneration from seeds, sprouts, advance regeneration, and other mechanisms.

Plant species are found where their growth patterns and regeneration cycles are compatible with the frequency and regularity of disturbances (Grime 1977; Johnson, 1981a,b; Runkle, 1982, 1984). A species does not survive where disturbances are too frequent for a nonsprouting species to mature and set seeds between them. Eastern hemlocks are not found in New England stands where frequent ground fires occur (Merrill and Hawley, 1924; Lutz, 1928). Red spruces are rare in frequently disturbed balsam fir-dominated stands because they compete best in stands of older ages (Sprugel, 1976). Too infrequent disturbances similarly exclude species which complete their life spans and disperse seeds which do not stay dormant in the soil before another disturbance allows these seeds to germinate. Red alders were probably excluded from natural upland stands in western Oregon and Washington before forest cutting because the times between natural disturbances exceeded the alders' life spans (Scott, 1980).

Spatial patterns of disturbances

Disturbances can be localized, isolated events when individual trees or tree groups die because of fire, wind, animals, or a combination of factors. Disturbances more often cover large areas and leave a mosaic of smaller areas with different levels of disturbance. For example, a single windstorm often creates many gaps throughout the forest, uniformly "thins" other areas by blowing over unstable trees—and may even blow down all trees in other areas and there acts as a stand replacing event. Similarly, many fires can burn across a landscape or large region at one time; or a single, large fire may burn over a very large area with a variety of intensities and leave a mosaic of forest conditions (Pierce, 1921; Boyce, 1929; Morris, 1935; Isaac and Meagher, 1938; Kemp, 1967; Lucia, 1983).. The disturbance to each area may appear as an isolated event at first; however, the events are often part of a landscape-scale disturbance caused by similar weather and stand conditions (Haines and Sando, 1969; Johnson and Wowchuck, 1993).

The landscape following a disturbance is often a mosaic of areas with different conditions. Some areas may not even contain snags or down logs from the previous stand if the disturbance is a repeated burn in the dry fuel several years after an initial burn (Morris, 1935; Isaac and Meagher, 1938; Kemp, 1967; Lucia, 1983), a windstorm followed by a hot fire (Henry and Swan, 1974), a volcanic eruption (Yamaguchi, 1986, 1993), or a very severe flooding. In some areas the soil can be quite depleted of organic matter and eroded. Other areas can contain snags and down logs if a fire killed standing trees; a windstorm left standing snags; or insects, floods, or siltation killed trees. Still other areas can have a dead overstory and a living understory where a very fast-moving fire completely skipped the forest floor (Isaac and Meagher, 1938), where a windstorm or avalanche blew over the overstory (Oliver et al., 1985), or where insect outbreaks killed the overstory (Osawa, 1986). Other areas contain dead understories but living overstories where ground fires (Segura and Snook, 1992; Covington and Moore, 1994), winds (Steinbrenner and Gessel, 1956), siltation, flooding, or insects selectively killed the smaller trees. Disturbances such as ground fires (Agee, 1993), insects (Haack and Byler, 1993; Lehmkuhl et al., 1994), and volcanic tephra (Segura et al., 1994) can be selective in killing tree species. Even quite severe disturbances frequently leave areas within the landscape which are relatively untouched because of topography, soils, chance weather behavior, or other reasons.

Following a disturbance, each spatially continuous area left with a similar structure can be viewed as a stand if it has similar soil and climatic conditions. Much of the knowledge and past management has focused at the individual stands; however, it is important to recognize that these stands are part of a broader landscape pattern.

The size and shape of a disturbed area influence the species composition and structure of each resultant stand in two ways: they determine how much of the stand is under the influence of the surrounding edge, and they determine how readily different parts of the stand are reached by seeds blowing in from surrounding trees.

Influences of surrounding edge. The edges of a stand created by a recent, stand-replacing disturbance is influenced by the surrounding stands (Chap. 14). This influence—and the disturbed area's influence within the surrounding stands—is referred to as *edge effect*. Some surrounding trees cast shade and extend roots into the opening, thus reducing growing space around the edges and creating forests there of different structures, species compositions, and growth rates. The extent of an edge's influence varies depending on the type of influence considered—shading, root competition, microclimate amelioration, animal habitat, or others.

Migration of seeds. Some species develop primarily from seeds entering an area after the disturbance. Wind-dispersed seeds have different flight characteristics (Fig. 4.2), depending on their weight and wing configurations and on wind conditions. Seeds of all species fall more numerously near their parents; at progressively farther distances the number of seeds declines, but not at the same rate for different species. Consequently, the centers of larger disturbances contain fewer individuals regenerating from wind-blown seeds, as well as different proportions of species than found at stand edges (Spring et al., 1974). Species with windblown seeds which do not travel far are often concentrated at stand edges (Thornburgh, 1969).

Depending on the species, seeds may be dispersed in the autumn, winter, or spring. Where wind directions are consistent for the season of seed dispersal, the windward edge of an opening contains more windblown seeds (Fig. 4.2A).

Configuration of disturbed area. The shape of a disturbed area is often characteristic of the causal agent, topography, and previous vegetation patterns. Fires usually start from a single point, fan outward with the wind or uphill, and cool down in valleys, leaving some trees alive there. Fires often stop at stand boundaries, which can reflect fuel, soil, or topographic changes (Hemstrom, 1979). Tornadoes generally leave a long, thin disturbed area in a north or northeasterly direction (Baldwin, 1973). Avalanches run in relatively narrow strips down mountains, and valley glaciers disturb the valley floor but leave trees growing on surrounding hills. Hurricanes and similar winds often knock over stands of tall trees on windward sides of hills but not on leeward sides.

The configuration of a disturbance dictates how much of the area is close to an edge, what the directional orientation is—and, hence, which and how many species invade. It also dictates what the microenvironment will be and how much growing space will be available—and, hence, which species have the competitive advantage and how fast they will grow. A disturbed area with a circular shape would have the smallest proportion of edge to nonedge. As the difference in length between longest and shortest axes increases, the percent of edge increases proportionately. Also, as the size of a disturbed area increases, the proportion of it under the strong influence of the surrounding edge decreases.

Characteristics of Different Disturbance Agents

Natural disturbances are quite common in all regions of North America and the world when viewed from a long-term perspective (Barrows, 1951; Washington State Library, 1975; Oliver, 1992d; Oliver et al., 1995; Botkin, 1995). Each area has characteristic frequencies and types of disturbances based on its climate, soils, vegetation, animals, and other factors.

Fires, windstorms, and other disturbances have specific behaviors and leave certain conditions for growth. Characteristics of some common disturbances are discussed below.

Fires

Fire occurrence, behavior, and effects have been intensively studied (Ahlgren and Ahlgren, 1960; Heinselman, 1970a,b, 1981; Wright and Heinselman, 1973; Artsybashev, 1974; Kozlowski, 1974; Pyne, 1982, 1984; Hendee et al., 1977; Johnson, 1979; Agee, 1981, 1993; Forman and Boerner, 1981; Romme and Knight, 1981; Romme, 1982; Chandler et al., 1983a,b). Fires covering tens and hundreds of thousands of hectares have occurred in forests in the tropics (Davis, 1984; Sanford et al., 1985; Lopez-Portillo et al., 1990), Asia (Salisbury, 1989; Uemura et al., 1990), American and Siberian boreal regions (Lutz, 1956; Barney, 1971; Viereck, 1973; Heinselman, 1981; Johnson, 1992; Wein and MacLean, 1983), inland and coastal western North America (Lutz, 1956; Heinselman, 1973; Van Cleve and Dyrness, 1983; Pyne, 1984; Teensma et al., 1991; Agee, 1993; Zybach, 1994), and elsewhere. Tropical rain forests previously assumed invulnerable to fires burn during periodic "dry times" caused by "El Nino" or other events (Davis, 1984; Lopez-Portillo et al., 1990).

Evidence of past fires has been found in nearly all forest regions and forest types (Walter, 1971; Pickett and Thompson, 1978). Forest fires are the result of certain fuel and weather conditions and an ignition source; therefore, the seasons, behaviors, and frequencies of fires vary dramatically between regions and between environments within a region (Fig. 4.3; Kessell, 1976; Agee and Huff, 1980; Hemstrom, 1982).

Recurrence interval and time of year for natural fires in the United States are shown in Table 4.1. Fires often occur in years and seasons of low humidity, a lack of snow on the ground, and/or a lack of overhead shade in deciduous forests (Pyne, 1984). Human activities have somewhat extended the fire season, have increased the frequency of fires in some regions, and have reduced the frequency in other regions. The greatest proportion of commercial forest land burned annually in the United States is in the Southeast, with about 0.4 percent, compared with slightly less than 0.4 percent for the Rocky Mountains, about 0.2 percent for the Northeast, and less than 0.2 percent for the Pacific coast (USDA Forest Service, 1982). The many wildfires in the southeastern United States are probably because the extensive road system and high rural population increase the contact of people and forests (Oliver, 1986b).

Fires can be roughly divided into *"crown fires,"* which burn large trees, and *"surface"* fires, which burn the forest floor and small vegetation but do not ignite the crowns of large trees. Crown fires are the result of dry weather conditions conducive to burning and high fuel buildups because of rapid growth, long times since a fire, low decomposition rates, and/or lack of removal of trees through logging. Crown fires generally burn foliage and small limbs of trees above the forest floor. Sometimes the forest floor vegetation and litter will not be burned by an active crown fire.

Surface fires usually occur where weather conditions are not conducive to intense burning or there is little fuel because of frequent fires or slow growth rates. Historically, they often created or maintained savanna ("park-like") structures of widely spaced large trees surrounded by forest floor vegetation. The structures were found in western ponderosa pine stands, (Agee, 1993; Johnson et al., 1993; Covington and Moore, 1994), southeastern longleaf pine stands (Chapman, 1926, 1944), and in many eastern hardwood stands (Bromley, 1935; Day, 1953; Thompson and Smith, 1970; Fissell, 1973) in the United States. Savannas were more prevalent in North America before active fire exclusion allowed younger trees to grow between the older ones. Ground fires also created savannas in pinyon pine stands in Mexico (Segura-Warnholz and Snook, 1989, 1992), and in dahurian larch stands in eastern Sibera (Osawa et al., 1994).

Whether the vegetation dies depends on characteristics of the fire and the species (Parsons and DeBenedetti, 1979; Thomas and Agee, 1986). Lingering fires on lee sides of thin-barked species can heat and kill the cambium (Toole, 1959, 1961; Viers, 1980a,b). Ponderosa pines, longleaf pines, loblolly pines, Douglas-firs, coastal redwoods, sequoias, and other species have thick bark which insulates and protects the cambium from fires when the trees are large. Thin-bark trees are much more susceptible to being scarred or killed by surface fires (Cole, 1977). Forest floor litter and tree roots can burn in surface fires. A lingering fire sometimes burns into the organic matter and becomes a *ground* fire. Surficially rooted trees fall over as ground fires burn their roots. Large, downed logs sometimes burn for long periods. Foliage, buds, and limbs in lower crowns of overstory trees can be scorched and killed even if not in direct contact with the flames.

Droughts lasting several months or longer predispose forests to fires (Plummer, 1912; Fahnestock, 1959; Haines and Sando, 1969; Heinselman, 1973; Hough, 1973; Rowe and Scotter, 1973; Hemstrom, 1979; Agee and Huff, 1980). Many fires start from lightning. Lightning strikes are most common in the southeastern United States (Fig. 4.1A) and become progressively less frequent toward the northwest (Pyne, 1984). Thunderstorms in the eastern United States are generally associated with rain, which reduces the incidence of lightning fires; the southwestern United States has the highest incidence of lightning fires (Fig. 4.1B). Native Americans were often quite unconcerned with fire control and contributed to the frequency of forest fires before European settlement (Thompson, 1916; Bromley, 1935; Isaac, 1940; Day, 1953; Thompson and Smith, 1970; Cole, 1977; Pyne, 1982).

Fuel buildup is necessary for forest fires, in addition to appropriate weather conditions and an ignition source. Fuels accumulate at different rates in different regions and are high during two periods of a stand's development: 1) Soon after a fire, windstorm, or similar disturbance, the injured and dead trees become dry and quite flammable. These fuels become moist and rotten and less conducive to fires as the new forest canopy grows. 2) As the stand ages, more limbs and trees die, creating more fuel in the understory and the stand again becomes susceptible. Sometimes the understory vegetation decreases the susceptibility to fires of very old stands.

The rate of fuel buildup varies with climate, site, species, and other factors which affect stand growth. Fuel buildup also depends on the rate of fuel decomposition; less fuel buildup with forest growth occurs in warm, humid climates, where microbes rapidly decompose dead organic matter (Barnette and Hester, 1930). Fuel buildup continues where weather or soils are not conducive to rapid decomposition. Insect outbreaks, diseases, or windstorms can also create large amounts of dead material and so lead to hot fires (Leiberg, 1904; Wellner, 1970; Henry and Swan, 1974; Dickman and Cook, 1989; Agee, 1993; Johnson et al., 1993; O'Laughlin, 1993). A fire can leave dead trees or weakened trees which later die, creating fuels which allow hot fires to reburn through the down, dead fuel during subsequent decades (Morris, 1935; Isaac and Meagher, 1938; Kemp, 1967; Lucia 1983). These reburns often kill regenerating trees and any remaining

TABLE 4.1 Estimated Recurrence Interval (before European Settlement) of Fires and Times of Year of Occurrence for Different Regions

Region	Natural recurrence interval	Time of year	Reference
Southeastern United States			
Coastal Plain			
Longleaf pine	2—3 years	Spring/fall	Chapman, 1926, 1944b; Pyne 1984
Lower Gulf Coast			
Sand pine	25—79 years*	Spring	Cooper, 1972; Wright and Bailey, 1982
Piedmont			
Loblolly pine	10 years	Sping/fall	Chapman, 1944b; Pyne,1984
Northeastern United States			
Southern New England	Very frequent	Spring/fall	Day, 1953; Bromley, 1935; Pyne, 1984
Central New England	150—300 years*		Henry and Swan, 1974
Northern New England	800 years*		Lorimer, 1977b
Eastern Canada			
Spruce/hemlock/ pine (includes human-caused fire)	230 years	Late spring	Wein and Moore, 1977
Upper-elevation con- ifer (includes human-caused fires)	1,000 years	Late spring	Wein and Moore, 1977
Lake States			
Jack pine barrens	15 years 60 years*		Heinselman, 1981
Birch/maple/hemlock	350 years*	Spring/fall	Stearns, 1951; Wright, 1974; Pyne, 1984
Aspen/birch	80 years*	Spring/fall	Heinselman, 1973, 1981; Pyne, 1984
Boreal forest			
Alaska	130—200 years*	Summer	Viereck, 1973; Barney, 1971; Pyne, 1984
Western Canada	50—200 years*		Johnson and Rowe, 1977; Heinselman, 1981
Eastern Canada	60—150 years		Heinselman, 1981

TABLE 4.1 Estimated Recurrence Interval (before European Settlement) of Fires and Times of Year of Occurrence for Different Regions (*Continued*)

Region	Natural recurrence interval	Time of year	Reference
Northern Rocky Mountains			
Ponderosa pine Douglas-fir/ lodgepole pine	6—120 years	Summer	Lemon, 1937; Arno, 1980; Houston, 1973; Davis, 1980; Laven et al., 1980; Martin, 1982; Pyne, 1984
Subalpine forests	100—300 years*	Summer	Romme, 1980; Hawkes, 1980; Pyne, 1984
Southwestern United States			
Ponderosa pine shrubs	2-10 years	Late spring, early fall	Dieterich, 1980; Ahlstrand, 1980; Leopold, 1924; Weaver, 1951; Pyne,1984
California			
Southern California Shrublands		November— December, summer	Pyne,1984
Sequoia, Ponderosa pine	3—40 years 300—1,000 years*	Summer	Hendee et al., 1977; Wagener, 1961; Kilgore and Taylor, 1979
Northern California			
Ponderosa pine	50 years	Summer	Veirs, 1980*a,b*
Coastal redwood	500—600 years	Summer	Veirs, 1980*a,b*
Pacific Northwest			
Olympic Peninsula	400—750 years*	Summer	Agee and Huff, 1980, 1987; Agee et al., 1981; Agee, 1981
West Cascade Douglas-fir/conifer	100—450 years*	Summer	Means, 1982; Hemstrom and Franklin, 1982
East Cascade			
Ponderosa/ lodgepole pine	2—169 years	Summer	Weaver, 1959, 1967; Martin, 1982
Spruce/fir/ lodgepole pine	50—400 years	Summer	Fahnestock, 1976

*Indicates crown fire, often leading to single-cohort stands.
SOURCE: Table follows Heinselman (1981), Martin (1982), and others.

seed source as well as harm the soil's productivity, and so allow the stand to grow as shrub and herbaceous communities for many decades (Smith, 1986). When fuel and weather are appropriate, a small ignition source is necessary to start a fire and even the best fire control efforts have been unable to prevent large fires. Appropriate weather occurs during severe droughts and large areas are burned periodically, rather than at evenly distributed times. Many old stands in western Washington began after droughts in about 1500 (Hemstrom, 1979; Agee and Huff, 1980) and millions of acres burned in the Inland West in 1889 and 1910.

Fires generally move in all directions after ignition (Fig. 4.2; Pyne, 1984). They move rapidly with the wind or uphill, where the headfire's flame is close to the forest floor and preheats the fuel. The fire's front advances more slowly on its flanks or in the rear, against the wind. First, a fire generally burns as a surface fire. If it gains in intensity, it creates its own advective weather column, sucking wind toward it and upward. The fire preheats the overstory canopy, causing it to burn rapidly as a crown fire—either a *passive* crown fire if it occurs behind the surface fire front, or an *active* crown fire if it advances without the surface fire's preheating the crown. The active winds take burning fuel upward in the convective column; where this fuel falls to earth beyond the head, it can create *spot fires* which expand and ultimately join the main fire. After the fire has advanced and burned the readily combustible fuels, the flame becomes discontinuous. Flames continue where readily combustible fuels have accumulated and in wind eddies—such as on the lee side of trees. Eventually, readily consumable fuels are burned and only scattered large fuels continue to smolder. The continuous flame zone can last in an area for several minutes, the discontinuous zone can burn for up to several hours, and roots have been known to smolder for 12 months.

Fires spread at different rates and in different directions depending on time of day, direction of wind, weather, and landscape. Fires are generally most intense during the day when humidity is low, winds are high, and temperatures are warm. Fires can shift directions with the wind and topography. Sometimes wind can change a head fire to a flank fire, and vice versa. It can also turn a headfire back into an area it recently burned—forcing it either to extinguish itself or to reburn the area. Cool, humid weather or rainfall can also extinguish a fire. A downslope, change in fuel characteristics of the vegetation, change in soil conditions, or other discontinuities of the landscape can also change the rate of burning of a fire—or limit the fire's spread. The result of such wildfires is often a landscape with a mosaic of unburned, partially burned, and severely burned areas, with the proportions being related to specific topographic, climatic, weather, and fire behavior characteristics (Heinselman, 1973; Morrison and Swanson, 1990; Camp et al., 1994; Camp, 1995).

The effects of fires on vegetation are generally correlated with fire intensity; however, low intensity fires of long duration also kill most vegetation. Fire intensity is generally expressed as energy released per length of fire front per unit of time. Forest fires can vary in intensity from 15 to 100,000 kilowatts per meter (kW/m) but are generally less than 50,000 kW/m. Intensity is dependent on both rate of spread and weight of fuel consumed. Heat of combustion of different forest fuels varies only slightly and so has

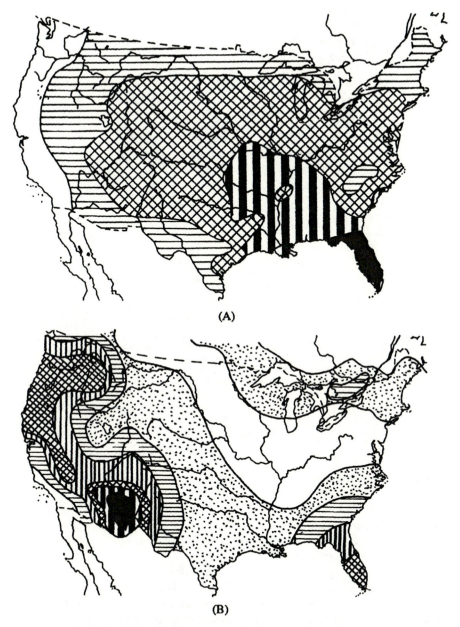

(A)

(B)

Figure 4.3 Distribution of thunderstorms (A; *from Baldwin, 1973*) and frequency of fires caused by lightning (B) in North America (*Schroeder and Buck, 1970*). More thunderstorms occur in the Southeast, but more lightning fires occur in the Southwest where rainfall often does not accompany thunderstorms. (A) Number of days per year with thunderstorms (1899—1938): clear = <10 days; horizontal stripes = 10—30 days; crosshatching = 30—50 days; vertical bars = 50—70 days; black areas = >70 days. (B) Average number of lightning fires per 400,000 ha per year (1 million acres): (clear = <1; dots = 1—5; horizontal stripes = 6—10; vertical stripes = 11—20; crosshatching = 21—40; vertical bars = 41—60; black areas = >60). (See "source notes.")

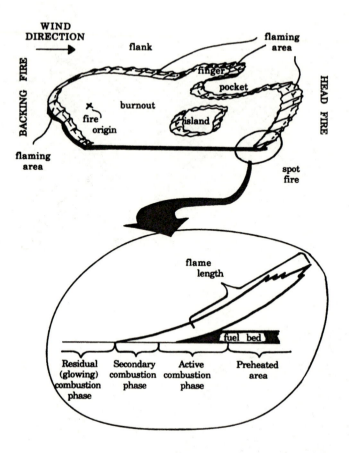

Figure 4.4 Basic terminology of a forest fire *(Pyne, 1984)*. (See "source notes".)

relatively little effect on intensity (Alexander, 1982). Unless soils are extremely dry, fuel is very concentrated, or fires burn in underground roots, the heat from fires remains primarily above the soil surface and temperatures within the mineral soil increase little (Hofmann, 1924; Chandler et al., 1983*a*).

Fires influence stand development patterns indirectly as well as by killing and scarring trees. Overstory and understory stems which were killed but not consumed by the fire remain in the stand, dry, and create combustible fuels for another fire. Standing snags also create partial shade and favor regeneration of some species by reducing wind, temperature, and humidity extremes; however, they buffer winds less than living trees do, so living trees in adjacent areas may blow over. Surviving trees—especially conifers— weakened by the fire often become susceptible to insect or disease infestations, which then spread to stands where the trees were not weakened. The fire may also consume many microorganisms deleterious to plant growth—such as the brown spot disease on longleaf pines (Lightle, 1960; Derr and Melder, 1970). Bark falls from areas at the bases of trees where fire killed the cambium, exposing a wound which can become infected by microorganisms, which proceed to decompose the tree's xylem (Toole, 1961). Most soil nutrients are released to the soil and become readily available for nutrient uptake or leaching to the groundwater (Chandler et al., 1983a). Some nitrogen in burning organic matter is readily volatilized (Grier, 1975; Klemmedson, 1976); other nitrogen becomes more available (Covington and Sackett, 1986). Where only the surface organic matter is consumed, the total carbon/nitrogen ratio of the soil can decrease and remaining nitrogen becomes more available, since surface layers usually contain the highest ratio (Biswell, 1972; Covington and Sackett, 1986). Nitrogen fixation by soil microorganisms can also increase after fires (Chandler et al., 1983a). The charcoal black soil surface and lack of shade after a fire increase soil temperatures and microbial activity—increasing decomposition and nutrient availability (Herman, 1963; Baker, 1968; Cochran, 1969). Accelerated erosion can occur immediately after a fire because of removal of forest floor litter and vegetation, a temporary water repellancy of the soil, and reduced root strength (Agee, 1993). Second burns can occur in dry logs on the forest floor several years after a fire. These burns can heat the soil and greatly reduce its productivity. Burns in the dry trees on the forest floor several years after a windstorm can similarly heat the soil.

Regeneration is affected in many ways by a fire (Ahlgren, 1959; Griffin, 1971; McQuilkin, 1989). The black soil surface color creates high temperatures and can kill germinants of some species until other species have provided slight shade (Shearer, 1974; Wells et al., 1979). Serotinous cones of some lodgepole and jack pines release seeds after the fire. Buried seeds of fire cherry, fireweed, and other species are resistant to being burned and germinate soon after a fire. Where surface fires have burned with or without crown fires, advance regeneration is usually killed to the root collar, or lower if the fire burned into the soil. Hardwood advance regeneration often sprouts from the root collar. Conifer advanced regeneration generally does not sprout and so dies. Other trees with sprouting ability can grow new stems from the root collars or roots. The fire can kill surficial roots of surviving overstory trees, making soil growing space available and allowing newly initiating understory plants to invade before it is completely reoccupied by the overstory.

Winds

Evidence of trees being uprooted by winds has been found in almost every forested region of the world (Lutz, 1940; Raup, 1941, 1964; Cline and Spurr, 1942; Jones, 1945; Goodlett, 1954; Stephens, 1956; Webb, 1958; Reiners and Reiners, 1965; Lyford and Maclean, 1966; Budowski, 1970; Chabreck and Palmisano, 1973; Henry and Swan, 1974; Whitmore, 1974; Sprugel, 1976; Halle et al., 1978; Stoneburner, 1978; Veblen and Ashton, 1978; Whitmore, 1978; Hartshorn, 1978, 1980; Thompson, 1980; Ashton, 1989; Oliver and Sherpa, 1990; Deal et al., 1991). Winds can cause both uprooting of the whole tree and snapping of the stem or branches. These winds need not be extremely severe. They need only be more severe than the tree's resistance. Since trees continually grow taller and develop wider crowns, they become increasingly more susceptible to wind.

Wind speeds over 85 km (50 mi) per hour occur in most areas of the United States, and over 160 km (100 mi) per hour in many areas (Baldwin, 1973). General directions of winds vary by region and by time of year. Prevailing winds on the Pacific coast are generally from the northwest, while they are from the northwest and north during January and February and from the south and southwest from May through August for most of the rest of the contiguous United States. As exceptions, Texas and Oklahoma receive winds from the south except in January and February, and Florida receives winds from the north in December and January and easterly winds at other times. Winds with speeds above 85 km (50 mi) per hour can occur at any time in coastal Alaska, and occasionally in other parts of North America. Strong winds are rare in Hawaii, but can occur locally. Periodicity of windstorms which destroy forests have not been studied extensively, but seem to vary by regions (Table 4.2).

Trees build up resistance to winds from the prevailing direction as they grow; therefore, windstorms which blow over trees either are unusually strong for that region, affect trees which have recently become exposed, or come from an unusual direction. Strong winds usually result from thunderstorms, frontal passages, tropical storms, tornadoes, or mountain winds. Thunderstorms produce winds from all directions. Thunderstorm frequency generally increases from the northwestern United States to the southeast (Fig. 4.3A).

Tropical storms come from the Atlantic and Caribbean; hurricanes and heavy winds of similar speed come to the Pacific coast from the southwest. The eastern ones generally begin heading west over the Atlantic or Gulf of Mexico, then turn north and enter land across the Atlantic or Gulf coastal states anywhere from Texas to Massachusetts, and then turn northeast. A storm is referred to as a *tropical depression* if its winds are less than 66 km (39 mi) per hour, a *tropical storm* between 66 and 125 km (39 and 74 mi) per hour, and a *hurricane* at higher speeds. Winds of hurricane force also reach the Pacific coast, especially during the autumn and winter months. True hurricanes are normally 340 km (200 mi) in diameter and advance at 17 to 34 km (10 to 20 mi) per hour when developing, are stationary when in full force, and move away (usually to the northeast) at speeds of 68 kin (40 mi) per hour (Baldwin, 1973). Hurricanes also generate tornadoes.

Although each one covers a relatively small area, tornadoes are the most intense storms, with surrounding wind speeds in excess of 340 km (200 mi) per hour. Tornadoes occur in all regions of the United States (Fig. 4.5) but are most frequent east of the Rocky Mountains in the central and southern Great Plains (Baldwin, 1973). About 650 tornadoes occur in the United States each year. Although they can happen during any month, high tornado frequency generally begins in the southern Great Plains during late winter and moves northward to the central and northern Great Plains, with maximum activity in April, May, and June.

Tornado funnels can reach the earth, creating a suction in their center surrounded by strong rains and/or hail, very high winds, and lightning. The funnel usually moves toward the east or northeast but can move in any direction. The usual speed is 42 to 68 km (25 to 40 mi) per hour; a tornado can remain stationary or change directions, and its speed can increase to 100 km (60 mi) per hour. The funnel can touch the ground, "skip up", and touch again nearby or at a distance. Tornado path widths can vary from a few meters to over a kilometer, and path lengths range from very short to over 340 km (200 mi); on average, tornado paths are about 250 m (800 ft) wide and 8 km (5 mi) long.

TABLE 4.2 Major Windstorms by Region, Showing Date and Estimated Property
Damage

Date	Type of windstorm	Place	Damage (in U.S. dollars)
Southeastern United States			
Sept. 8-11, 1900	Hurricane	Texas	$25,000,000
Sept. 20-21, 1909	Hurricane	Louisiana, Mississippi	5,000,000
Aug. 16-23, 1915	Hurricane	Texas, Louisiana	50,000,000
Sept. 29-Oct. 1, 1915	Hurricane	Mid-Gulf coast	13,000,000
Sept. 14-15, 1919	Hurricane	Florida, Louisiana, Texas	22,000,000
Sept. 20-22, 1926	Hurricane	Florida, Alabama	73,000,000
Apr. 12, 1927	Tornado	Texas	1,230,000
May 9, 1927	Tornado	Arkansas, Mississippi	2,300,000
Sept. 16-20, 1928	Hurricane	Florida	25,000,000
Sept. 2-6, 1935	Hurricane	Florida	6,000,000
Apr. 5, 1936	Tornado	Mississippi	35,000,000
Apr. 6, 1936	Tornado	Georgia	13,000,000
Nov. 24-26, 1950	Cyclones	Eastern United States	100,000,000
Mar. 21-22, 1952	Tornado	Arkansas, Missouri, Tennessee	14,000,000
May 11, 1953	Tornado	Texas	41,000,000
Dec. 5, 1953	Tornado	Mississippi	25,000,000
Aug. 30-31, 1954	Hurricane	N. Carolina to New England	439,000,000
Oct. 15, 1954	Hurricane	S. Carolina to New York	200,000,000
Aug. 16-20, 1955	Hurricane	N. Carolina to New England	832,000,000
June 26-29, 1957	Hurricane	Texas to Alabama	138,000,000
Sept. 9-13, 1960	Hurricane	Florida to New England	426,000,000
Sept. 11-14, 1961	Hurricane	Texas	408,000,000
Mar. 6-9, 1962	Cyclone	Florida to southern New England	200,000,000
Sept. 7-12, 1965	Hurricane	Florida to Louisiana	1,420,000,000
Mar. 3, 1966	Tornado	Mississippi, Alabama	18,500,000
Sept. 19-22, 1967	Hurricane	Texas	208,000,000
Aug. 17-20, 1969	Hurricane	Mississippi, Louisiana, Virginia	1,420,000,000
May 11, 1970	Tornado	Texas	135,000,000

TABLE 4.2 Major Windstorms by Region, Showing Date and Estimated Property Damage *(Continued)*

Date	Type of windstorm	Place	Damage (in U.S. dollars)
Aug. 3-5, 1970	Hurricane	Texas	453,800,000
Feb. 21,1971	Tornado	Mississippi	17,200,000
June 18-25, 1972	Hurricane	Florida to New York	2,103,600,000
Sept. 22, 1989	Hurricane	S. Carolina, N. Carolina	3,000,000,000
Northeastern United States			
Sept. 21-22, 1938	Hurricane	Long Island, southern New England	306,000,000
Nov. 24-26, 1950	Cyclone	Eastern United States	100,000,000
June 9, 1953	Tornado	Massachusetts	52,000,000
Aug. 30-31, 1954	Hurricane	N. Carolina to New England	439,000,000
Oct. 15, 1954	Hurricane	S. Carolina to New York	200,000,000
Aug. 16-20, 1955	Hurricane	N. Carolina to New England	832,000,000
Sept. 9-13, 1960	Hurricane	Florida to New England	426,000,000
Mar. 6-9, 1962	Cyclone	Florida to southern New England	200,000,000
June 18-25, 1972	Hurricane	Florida to New York	2,103,600,000
Midwest			
Mar. 23, 1913	Tornado	Nebraska	3,500,000
June 28, 1924	Tornado	Ohio	12,000,000
Mar. 18, 1925	Tornado	Indiana	17,000,000
Sept. 29, 1927	Tornado	Missouri	22,000,000
Apr. 11, 1965	Tornado	Indiana, Ohio, Michigan, Wisconsin, Illinois	200,000,000
June 8, 1966	Tornado	Kansas	100,000,000
Lake States			
June 2, 1919	Tornado	Minnesota	3,500,000
June 8, 1953	Tornado	Michigan	19,000,000
Pacific Coast			
Jan. 29, 1921	Windstorm	Washington	150,000,000
Oct. 12, 1962	Cyclone	Washington, Oregon	150,000,000

SOURCE: Baldwin, 1973, and Washington State Library, 1975.

Tornado winds are quite different from most destructive winds. They spiral and lift, causing tree stems to be twisted and shattered, with the tops raggedly severed from the stumps and scattered in all directions. After tornadoes, the fallen tree stems are not parallel and are so broken that they have little salvage value for wood products.

Mountains can create wind disturbances in several ways. Where winds are funneled between mountains or over passes, they accelerate and create areas of high potential for trees' being blown over. Trees on windward sides of ridges can be desiccated by excessively strong, cold, hot, or dry winds. Lee slopes are sometimes affected by violent downslope winds created by wave motions of wind moving over the summit (Barry, 1981). Upslope (up-valley) winds in the afternoon and downslope (down-valley) winds at night are caused by differential heating and air expansion in mountains and valleys; these winds can cause rapid changes in temperature and fire behavior.

Winds moving over mountains and into valleys first cool and release moisture as they rise and then warm and absorb moisture as they descend. In winter these *fall winds* can have extreme effects. Warm, dry winds blowing to valleys are known as a *chinooks* (also, as *fohns* and *santa ana winds*), and cool winds are known as boras (Barry, 1981). Such winds have caused air temperatures to rise 27°C (49°F) to an ambient temperature of 7°C (45°F) in 2 minutes (min) in Spearfish, South Dakota, Jan. 20, 1943; to fall 26°C (47°F) to -13°C (8°F) in 15 min in Rapid City, South Dakota, Jan. 11, 1920; and to fall 56°C (100°F) to -49°C (-56°F) in 24 hours in Browning, Montana, Jan. 23-24, 1916 (Baldwin, 1973).

A tree can become uprooted (windthrown), or its stem can break from the wind (Putz et al., 1983). Trees which grow at wide spacings and in a windy area can develop resistance to wind by growing strong stems and strong, spreading root systems. The constant motion of a tree growing in a windy area will cause reaction wood—tension wood in hardwoods and compression wood in conifers—to develop along the stem and supporting roots. Trees in windy places such as exposed areas, wide spacings, or stand edges develop this wind resistance in the direction of the prevailing wind. Trees at tight spacings on the interior of stands are protected from the wind and will not develop the resistant stem or roots. An individual tree bending with the wind compresses the xylem at the junction of the stem and roots on the leeward side, and puts tension on the xylem and root junction on the windward side. Green wood—xylem—is much stronger in tension than compression; consequently, wood is crushed at the root junction on the lee side, and tension on roots on the windward side keeps the tree from overturning. A tree falls over more easily if a large rock, an anaerobic microsite, or other factor has kept roots from extending on the windward side. Trees are often buffeted by many gusts in a windstorm, causing them to sway in an arc of 30° or more. If enough force is exerted on the stem, the leeward roots will be crushed, the windward roots will be pulled out of the ground, and a root mound will be formed as the tree falls over. On lee sides of hills or eddies where winds come from several directions, trees can be pushed in opposite directions during a storm; as trees sway in strong winds, the roots are crushed on both the windward and leeward sides in the directions of the sway; finally, many root fibers are crushed on the windward side, these roots are no longer strong enough to hold the tree upright, and the tree falls in the direction of the wind (Mergen and Winer, 1952; Mergen, 1954). Conifers also develop stronger roots on their downhill side. Consequently, wind

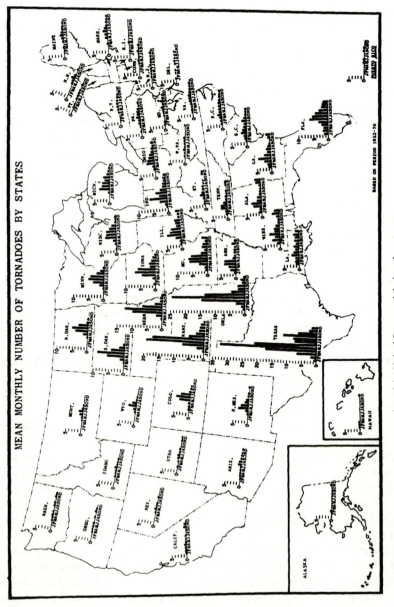

Figure 4.5 Tornadoes can occur in every region of the United States and during almost every season, but are most common in the lower Great Plains (*Baldwin, 1973*). (*See "source notes."*)

pressure from the uphill side can more easily cause trees to blow over (Steinbrenner and Gessel, 1956). Trees can fall over in a single gust during a storm or can be pushed farther with repeated gusts.

The first trees blown over or broken in stands, or the ones blown over when only some trees are destroyed, are generally trees of lower crown classes or ones infected with rot (Wiley, 1965). Trees at narrow spacing or of lower crown classes are thin and have weak roots connecting the stems and so readily buckle or tip over. Shallow soils do not allow deep penetration of roots and so make a tree susceptible to windthrow (Croker, 1958), although wide spacings allow the roots to spread more and help the tree become more stable. On the other hand, a tree growing at a wide spacing develops a large crown which provides more surface to the wind and consequently more force to push the tree over. Stems or roots infected with rot are less resistant to breakage or uprooting.

Stands vary in the time at which they become susceptible to windthrows depending on their structures, species compositions, and sites (Ruth and Yoder, 1953). At older ages, after competition-related mortality is less common, windthrow can be a major cause of death of individual trees (Harcombe and Marks, 1983; Franklin et al., 1986). Trees are more susceptible to windthrow just after thinning a stand, since newly released trees do not offer the mutual support and wind protection of the tight stand. Stem rot—which usually leaves sound wood around the perimeter of the stem—increases a tree's susceptibility to breakage because less strong wood is available for support. Stems often break at places of rot, forks, or crooks formed after the release of suppressed trees. Trees growing rapidly in height or crown spread become proportionally more susceptible to the wind. Wind exerts equal pressure for uprooting a tree twice as tall as another if the shorter has twice the crown surface area exposed to the wind. The taller tree is more susceptible to stem breakage, however, even if its stem diameter is not reduced by its small crown. Evergreen trees are more susceptible to winter windstorms than deciduous trees, since the retained foliage offers more surface area to the wind. Sites and climates favorable to tree growth encourage rapid height growth, making the trees increasingly susceptible to winds.

Trees on stand edges are generally first hit by strong winds, reduce the wind velocity within the stand, and thus buffer the rest of the stand. Having been exposed to other winds, edge trees are resistant to windthrows. Edge trees also have longer crowns extending down their exposed side, favoring diameter growth and hence stability in windstorms. If the edge trees are destroyed by logging, insects, fire, disease, or some other disturbance, wind will enter the stand and blow over more sensitive, interior trees. "Walls" of trees along stand edges facing the direction of prevailing windstorms are more susceptible to windthrow than are edges at angles to the wind (Alexander and Buell, 1955; Gratkowski, 1956). Windblown trees generally fall parallel to the wind direction. Different storms can sometimes be distinguished by the direction of tree fall (Stephens, 1956). Where the soil is dry or frozen, uplifting of a tree's roots is more difficult and trees more commonly break in the stem or remain standing. When the soil is moist and friable, root uplifting is more common.

A severe windstorm can leave a mosaic of stand conditions across the landscape based on topographic and soil features (Harris, 1989), stand structures, and specific characteristics of the windstorm. For example, a severe windstorm of 1921 which overturned approximately 25 thousand ha (60 thousand acres) of forests on the Olympic Peninsula of Washington left the forest landscape in a mosaic of stand conditions. Some areas over a square kilometer in size had very few trees remaining upright, while other areas were essentially untouched (Boyce, 1929; Pierce, 1921). Hurricanes in the southeastern United States (Marsinko et al., 1993) and windstorms in tropical rainforests (Nelson et al., 1994) similarly produced a varied landscape pattern.

The immediate effects of winds are to make growing space available by removing shade and, where they overturn nonsprouting trees, by killing tree roots. Even small winds abrade lateral branches, so that more light passes to lower parts of the canopy. Winds can also blow over individual trees, releasing large amounts of growing space. Except for areas crushed by falling trees or upturned by windthrow mounds, winds generally do not damage the soil or forest floor vegetation.

Wind abrasion among tree crowns can change competitive advantages among species. One species' terminal or lateral branches can be knocked off by the tougher limbs of another species. Winds also disperse seeds over long distances.

Winds have indirect effects on stands. Overturned trees create hummocky soil conditions (mound-and-pit microrelief) and mix soil horizons (Lutz, 1940; Stephens, 1956; Bowers, 1987; Ugolini et al., 1989). The mounded microrelief then creates raised, mineral seedbed microsites suitable for germination of light-seeded species, as well as nearby "pits" unsuitable for tree growth. These microsites help determine the spatial patterns of trees in the next stand; the raised microsites may even be necessary for growth of certain species in some areas—such as Sitka spruce in Alaska (Deal, 1987). The mixing of the soil may also reverse podzolization and favor organic matter decomposition, making the soil more productive for tree growth by releasing tied-up nutrients (Bowers, 1987; Ugolini et al., 1989). Except where it is crushed, advance regeneration is essentially untouched in wind-disturbed stands. Advance regeneration often grows rapidly into growing space released by the windstorm, and these trees dominate the next stand. Sprouts also grow from overturned tree roots of some species, provided part of the plant retains contact with the soil, and from the stems of broken trees. Buried seeds germinate and begin growing; water sprouts appear on those trees which survive the winds. Cracks in roots allow diseases to enter (Hintikka, 1972).

After several years, the trees blown over and killed by the wind dry out, increasing the fuels and possibility of a severe fire. Such fires were probably common before recent salvage and fire suppression efforts (Henry and Swan, 1974). Fires of tens or hundreds of thousands of acres would probably have occurred in the southeastern United States following Hurricane Hugo (Marsinko et al., 1993), in the northeastern United States following the 1938 hurricane (Clapp, 1938), in Hokkaido Island of Japan following the 1954 typhoon (Osawa, 1992), and in the Pacific Northwestern United States following the 1921 windstorm (Pierce, 1921; The Timberman, 1921; Boyce, 1929) were it not for active efforts by people. Where such fires occur, they are generally very hot because of the large amount of dry fuel close to the ground. They burn all advance regeneration, sprouts and even root collars, giving a growth advantage to light-seeded species which blow in afterward.

Floods

Flooding, erosion, and siltation are normally associated with heavy rains. Flooding occurs in every region of the United States at some time of the year (Fig. 4.6), while erosion and siltation occur under certain soil, topography, and land use conditions.

Floods generally result from very heavy rainfalls or snowmelt. Consequently, floods usually occur during the spring snowmelt in northern areas or in the Mississippi River, which drains northern areas. In mountainous areas floods occur in winter when heavy rains melt accumulated snow—especially along the Pacific coast. Throughout the southern United States, floods result from rains and often occur after hurricanes or local thunderstorms. Streams and rivers originating from glaciers often have maximum daily flows in late afternoon as the glacier is melting, although maximum annual flows depend on precipitation distribution and other weather conditions.

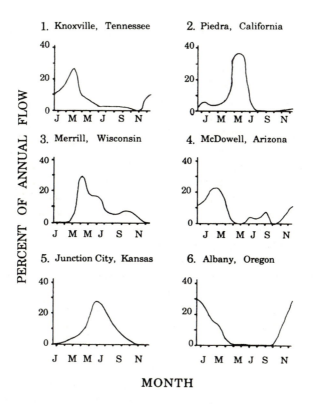

MONTH

Figure 4.6 Floods occur in each drainage with greater probability during certain seasons. The periods of flooding are dependent on the times of rainfall and snowmelt. 1. Tennessee River at Knoxville, Tenn., 169 floods in 51 years; 2. Kings River at Piedra, Calif., 211 floods in 39 years; 3. Wisconsin River at Merrill, Wis., 130 floods in 31 years; 4. Verde River at McDowell, Ariz., 88 floods in 35 years; 5. Republican River at Junction City, Kan., 117 floods in 27 years; 6. Willamette River at Albany, Oreg., 166 floods in 50 years. (*Wisler and Brater, 1959.*) (See "source notes.")

Sizes and effects of floods are dependent on local weather and drainage basin conditions (Hack and Goodlett, 1960). Probabilities of floods of a predetermined magnitude can be estimated for a given river. Floods are generally restricted to more predictable areas than are fires and windstorms. Floods occur primarily around streams and rivers (Shelford, 1954). The flat, wooded floodplains surrounding rivers in the southeastern United States are the most extensive areas flooded—some of them being tens of miles across [before it was restricted, the "Delta"—a floodplain of the Mississippi River in western Mississippi—was about 170 km (100 mi) wide and 425 km (250 mi) long]. Narrow mountain valleys such as in the Colorado Front Range can suffer sudden, periodic, and catastrophic floods and extreme—if localized—damage.

Floods cover trees to different heights, they carry pieces of wood which float on the water surface and batter stationary vegetation, and they deposit silt and sand on the soil surface. Floods can wash excess salts from the soil and so increase the soil productivity. After floods, water stays impounded in *low flats* until it evaporates or slowly drains. Fish which washed in with the water can die and rot there. Moving floodwaters often change stream and river channels (Coleman and Gagliano, 1964; Coleman and Wright, 1975; DeLaune et al., 1983), in spite of the best human efforts to dictate a stream's channel. The most notable example is the constant tendency of the Mississippi River to divert to the west near Nachez and flow down the Atchafalaya River channel, thus abandoning its channel through New Orleans (McPhee, 1990).

During the last glacial event, ice dams impounded water and created lakes as large as lake Michigan in eastern Washington and northern Idaho. When the ice dams broke or floated, water flooded across central Washington and finally drained to the Pacific Ocean through the Columbia River (Baker, 1978). The affected landscape still shows evidence of deep erosion and other effects of these catastrophic floods. These long-ago events have affected the present soil conditions and thus growth potential of the site. Similar *jokulhlaups*—outbursts of water from glaciers or glacier-dammed lakes—have been observed in many parts of the world (Meier, 1964).

Species vary in how long they survive and grow in waterlogged soils or when completely submerged (Ahlgren and Hansen, 1957; Hosner, 1960). Some species survive less than 4 weeks during the growing season, while others survive more than a full growing season. Seeds can also remain viable for different periods when submerged (McDermott, 1954; Broadfoot, 1967; Kennedy and Krinard, 1974; Keeley, 1979; Walters et al., 1980; McKnight et al., 1981). Running water can mitigate temperature fluctuations in the soil and maintain warm temperatures suitable for plant growth in winter. Most temperate species—especially hardwood—can tolerate standing water in winter if it is drained off before springtime. Species distributions in commonly flooded areas are often based on their flood tolerances (Demaree, 1932; McDermott, 1954; Hall and Smith, 1955; Hosner, 1957, 1960; Hosner and Boyce, 1962; Broadfoot and Williston, 1973; Koevenig, 1976; Baker, 1977; Teskey and Hinckley, 1977, 1978; Hook and Scholtens, 1978; McKnight et al., 1981; Malecki et al., 1983; Pezeshki and Chambers, 1985).

Erosion

Erosion usually occurs at times of flooding—during heavy rains or snowmelt. Some erosion occurs as a normal part of stream movement and dissipation of stream energy. The extent of erosion varies with soil textures, topography, and disturbance history—especially land use activity. Extensive erosion occurred in the hilly, clay Piedmont soils of the southeastern United States before the 1950s. Poor farming practices removed the roots and other organic matter holding soil particles together and allowed water to gain momentum by flowing downhill unobstructedly. Erosion occurred in the Rocky Mountains in the late 1800s because extensive grazing and soil compaction by animals destroyed roots which held soil particles together. In the Pacific coast region, localized erosion occurred in the late 1800s and abated after about 1950. Poor forest road construction and, to a lesser extent, poor logging practices caused much of the erosion. In California, hydraulic washing of hillsides to obtain gold eroded large mountains. Relatively little erosion occurs in glaciated regions of the northern United States because glacial activity has generally left soils of coarse texture (DeWiest, 1965).

Erosion occurs when water moves rapidly over the soil surface, gaining momentum as it flows downhill. The water picks up soil particles as it moves and transports them. The size (d) and amount (Q_s) of soil particles moved is proportional to the amount (Q) and slope (S) of water flow, as expressed in *Lane's relationship* (Garstka, 1969):

$$QS \propto Q_s d$$

This relationship is a proportionality, not a mathematically rigorous relationship; however, it illustrates that change in any one variable affects the others. High rains and steep slopes increase the Q and S relations, respectively, and thus the amount of erosion (Q_s) for a given size of soil particle. Smaller soil particles (finer textures of soil) produce more erosion for a given Q and S.

Vegetation reduces erosion in several ways. Until the vegetation is completely wetted and the soil infiltration rate is exceeded, vegetation reduces the amount of water flow (Q) both by intercepting it on the foliage, stems, and litter and by producing soil pore spaces for water to infiltrate into the soil rather than run over the surface. Vegetation also interrupts channels in the soil, reducing the rate and slope (S) of soil movement. Particle sizes (d) of fine textured soils are increased by vegetation, since roots and accompanying fungal mycelia bind particles together.

When water flows overland through a forest, it picks up (erodes) soil particles and carries them downslope. If allowed to continue, the water creates a channel and begins removing soil particles within and beneath adjacent vegetation, thus killing the roots and eventually undermining the vegetation and causing it to fall over. As the soil washes away, growing space is reduced, vegetation grows more slowly, and the preventive effects of vegetation on erosion are diminished. Erosion then proceeds even faster.

Erosion occurs primarily where vegetation is destroyed or prevented from growing by grazing, agricultural practices, a lack of moisture, or passage of a steep channel of water through the soil, such as after poor road construction. Disturbances create conditions for erosion when they destroy enough roots or soil organic matter to allow rapid overland water flow or to diminish the soil structure and infiltration capacity. Heavy fires favor erosion in arid regions where regrowth of vegetation is slow. A fire-erosion cycle can also occur in dry regions, creating a *gully* drainage network where the better soils produce more organic matter, are burned hotter, and erode more severely because of their depletion of soil organic matter (Segura-Warnholtz, 1989). Removal of vegetation in warm, humid areas where the dead roots and other soil organic matter decompose rapidly also predisposes the soil to erosion.

Efforts to stop erosion in one place often exacerbate it farther downstream. A known volume of water moving at a specific rate can carry a certain amount and size of sediment. If the amount of sediment is reduced by erosion-control activities without reducing the slope or amount of water, the water increases in force and removes sediment from elsewhere. Similarly by Lane's relationship, increasing water speed (slope, S) by straightening the channel, paving it, or removing obstacles such as logs and rocks can lead to more erosion.

Besides the direct effect of killing vegetation, erosion has the indirect effect of removing soil growing space by reducing its nutrients, water-holding capacity, and even the physical pore space for root penetration. At its extreme, erosion removes all organic matter, advanced regeneration, stumps, roots, rhizomes, and buried seeds. Erosion also changes an area's topography. Those species which can survive extremes of soil conditions and regenerate from seeds blowing into an area (or from stolons moving along the soil surface) are generally favored on eroded landscapes (Fisher et al., 1982).

Siltation

Siltation is the deposition of fine soil material, usually by water or wind action. Where siltation occurs on the forest floor, it can damage or kill trees. Some siltation occurs with normal flooding, and extreme siltation occurs downstream from eroded areas. Most alluvial floodplains contain soil of a few meters to hundreds of meters deep deposited by siltation (Shelford, 1954).

Siltation can be understood by Lane's relationship, described previously. The speed (slope, S) of water declines as it reaches a flat area or a wide channel, and the amount of sediment (Qs) it can carry is reduced. The excess sediment is deposited. Large particles are deposited first, and progressively smaller particles are deposited as the water slows. Water-deposited soils usually are deposited in relatively flat layers (*lenses*) of similar textures. A stream or river slows when it overflows its bank, depositing sand along the bank—often as elevated levees—and silt and clay progressively farther from the channel as the floodwater slows. Soils thus deposited by floods contain distinct areas of uniform soil particle size roughly parallel to the channel and create a mosaic of site conditions in river floodplains (Bell and del Moral, 1977).

Siltation buries the soil profile with less than 1 cm to several meters of new soil material, reducing the oxygen to the former rooting zone. Some species are killed by siltation, while others survive various amounts of overburden. Some species survive because their roots grow upward to the new soil surface. Cottonwoods (Everitt, 1968), aspens, redwoods (Fritz, 1934; Stone and Vasey, 1962, 1968), and some other species survive because their stems produce adventitious roots which grow into the new soil surface.

Freshly deposited alluvium varies from coarse sand (or even boulders at an extreme) to fine silts and clays. Where the texture is fine enough and moisture is available, seeds entering by water or wind after a disturbance germinate and rapidly refill the available growing space (Nanson and Beach, 1977). Where the new soil is of fine texture, it increases the total growing space of the site, although it often brings nutrients and pH relations characteristic of the area from which the erosion occurred. For example, the soil pH of the lower Mississippi River floodplains is generally higher than the pH of floodplains of nearby local rivers because the Mississippi River alluvium originates in the more basic soils of the Great Plains.

Trees weakened by siltation become predisposed to insect or disease attack, which then spread to adjacent healthy areas. Trees killed by siltation also become dry and susceptible to fires.

Landslides

A landslide is the sudden translocation of soil material from a higher to a lower place under the direct influence of gravity. Landslides occur in most hilly or mountainous regions of the United States—the southeastern Piedmont, Great Smokey Mountains (Bogucki, 1970), Adirondacks (Bogucki, 1977), White Mountains (Flaccus, 1959), Rocky Mountains (Langenheim, 1956), Pacific Northwest (Washington State Library, 1975)—as well as elsewhere in the world (Veblen et al., 1980; Garwood et al., 1979; Glynn, 1988; Bruenig and Y-w Huang, 1989; Gentry, 1989; Irion, 1989; Junk, 1989; Salo and Rasanen, 1989).

Soils usually rest on slopes below a maximum stable *angle of repose* characteristic of the particular soil and based on the particle size, chemical and physical characteristics, climate, structure, and method of formation or deposition. Above this angle of repose, the soil is unstable and can slide or slump as a *rotational failure* (Sidle et al., 1985). In areas of geologic uplift, slopes constantly increase in steepness, eventually exceed the angle of repose, and become subject to failure. Tree roots can increase the stable angle of repose. After the trees are killed (e.g., by fires, windstorms, insects, or forest harvesting), a steep slope can remain stable until the fine roots have rotted, after which the slope often fails.

Earthquakes and heavy rains can precipitate landslides by reducing the friction between soil particles, thus reducing the maximum stable angle of repose. The soil then moves downhill to a new stable—or temporarily stable—angle (Garwood et al., 1979). Excavations—such as for road construction—along a slope can increase the water infiltration beneath the excavation and increase the angle of repose above the slope, causing slope failure above and below the road. Glacial lateral and terminal moraines are often prone to landslides because the soil (or rock) slopes are deposited against ice. After the ice melts, the slope is often greater than the maximum stable angle of repose and rockslides or landslides occur with little stimulation.

Landslides generally remove all buried seeds, advance regeneration, stumps, and other organic matter and expose mineral soil material or bare rock in the area of denudation. The result is a general decrease in soil growing space and a favoring of species which germinate from windblown seeds or grow from stolons. The former soil profile is generally buried except in isolated chunks; and the newly arrived soil is so mixed in the accumulation area that little regeneration occurs from pre-existing advance regeneration, buried seeds, or stumps. Species which initiate growth from windblown seeds are generally favored (Flaccus, 1959). Unlike the area of denudation, growing space may be increased in the area of accumulation, since the soil is mixed and the total depth increased.

Mudflows occur where internal heating of volcanic mountains causes saturated soil material to move rapidly downhill (Crandell, 1971). One such mudflow was begun by an ice avalanche and descended at a rate of 105 km (65 mi) per hour. The effect of a mudflow is similar to that of other landslides described above, and accumulation areas can bury the original soil profile with several meters of soil material (Gartska, 1964).

Rockfall areas often shatter existing vegetation, giving growth advantages to species which are very short-lived and set seeds rapidly or can regenerate readily from stump or root sprouts, bulbs, tubers, rhizomes, or stolons.

Avalanches

Snow or ice avalanches occur in snowy mountain regions throughout the world. They occur either as loose snow or as slabs of snow, with each having unique characteristics depending on the amount of liquid water present. Dry-powder snow avalanches are generally the fastest, with speeds over 320 km (200 mi) per hour on slopes of 44°. One wet-snow avalanche on a similar angle was estimated at 27 km (17 mi) per hour. Ice avalanches occur when ice breaks from a glacier (Gartska, 1964).

Avalanches generally occur during the 1 or 2 days following 10 or more inches of new snow accumulation; during warm, fair weather following a snowstorm; during windy periods; or after a rain in snow. Avalanches follow repeated tracks or create new paths through stands which have not been previously disturbed for many centuries. Avalanches generally fall downhill, although their momentum often carries them across valleys and up opposite slopes.

The upper part of avalanche tracks (the *start zone*) is generally scoured as the slipping snow removes the vegetation (and sometimes soil; Flaccus, 1959; Cushman, 1976, 1981). Below this is a *runout zone,* where snow from above accumulates. Preceding the moving snow is often a strong wind which blows down or breaks off trees or other objects protruding above the snow level in the runout zone, creating broken or uprooted trees parallel to each other on the ground, but otherwise not harming the forest floor or advance regeneration in patterns similar to other windstorms. An avalanche track offers a gradient of disturbances from very severe in the upper start zone to not severe in the lower runout zone, with varying regeneration mechanisms favored along the gradient. Advance regeneration species are often favored in the runout zone, sprouting species in the central zone, and light-seeded species in the upper zone (L. Smith, 1973; Burrows and Burrows, 1976; Cushman, 1976, 1981).

Glaciers

New England, the northern border of the contiguous United States, and parts of Southeast Alaska were covered by a continental glacier for a long period before 12,000 years ago. Within the past 2 million years, a continental glacier has extended approximately to the same point and retreated several times, keeping most of Canada under ice during this period for longer than it has been exposed (Fig. 4.7). Valley glaciers extended in mountainous areas south of the continental glacier, and are still present in many mountains west of the Great Plains (Flint, 1971).

Glaciers generally advance and retreat with long-term fluctuations of the weather (Skinner and Porter, 1987); however, internal dynamics can cause them to advance or retreat abruptly in contrast to the weather changes (Gartska, 1964). Valley glacier termini have been estimated to advance over 100 m (300 ft) per day during short intervals. Usually, however, these glaciers advance and retreat only 0.5 m (1 or 2 ft) per day (Sharp, 1960).

When glaciers advance rapidly, they physically break over trees and push away soil material—acting as a giant bulldozer. When they retreat—weeks, months, centuries, or millennia later—their meltwaters wash away the organic matter and fine soil material, generally leaving a landscape completely bare of vegetation and consisting of sands, silts, rocks, and boulders distributed in characteristic patterns of glacial till and outwash and terminal and lateral moraines. These patterns then form the basis for variations in growth potential of the soil which develops on them. The retreating glacier creates secondary disturbances such as rockslides, landslides, avalanches, frost actions, and floods which continue for hundreds of years after the glacier has retreated—well after a forest stand has become established on the exposed area. An isostatic rebound is still occurring in Canada where the land is rising now that the weight of the continental ice sheet has disappeared. Similarly, sea level is estimated to have risen 110 m (350 ft) from its extreme when much of the water was frozen in the extended polar icecap (Emery, 1967).

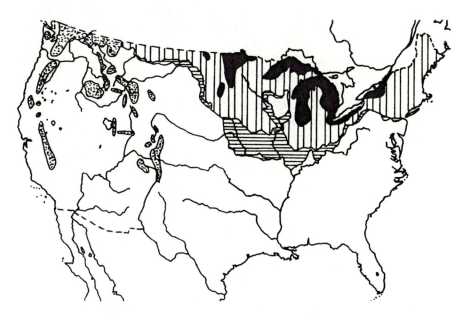

Figure 4.7 Areas covered by glacial activity in the United States: (clear = unglaciated; horizontal stripes = old drift; vertical stripes = young drift; black areas = glacial lakes; dots = mountain glacial activity. (*Shimer, 1972*.) (See "source notes.")

The local climate surrounding the continental glaciers was probably much different from that surrounding recent valley glaciers. Frost heaving and other effects of very cold climates can be found around the continental glaciers' margins (Stout, 1952; Goodlett, 1954). By contrast, except for the first 30 to 90 m (100 to 300 ft), the microenvironment in front of a retreating valley glacier is not strongly affected by the glacier. The glacier's presence does not affect the species composition of the invading community markedly (Oliver et al., 1985). Species are favored which reach the area by windblown seeds— either from vegetation advancing before the retreating glacier or from vegetation on hills above the glacier. Even on extreme sites recently uncovered by the retreating glacier, many tree species are able to invade and survive without other species' first modifying the soil and aboveground environment (Sigafoos and Hendricks, 1969; Oliver et al., 1985).

Volcanic activities

Volcanoes occur near contacts of continental plates such as along the Sierra and Cascade Ranges of California, Oregon, and Washington. Some (*mafic*) volcanoes spread hot lava on the immediately surrounding area; lava creates fires and eventually cools to basalt rock. *Felsic* volcanoes produce explosive eruptions which blast trees over with strong winds and emit particulate *tephra* (a mixture of volcanic ash and pulverized rock from the original volcanic cone) into the atmosphere (Strahler and Strahler, 1978). Some of this particulate matter falls locally; and some travels long distances with the wind, eventually falling to earth. Soils containing volcanic tephra are found locally around the Pacific coast volcanoes and as far away as Idaho and Montana where the tephra was blown after eruptions of Mount Saint Helens in western Washington, Mt. Mazama (now Crater Lake) in western Oregon, and others. Finer particulate matter can reach the upper atmosphere and circulate around the globe, reducing sunlight intensity for several years and causing cooler global climates.

The immediate effects of basalt lava flows are to kill existing vegetation and reduce the soil growing space by covering previous soils with rock. Basalt decomposes rapidly compared with many rocks, and small plants capable of tolerating the site extremes can be found in crevices on basalt within a few decades (Eggler, 1971). Basalt will weather to relatively fertile soils within a few hundred years in favorable climates.

Most volcanic blasts are directed upward and do not affect vegetation except immediately around the volcano's rim. The 1980 blast of the Mount Saint Helens volcano was an exception, since a landslide immediately before the blast directed it sideways. The immediate effect of a volcanic blast is similar to a wind, except the hot air and gases also kill any exposed vegetation on the forest floor—but will not necessarily harm buried seeds, rhizomes, tubers, or roots (Griggs, 1919).

Tephra deposition has varying immediate effects depending on characteristics of the tephra, amount deposited, and time of year it is deposited (Griggs, 1919; Veblen et al., 1980; V. A. Seymour et al., 1983; Hinckley et al., 1984; Adams and Adams, 1987; Bilderback, 1987 Yamaguchi, 1986, 1993; Segura, 1991; Segura et al. 1994). Fine tephra can coat leaves, restrict photosynthesis, and eventually kill trees. Coarse pumice batters the limbs and buds of trees, thus killing them. If tephra is deposited in winter, forest floor vegetation is buried beneath snow and escapes immediate injury. If fine tephra is deposited before budbreak in spring, newly emerging leaves may photosynthe-size enough to keep the tree alive and annual and perennial plants may emerge through the tephra from tubers, rhizomes, and roots—providing that the tephra layer is not too thick. Tephra fall can have a mulching effect on soils, which increases tree growth where it does not immediately suffocate the roots or leaves. If the deposition occurs later in the spring, the same plants are smothered and killed. Vegetation favored by a vol-cano, therefore, varies with specific characteristics of the volcanic ejecta; however, advance regeneration is probably not favored unless it can resprout from the root collar; and light, windblown seeds germinating on the tephra are favored where the blast removed or obliterated the previous soil horizon and where the tephra layer is too thick to allow buried seeds or sprouts to penetrate upwards.

Trees killed by a volcanic eruption may eventually dry and create fuels, making the area susceptible to fires. Trees weakened by the falling tephra can become susceptible to insects and diseases, which then build their populations and invade otherwise healthy stands. The long-term effects of volcanic tephra depend on its local characteristics. Most of the tephra layers in soils in eastern Oregon are several centimeters to 1 or 2 meters thick. As a general rule, tephra is very porous, easily weathered, and easily com-pacted. Tephra layers in soils generally have a high water-holding capacity. Actual development of tephra into soils depends on local climatic and biological conditions.

Ice storms

Ice storms—freezing rain or "glaze"—occur generally from November to March in the contiguous United States. They occur mostly in the northern part of the United States—averaging 5 to 15 days per year, compared with less than 1 day per year in the extreme southwestern United States, along the Gulf of Mexico, and in southern Georgia. Glaze practically never occurs in southern Florida (Baldwin, 1973). Although they are most prevalent in Minnesota, Wisconsin, Michigan, Pennsylvania, New York, and New England, they are often most destructive farther south.

Ice storms occur when warm, moist air produces rain, which then falls into a subfreez-ing air layer at ground level. The supercooled rain freezes on tree limbs, foliage, and other structures. The ice usually coats the objects as a thin film or builds to 3 cm (1 in) thick, but has been known to be 20 cm (8 in) thick.

Trees are usually hardened to the cold, and most damage is from breakage because of the ice's weight. Some species withstand the excess weight, remaining balanced upright, and others shed the glaze by allowing needles and limbs to break off. Others cannot withstand the glaze; their stems bend, buckle, or often break. Evergreen trees contain more surface area for adherence of ice and, so, are more heavily weighed down. For evergreen conifers, this excess weight may be offset by their more conical crown shapes. An evergreen tree 15 m (50 ft) tall with an average crown width of 6 m (20 ft) can con-tain 4.5 metric tons (5 tons) of ice during a severe storm. Winds while trees are coated with ice further unbalance the trees and greatly increase the amount of breakage.

Severe damage occurs from breakage of limbs or whole trees (Downs, 1938; Carvel et al., 1957; Lemon, 1961; Baldwin, 1973; Siccama et al., 1976). Some species are less adapted to glaze storms than others (Rogers, 1922, 1923). One glaze storm of February 15-17, 1969, destroyed $60 million of forest timber in central South Carolina (Baldwin,

1973)—primarily by buckling and breaking slash pines which had been planted north of their natural range during the preceding 2 decades. Glaze storms can cause indirect damage by breaking limbs and allowing pathogens to enter stems, and by breaking tops from some trees and allowing others to overtop them. Sprouting species are generally favored by these storms. Trees killed by glaze often attract insects, which build up their populations and then attack otherwise vigorous trees.

Heavy, wet snows cause breakage in conifers in the Pacific northwestern United States (Thornburgh, 1969) but do less damage to leafless deciduous trees in other regions (Blum, 1966).

Mammals

Mammal damage is usually from grazing or browsing (Apsley et al., 1985), from eating of seeds (Jensen, 1982) and tree bark, from physical breakage, or from indirect activities such as flooding or burrowing (Platt, 1975). Deer and rabbits browse small trees in many areas (Anderson and Loucks, 1979; French and Lorimer, 1985), and beavers are becoming locally abundant in the United States after being greatly reduced for almost a century (Hill et al., 1978). Buffalos and woodland bison were previously important but have become less so (Larson, 1940). Other areas contain specific animal problems—mountain beavers and bears in the Pacific northwestern United States, for example. In regions of Africa, elephants have become more numerous, increasingly damaging vegetation (Buechner and Dawkins, 1961). Some vegetation types in North America may have resumed the browsed conditions after domestic livestock introduction; this browsing had been absent since the elimination of many browsing mammals during the Pleistocene extinction about 10,000 years ago (Janzen, 1982; Martin, 1982; Pielou, 1991).

Wild boars have damaged forests in Europe, Asia, and America where introduced (Bratton, 1975). Introduced deer and possums in New Zealand are browsing the forest vegetation and significantly altering its structure (Veblen and Stewart, 1980, 1982b).

Browsing usually occurs on small twigs and buds below about 1.5 m (5 ft) tall. Where browsing animals—deer, elk, sheep, goats—are numerous, they inhibit development of some or all woody vegetation (Sparhawk, 1918; Stoeckler et al., 1957; Marquis, 1975b, 1992; Amaral, 1978; Hanley and Taber, 1980; Irwin et al., 1994). Excessive grazing can also reduce soil growth potential through compaction (Steinbrenner, 1951). Some competing brush is more palatable than tree seedlings, so grazing may sometimes promote tree regeneration (Sparhawk, 1918; Nakashizuka and Numata. 1982b). Some new twigs of woody plants contain undesirable chemicals, and these parts are discarded after the animal has eaten older twigs (Coley et al., 1985). In other cases, newly emerging spring growth is most desired, and older growth is not eaten. Where the animal population is not so high that all palatable vegetation is eaten, an animal will generally "tire" of eating one plant species and will diversify its diet as much as possible. Consequently, an infrequent species can be eliminated from a plant community, and the competitive advantage of the most prevalent species is maintained (Habeck, 1959; Beals et al., 1960; Halls and Crawford, 1960; Crouch, 1974; Hanley and Taber, 1980). Browsing often gives a competitive advantage to distasteful species and to sprouting species (Cates and Orians, 1975). Palatable seeds are eaten by rodents, deer, birds, or other animals where a high animal population exists (Krajicek, 1955; Duvendeck, 1962), and a new stand of non-palatable species may result.

The inner bark of trees is often eaten by beavers, porcupines, mountain beavers, and bears before the bark becomes too thick—generally in trees less than about 30 cm (12 in) in diameter (although beavers often chew much larger trees). Attacked trees can be either completely girdled and killed or partly girdled and made susceptible to pathogens which decay the xylem.

Physical breakage occurs when such animals as bears, elephants (Buechner and Dawkins, 1961), or the now-extinct passenger pigeons weigh down and break limbs and stems (Lawson, 1709; Savage, 1970).

Flooding of forests by beavers—with the subsequent killing of trees—is common in most of the United States (Ives, 1942). The level topography of the southeastern Coastal Plain and parts of the Minnesota-Wisconsin area makes a single dam capable of flooding and killing large areas of forest. Small animals and insects can have indirect effects of transporting disease-causing fungi as well (Maser et al., 1978).

Heavy grazing often keeps growing space unoccupied. A sudden cessation of grazing can allow light-seeded tree species to invade and grow vigorously. Intensive grazing to reduce grass competition, followed by broadcast seeding and excluding grazing, has been used to establish stands of Cedar of Lebanon in the mountains of southern Turkey (Boydak, 1994). The dramatic reduction of grazing for economic and policy reasons caused dense stands of conifers to grow in much of the Inland West, creating fire problems as the stands grew overly dense (Covington and Moore, 1994; Oliver et al., 1994b).

Insects and diseases

Insect and disease damage has also been common in many places (Craighead, 1924; Swaine, 1933; Turner, 1952; Pomerleau, 1953; Blais, 1954, 1965; Ghent et al., 1957; Toole and Broadfoot, 1959; Brandt 1961; Anderson and Anderson, 1963; Blais, 1964; Macaloney, 1966; Childs and Edgren, 1967; Bazter, 1969; Brown, 1970; Baskerville, 1971; Davis, 1976; Amman, 1977; Lorimer, 1977b; Filip and Schmitt, 1979: Cole and Amman, 1980; MacLean, 1980; Cook 1982; Tinnin et al., 1982; Royama, 1984; Osawa, 1989, 1994; Petersen, 1992; Johnson et al., 1993; Hessburg et al., 1994; Lehmkuhl et al., 1994; Sampson et al., 1994), although mostly studied in conifers. Native insects often remain at low (endemic) levels within many stands. When the trees become weakened by climatic changes, fires, overcrowding, or other conditions, the insect populations can build to epidemic proportions. Insects and diseases are most damaging when trees are otherwise weakened by overcrowding, fires, drought, age, or other agents (Cole and Amman, 1969; Alexander, 1975; Amman, 1978; Geiszler et al., 1980; Waring and Pitman, 1983; Harvey, 1994). Weakened trees become less resistant to insects and diseases, which then build up populations in these weakened trees and attack healthy trees as well.

Occasionally, an introduced insect or disease can kill even vigorous trees, as occurred when the American chestnut was attacked by the chestnut blight [*Endothia parasitica* (Murr.) Anderson] in the early decades of the twentieth century (Braun, 1942; Woods and Shanks, 1959; Hepting, 1971). The Dutch elm disease is currently killing the

American elm throughout its range in North America (Hepting, 1971); and the white
pine blister rust, which affects five-needle pines in North America, is drastically reducing the range of western white pine (Miller et al., 1959; Hepting, 1971).

Tree Responses to Disturbances

Tree responses to disturbances vary with the characteristics of the species and the disturbance. Trees surviving minor disturbances often are scarred or otherwise changed.
These changes often leave evidence of the disturbance for many years. Trees regenerate
by different sexual and asexual mechanisms which give different species competitive
advantages following both minor and major disturbances.

Trees surviving minor disturbances

Species have different physiological and anatomical features allowing them to survive
specific minor disturbances. The responses often leave visible features on trees which
can be used to reconstruct their previous disturbance history.

Scars. Fires (Toole, 1961), falling trees, mammals, insects, and logging activities can
kill part of a stem's cambium (Kauffert, 1933; Toole, 1959; McBride and Laven, 1976;
Stuart et al., 1983). The age of the disturbance can be dated on an increment core or
wedge cut from the tree by counting the annual rings since scar formation began.
Sometimes pathogens will invade the exposed wood of the scar and cause decay (Wright
and Isaac, 1956; Shea, 1960a,b). Many tree species "encapsulate" certain kinds of rot.
Wood created after the disturbance will not be decayed, and the time of disturbance can
be dated (Shigo and Larson, 1969). Scars to the roots or stem do not necessarily reduce
diameter growth but do make the tree more subject to windthrow and stem breakage
(Shea, 1967).

Adventitious roots. Landslides and siltations kill most tree species by suffocating their
roots. Only a few inches of additional soil will kill the roots of most species. Populus
species, willows, and sequoias survive by developing new roots into the new soil from
the buried tree stem. Separate trees have been found actually to be different limbs of a
single tree which had previously been partly buried (Everitt, 1968). Black spruce stems
similarly produce adventitious roots as their old roots become covered with layers of
moss (LeBarron, 1945).

Weakened trees and insects and diseases. Photosynthetic energy is allocated for various uses within a plant. A tree weakened by a minor disturbance may not have enough
energy to allocate to insect and disease resistance (Chap. 3; Furniss and Carolyn, 1977),
and it may be attacked and possibly die.

Freezing damage. Temperate species become physiologically "hardened off" to winter
frost after being subjected to increasingly cooler temperatures above freezing. Suddenly
cold temperatures before trees are hardened off kill trees or parts of trees. Freezing can
kill part of the cambium, and a scar will form. Where part of the crown is killed, the surviving crown responds as to a physical injury. In very cold weather, tree boles can split

vertically, creating a loud sound and a "frost crack." Long vertical scars then develop, and the same crack can reopen during subsequent years. Under sunny and sometimes windy conditions in winter, conifer crowns can evapotranspire, but the roots systems may not provide replacement water if the soil is frozen—especially where snow has not covered the soil to keep it from freezing or enable it to thaw. The needles become dehydrated and turn brownish red, creating *winter burn*. Because such "burns" occur at certain elevation and temperature conditions, they appear in horizontal bands on mountains, and when covering large areas are referred to as *red belts*.

Sun scald. Temperatures on bark surfaces can be much hotter than the surrounding air and are often lethal to living cells. Stem parts exposed to direct sunlight develop thick, insulating bark; however, stems previously in shade but suddenly exposed to full sunlight can develop scars on the sunny sides where phloem and cambium tissues are killed. Where the cambium is killed, permanent scars are found in the xylem. These scars appear similar to fire, logging, and other scars and can be at all heights in trees, but usually on exposed, sunny sides.

Leaf appearances. Shade leaves (or needles) were in shade when they first developed during budbreak and are poorly adapted to growing in full sunlight. When exposed to full sunlight by minor disturbances, they may die or become weakened. This leaf death can be distinguished from normal leaf fall, since dead leaves do not develop an abscission layer and so remain attached to the tree. New leaves developing under full sunlight are more vigorous, smaller, thicker, and oriented at steep angles to the sun.

Release from overhead shade. Following a minor disturbance which releases a tree from high shade, the growth rate and shape of the leader changes (Fig. 3.7). The terminal grows faster relative to the laterals and a more pointed crown shape develops in trees with strong epinastic control for which a terminal is maintained. Several lateral branches may turn upward, forming new terminals where the terminal became a lateral branch (weak epinastic control) in the shaded condition. The upward-turning laterals can lead to crooks or forks, depending on whether one or several branches stay vertical.

Crown breakage. Stem breakage in trees appears as a crook where the branch stem created an irregularity as it turned upward, or as a fork where two or more branches assumed the terminal position. Broken lateral branches also form crooks or forks where previous breaks occurred.

Water sprouts. After a minor disturbance, new branches develop from adventitious or epicormic buds (Blum, 1963; Kormanik and Brown, 1964; Kormanik, 1968) along stems and branches (Fig. 3.5). These new branches are known as *water sprouts* and can be caused by many alterations of the tree's environment. Dominant, vigorous trees are less likely to produce water sprouts than less vigorous trees are. Sudden release of a tree from competition such as by thinning can create water sprouts. The number of sprouts is directly related to the degree of release of the tree (Marquis, 1986). Water sprouts also develop on suppressed trees of some species even when they are undisturbed (Krajicek,

1959). The tendency to produce water sprouts varies by species. Most pines do not produce water sprouts. Hardwoods generally do; hardwood species such as white oaks produce them more readily than others such as white ashes (Bruner, 1964; Smith, 1966). Douglas-firs, sitka spruces (Herman, 1964), and true firs produce them, but less readily than hardwoods do. Following a disturbance, water sprouts within a crown increase light interception rapidly, increasing the tree's growth and decreasing the sunlight passing through the crown. Water sprouts can develop on boles beneath crowns and reduce the amount of desirable, knot-free wood.

Changes in stem growth. Diameter growth and annual ring width generally increase in trees surviving a minor disturbance. This increase may not be immediate if the physiological shock of growing under different conditions causes a temporary growth decline (Staebler, 1956a,b; Bradley, 1963; Harrington and Reukema, 1983). Reduction in growth can also be caused by broken branches or roots which reduce the amount of growing space occupied.

Changes in root growth. Roots can be injured or broken by browsing animals, feeding animals, frost, and wind movement to the tree stem (Hintikka, 1972). Broken woody roots develop multiple roots at the point of breakage, creating acute forks distinct from normal root branching patterns (Wilson, 1970a; Lyford, 1980). These forked roots rapidly increase a tree's root absorption area, although the tree's resistance to windthrow may be more slowly regained.

Disturbances and regeneration mechanisms

Disturbances can be placed along a gradient depending on how much of the forest floor vegetation and soil is removed (Fig. 4.8; Oliver, 1981; Runkle, 1982, 1984). Winds—such as hurricanes and tornadoes—often do little damage to the forest floor except where upturned trees expose mineral soil or where falling trees crush the vegetation. Snow avalanches do little damage to the forest floor vegetation in the runout zones—especially when this vegetation is under several meters of snow. Except for roads and skid trails, forest cutting does even less damage than windstorms and avalanches, since tree roots are less disturbed. Fires generally kill aboveground vegetation and eliminate the forest litter layer; however, they generally do not kill root tissues more than a few millimeters beneath the soil surface and generally do not alter the soil profile unless very hot. After glacial activity or severe erosion, the soil organic matter can be lost and the horizons greatly modified.

Some regeneration methods—e.g., seeds and advance regeneration—involve genetic recombinations and produce genetically unique individuals (*genets*; Harper, 1977). Other regeneration methods such as sprouting from stumps or roots produce new individuals without a genetic recombination (*ramets*; Harper, 1977). Little work has been done to determine how frequently each method of regeneration is utilized by various species. Although a given species is capable of producing genetically new individuals after each disturbance, actual genetic recombinations probably occur with less frequency in species with asexual regeneration mechanisms. Whether a sexual or an asexual regeneration mechanism is favored depends on the species and the type of disturbance. Asexual regeneration mechanisms in certain species imply that evolutionary change occurs at a very slow rate, thus providing further evidence that tree species do not rapidly develop coevolved "superorganisms" with other tree species.

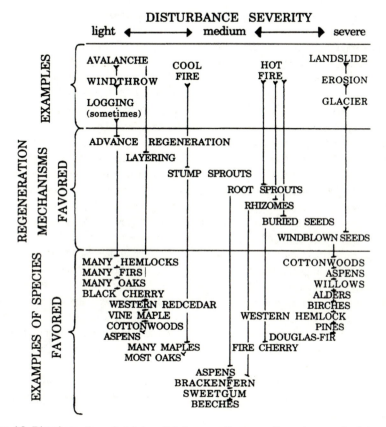

Figure 4.8 Disturbance "severity" is here listed as a gradient according to how much of the understory, forest floor, and soil is destroyed. Different forms of sexual and/or asexual regeneration have competitive advantages depending on the severity of the disturbance.

Each regeneration method is most successful when a specific amount of the forest floor has been removed. Consequently, one or more regeneration methods will be favored after each disturbance type (Fig. 4.8; Bazzaz, 1983; Kanzaki, 1987; Yamamoto, 1988; Abrams et al., 1985). Disturbance severity—how much forest floor is removed—is one gradient defining a species' niche (Chap. 2). Most species can regenerate by more than one mechanism, and there are several situations along this gradient for which a single species may have a competitive advantage. Once the advantage is gained by one or several species, they can occupy the growing space and dominate the site for a long time—often until the next major disturbance. The vigor of each regeneration mechanism sometimes varies with season during which the disturbance occurs (Franklin et al., 1985). Consequently, time of year of a disturbance may be considered another component of the disturbance gradient.

Regeneration mechanisms vary among species and varieties within species. The vigor of plants growing from each regeneration mechanism can also vary with species, site, and vigor of the parent. The various regeneration mechanisms will now be discussed. Their competitive advantage along the gradient from the most severe disturbance to the least severe will be stressed (Fig. 4.8).

All natural tree species can regenerate from seeds. Seeds of some species invade after a disturbance. Others can be present but dormant in the site before the disturbance and germinate afterward. Advance regeneration develops from seeds which germinate many years before a disturbance but grow little until afterward. Species vary in their seed distribution behavior. Some species produce few seeds but give each seed large food reserves. Such seeds can grow on stored food for an extended period and to a large size before they must support themselves by photosynthesis. Although few seeds are produced, the probability of survival of each seed is quite high, since it does not need to grow into available growing space immediately to survive. Other species produce many small seeds which contain little stored food reserves. The probability of survival of each seed is low because it must quickly occupy growing space and begin photosynthesizing or it will die. Alternatively, the large number of small seeds produced by these species ensures that at least some individuals will reach suitable sites and grow (Janzen, 1969; Harper et al., 1970; Gadgill and Solbrig, 1972). In all cases, the rate of successful establishment of trees from seeds is low, and it is the rare individual which survives.

Asexual regeneration results when a new stem develops from a dormant or newly formed bud on a parent stump, root, or other tissue. Species vary in their ability to regenerate from various asexual mechanisms, and in the vigor of the new plant. Some species, such as the American chestnut since the chestnut blight, regenerate almost exclusively through sprouting (Korstian and Stickel, 1927; Jaynes et al., 1976).

Seeds invading a disturbed area. Some species' seeds are especially adapted to travel into a disturbed area, germinate, and grow. Seeds generally travel by wind, animals, and water.

Wind. Seeds disseminated by air are usually light and contain aerodynamic adaptations for being carried by wind. Species such as cottonwoods, aspens, and willows contain cottonlike or other fine structures to increase their surface area so that they can travel in wind. Others, such as maples, ashes, pines, firs, spruces, hemlocks, and birches, have attached "wings" or other air-catching structures which help them travel (Schopmeyer, 1974).

The distances seeds travel vary with species and are related to wind velocity and tree height and are inversely related to seed weight (Fig. 4.2B). Density of seeds (number per ground area) usually decreases logarithmically with distance from parent tree. Most seeds are disseminated in fall, winter, or spring, and the time and intensity of prevailing winds during seed dispersal strongly affect the distribution patterns (Tanaka, 1995).

Light seeds of poplars and aspens can travel many miles and invade distant areas downwind. Alternatively, heavy seeds of oaks and hickories generally travel less than 100 m (300 ft). The heavy, winged seeds of Pacific silver firs are usually found on the edges of disturbed areas immediately downwind from parent trees (Thornburgh, 1969). Birch seeds are blown for long distances over crusted snow when the seeds fall in winter.

Animals. Some seeds are carried by animals (Janzen, 1969; Harper, 1977; Thompson and Wilson, 1978; Kominami et al., 1995). The accessible, succulent forage regenerating after a disturbance and the variety of spatial niches encourages many animals into newly disturbed areas (Ahlgren, 1966). These animals then deposit seeds which further colonize the areas. Seeds of some species are specialized for transport by animals (Krefting and Roe, 1949). Birds eat the fleshy fruit of juniper, cherry, rubus, and vaccinium species, swallowing the seeds and, later, (distally) dispersing the seeds within their fecal droppings (Phillips, 1910; Van Dersal, 1938; Marks, 1974). Germinants of these species can sometimes be found in patterns of bird roostings, such as along fences.

Squirrels and other rodents collect and bury large fleshy seeds such as acorns, pecans, hickory nuts, and pine nuts for future eating (Seton, 1912; Smith, 1970). The seeds often germinate in these stashes if the squirrel dies, forgets the cache, or otherwise does not eat them (Abbott and Quink, 1970).

Some seeds become attached to animal fur and later are shed at far distances from their parent.

Water. Some seeds are distributed by running water. Cottonwoods, alders, willows, sycamores, and birches can be washed from running water onto moist stream or river banks, where they germinate readily, forming *willow breaks*. River floodplains often contain standing water in winter which carries away floating acorns and other nuts, depositing them in clusters where they accumulate in eddies against downed logs as the water recedes. Erosion can wash seeds away from hillsides to areas of deposition, leaving an irregular pattern to the seed distribution.

Species variations. Seed production varies with species and tree age. Most temperate tree species have years of abundant seed production interspersed with years of moderate or low seed production. Red alders, hemlocks, grey birches, aspens, many maples, and some other species produce numerous seeds every year beginning before they are 2 decades old (Frissell, 1973). Other species—Douglas-firs, loblolly pines, most oaks, beeches, and ponderosa pines, for example—produce few seeds in most years and abundant crops infrequently, possibly in response to certain weather conditions (Lamb, 1973).

Seeds of some species—e.g., cottonwoods and aspens—are disseminated in the spring and must immediately land on favorable germination sites or they will die. White oaks disseminate in the fall, germinate immediately, and over winter as new germinants, whereas northern red oaks remain dormant on the ground during winter and germinate in the spring. Sometimes, a seed crop will vary in its germination rate, with some seeds germinating immediately, some germinating later in the season, and some delaying germination for one or more years. Other species' seeds will remain dormant where they land until induced to germinate by some environmental stimulus (Eyre, 1938; Ahlgren, 1959).

Disturbances can stimulate surviving trees to produce seeds. Weakened plants sometimes produce abundant *distress crops* of seeds. These seeds may not always be vigorous (Sundstrom and Pezeshki, 1988). Varieties of lodgepole and jack pines produce serotinous cones which contain viable seeds; fires stimulate opening of these cones and the seeds are disseminated in growing space made available by the fire.

Seeds remaining dormant until after a disturbance. Disseminated seeds can also land in closed forests. Depending on the species, they can either germinate or remain dormant in the forest floor litter (Olmsted and Curtis, 1947; Livingston and Allessio, 1968; Marquis, 1975a; Graber and Thompson, 1978; Osawa, 1986; Fried et al., 1988). Those which germinate either soon die or remain as advance regeneration.

Seeds which do not germinate can remain dormant for years, decades, or even several hundred years in plants such as lotus species (Exell, 1931; Libby, 1951; Mizunaga, 1995), *Chenopodium album,* and *Spergula arvensis* (Quick, 1961; Odum, 1965). Pin and black cherries remain viable for decades (Wendel, 1977; Marks, 1974), while yellow poplars and other species remain viable for shorter periods (Olmsted and Curtis, 1947; Clark and Boyce, 1964; Livingston and Allessio, 1968). With time the dormant seeds become buried within the organic or mineral soil. The seeds are protected there from many disturbances. In their dormant condition, they can withstand extreme heat, cold, drying, and physical abrasion.

Viability of buried seeds varies within a species and with climate (E. A. Johnson, 1975; Harper, 1977). Dormancy is broken and the seeds germinate by various complex, interacting stimuli (Went et al., 1952; Lang, 1963; Barton, 1965; Schopmeyer, 1974). These stimuli are often associated with a disturbance or availability of growing space. For example, increases in red light associated with canopy openings can be detected by the phytochrome pigment in such species as Virginia pine (Toole et al., 1961), longleaf pine (McLemore and Hansbrough, 1970), red maple, yellow birch, paper birch (Marquis, 1975b), and various European and Asiatic pines, birches, firs, and spruces (Krugman et al., 1974). Seeds of these species then germinate when temperature and moisture are also appropriate (Pollock and Toole, 1961; Quick, 1961; Stanley and Butler, 1961; Toole and Toole, 1961). Increases in heat and other stimuli caused by fires or direct sunlight can indirectly stimulate species such as raspberry, fire cherry, and fireweed to germinate (Auchmoody, 1979).

Root sprouts. New stems can grow from newly stimulated buds along roots of certain species (Fig. 4.9). Roots can survive disturbances which destroy the forest floor and living tissue closer to the soil surface.

Species vary in the degree to which they produce root sprouts and in the stimulation needed to induce sprouting. American beech (Ward, 1961; Bormann and Buell, 1964) and oriental beech produce root sprouts, for example, while closely related European beech does not. Sprouts from sweetgum roots grow from a vigorous parent when growing space is available (Kormanik and Brown, 1967), such as in an adjacent opening. Aspens and other poplars produce root sprouts when the parent is weakened by a disturbance (Hook et al., 1970; Zimmerman and Brown, 1971). Still other species are stimulated only under extreme conditions. Root sprouts of varying vigor have been observed in some healthy, released poplars, aspens, ashes, dogwoods, acacias, and yellow poplars; in unhealthy or fallen trees with uninjured roots in some alders, cherries, and maples; and in unhealthy or fallen trees with injured roots in some elms, chestnuts, horse chestnuts, plane trees, walnuts, and birches (Busgen and Munch, 1929). Some oaks and other species of ash, maple, alder, and willow have little tendency to produce root sprouts (Busgen and Munch, 1929). Honey locusts, sumacs (Zimmerman and Brown, 1971), black spruces (LeBarron, 1945), and possibly evergreen magnolias can also produce root sprouts.

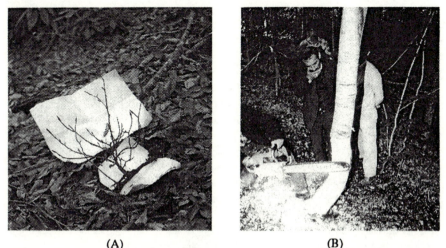

(A)   (B)

Figure 4.9 Root sprouts occur in some species, such as aspens, beeches, and sweetgums. (A) Root sprouts of oriental beeches beneath a shelterwood (*Zonguldak, Turkey*); (B) Nine-year-old root sprout of oriental beech cut open along the pith, showing that rot did not enter the stem from the base.

The root sprouts of some species will sever contact with the parent as it grows, as observed in sweetgums (Zimmerman and Brown, 1971), aspens, and poplars. This severance probably reduces stability of the parent tree. Most sprouts generally grow vigorously when released from overhead competition. Rot does not readily develop in the lower stem (Fig. 4.9B) because of compartmentalization (Shigo and Larson, 1969). Trees growing from root sprouts have advantages over trees of the same species growing from seeds, since root sprouts begin growth immediately after the disturbance while seeds may not invade and germinate until several years later. Also, shoots from root sprouts generally have faster initial height growth because their preestablished root systems and food reserves allow growth to be concentrated in the stem.

Some herbaceous plants and woody shrubs have underground stems—rhizomes—which behave in an ecologically similar fashion to sprouting roots (Tappeiner, 1971; Marsh and Koerner, 1972). These rhizomes live within the mineral soil and allow the plant to endure disturbances to the forest floor, after which they reemerge from the rhizome. Rhizomes enable plants to disperse asexually. The dispersion can be geometri-cally regular in all directions, as occurs in Indian cucumber (Bell, 1974). Bracken fern *(Pteridium)* species can develop up to 9 meters (30 ft) of rhizomes beneath each square meter (10 ft^2) of soil, allowing the species to expand rapidly into available growing space within its territory. Species vary in the persistence of rhizome connections. One Vaccinium species was found to remain connected for 34 years. A rhizome of Pteridium was found to remain attached underground for 35 to 72 years (Watt, 1970); single clones may be 1,400 years old (Oinonen, 1967; Harper, 1977) and over 500 m (1500 ft) long, and weigh hundreds of tons. It is unclear if rhizomes can remain dormant beneath a closed forest and be stimulated to grow after a disturbance. Certainly rhizomatous species grow very vigorously in disturbed areas.

Various bulbous herbaceous perennials similarly can endure disturbances, since their bulbs remain protected within the soil and resprout even if the disturbance destroys their aboveground tissue.

Stump sprouts. Most hardwood species and some conifer species produce sprouts from the stumps of severed trees (Wood, 1939; Clark and Liming, 1953; Roy, 1955; Johnson, 1964; Hook et al., 1967; Solomon and Blum, 1967; Wilson, 1968; Arend and Scholz, 1969; P. S. Johnson, 1975; Tappeiner et al., 1984). These sprouts rapidly reoccupy the available growing space. Vigor of sprouts varies by species and condition of the parent stump. Hornbeams, basswoods, and some beech species (Ohkubo et al., 1995) develop sprouts from the base of vigorous, intact parent trees; and oaks and cottonwoods some-times develop sprouts from the base of suppressed or weakened, intact parent trees. Most species, however, develop sprouts only when the stem is severed. Most hardwoods readily develop stump sprouts, although the vigor of the sprouting is generally less if the stump is very large (Roth and Sleeth, 1939; Roth and Hepting, 1943).

The vigor of sprout shoots is less when trees are cut at the peak of the growing season (Frothingham, 1912; Clark and Liming, 1953; Buckman, 1964). During this time, photo-synthetic reserves have been reduced by shoot elongation, and the tree has less energy to put into new sprouts (Zon, 1907; Mattoon, 1909; Roth and Sleeth, 1939; Buell, 1940; Roth and Hepting, 1943). Coastal redwoods, radiata pines, and pitch pines readily sprout from the root collar, although conifers generally do not sprout. Western hemlocks and lodgepole pines often appear to sprout when young; actually, a living branch on the stump base grows upward as a new stem.

Following severing of the parent tree, shoots can develop from preformed basal buds, sometimes concentrated in a burl (Stone and Cornwell, 1968; Tappeiner and McDonald, 1984), or from buds on the stem below the point of cutting. A realignment of xylem and phloem tissue occurs (Wilson, 1968), and sprout shoots pierce outward through the bark. New shoots are often constricted and weakened where they penetrate the old bark; such sprouts, although vigorous, may lack stability and blow over. Many sprout shoots can develop on a stump. Generally, one—or several on a large stump—will gain dominance, and the other sprouts will die, causing an apparently rapid, early mortality to stump sprouts (Roth and Hepting, 1943). Sprouts from low on the stump are often considered favorable because rot from the parent stump less readily enters the new stem (Roth and Sleeth, 1939; Roth and Hepting, 1943).

The vigor of trees growing from sprouts varies with species, size of stump, and vigor of the parent stump (Solomon and Blum, 1967). Larger stumps often produce less vigorous sprouts. Stumps from parent trees which were suppressed or which were weakened by pathogens also may not produce vigorous sprouts. Sprouts from vigorous stumps can grow more rapidly than trees of the same species germinating from seeds, since the sprouts begin growth immediately after the disturbance and their preestablished food resources and root systems allow growth to be concentrated in the stem. Rot may enter the sprout through the heartwood of the parent stump in oaks (Roth and Sleeth, 1939; Roth and Hepting, 1943). In sugar maples, beeches, and yellow birches, rot generally does not enter the new sprout unless the parent stump is reinjured after the sprout has been established or (perhaps) if the parent stump is very decayed (Shigo and Larson, 1969).

Sprout growth in general has received an undeserved bad reputation in silviculture as a result of continuous coppice (or stump sprout) management of hardwood stands for hundreds of years in parts of the eastern United States (Scholz, 1931) and in central Europe. Continued coppice management has produced rotting, weakened stumps, and, consequently, poorly growing stems (Fig. 4.10A). In spite of these extreme cases, stump and root sprouts can be successfully managed without loss of tree vigor and without rot in the new stems especially when the sprouts are from root stock which has not been repeatedly coppiced (Fig. 4.10B; Frothingham 1912; Trimble, 1968b).

Seedling sprouts (Fig. 410C) are arbitrarily distinguished from stump sprouts (Liming and Johnston, 1944; Society of American Foresters, 1950) and defined as sprouts arising from stumps with diameters of less than 2.5 cm (1 in). Generally one sprout shoot arises from these very small stumps. This shoot quickly overgrows the old stump, and the tree grows vigorously, resembling a seedling (Boyce, 1961; Arend and Scholz, 1969; Sander and Clark, 1971).

Layering. Layering occurs when a stem or branch touches the ground or becomes buried by litter or soil. The buried section develops roots and produces another tree either from an existing or a newly sprouting shoot. Layering occurs in many forms, but only among species whose stems develop roots when placed in dark, moist conditions.

Layering occurs in some spruces (Cooper, 1911; Fuller, 1913; Gates, 1938), in hornbeams (Busgen and Munch, 1929), and to a lesser extent in western redcedars when a living limb droops to the ground, becomes covered with litter, and develops roots. The distal (from the base) part of the branch then turns upwind and grows as a separate tree.

When vine maples becomes partly buried by disturbances such as logging, these buried stems, too, develop roots. The existing stem stays alive and new shoots sprout vertically (Oregon State University, 1974). Often, many "sprout clumps" of vine maples are still connected by a living, but partly buried, stem as long as 60 years after the stem was first buried.

Roots and shoots also sprout from cottonwoods and aspens which are overturned and partly buried while still alive. Here, a straight row of genetically identical poplar trees can grow from the old stem.

Figure 4.10 (A) Oak stump from what was formerly a coppice forest in central Europe. This stump underwent several centuries of cutting (and many generations of sprouts). Note the enlarged stump and many rotten stump holes. Experiences with continuous coppicing have probably led to the poor reputation of sprout regeneration. (B) Vigorous, nonrotten oriental beech growing from a stump sprout. (C) Seedling sprout of cherrybark oak, rapidly growing over exposed stump.

Advance regeneration. Trees of many species germinate in the restricted growing space beneath a forest canopy—usually in the *understory reinitiation stage* (Fig. 5.2)— and survive, but grow very little for an extended time. At an extreme, Pacific silver fir seedlings beneath a forest canopy can be less than 1 m (3 ft) tall and over 100 years old (Fig. 4.11A; Utzig and Herring, 1974). This form of "suspended growth" of forest floor trees is known as *advance regeneration.* Many Holarctic *Abies* (Blum, 1973), *Quercus* (Merz and Boyce, 1956), and other species; many Paleotropic *Dipterocarpaceae* (Ashton,P.M.S., 1992); and many Neotropic species (Uhl et al., 1988; Uhl and Guimaraes, 1989; Devoe, 1989) have the advance regeneration form. When released from overstory competition, advance regeneration can grow rapidly and form the new stand. The advance generation often retains immature morphological characteristics when suppressed and, when released, "act their size, not their age." Consequently, the age of trees which began as advance regeneration is considered from time of release rather than from time of germination for stand dynamics studies (Morris, 1948; Sprugel, 1974; Oliver and Stephens, 1977; Oliver, 1978a, 1981). Disturbances such as avalanches, windstorms, or logging operations which do not destroy the forest floor allow these trees to gain a competitive advantage, since they are already present and have established roots and foliage. They often exhibit more rapid growth and can better survive competition after release than trees beginning from seeds or seedlings (Hough, 1937; McGee, 1967; Sander, 1972; McGee and Loftis, 1986; O'Hara, 1986). Advance regeneration's ability to grow after release is closely related to vigor before release (Sander, 1972; Herring and Etheridge, 1976; Marquis, 1981a,b; Ferguson, 1994).

Species vary in how long they endure shaded conditions as advance regeneration and in what minimum light levels they can survive. Intolerant southern pines live only one or two growing seasons beneath a full canopy; Japanese beeches can live about 10 years beneath a dense overstory (Nakashizuka and Numata, 1982); cherrybark oaks can live for about 4 years as advance regeneration; while northern red oaks, many fir species, red spruces, and western hemlocks can survive as advance regeneration for many decades. Advance regeneration trees develop long lateral branches and short stems, giving them a flat-topped appearance characteristic of trees growing beneath high shade (Fig. 4.11B). Depending on the species, the tree may or may not lose a central stem.

Advance regeneration can be divided into three classes, based on behavior: tolerant conifers, "grass stage" pines, and hardwoods.

Many spruces, firs, and hemlocks can live as advance regeneration for extended times beneath a shaded overstory (MacBean, 1941; Ghent, 1958; Harris, 1967; Thornburgh, 1969; Blum, 1973; Seidel, 1977, 1980; Williamson, 1978; Kohyama, 1980; Jaeck et al., 1984; Seymour, 1992). Some other conifers also exist for limited times beneath a less dense overstory. Conifer advance regeneration generally does not sprout, so fires which kill the stem will destroy the entire plant. When released, advance regeneration will either die, stay alive but grow very little, or grow rapidly depending on the site and seedling vigor (Aho, 1960; Hatcher, 1964; Johnson and Zingg, 1968; Gordon, 1973; Utzig and Herring, 1974; Herring and Etheridge, 1976; Ferguson and Adams, 1980; Seidel, 1980; Ferguson et al., 1986). Younger seedlings with more photosynthesis area than respiring surface will generally grow more readily upon release. Pacific silver fir advance regeneration up to 100 years old was found to grow rapidly after the overstory was removed; however, most vigorously growing trees were from advance regeneration less than 30 years old (Long, 1976; Wagner, 1980). Indian paint fungus (*Echinodontium tinctorium*) can enter the stem through dead twigs of hemlock and fir advance regeneration. This pathogen may not cause rot in the stem of the small seedling, even upon release. If the tree is scarred many years later after it has grown large, oxygen can enter the stem and the pathogen becomes activated and quickly rots the stem (Aho, 1982).

Some fir species can live very long as very small seedlings. When slightly more growing space is available, they grow larger and appear as the usually noticed advance regeneration (Kohyama and Fujita, 1981; Kohyama, 1983). Consequently, there may be a condition of both *"seedling advance regeneration"* and *"sapling advance regeneration."* In conditions too shaded for the larger (sapling) advance regeneration, many small Pacific silver firs were found near the forest floor. These seedlings were up to 15 years old and less than 5 cm (2 in) tall, and generally contained less than 20 needles (Oliver et al., 1985). These two advance regeneration forms have been reported in depth for *Abies mariesii* Mast. and *Abies veitchii* Lindl. in Japan (Kohyama and Fujita, 1981; Kohyama, 1983.)

Longleaf pine, south Florida slash pine, Caribbean pine, *Pinus montezumae*, *Pinus michoacana*, *Pinus tropicalis*, *Pinus engelmannii*, and some varieties of *Pinus pseudostrobus* exhibit an unusual grass stage; it is most pronounced in longleaf pine (Mirov, 1967). Longleaf pines germinate but do not produce lateral branches or grow over 5 cm (2 in) tall for a period ranging from 2 to over 25 years. These trees resemble tufts of grass, hence the name. Unlike other conifer or hardwood advance regeneration, grass stage pine seedlings can not survive in shade. They are adapted to survive droughty soils subject of frequent ground fires (Wahlenberg, 1946). During the grass stage the seedlings are resistant to fires and grow extensive root systems. Competition which reduces their vigor will lengthen the time these seedlings are in the grass stage. When vigorous enough—usually when the root collar is about 2.5 cm (1 in) in diameter—they grow rapidly in height; their sensitive buds grow above the usual height of ground fires within 2 or 3 years after emerging from the grass stage (Wahlenberg, 1934; Croker and Boyer, 1975).

Many hardwood species exist as advance regeneration for varying times and under varying light intensities (Tryon and Carvell, 1958; Phillips, 1963; Bowersox and Ward, 1972; Marquis, 1981a,b; Nakashizuka, 1983). As an extreme exception, cottonwoods, yellow poplars, white birches, and aspens do not grow well if shaded for any length of time. Generally, progressively more tolerant hardwood species survive as advance regeneration for longer times and under deeper shade. Shade-tolerant sugar maples, beeches, and hickories are often found as advance regeneration beneath dense canopies. Less tolerant black birches, white ashes, red maples, yellow birches, and black cherries can live beneath less dense canopies and beneath dense canopies for a short time. Intolerant paper birches, grey birches, yellow poplars, and sweetgums are rarely found as advance regeneration; they can germinate under dense shade but rarely live more than one year. White, black, northern red, and many other oak species can survive in the shade as advance regeneration (Merz and Boyce, 1956; Carvell and Tryon, 1961; Bey, 1964; Sander, 1966, 1971, 1972; Clark, 1970; Crow, 1988), even though some species

(A)

(B)

Figure 4.11 Advance regeneration of (A) Pacific silver fir, (B) fir (*Abies bornmuellerana*) and oriental beech, and (C) southern red oak. These trees often (but not always) grow vigorously after release; however, conifer species with advance regeneration often do not sprout if the top is destroyed (e.g., by fire). Note the flat-topped appearance similar to that found in Fig. 3.7. (A) Pacific silver fir advance regeneration in the foreground is over 60 years old (based on later destructive analysis). (B) Flat-topped appearance of both fir with strong epinastic control and beech with weak epinastic control is seen. (*From Zonguldak, Turkey.*)

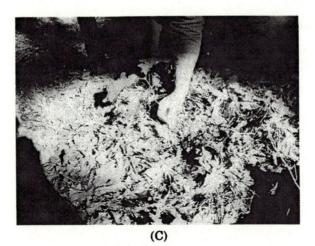

(C)

Figure 4.11 (*Continued*) (C) Southern red oak and some other species can survive in the under-story, grow little until released, and (in some oak species) die back to the root collar and resprout in the understory. After release, they can grow rapidly with a straight, vigorous stem. This shoot is 13 years old. Note the stubs of former sprouts near the root collar.

such as northern red oak are not very shade-tolerant (Liming and Johnston, 1944; Bey, 1964). Oak advance regeneration of some species grows very little for 10 to 15 years, dies back to the root collar, and sprouts a new stem, which again grows for 10 or 15 years before again dying, to be followed by a new sprout (Fig. 4.11C). This oak advance regeneration can have a root collar over 30 years old and a much younger shoot (Merz and Boyce, 1956).

Unlike conifer advance regeneration, hardwood advance regeneration can survive light fires which kill the aboveground shoots, since the hardwoods sprout from the root collar. Sometimes, hardwood advance regeneration shoots actually die upon release, and new, vigorous, straight shoots sprout from the root collar and grow rapidly in the altered environment (Sander, 1971).

Applications to Management
Silvicultural activities can help prevent disturbances from occurring at undesirable times. Control of spacing by planting, weeding, thinning, or underburning can make stands less susceptible to insect and disease buildups or windthrows and stem breakages in storms. Similarly, underburning or removal of trees expected to die can prevent fuel buildup and subsequent wildfires. Managed forests can also prevent damage to other areas and are often used for erosion and avalanche control and wind abatement.

Natural disturbances are not completely controllable even though the probability and severity of damage can be reduced. Many disturbances are the result of well-established weather patterns and can be anticipated by examining local history or climate data. Many weather-related disturbances occur infrequently and destroy a forest stand at a critical stage just as it is becoming economically mature. Even if the trees are large enough to be salvaged after the disturbance, with many landowners trying to salvage their trees, the market will be saturated, pushing down the price and creating a shortage of loggers.

Stands managed on areas vulnerable to blowdown should be considered high risks for investment. Activities such as expensive site preparation or conversion, pruning, and fertilization which occur long before the anticipated harvest age should generally be avoided. Thinning to keep stable, windfirm trees can be worthwhile. Harvesting should be planned so that abrupt edges of susceptible stands are not left exposed to the usual storm winds. The forester should also stay aware of the risk of a catastrophic fire. Management activities often depend on the stand's susceptibility to fire. Natural disturbances often occur during abnormal weather patterns; consequently, preparation for natural disturbances involves anticipating possible extremes of weather and their effects— not simply preparing for conditions common during normal weather patterns.

The forester needs to predict the structure and species composition which will develop after each type of disturbance. Clearcutting in the northeastern United States which removes small-diameter red maples will result in vigorous red maple sprouting. Clearcutting pine stands in the southeastern United States often results in hardwood stands because of the advance regeneration hardwoods beneath the pines. These original pine stands had grown on abandoned fields, where they had seeded in before hardwoods had become established. Only with much silvicultural intervention can a pine stand be reestablished on most sites of previously clearcut pine stands.

Many large red oaks currently in the northeastern United States developed from advanced regeneration after the cutting of white pine stands. The pine stands grew following agricultural land abandonment, as later occurred in the Southeast. Growth of the oaks was aided by the death of chestnuts which otherwise would have dominated the oaks.

Knowing the regeneration mechanisms of all desired and undesired species in a stand is necessary to regenerate the stand efficiently.

In Maine, most stands of spruce and fir arose from advance regeneration. Future stands may contain different species and/or structures where the spruce and fir have been cut or have died before advance regeneration has become established. Similarly, many young stands of Pacific silver fir at upper elevations in western Washington are growing from advance regeneration. If these are harvested or blow over before advance regeneration is again established, forests or shrublands of different species may become established.

Most silvicultural activities—timber harvest, site preparation, weed control, regeneration, thinning, and others—resemble specific natural disturbances inherent to forests and therefore do not create wholly unnatural stands. Applications of chemicals—herbicides, insecticides, and fertilizers—may be exceptions. While each managed stand may resemble a natural pattern, the main impact of human activity is to change the frequency with which different types of disturbances occur.

Changes in populations, cultures, and rural policies can shift disturbance patterns. These shifts alter the growth and relative advantages of tree species and lead to changes in stand structures which become apparent many decades later. Presently high populations of grand firs and Douglas-firs in parts of the Inland West and red maples in parts of the eastern United States are caused by fire exclusion and partial harvesting (high grading) which began decades ago. True firs (*Abies* species), Douglas-firs, and red maples are more shade tolerant but less fire resistant than the previously dominant pines and oaks. Similarly, decades-old policies of excluding fires in the Inland Western of the United States (Steen 1976) have led to extreme insect outbreaks and fire hazards over millions of acres (Sampson and Adams 1994).

Those wishing to maintain all native animal and plant species in a forested area some-times propose setting aside "reserves," where most or all human activities are avoided (Hunter, 1990; FEMAT, 1993; Franklin, 1993). Several considerations with this approach are:

—the forests have always changed in response to disturbances, species migrations, and climate changes (Chapter 2); therefore, there never was nor will be a single, stable condition which can be maintained by setting aside such reserves;
—quite large and catastrophic disturbances can occur to forests unless conditions are mitigated by people. The forest's susceptibility to these disturbances can usually be predicted and mitigated most efficiently long before they occur. Where conditions were not mitigated, these disturbances have led to extreme fluctuations of animal pop-ulations over large areas (Oliver et al., 1995) and have harmed people living in the dis-turbed area;
—the size of natural disturbances and other processes means that very large areas (prob-ably millions of hectares) will need to be reserved to ensure one or a few disturbances (and regrowth) do not exclude stand structures needed by some species (Baker, 1992). If human activities are excluded in some areas, they will have to be crowded into other areas where they will use resources more intensively (Peres-Garcia, 1993). Consequently, the "reserves" approach to conservation of species involves making a choice of preserving some areas at the expense of other areas (Lippke and Oliver, 1993; Oliver and Lippke, 1995).
—people have impacted forests in nearly all parts of the world for thousands of years. Even in North America, where people may have lived for only about 12,000 years, they have dramatically altered the disturbance patterns (Pielou, 1991). Each new cul-tural adjustment has again altered the disturbance patterns. A sudden cessation of all human activities may cause disturbances of magnitudes not previously common—and so cause very unusual shifts in populations, with probable extirpations and extinctions of some species as well. For example, Native American and early European American burning maintained savannas in parts of North America for thousands of years. Active fire exclusion policy allowed these areas to grow to dense forests. In some areas, species are becoming endangered, extirpated, or extinct as a result (Chapter 16). Sudden cessation of fire protection would allow these forests to burn much hotter than they did during most of the previous few thousand years; all trees would be killed in many burned forests; and a savanna forest would not return for many decades or cen-turies afterward.

Alternatively, the various disturbances can be imitated, avoided, and recovered from on a smaller, controllable scale through various silvicultural manipulations. Silvicultural activities have generally imitated a relatively narrow range of possible disturbances;

however, more intensive management could allow imitation of a wider range at a controllable spatial scale. The extreme effects of some previous disturbances such as glaciation may make it undesirable to imitate these. A concern is determining the target range of conditions for management (Morgan et al., 1994). This target range will be discussed in the "Applications for Management" section of Chapter 5.

CHAPTER 5

OVERVIEW OF STAND DEVELOPMENT PATTERNS

Introduction

Similar stand development patterns and processes appear after disturbances and change in similar ways with time in many parts of the world in stands of diverse species and distant locations. The same tree species dominate an area following a given type of disturbance within a geographic area, and the same species grow together and change the stand structure in similar patterns following the disturbance. These patterns and changes in stand physiognomy and species dominance result from competitive advantages gained by the new species following the disturbance and events occurring after the disturbance

These patterns and processes are expressed as stand structures (for definition, see Chapter 1). Two approaches have been recently used to study and manage forests: one is to concentrate on the processes of change following major (stand-replacing) and minor disturbances; the other is to concentrate on the stand structures which provide various values. Confusion between these two approaches has led to problems in both forest policy and management.

145

This chapter gives a descriptive overview of the general patterns and processes of forest development following both major and minor disturbances. It then describes the approach of concentrating on the various stand structures. Subsequent chapters then give more detailed, mechanistic explanations and quantitative descriptions of the patterns.

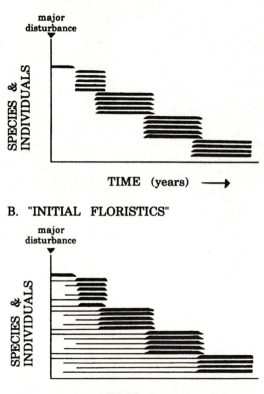

Figure 5.1 Schematic of two patterns assumed to occur in stand development. (*After Egler, 1954.*) (A) Traditionally, a "relay floristics" pattern has been assumed to occur, with one species or group invading and being replaced by successive species or groups. (B) An "initial floristics" pattern is actually more prevalent, whereby all species invade at approximately the same time after a disturbance but assert dominance at different times. The type of disturbance acts as an "environmental sieve" (*Harper, 1977*), giving some species a competitive advantage. (See "source notes.")

Development Patterns following Major (Stand Replacing) Disturbances

Successive changes in species dominance after a major disturbance have long been recognized by ecologists. Early interpretation of the pattern was that one or a few species invaded a disturbed area and predominated; as these altered the environment, other species invaded which achieved dominance, altered the environment, and allowed still other species to invade and eventually predominate (Clements, 1916; Cline and Spurr, 1942; Daubenmire, 1952; Oosting, 1956; MacArthur and Connell, 1966). This progressive change in species was referred to as *succession* and was believed to occur in an all-aged manner (Drury and Nisbet, 1973). The concept of one species invading after another in a "relaylike" manner is known as *relay floristics* (Fig. 5.1A). Eventually a species or group of species invades which predominates and replaces *itself* rather than being replaced by other species, creating a stable end point to succession ("steady state"), known as the *climax* (Cowles, 1911; Clements, 1916, 1936; Braun, 1950; Oosting, 1956; Dansereau, 1957; Odum, 1959; Daubenmire, 1968).

Other ecologists believed most plants existing in a stand invade shortly after a disturbance (Egler, 1954; Stephens, 1956; Raup, 1957, 1964; Olson, 1958; Johnson 1972; Drury and Nisbet, 1973; D. M. Smith, 1973; Henry and Swan, 1974; Niering and Goodwin, 1974; Oliver, 1978a, 1981), rather than throughout the life of a stand. According to this *initial floristics* pattern (Fig. 5.1B), species which predominate later have been present since soon after the disturbance but are often overlooked because of their small size, relatively few numbers, and, "to some extent, the preoccupation of botanists with the dominant species" (Drury and Nisbet, 1973). Elements of both initial and relay floristics are characteristic of forest development; however, the invasion pattern after a disturbance predominantly follows the initial floristics pattern.

Single- and Multiple-Cohort Stands (Even-Aged and Uneven-Aged Stands)

Stands developing after major disturbances have been described as "even-aged" stands, since all component trees have been assumed to regenerate shortly after the disturbance. In fact, trees may continue to regenerate for several decades where growth is slow before the available growing space becomes reoccupied, resulting in a wide age range in the stand.

The term *even-aged* is misleading for stands where there is a wide age range following a single disturbance. The group of trees developing after a single disturbance— whether major (stand-replacing) or minor—has been referred to as an *age class*, a *cohort* (Bazzaz, 1983), or a *generation* (Smith, 1984). In this book, a group of trees regenerating after a single disturbance will be referred to as a "cohort." The age range within a cohort may be as narrow as 1 year or as wide as several decades, depending on how long trees continue invading after a disturbance. Stands developing after a minor disturbance will be referred to as *single-cohort stands* instead of "even-aged stands." Stands where component trees arose after two or more disturbances (all but the first of which would be minor disturbances) will be referred to as *multiple-cohort* (or *multicohort*) *stands,* instead of the usual designation of "uneven-aged" or "all-aged" stands.

More is known about single-cohort stands than multicohort stands, probably because they are less complicated, they have been considered as part of "secondary succession" by ecologists, and they are more commonly managed by foresters.

Stages of Stand Development

Patterns of species dominance and changes in stand structures are not the result of oblig-atory laws which forest stands must follow. While they can be somewhat—but not com-pletely—anticipated, they are simply the result of interactions of plants and are emergent properties of the tree interactions. For purposes of this discussion, the development pat-terns following a disturbance will be divided into stages (Fig. 5.2; Oliver, 1981). These stages are similar to the descriptions of Isaac (1940), Jones (1945), Raup (1946), Watt (1947), Bloomberg (1950), Horton (1956), Daubenmire and Daubenmire (1968), Reiners et al. (1971), Day (1972), Harris and Farr (1974), Bormann and Likens (1979), Crow (1980), Hartshorn (1980), Wallmo and Schoen (1980), Peet (1981), Alaback (1982a,b, 1984b), Brady and Hanley (1984), Felix et al. (1983), Nakashizuka (1984a,b), Whitmore (1975, 1984), and Peet and Christensen (1987).

Stand initiation stage. After a disturbance, new individuals and species continue to appear for several years.

Stem exclusion stage. After several years, new individuals do not appear and some of the existing ones die. The surviving ones grow larger and express differences in height and diameter; first one species and then another may appear to dominate the stand.

Understory reinitiation stage. Later, forest floor herbs and shrubs and advance regen-eration again appear and survive in the understory, although they grow very little.

Old growth stage. Much later, overstory trees die in an irregular fashion, and some of the understory trees begin growing to the overstory.

How rapidly a forest changes from one of the above-described stages to another varies greatly. In some cases—as in balsam fir/red spruce stands—a disturbance may intervene before the later stages are reached (Sprugel, 1976; Osawa et al., 1986). Where there is a partial overstory the pattern may differ slightly.

Development of single-cohort stands growing after major disturbances will first be discussed because it is the least complicated pattern. The development of multicohort stands created by minor disturbances will then be discussed. Major disturbances are not necessarily more prevalent, and in some regions they may rarely occur; however, for purposes of this discussion, they allow a "starting point" for describing forest develop-ment.

Stand initiation stage

A major disturbance kills all large trees but may not destroy the forest floor herbs, shrubs, advance regeneration, buried seeds, and roots—depending on the type of distur-bance.

A major disturbance radically changes the forest floor and soil environments, even if it does not destroy them. There are losses of evapotranspiration and litter input as well as great changes in pedochemical and biological processes. Many pedologic changes such as alteration of the soil profile (Lyford, 1964b) or releasing of nutrients (Bormann and Likens, 1981) from the upper soil are almost immediate; however, others occur over several years as the newly dead roots and other organic matter decompose and reduce the cation exchange sites and alter the soil structure. Such disturbances do not necessarily reduce the soil's productivity. The disturbance may actually have a rejuvenating effect in soils undergoing podzolization (Ugolini et al., 1989). The rate of these pedologic changes after a major disturbance depends on the climate, soil texture and structure, nature of the overstory, nature of the disturbance, and other factors (Bormann and Likens, 1981).

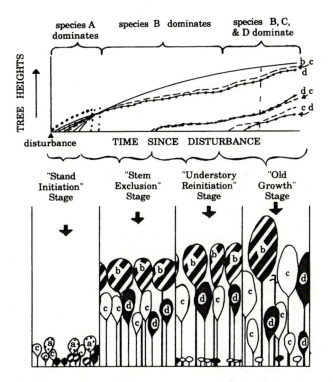

Figure 5.2 Schematic stages of stand development following major disturbances. All trees forming the forest start soon after the disturbance; however, the dominant tree type changes as stem number decreases and vertical stratification of species progresses. The height attained and the time lapsed during each stage vary with species, disturbance, and site. (*Oliver, 1981.*) (See "source notes.")

New plants—trees, shrubs, and herbs—grow from seeds, sprouts, advance regeneration, and other mechanisms into the growing space made available by the death of the previous overstory (Dyrness, 1973; Johnson, 1981a,b; Canham and Marks, 1985). Only certain individuals are able to grow on a given area, depending on the type of disturbance, competing vegetation, availability of seeds from surrounding areas, suitability of microsites for growth, and predation by animals. Those which grow rapidly at first reoccupy the available growing space. Their advantage over other species which invade later often allows them to dominate and/or exclude other species for many years or centuries—often until the next disturbance (Nichols, 1935; Hough and Forbes, 1943; Chapman, 1942; Spurr, 1954; Niering and Egler, 1955; Raup, 1964; Dyrness, 1973; Strang, 1973; Henry and Swan, 1974; Niering and Goodwin, 1974; Whitmore, 1974; Hanley et al., 1975; Ashton, 1976; Keeley, 1979; Veblen et al., 1980; Strang and Johnson, 1981; Hibbs, 1983; Felix et al., 1983; Frelich and Lorimer, 1985; Thomas and Agee, 1986; Aplet et al., 1988). As a result, the species composition of each stand is largely the result of the type of disturbance which initiated it; i.e., the disturbance determines which species have the initial advantages immediately after the disturbance (Ahlgren, 1960; Buckman, 1964; Pase, 1971; Tappeiner, 1971; Drury and Nisbet, 1973; Cattelino et al., 1979; Johnson, 1981a,b; Oliver, 1981; D. M. Smith, 1982; Tappeiner and McDonald, 1984; Adams and Adams, 1987). When germinating seeds are favored by a disturbance, the species composition of the ensuing stand will also depend on which species are producing seeds during the years immediately following the disturbance (Frissell, 1973). Soil, climate, and other gradients also influence which species have a competitive advantage. The concept that a single area can be occupied by different, relatively stable groups of species (Botkin, 1979; Sprugel, 1991) is different from early assumptions that each area returns to a single equilibrium vegetation composition soon after disturbance (Clements, 1916).

Also contrary to older interpretations of forest development, tree species often invade newly disturbed areas without other, nonwoody species having preceded them to modify the soil and microclimatic conditions (Niering and Egler, 1955; Hack and Goodlett, 1960; Ohmann and Ream, 1971; Purdie and Slatyer, 1976; Johnson, 1981a,b). Tree species have even been found invading newly available growing space within a few years after the soil was uncovered by a retreating glacier (Sigafoos and Hendricks, 1969; Oliver et al., 1985).

A given environment permits only certain individuals to grow on an area. This process of species limitation is referred to as an environmental sieve (Harper, 1977), and the composition of the stand is restricted by those species able to "pass" the environmental sieve. Disturbance types, microclimates, and soil conditions may all be considered environmental sieves. The environmental sieve may change slightly during stand initiation, since some plants slightly modify the environment—such as by creating slight shade which allows other plants to germinate and grow which could not become established in direct sunlight.

Some invading plants grow rapidly at first and complete their life cycles before other plants crowd them out (Marks, 1974; Cook, 1978; Zamora, 1982; Alaback, 1982a,b; Oliver et al., 1985). Annuals and biennials have a competitive advantage immediately after a disturbance where time since disturbance is considered as a gradient. Tree species such as fire and bitter cherries (Marks, 1974; Stubblefield, 1978) invade early and set seeds after several years. Douglas-firs, oaks, and some other species can begin growth at the same time as the rapidly growing ones, but are not so numerous, do not grow so rapidly at first, and do not produce seeds until they become older (Oliver, 1978a; Wierman and Oliver, 1979). Consequently, they do not dominate the site at first.

Still other plants—such as red alders, tulip poplars, and vaccinium shrubs—grow rapidly at first and maintain a dominant position for a long time (Stubblefield and Oliver, 1978; O'Hara, 1986). Some trees such as western hemlocks can either predominate the site soon after a disturbance or become dominant later in their lives, depending on the site and the other plants on the area (Franklin and Dyrness, 1973; Atkinson and Zasoski, 1976). Individuals will continue to invade an area so long as growing space is available. The invasion period can vary from a few years to many decades, depending on how rapidly the newly growing stems reoccupy the growing space. Factors leading to a short invasion period are soil and site conditions favorable to rapid growth; a rapid appearance of new plants in the disturbed area—either because the seeds, root collars, or advance regeneration survived the disturbance or because seeds came in rapidly from adjacent areas; the presence of species which naturally grow rapidly and in directions which occupy the most growing space when young (e.g., lateral expansion of branches is often more important than height growth during this stage); a high frequency of plants so that the growing space is filled by each plant growing only a small amount; and the absence of animal predators or harsh conditions which kill invading plants.

The age range of invading plants, therefore, is limited by the time it takes the growing space to become reoccupied. Actually, the age range of the cohort may become somewhat narrower as the stand grows older if the last arriving plants are outcompeted and killed by more dominant ones. In forest development each individual is considered, regardless of whether it is a genetically different individual (a genet) or simply a clone (a ramet) of another or previous plant. Consequently, age is considered as time of initiation of sprouts rather than time of germination of the original seed, which may have occurred several generations earlier. Advance regeneration is also commonly aged from the time it is released rather than when it germinated, which may have been over 50 years earlier.

The stand initiation stage—before the growing space is fully reoccupied and new stems quit initiating—is the time of very high numbers of plant species (Isaac, 1940; Alaback, 1982a,b, 1984b; Zamora, 1982; Felix et al., 1983; Hibbs, 1983; Oliver et al., 1985). Long-lived tree species have not had time to outcompete other species and reduce the diversity. It is also a period when many animals are found, since the variety of plants and seeds provides an abundance and diversity of food. The palatable, succulent upper canopy of young plants is still accessible and provides hiding cover close to the ground and a variety of spatial patterns for territorial establishment.

There may be elements of relay floristics during plant initiation. Some species grow only on extreme sites after other species have altered or increased the growing space (Connell and Slatyer, 1977; Agee and Dunwiddie, 1984; Peet and Christensen, 1987). Also, some short-lived herbaceous and woody plants may die and relinquish their growing space before a disturbance kills them (Johnson, 1981a,b; Hibbs, 1983). Other species invade in a relaylike manner when this growing space is made available. This relinquishing of growing space is similar to the death of overstory trees in the old growth stage. Herbaceous and shrub communities may at times pass through initiation, exclusion, reinitiation, and senescent processes similar to a forest stand, but over the course of a few months or years during the stand initiation or understory reinitiation stages of a forest stand.

Stem exclusion stage

Before the available growing space is reoccupied, plants are expanding in an *open growth* condition, unhampered by competition with other individuals. After the available growing space is reoccupied, new individuals do not become established successfully. Those plants with a competitive advantage in size or growth pattern are able to expand into growing space occupied by other plants and reduce their growth rate or kill them. The stage when this occurs is referred to as the *stem exclusion stage* both because new stems are prevented from successfully invading and because some existing stems die and thus are excluded from the stand.

Parts of a stand may be in the stem exclusion stage, with some trees dying from competition, while other, more open areas are still in the stand initiation stage because of spacing, species, and site irregularities. In fact, a given stand may take several decades before all parts make the transition from the stand initiation stage to the stem exclusion stage.

As the plants begin competing with each other, they form a layer of living foliage several meters thick in which their leaves are able to obtain sunlight (Assmann, 1970). Being in a slightly lower portion of this layer—in low shade—does not put these plants at a distinct disadvantage when young. At first there is not a pattern of height by species within the foliage layer. This period with all trees within the same layer at the beginning of the stem exclusion stage is called the *brushy stage* (Gingrich, 1971).

The foliage layer rises as the trees grow taller, and leaves cannot survive in the diminished sunlight beneath it. Plants which cannot grow tall enough to stay within this foliage layer often die (Isaac, 1940; Alaback, 1982a,b; Oliver et al., 1985). The shaded forest floor becomes devoid of living plants and consists of brown, dead leaves, twigs, and stems.

During rising of the foliage layer, growth characteristics of the different species manifest themselves, causing varied stand development patterns. In some cases, early dominance is a strong indication of which species will maintain dominance in the stand (O'Hara, 1995a; O'Hara et al., 1995). In other cases, a species can become dominant even if it begins at a relatively low canopy position in the "brushy stage" (Oliver, 1978). For simplicity, the various growth patterns of stands with only one species will first be described, although perfectly pure stands rarely occur. Afterward, mixed species stands will be described.

Single-species stand development patterns. When growing space is available in the stand initiation stage, all trees would expand at an equal rate in an ideal stand with trees of the same genetic potential and age, on a uniform site, and with uniform spacing between trees so that each tree occupies the same growing space. Each tree would continue to increase in size in the stem exclusion stage, but would not increase in amount of growing space it occupies. Because all trees would be identical, none would have a competitive advantage which would allow it to expand by taking growing space away from a neighbor. Each tree would eventually expand until all photosynthate was consumed by its own respiration to keep its increased size alive, and it would no longer grow. If all trees behaved uniformly in this manner, the stand would eventually quit growing and become a *stagnant stand.* Absolute stagnation does not occur in nature because stands do not exist as ideally as described above. Uniform, crowded stands of some species—such as lodgepole pines—can approach stagnation and be only 2 m (3 ft) tall at 18 years while adjacent, less crowded stands of the same age can be almost 4.5 m (15 ft) tall (Mitchell and Goudie, 1980). The stagnated condition seems to be more pronounced on poor sites and in stands with high numbers of trees, although more widely spaced stands on more productive sites can also stagnate. The uniform age and spacing of forest plantations may be creating stands which approach stagnation more commonly than do naturally grown forests.

Actually, some trees outgrow others in a stand because even single-species plantations are rarely uniform in site, tree age, spacing, or genetic potential. Even in apparently stagnated lodgepole stands studied in eastern British Columbia, some trees were found to be outgrowing others—albeit at a rate of less than 1 cm/year. As some individuals outgrow others, they take away available light space by shading foliage and physically abrading foliated branches of adjacent trees. The reduced growing space of the tree being outcompeted forces it to put less of its reduced photosynthate into growth, and it declines in growth while the dominating one may not. The result is crown *differentiation,* with some trees becoming larger at the expense of others (Guillebaud and Hummel, 1949).

Trees in pure, single-cohort stands usually have their crowns within a single horizontal layer, although the more vigorous trees' crowns can extend higher within the layer and the foliage of suppressed trees may live lower in the layer. In mixed-species stands, there are often several distinct layers of tree crowns. Each horizontal layer is referred to as a *stratum.* Single-cohort, single-species stands generally contain a single stratum when in the stem exclusion stage. The degree of dominance and suppression within each stratum has been described by classifying trees into four crown classes (Kraft, 1884; D. M. Smith, 1986). These crown classes are distinct from canopy strata discussed later (Fig. 5.3).

Dominant. Crowns extend above the general level of crown cover of other trees of the same stratum and are not physically restricted from above, although possibly somewhat crowded by other trees on the sides.

Codominant. Crowns form a general level of crown stratum and are not physically restricted from above, but are more or less crowded by other trees from the sides.

Intermediate. Trees are shorter, but their crowns extend into the general level of dominant and codominant trees, free from physical restrictions from above, but quite crowded on the sides.

Suppressed (overtopped). Crowns are entirely below the general level of dominant and codominant trees and are physically restricted from immediately above.

Trees in single-species stands expand their crowns horizontally and intercept more sunlight as they become larger. Consequently, they constantly compete with neighbors and either dominate (and eventually kill) them, or are dominated (and killed) by them.

This death from suppression is referred to as *natural thinning* or *self thinning* (Peet and Christensen, 1987). The dominant (survivor) then grows until it comes into contact with the next-closest tree, these trees compete, and eventually one dominates (and survives). Trees tend to become somewhat regularly spaced in a stand as they grow older. Stands sometimes have irregular waves of mortality separated by times of relatively little mortality of the remaining trees.

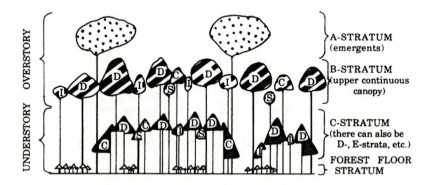

Figure 5.3 The relative positions of canopy (or crown) strata and crown classes. Canopy strata are the horizontal layers of tree crowns which are separate from higher and lower layers. Crown classes refer to trees within each stratum [D = dominant; C = codominant; I = intermediate; S = suppressed (overtopped), as defined in text]. Trees of each crown class can be found in each stratum. (*After Richards, 1952; Oliver, 1978a; D. M. Smith, 1984, 1986*).

Mixed-species stand development patterns. Most stands in North America contain more than one species, even where a single species has been planted. The growth patterns of mixed stands vary greatly, depending on environmental conditions and the physiological predispositions of the interacting species. Some general patterns emerge, however (Larson, 1992; Kelty et al., 1992).

Some tree species grow rapidly when young and dominate the stand. Species growing in this way are referred to as *pioneers*. A species such as western hemlock can either act as a pioneer species or dominate a stand much later after a major disturbance (Franklin and Dyrness, 1973; Atkinson and Zasoski, 1976). Other species—such as yellow poplars (Beck and Della-Bianca, 1981) or Douglas-firs (Isaac, 1940)—act as pioneers and continue to dominate the stand for many decades or centuries. Quaking aspens and grey birches behave as pioneers but have short natural life spans; they die even before becoming dominated by other species—having already set seeds for perpetuation of the species (Oliver, 1978a). Bitter, pin, and fire cherries behave as pioneers but are soon overtopped by other species and die in the understory after an initial period of domination (Marks, 1974; Stubblefield, 1978; Hibbs, 1983). Red maples and black birches can act as pioneers in mixed stands with red oaks and dominate the stand for the first few decades; then they become overtopped and continue to live for many decades in subordinate positions (Olson, 1965; Stephens and Waggoner, 1970; Oliver, 1978a). Where species are tolerant enough of shade to survive in subordinate positions completely over-topped by other species, the forest can have a *vertical stratification* of foliage layers (Richards, 1952; D. M. Smith, 1962, 1986; Grubb et al., 1963; Walter, 1971; A. P. Smith, 1973; Mueller-Dombois and Ellenberg, 1974; Beard, 1978; Oliver, 1978a; Yamakura, 1987).

Sometimes a species grows so much taller than the others in a mixed stand that it appears to emerge above them. Such stems are referred to as *emergents* (Fig. 5.3). Emergent trees are also found in tropical forests (Richards, 1952) and probably devel-oped similarly as in temperate forests. The emergent stratum in a forest is also referred to as the *A-stratum* (Smith, 1984; Oliver, 1978a). An individual which acts as an emer-gent when intermixed with slower-growing species may appear in the upper continuous stratum—the *B-stratum*—when grown with more trees of the same species. Pioneer species such as aspens can be described as emergents before they die when they are iso-lated among shorter species. Isolated white pines in eastern hardwood forests often appear as emergents (Oliver, 1978a), as do isolated western larches in mixed coniferous stands in the northern Rocky Mountains (Cobb, 1988). Sometimes, isolated western hemlocks or Douglas-firs can act as occasional emergents in mixed stands dominated by red alders (Stubblefield and Oliver, 1978).

Two species can have similar enough growth patterns that they interact as a single species, with each asserting dominance and killing the other depending on subtle differ-ences in ages, microsites, and spacings which give one or the other a competitive advan-tage. Loblolly and longleaf pines behave this way on certain sites in the southeastern United States. A slightly greater shade tolerance of the loblolly pines allows them to maintain fuller crowns and larger limbs and to grow larger in diameter than surrounding longleaf pines.

Different species growing together can also have quite different growth rates, result-ing in a predictable pattern in which one species overtops the other. When the over-topped species is not able to live in the shade of the other, it dies (Lorimer, 1981). Douglas-firs usually become overtopped by red alders when the two are grown closely together. Since Douglas-firs cannot live in the shade of alders, the Douglas-firs die. Consequently, Douglas-firs are rarely found beneath an alder canopy (Stubblefield and Oliver, 1978).

If the overtopped species is able to survive in the reduced light beneath the dominat-ing species, it can form a lower canopy stratum. Such stands are commonly referred to as *stratified, mixed-species stands*.

A dominating species can relegate other species to lower strata by shading them with high shade so that their height growth slows down, and by physically abrading the oth-ers' terminals during windstorms (Oliver, 1978a; Wierman and Oliver, 1979; Kelty, 1986; Larson, 1986). Once in the reduced light beneath the main canopy, the tree may survive but generally grows very slowly. Eventually the continued growth of the upper stratum trees and the curtailed growth of the lower make the dominating trees much larg-er (Fig. 2.1; D. M. Smith, 1962, 1986; Gibbs, 1963; Stubblefield and Oliver, 1978; Wierman and Oliver, 1979). Such stratified, single-cohort, mixed stands have been mis-taken for uneven-aged stands which grew in a relay floristics pattern (Oliver, 1980; Johnson, 1981a,b). The smaller trees in the lower strata have been incorrectly assumed to be younger and in the process of overtaking the larger, apparently older stems in the upper stratum (Meyer and Stevenson, 1943; Phillips, 1959; Minckler, 1974; Mueller-Dombois and Ellenberg, 1974).

Progressively lower strata are referred to alphabetically (Fig. 5.3):

A-stratum (emergents). Trees above the highest continuous canopy

B-stratum. Upper continuous canopy

C-stratum, D-stratum, etc. (understory strata). Progressively lower strata beneath the B-stratum, although each one is not necessarily as continuous as the B-stratum; these strata may not be present

Forest floor stratum. Trees and shrubs very close to the soil surface; usually not more than 2 m (6 ft) tall.

During the initial brushy stage, all trees exist within the same stratum. They first appear to differentiate into dominant, codominant, intermediate, and suppressed crown classes (O'Hara et al., 1995). The crowns of some species can live beneath others. In these cases, the overtopped trees do not die; instead, they form lower strata beneath the B-stratum as differentiation proceeds (Smith, 1984).

Within each stratum, the trees tend to differentiate further into crown classes. It is important to distinguish between crown classes and strata. Crown classes can occur within each stratum as the trees differentiate. The behavior of each tree varies greatly depending on both its stratum and its crown class. For example, a suppressed shade-tolerant tree such as hemlock in the B-stratum of a pure hemlock stand will behave much differently upon release than would a dominant hemlock in the C-stratum of a mixed stand.

A stratum can contain one species or several species of similar height growth rates and shade tolerances. In both single and multicohort stands, trees of a single cohort can be found in more than one stratum. In a multicohort stand, a stratum can contain trees of one cohort or of several cohorts.

There are, of course, many variations of the stratification patterns. Sometimes a species has such a height growth advantage that it can stratify above other species under almost any conditions, as tulip poplars outgrow white oaks (O'Hara, 1986). In other cases—such as in mixtures of cherrybark oaks and sycamores (Clatterbuck et al., 1987; Oliver et al., 1989) or Douglas-firs and red alders (Stubblefield and Oliver, 1978)—rapid early height growth of one species can be offset by later sustained height growth of the other species. Which species dominates depends on whether the initial spacing is wide enough for the initially slow growing species to avoid being overtopped before it outgrows the other. Relatively few species occupy the upper strata even in stands containing many species, as has been found in deciduous forests in eastern North America (Putnam et al., 1960), as well as in tropical humid forests of the Amazon (Hubbell and Foster, 1986; Balslev and Renner, 1989; Hartshorn 1989b) and southeastern Asia (Ashton, 1989).

In a few cases a species can grow upward from a lower stratum in mixed, single-cohort stands after the structure of the stand appears stabilized. In these cases, the species emerging into the overstory generally is very tolerant of shade and has tough lateral branches. As winds cause the tree to sway, its branches rub and batter against those of other species and break them.

The numbers of species and individuals decline during the stem exclusion stage. This stage contains the fewest plant species (Alaback, 1982a,b; Oliver et al., 1985). Comparatively few species of animals are found in forests in this stage because there is very little living foliage near the forest floor to serve as food, cover, and habitat diversity. Some animals specifically live in these stands, however. The Kirtland's warbler makes its nests in these young jack pine stands in the Lake States (Walkinshaw, 1983).

Understory reinitiation stage

As the overstory grows older, new herbs, shrubs, and/or trees appear in the forest floor (McKinnon et al., 1935; Sprugel, 1976; Alaback, 1982a,b; Henderson, 1982; Oliver et al., 1985). These are usually species capable of living under low intensity, high shade and can be the same as or different from species in the overstory and those shrubs and herbs which grew during stand initiation. Less is known about the understory reinitiation and old growth stages.

Time of this understory reinitiation varies from only about 60 years in shade-intolerant southern pines (Peet and Christensen, 1987) or stressed stands to over 150 years in stands of shade-tolerant conifers (Oliver et al., 1985).

Why understory stems appear later in a stand's age and not earlier has not been extensively studied. Foresters have generally recognized it as a time when the overstory "loses its grip on the site" but have not explained the causal mechanisms. In some cases, more light reaches the understory as the suppressed overstory trees die and/or as the remaining trees become taller, sway more in wind, collide with neighboring trees, and abrade their outer branches (Sprugel, 1976). A higher carbon dioxide concentration near the forest floor may also increase the photosynthetic efficiency of understory plants. Pedogenic processes may create more growing space near the forest floor.

Some species seed continually into an understory even during the stem exclusion stage. The seeds may germinate, but the seedlings grow very little and are so small that they are not readily visible. These seedlings may die within a few years, creating an ephemeral understory of seedlings and creating wavelike populations as they seed in at periodic intervals of good seed years and stand conditions (Hett and Loucks, 1968, 1971, 1976; Hett, 1971).

Still other species have seeds which stay dormant in the soil for many years. These seeds can germinate after subtle changes in growing space during the understory reinitiation stage (Krugman et al., 1974; Marquis, 1975a; Osawa, 1986). Many shrub and herb stems initiate from rhizomes and root sprouts during the understory reinitiation stage.

Plants in the forest floor stratum in the understory reinitiation stage do not grow rapidly in the small amount of high-shade sunlight reaching them for photosynthesis. They grow even less in height relative to their total biomass accumulation than would trees growing in full sunlight. The reinitiated understory beneath a forest stays quite small, and thus visibly distinct from older trees in upper strata, for many years.

Lower strata of a single-cohort, mixed species stands can contain the same tolerant species as the advance regeneration. The stand can be mistakenly assumed to be all-aged, with each successively lower strata being younger because the trees are smaller; and the advance regeneration are sometimes assumed to be climax trees during the understory reinitiation stage, since these trees are expected to grow to the overstory (Braun, 1950; Scott, 1962).

In a literal sense, the stand would no longer be a single cohort after the forest floor stratum develops; and the development of the forest floor vegetation could be interpreted as "relay floristics" (Chapter 5). In practice, however, a stand is still considered to have a single cohort until younger trees from the forest floor grow much larger.

Barring intervening disturbances, the understory reinitiation stage can conclude within the first 100 years of the stand on highly stressed sites or can end when the stand is more than 500 years old if the overstory is shade-tolerant and vigorous.

The understory reinitiation stage generally contains more animal species than does the stem exclusion stage, but fewer than the stand initiation stage. Understory plants generally contain less starch nutrition for animals than those growing in full sunlight. Many resident animals are particularly adapted to growing in this stage and in the subsequent old growth stage, during which they utilize rotting wood and hollow trees for food and shelter. A large animal population can prolong the time before the understory reinitiation stage appears by browsing forest floor vegetation while it is very small.

Old growth stage

Older stands are often destroyed by periodic catastrophic disturbances. Where such disturbances do not occur, the overstory trees die from various agents (Peet, 1981; Peet and Christensen, 1987), such as winds or the accumulated effects of pathogens, droughts, or insects, which become more influential as trees age. The maximum life span of stand-grown trees can vary from less than 100 years for black willows to over 1,000 years for Douglas-firs and over 1,500 years for coastal redwoods. As overstory trees die, trees from the forest floor stratum of the understory reinitiation stage grow slowly into the overstory. Death of overstory trees and growing of understory trees are not caused by external factors. The autogenic process achieved when the trees regenerate and grow without the influence of external disturbances is referred to as the *old growth condition* (Thomas, 1979; Oliver, 1981; Runkle, 1981). This definition of old growth describes a process occurring by a unique development sequence. When the trees which invaded immediately after the disturbance have all died, the stand enters a *true old growth* stage. Young trees at first grow into the overstory, while some relic trees from a previous allogenic disturbance remain alive. Stands in this condition are referred to as *transition old growth* in this book.

As occurs in the transitions between other stages, the change from the understory reinitiation to the old growth stage may be gradual and spatially varied. Some areas may maintain distinct upper canopy and forest floor strata, while nearby areas where overstory trees have died can have a gradation of trees of different heights and diameters. The result is often a large variety of structures within a stand (Peet and Christensen, 1987).

The death of dominant trees in very old stands can be viewed as either allogenic or autogenic. Death of one or a few overstory trees acts like a small minor disturbance and permits a small, single-cohort stand to grow from advance regeneration and other regeneration mechanisms. Development of a forest by such minor disturbances has been referred to as *gap phase development* (Watt, 1947; Veblen and Stewart, 1982*a*; Pickett and White, 1985).

When part of the overstory begins to die during the transition from understory reinitiation to old growth stages, other overstory trees may weaken and the residual overstory may not last many more decades. Eventually, no trees are left alive which are relics of past disturbances other than minor disturbances caused by the deaths of individual overstory trees. At that time, all existing stems have begun from autogenic origins. Such a condition—a true old growth stand, in which the tree species live a long time—has very rarely been reached, because the probability of a large disturbance increases as the stand becomes older. Very old stands in the cold climates of Alaska and northern Canada may develop into organic bogs (muskegs) if they persist without a disturbance. Gradually, the trees die as forest floor organic matter increases, tying up nutrients, moisture, and oxygen (Heilman, 1966, 1968; Reiners et al., 1971; Vitousek and Reiners, 1975; Ugolini and Mann, 1979; Strang and Johnson, 1981). Such conditions may require over 1,000 years without a disturbance. In warmer climates with more rapid decomposition of organic matter, old growth stands maintain a variety of tolerant and intolerant tree species.

The term "old growth" has also been used to describe stands of specific structural characteristics, regardless of the autogenic or allogenic processes which led to these structures, as discussed later. Structural features include large, living old trees; large, dead standing trees; massive fallen logs; relatively open canopies with foliage in many layers; and diverse understories. Such structures are achieved by a variety of major and/or minor disturbance patterns in single- or mixed-species stands; therefore, they do not represent a unique stage of forest development

The old growth structure probably has the greatest horizontal and vertical variation in structure, with both large and small trees growing in separate and intermixed patches (Alaback, 1982*a,b*). The number of plant and animal species found is generally more than in most other structures (Manuwal and Huff, 1987) but less than in the open structure of the stand initiation stage. Some plants and animals are dependent on the rotting wood or other conditions found exclusively in these old growth forests for their survival. The spotted owl in the Pacific northwestern United States and the red-cockaded woodpecker in the Southeast, for example, seem to exist predominantly in forests with old growth structures.

Development Patterns following Minor Disturbances
Overview

Minor disturbances that injure or kill some of the trees in a stand can occur between or instead of major disturbances. Like major disturbances, their type and frequency are determined by both allogenic and autogenic factors. Minor disturbances may reduce the frequency of major disturbances. For example, understory fires which reduce litter buildup prevent large conflagrations which kill all trees in a stand (Agee, 1981).

Windstorms which blow over some trees allow residual ones to grow more windfirm and hence more resistant to winds (Wiley, 1965).

(A)

(B)

Figure 5.4 Minor disturbances can create (A) multicohort stands of various structures or may even result in (B) a single-cohort stand. (A) This multicohort stand began after repeated, small disturbances—primarily cuttings—in northern Idaho. (B) This stand in northern Connecticut began in 1897, but a hurricane in 1938 blew down some trees and tilted others, releasing growing space (photo taken in 1972). The minor impact of the hurricane and the vigor of residual trees allowed the residual trees to grow and reoccupy the growing space, excluding a new cohort.

Minor disturbances create multicohort stands when new trees regenerate following the disturbance. A single-cohort stand can be maintained if a single-cohort stand was disturbed but the residual trees reoccupied the released growing space and excluded a new cohort from initiating or surviving (Fig. 5.4B). A minor disturbance can create a multi-cohort stand where the surviving trees are not vigorous or the disturbance is too severe to exclude a new cohort from entering (Fig. 5.4A).

Response of a stand to a minor disturbance depends on several factors: type of disturbance—fire, windstorm, human cutting, animal damage, or other; severity of the disturbance; location of damage in the stand—in the upper (A- and B-) strata, middle strata, or forest floor stratum; and vigor of the residual trees—their ability to grow after the disturbance. These factors are partly the result of stand development patterns and partly the result of influences external to the stand.

Species vary in ability to endure different disturbances and in ability to regenerate soon afterward. The different types of disturbances act as "environmental sieves," determining which species continue to constitute the stand. A light fire in an eastern mixed hardwood stand eliminates thin-barked species and thus increases the predominance of oaks and pines; in mixed conifer stands in the western United States, a light fire kills thin-barked spruces, firs, and hemlocks but allows Douglas-firs and pines to grow. Insect and disease outbreaks are selective for certain species; e.g., spruce budworm kills fir species (Blais, 1954); the chestnut blight killed chestnuts; and the white pine blister rust attacks white pine species (Hepting, 1971).

Disturbances lead to changes in a stand's structure, species composition, and tree shapes depending on which trees are left. Minor disturbances kill trees in the upper (A- and B-) strata, the middle strata, or the forest floor stratum. Surface fires kill forest floor trees and, when more intense, kill trees in higher strata (Viers, 1980a,b, 1982; Means, 1982). Lightning often kills the tallest trees—in the A- and B-strata. Insects and diseases which attack weakened trees often kill intermediate and suppressed trees in upper strata and more dominant crown classes in lower strata.

A disturbance's severity and the vigor of the surviving trees determine how many and what trees are killed, and thus how much growing space becomes available (Fig. 5.5). If a light, minor disturbance releases very little growing space, vigorous residual trees can refill it rapidly and prevent a new cohort from becoming established (Fig. 5.4B; Woods and Shanks, 1959; Hibbs, 1982). Such disturbances only stimulate a change in the growth rates of trees which survived. A slightly more severe minor disturbance may create a new cohort which soon becomes so suppressed that the trees remain in the forest floor stratum as advance regeneration. More severe minor disturbances release more growing space, allowing a new cohort of trees to grow as a major component of the stand. Vigor of surviving trees determines how rapidly they expand into released growing space. If they expand rapidly, the invasion or survival of a new cohort is prevented. Vigor is dependent on the trees' ages, their canopy strata and crown classes, their crowding, and the site conditions (Awaya et al., 1981).

In young stands in the early stem exclusion stage, most residual stems are vigorous enough to grow rapidly after a minor disturbance and reoccupy released growing space completely, maintaining a single-cohort stand. A partial disturbance in the stand initiation stage prolongs the time of plant invasion (Oliver and Stephens, 1977; Franklin and Hemstrom, 1981); however, it is difficult to determine which stems invade because of the second disturbance, and a second cohort is generally not recognized here. Dominant and codominant trees in any stratum generally respond more rapidly to minor disturbances than trees in lower crown classes do (Oliver and Murray, 1983). Consequently, ground fires or silvicultural thinnings from below tend not to create a new cohort, whereas "high-grade" logging, which eliminates the more dominant trees, favors establishment of a new cohort.

A. SINGLE-COHORT STAND MAINTAINED

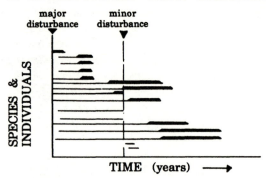

B. MULTI-COHORT STAND CREATED

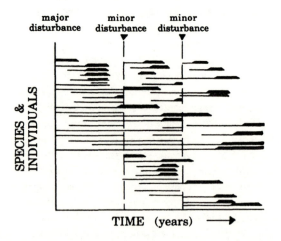

Figure 5.5 Minor disturbances (which do not kill all trees above the forest floor) may (A) allow any remaining trees to grow larger as they reoccupy the released growing space or (B) create new cohorts. New individuals enter primarily following disturbances, similarly to the initial floristics pattern of Fig. 5.1.

Development of multicohort stands

Development of multicohort stands is determined by the behavior of stand components following each minor disturbance. Following a minor disturbance, surviving trees expand and newly initiating ones invade and expand. There is a period of "free growth" during which both new and old plants grow into unoccupied growing space. Eventually, all growing space is reoccupied by existing stems and newly invading ones, even newer stems are excluded, and some existing stems are eliminated or severely suppressed through competition. In this way, there are also "stand initiation" and "stem exclusion" stages following minor disturbances.

Growth of new stems. After a minor disturbance, new stems begin growth from advance regeneration, stump and root sprouts, and buried and invading seeds, depending on the type of disturbance. Many shade-tolerant and -intolerant species initiate growth by utilizing stored energy in the seeds, roots, or stems of germinating, sprouting, or releasing trees.

As these new plants grow, they and the surviving older plants reoccupy the available growing space and begin competing. High shade from older trees slows height growth of the understory trees (Minckler and Woerheide, 1965). Shade cast by the overstory gives a competitive advantage to those species capable of surviving and growing most rapidly under partial, high shade (Toumey and Kienholz, 1931; Lutz, 1945; Bray 1956; Franklin and DeBell, 1973; Veblen et al., 1980).

Where numerous large trees survive the disturbance, regenerating species capable of living under deep high shade survive and have the growth advantage. The shade cast and the root competition increase as the overstory trees reoccupy newly available growing space. The understory trees first continue to slow in height growth. Then they begin to lose epinastic control and become flat-topped (Fig. 3.7). Less shade-tolerant species are soon eliminated from a new cohort when the residual overstory casts much shade. These intolerants generally do not survive the low light intensity even though they were able to begin growth. Finally, all understory trees die if the overstory excludes too much sunlight.

Response of the understory strata. Older trees in the lower (C- and D-) strata generally have flat-topped crowns characteristic of trees living in high shade (Fig. 3.7). After release, the terminal reasserts its dominance, producing a branching pattern characteristic of suppression and release—a cluster of branches, a crook, or a fork (Trimble, 1968a). Growth after release varies by species and depends on the tree's ability to withstand exposure and on the size of its crown (Marquis, 1981a,b). Dominant and codominant trees in lower strata often grow rapidly, since they have large living crowns capable of rapidly increasing their foliage. More suppressed trees continue growing slowly for a long time, until their photosynthetic surface area becomes large.

Growth of overstory trees. There is little visible effect on the upper strata if a minor disturbance injures or kills trees in the lower or forest floor strata, since the disturbance does not provide extra room for crown expansion. There is generally some increase in growth of these trees, since soil growing space becomes available.

A disturbance to some overstory trees can cause water sprouts and sun scalding on the remaining ones. Soon after release, foliage increases throughout the crowns of dominant and codominant trees. Diameter and height growth are also rapid. Intermediate and suppressed trees in all strata increase foliage and height and diameter growth relatively little, since most photosynthate is used for maintenance respiration.

The limbs of species with weak epinastic control rapidly grow outward into sunlight if a neighboring tree dies. Species with strong epinastic control maintain symmetric crowns, and branch growth does not increase. Even when overstory branch growth does not increase after a disturbance, light reaching lower strata dramatically declines as foliage increases within the overstory crowns. In temperate forests sunlight never comes from directly overhead, and long, dense crowns reduce light to the understory even when adjacent crowns are not touching.

The understory can reduce the vigor of overstory trees when a minor disturbance allows a vigorous understory to develop beneath a scattered overstory. The vigorous understory often outcompetes the overstory for root growing space—moisture and nutrients. This mechanism may be responsible for the susceptibility of loblolly pines to the southern pine beetle (*Dendroctonus frontalis*) on clay soils in the southeastern United States as hardwoods grow in the understory. Similarly, large areas of mixed conifers are dying from *Armellea obscura* in mixed conifer forests in the eastern Cascades of Washington. *Armellea* attacks weakened trees. Exclusion of ground fires in recent decades has allowed grand firs to grow into the understories of these stands and possibly weaken the overstory trees.

Overview of Forest Development Patterns

Forests are frequently subjected to disturbances. Trees in young, vigorous stands can respond to light, minor disturbances by rapidly expanding into available growing space and thus maintain a single-cohort stand. Stands resembling the understory reinitiation stage may begin after a minor disturbance if the existing trees are too old to exclude the new stems but vigorous enough to prevent them from growing large. Trees in the old growth stage behave like those responding to minor disturbances.

Both initial and relay floristics patterns occur at various times in forest development. The shifting dominance among species during the stem initiation, exclusion, and reinitiation stages is not the result of relay floristics. Rather, the initial floristics pattern occurs after a disturbance, with all species and individuals invading during a relatively short interval. Once established, they exclude plants arriving later. Later development of shade-tolerant species in the understory during understory reinitiation follows the relay floristics pattern. This invasion could be interpreted as directional development of a stand toward specific species compositions, since trees invading during understory reinitiation are generally shade-tolerant and only a limited number of such species exist in each region. Actually, the patterns are the casual result of competitive interactions rather than any grand design or external force on forest development.

Structural vs Process Approaches
to Stand Classification

Much of this book describes the processes of change. For ease of discussion, the processes have been classified into discrete development stages where different interactions are dominant. The processes can lead to various stand structural features.

Various stand structures have also been observed and classified as discrete conditions—or structures—independent of the processes leading to the structures. Because each structure is generally, but not always, the result of processes occurring in each development stage, the same terms have been used to classify both processes and stand structures (e.g., Oliver, 1992a,b,c).

Confusion has arisen because each of the various stand structure classes are wrongly assumed to be *uniquely* achieved in a single development stage; and the confusion is especially strong when development stages and structures have the same names. Specific, defined stand structures can often be achieved by several processes and in several development stages; and in some cases, a stand structure with a given name can not be achieved in a developmental stage with the same name. For example, "old growth" is defined both as a process (Oliver, 1981) and a development stage (Franklin et al., 1981, 1986; Hunter, 1990). As a development stage, it defines a stand's condition when disturbances have been absent for so long that none of the trees which invaded during the "stand initiation" stage following the disturbance are still alive. As a structure, "old growth" has different definitions and usually describes large trees, multiple species, and some component of rotting wood, snags, and logs. In many cases, the "old growth" *structure* would not exist in stands in the "old growth" *development stage.* For example, "old growth" longleaf pine stands (structural definition) suitable for the red cockaded woodpecker habitat are generally found in stands in the "understory reinitiation" development stage or in stands impacted by minor disturbances. If longleaf stands were allowed to grow to the "old growth" development stage, they would probably not be suitable habitat for the red cockaded woodpecker. Similarly, most stands with the "old growth" *structure* in the Pacific Northwest (Franklin et al., 1986) are not in the "old growth" *development stage*; rather, they are the result of minor disturbances or specific growth patterns in much younger stands (Newton and Cole, 1987; McComb et al. 1993; Oliver et al. 1994b).

This book will follow the general practice of referring to such terms as "stand initiation," "stem exclusion," "understory reinitiation," and "old growth" as both structures and development stages. It is important to realize that the two are not the same, and this book will distinguish between the two wherever there appears to be ambiguity. Other structural classifications have also been used, usually with more structures defined. Which and how many structures are defined depends on the objectives of management.

Forests generally are suited for various commodity and non-commodity values based on their structural condition, rather than the development stage they are in. For example, large trees with knot-free wood are valuable regardless of whether they are found in stands in the "stem exclusion," "understory reinitiation," or "old growth" development stage or in a multicohort stand. Certain animal species are found where there is a large vertical distribution of foliage, which can be found in stratified, mixed species stands in the stem exclusion development stage, as well as in older stages.

Disturbance Frequency and General Forest Type

Disturbances generally do not occur randomly within a region. Instead, they often occur in waves during times of unusual weather conditions, insect outbreaks, fuel buildups, or economic pressures (in cases of human disturbances). Within a region, each severe past disturbance has created many stands or cohorts (within multicohort stands) which grow and become predisposed to later disturbances (Fig. 4.1) of a given magnitude at approximately the same time. When a later regional disturbance occurs, the stands or cohorts create another wave of one cohort distributed throughout many single and multicohort stands. Consequently, ages of stands or cohorts are often not randomly or uniformly distributed throughout a region. Rather, stands and cohorts are often of characteristic ages dating from regionwide disturbances.

Different structural features are commonly found across a landscape in certain topographic positions (Camp, 1995) because of the effects of topography on disturbance patterns and growth rates. Variation in disturbance impacts across the landscape has generally created a variety of stand structures following each disturbance. Some areas were sheltered for topographic or edaphic reasons and were subject to a minor disturbance, if disturbed at all; other areas were frequently subjected to stand-replacing disturbances. Following the disturbance, there was a variety of stand development patterns occurring as different areas grew to single and multicohort stands (Yamamoto, 1992, 1993; Cao, 1995; Mizunaga, 1995), . The variety of structures became more diverse with time because stands grew differently on soils with different productivities, even if the disturbance had left them with initially similar structures.

The types of disturbances are also somewhat characteristic of each region, favoring particular regeneration mechanisms and certain forest species compositions within each soil and climatic area. The common species compositions and ages of stands and cohorts over regions give them similar physiognomies which change slowly with time. Before recent, rapid communication and accumulation of historical records proved otherwise, forest types were assumed to be permanent and stable on an area, because changes occur so slowly.

Figure 5.6 shows how types and frequencies of large disturbances can determine the stand patterns over broad forest regions of similar soil and climate. Figure 5.2 shows the relative time after a major disturbance during which each species (or group of species) is dominant. Figure 5.6A shows the pattern of dominance of each species in an area where major disturbances occurred at equal time intervals. If the disturbance was of a given severity (shown as "X" in Fig. 5.6), the species shown in Fig. 5.2 would be favored. If a more severe or less severe disturbance occurred ("Y" in Fig. 5.6), different species would be favored and the forest composition would be different until the next disturbance. If disturbances were more frequent (Fig. 5.6B), the predominant forest types would shift to species which dominated when the stands were younger. The probability of finding a particular forest type in a large area, and hence the dominant type of the area, is therefore directly related to both frequency and type of disturbance. Minor disturbances would similarly influence the pattern, depending on their type and severity, except not all trees would be eliminated at each disturbance. Consequently, each cohort would be characteristic of the type and frequency of the disturbance, as well as its magnitude.

Figure 5.6 also shows how different stands can achieve a targeted structure at different times and in different development stages. If stands with a large vertical distribution of canopy strata were desired, the structure can be achieved quite readily during the stem exclusion development stage and maintained thereafter in stands containing species "A" through "D." On the other hand stands containing species "E" and "F" may have only one stratum and may remain in the "stem exclusion" and "understory reinitiation" development stages for much longer; consequently, the targeted stand structure would not be achieved in these stands for a very long time--if at all.

Where disturbances are randomly scattered in space over a large area, travel across the region can be substituted for time in determining the probability of occurrence of a given forest type. Pulselike patterns have also been suggested by others (Odum, 1969; Loucks, 1970; Heinselman, 1973; Wright, 1974; Whittaker, 1975; Connell, 1978; Oliver et al., 1985).

Applications to Management

Management of stands does not alter a "natural direction" of forest development, since no single obligatory direction occurs. Repeatable patterns which emerge from similar disturbance regimes on similar sites are consequences of plant interactions, not driving forces.

Most commodity and non-commodity management objectives are better expressed in target structures, rather than in processes. For example, if an "old growth" structure is desired, this *structure* is better explicitly stated, rather than stating a desire to follow the *process* of avoiding all disturbances to a stand—thus achieving (or trying to achieve) the old growth developmental stage. As discussed above, following the process of avoiding disturbances may not achieve the old growth structure—and usually not as rapidly as through other means. Understanding the processes of development, the development stages, and how different processes can lead to different structures will allow the silviculturist to design management activities which will achieve the desired structures as efficiently as possible.

Silvicultural operations are tied to the stages of stand development. Applying operations during an inappropriate stage makes them much more difficult and costly or virtually impossible to implement successfully. Some operations may not be necessary if the natural stages of development are known and anticipated. For example, since advance regeneration becomes established during the understory reinitiation stage, planning on this form of regeneration of a desirable species can avoid costly planting of seedlings (e.g., Seymour, 1992).

Within limits, however, the time a stand appears in any stage can be prolonged or shortened with appropriate management. Single cohort stands can be moved rapidly through the initiation stage by planting trees and controlling weeds to stimulate tree growth, and this stage can be prolonged by removing some trees. The stem exclusion stage can be prolonged with judicious, light thinnings of intermediate and suppressed overstory trees, and shortened by heavy thinnings which allow an understory to develop.

168   Chapter Five

The understory reinitiation stage can also be prolonged by light thinnings which stabilize the overstory, or shortened by heavy thinnings which allow the advance regeneration to grow upward, creating a structure resembling old growth.

Operations such as weeding and planting or interplanting to alter the species composition of the new stand are most efficient during the stand initiation stage. Trying to adjust the species composition by adding species during the stem exclusion stage is nearly impossible. If the desired species stratifies above the undesired one, thinning operations which adjust to the species composition which occurs naturally during the stem exclusion stage may not be necessary.

A. DISTURBANCES AT LONG INTERVALS
("old growth" stage reached)

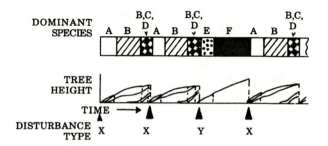

B. DISTURBANCES AT SHORT INTERVALS
(before "old growth" stage reached)

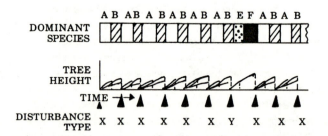

Figure 5.6 Changes in forest composition over many centuries, based on development patterns in Fig. 5.2 and following periodic disturbances of type X. Sloping lines represent tree height growth patterns shown in Fig. 5.2. (A) The forest is dominated with species B (pure or mixed with C and D) if disturbances are infrequent enough to allow all four development stages. The forest shifts to being dominated by species E and F after a different disturbance (type Y) favors other species. (B) The forest is equally dominated by species A and B when disturbances of type X occur more frequently. Again, a type Y disturbance is followed by a predominance of species E and F. (*Oliver, 1981.*) (See "source notes.")

Stand structures reached at different development stages are susceptible to certain disturbances. The forester can anticipate dangerous conditions at each stage of development. Both fires and insect attacks tend to occur during the stem exclusion stage; activities such as thinning can reduce or increase the risk of fires or insects. Thinning which is planned for a time early in the stem exclusion stage may need to incorporate extra slash disposal measures which may not be necessary later in the development of the stand. Progressively older stands become more susceptible to windthrow as the trees grow taller.

Each structure generally has both commodity and non-commodity values (Oliver 1992*a*). For example, mushrooms and berries can be plentiful in the stand initiation structure; rapid tree volume growth (per hectare) occurs in the stem exclusion structure; advance regeneration, mushrooms, floral greens, some medicinal plants, and high quality timber are often found in the "old growth" structure. In addition, each forest structure is necessary for some species to survive. Some structures are more or less aesthetically pleasing and affect the recreational value of the stand.

Other stand structures have value by their contribution to the habitat of certain wildlife species. Different wildlife species require different stand structures, so different species are more abundant in each development stage. Some wildlife require a group of stands in different stages, and some development stages may or may not be desirable depending on the structure of surrounding stands in the forest. Because there is no single structure which will provide all values at all times, the various values can be maintained by coordinating stands across a landscape so that a proportion of the landscape is maintained in each desired structure at all times (Boyce, 1985, 1995; Oliver, 1992*a,b,c*; Oliver et al. 1992). Maintaining the various structures requires a coordination of silvicultural activities among stands and through time as the structures change through growth and disturbances. Several technical tools have been developed to allow this coordination to occur (Boyce, 1985, 1995; Sessions and Sessions, 1992; Oliver and McCarter, 1995), and various incentives (Lippke and Oliver, 1993) and portfolio approaches to management (Oliver, 1994*a,b*) have been developed to make landscape management as efficient as possible.

Managing landscapes to provide and maintain all structures using silvicultural activities will generally cause less extreme shifts in structures and species populations than would occur if large areas were set aside as reserves and excluded from human activities (discussed in Chapter 4).

Forests can develop in only certain patterns, depending on their present conditions. A stand in the stem exclusion structure may take many decades to appear as an old growth structure, even with appropriate silvicultural manipulations; a single-cohort stand in the stem exclusion stage takes many decades to develop as a stand with three or more cohorts; and a stand with several cohorts takes many years to revert to a single-cohort stand in the stem exclusion structure.

Various studies have proposed efficient ways to achieve desired structures (Newton and Cole, 1987; McComb et al., 1993; Oliver 1994d; Oliver et al., 1994a). The silviculturist can take advantage of different growth rates on different sites to achieve the desired structures as rapidly as possible.

Thinning, pruning, harvest, prescribed fire, or other silvicultural activities often detract from the stand's appearance for several years afterward, until the responding growth begins creating the desired structure. Because of the commonly unattractive early appearance, the silviculturist needs discipline and diagnostic tools to help ensure that the correct actions have been done to achieve the results. Some of these tools are described in Chapter 15.

Stand structures can be defined in many ways, depending on the objectives, the region, and other biological and non-biological factors. They have been defined several ways:
—tree size classes (e.g., seedling-sapling, pole, small sawtimber, large sawtimber);
—modifications and variations of the development stages of Figure 5.2;
—other definitions (e.g., "savanna," "open," "ancient," and others).
Other classifications are currently being developed as more emhasis is being placed on managing for a variety of objectives. It is premature and probably inappropriate to set standard structures for universal use. Eventually, coventional classification systems may become accepted for various regions, objectives, and/or management organizations.

Because structures often take time to develop, silvicultural practices often focus on actively encouraging stands to develop toward those structures in short supply, recognizing that the stands will become increasingly more valuable as they accumulate attributes of the desired structure.

The precise target amount of the various structures across the landscape is often not an immediate concern. Historical records and some understanding of the structure needs to provide the different values can give a preliminary indication of *which* structures are to be maintained. Where there is too little of one stand structure, there will be too much of other structures; and most forested landscapes in the world have quite disproportionate areas in the different stand structures. For several decades, silvicultural activities can focus on converting the stand structures which are extremely plentiful to those in extremely short supply; during this time, more detailed studies and adaptive management approaches (see "Applications to Management," Chapter 15) can determine more precisely the target amounts of each structure needed. Fractal geometry and other geometric tools may prove useful in managing different spatial patterns at different scales (Lorimer et al., 1994; McQuillan, 1995).

Disturbances such as wind or agricultural land abandonment often cover large areas, creating many stands (or cohorts within stands) in very similar stages of stand development, with similar structures, and with similar species compositions. Where stands are similar over large areas, returns from the forests and the silvicultural operations needed are limited by the little variation in development stages. There is a danger that silvicultural innovation can become stifled because the development of these similar stands occurs over a long time. As a result, foresters tend to do what has been done in the past and what has worked in the past. As stands reach more advanced stages of development or are harvested and stands with new development patterns emerge, new silvicultural problems and opportunities will exist. Foresters must adapt to the changes and devise prescriptions which reflect the changes in stand development.

CHAPTER 6

TEMPORAL AND SPATIAL
PATTERNS OF TREE INVASION

Introduction

Trees do not all invade a disturbed area in a single year, nor do they become established in a uniform spatial pattern. The age range of stems is determined by how long new plants initiate after the disturbance. The spatial distribution is determined first by the initial plant spatial pattern after the disturbance, and then by ability to survive the ensuing competition. Survival is not random. It is possible to anticipate which trees will survive and dominate a particular site based on the species, regeneration mechanisms, relative times of invasion, and spatial positions.

Once established, the location of every tree in a stand is fixed. Very many trees die during the first decades of a stand (Peet and Christensen, 1987). Surviving trees gain an advantage because of, or in spite of, the microsites on which they grow. A favorable microsite may be relatively permanent, such as a raised hummock where many generations of dominant trees grow from the same place. It may be only temporary, but still give a tree an advantage—such as when a dead tree's fallen crown or stem protects a newly growing tree from frost, heat, or browsing. A tree may survive and dominate even if it grows on an unfavorable microsite, providing other factors give it a competitive advantage—such as a rapid early height growth rate, an appropriate regeneration mechanism, or a physiological resistance to adverse climates and soils. A tree also influences the locations of trees in future stands in the same place, since its shade, soil changes, and roots influence the locations of competing herbs, advance regeneration, and sprouting tree stumps. A mound created by a windthrown dominant tree can leave a mineral soil microsite, which gives a competitive advantage to a tree initiating from a windblown seed and allows this tree to dominate in the same location.

Age Distributions in Single- and Multiple-Cohort Stands

Tree age is calculated from the time of germination of seeds, initiation of stump or root sprouts, or release of advance regeneration (Morris, 1948; Sprugel, 1974; Oliver and Stephens, 1977; Oliver, 1978a). The tree age distributions within stands vary from all stems being in a single cohort, to stems distributed uniformly in a few or many cohorts, to stems distributed unequally in a few or many cohorts. The frequency and magnitude of disturbances dictate the age distributions, and a single-cohort stand is simply the result of a special case of the disturbance pattern. The age distribution can change as a stand ages. A single-cohort stand can have an all-aged tree distribution during the stand initiation stage (Fig. 5.2), a single age class during the stem exclusion stage, and two age classes during the understory reinitiation stage if the advance regeneration is considered a separate age class. A true old growth forest, in which all trees established after the initial disturbance have died, has a broad range of tree ages.

Disturbances occur in a continuum of intensities from very slight to very great (Jones, 1945; Whitmore, 1974). The amount of overstory removed by a disturbance and the vigor of remaining stems influence the age distribution and growth of establishing stems (Fig. 6.1A; Hough and Forbes, 1943; Larson, 1982; Marquis, 1992; O'Hara, 1995). A single-cohort stand can follow either a disturbance which removes all previously existing stems or a disturbance which removes so few stems that the remaining ones rapidly reoccupy the growing space and preclude a new cohort's becoming established (Veblen and Stewart, 1982a). Multicohort stands with trees predominantly in one cohort exist if only a few trees survive the disturbance or if only enough growing space is released for a few young trees to become established (Deal, 1987). A multicohort stand with trees distributed evenly in all cohorts can also exist (Oliver and Stephens, 1977; Viers, 1980a,b, 1982; Deal, 1987).

The frequency of disturbances relative to the life spans of component trees also determines the age distribution of stems (Fig. 6.1B). Rare disturbances lead to old stands containing a very broad range of ages. Slightly more frequent disturbances allow the stand to grow old and predisposed to major disturbances, and then be entirely destroyed, creating single-cohort stands. More frequent disturbances allow some trees to withstand each disturbance, with multicohort stands being formed. These stands can have either regular or irregular age-class distributions, depending on the disturbance patterns. Very frequent, small disturbances can create all-aged stands if successive disturbances occur while trees are still initiating from the previous one.

Physiognomy and growth pattern differences between some multicohort stands may be greater than between some single-cohort and multicohort stands. A stand with unequal tree numbers in each cohort (e.g., stand II, Fig. 6.1A) can behave much like a single-cohort stand but quite differently from a multicohort stand in which trees are distributed evenly among the cohorts. Stands are often classified as either single- or multicohort; however, a gradation exists between the extremes of stands with a single age class and those with trees of all ages.

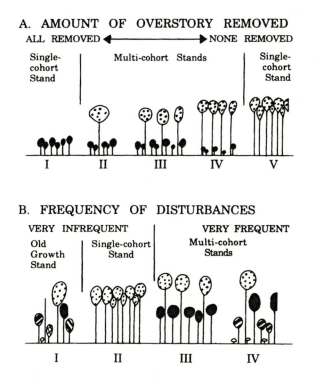

Figure 6.1 Schematic of the relation between disturbances and stand cohort (age-class) distributions. Stands can range from containing trees solely in one cohort, to containing trees predominantly in one cohort, to containing trees balanced in two or more cohorts depending on such factors as (A) amount of overstory (above the forest floor) removed and (B) frequency of disturbances. Different crown shadings represent different cohorts. (A) Amount of overstory removed. Disturbances which remove all or none of the overstory leave single-cohort stands. Multiple-cohort stands with a gradation of structures occur when intermediate amounts of overstory are removed. (B) Frequency of disturbances. Very infrequent disturbances allow old growth stands to develop, with a variety of tree ages. Progressively more frequent disturbances create single-cohort and then multiple-cohort stands with progressively more age classes

Age distributions in single-cohort stands
Trees invade a stand for a few years or for many decades before the growing space is occupied by the existing trees and additional trees are excluded (Table 6.1;

TABLE 6.1 Age Ranges of Naturally Invading Stems following Various Natural Disturbances

Stand condition	Disturbance type	Time since disturbance, years	Tree age range, years	Source
Douglas-fir; dry site; Oregon west Cascades	Natural fire	540	200+	1
Douglas-fir, incense cedar; dry site; Oregon west Cascades	Natural fire	Various times	60-150	10
Douglas-fir; mesic site; Washington west Cascades	Volcanic ash	680	70	2
Douglas-fir; mesic site; Washington west Cascades	Volcanic ash	500	180	2
Douglas-fir; mesic site; Washington west Cascades	Natural fire	750	100+	11
Douglas-fir; mesic site; Washington west Cascades	Natural fire	92	92	11
Douglas-fir; dry site; Washington east Cascades	Natural fire	90	60	3
Ponderosa pine, maple, Douglas-fir, mesic/dry site; Washington east Cascades	Natural fire	90	50	3
Lodgepole pine, maple, Douglas-fir, willow; mesic site; Washington east Cascades	Natural fire	90	40	3
Western hemlock; mesic site; coastal Washington	Shelterwood harvest (first cut 17 years; final cut 9 years)	17	9	4
Douglas-fir, hemlock; mesic site; coastal Washington	Clearcut and burn	Several stands: 35-90	16	5

TABLE 6.1 Age Ranges of Naturally Invading Stems following Various Natural Disturbances *(Continued)*

Stand condition	Disturbance type	Time since disturbance, years	Tree age range, years	Source
Alder, Douglas-fir, hemlock, redceder; mesic site; Washington west Cascades	Clearcut and burn	52	22	6
Fir, hemlock; upper-elevation mesic site; Washington northwest Cascades	Glacial retreat	60	19	8
Fir, hemlock; upper-elevation mesic/dry site; Washington northwest Cascades	Glacial retreat	105	44	8
Fir, hemlock; upper-elevation dry site; Washington north-west Cascades	Glacial retreat	63	63+ (still invading)	8
Fir, hemlock; upper-elevation mesic site; Washington north-west Cascades	Avalanche (runout zone)	100	10	8
Fir, hemlock; upper-elevation mesic site; Washington north-west Cascades	Avalanche runout zone)	480	185	8
Oak, maple, birch, hemlock; mesic site; Massachusetts	Hurricane	40	10 (except hemlock, which continued)	7
Oak, maple, birch; mesic site; Connecticut	Hurricane	35	25	9
Oak, maple, birch; mesic site; Connecticut	Clearcut	65	35	9
Oak, maple, birch; mesic site; Connecticut	Clearcut	45	25	9

NOTE: Ages are often reconstructed from living trees and thus may reflect underestimation of the total age range.
SOURCES: 1 = Franklin and Waring, 1979; 2 = Yamaguchi, 1986; 3 = Larson and Oliver, 1982; 4 = Jaeck et al., 1984; 5 = Wierman and Oliver, 1979; 6 = Stubblefield and Oliver, 1978; 7 = Hibbs, 1983; 8 = Oliver et al., 1985; 9 = Oliver, 1978*a*; 10 = Means, 1982; 11 = Hemstrom, 1979.

Bazzaz, 1983). The age range is narrow when the first invading stems rapidly occupy the growing space and exclude later arriving trees. Narrow age ranges occur where sites are productive, where species and regeneration mechanisms promote rapid growth, or where many trees begin soon after the disturbance.

Plantations usually have a narrower age range, both because planted seedlings grow vigorously and because seed sources and other regeneration mechanisms are limited. In one study, new stems quit invading a western hemlock stand the year after removal of a shelterwood overstory which had been left for 8 years (Jaeck et al., 1984). The short time interval was probably because the productive soils, nearby seed sources, good seedbed conditions, and rapid growth of the advance regeneration and germinating hemlocks allowed them to reoccupy the growing space rapidly and exclude younger stems. On the other hand, stands beginning after clearcuttings or windthrow of old field white pine stands in New England allowed stems to continue entering for 35 years, even though most trees entered during the first decade (Fig. 6.2; Oliver, 1978a). Apparently the vigorous growth of sprouts and advance regeneration did not adequately exclude new stems during a shorter interval.

Broad age ranges develop if the initially invading stems grow slowly and so do not exclude later stems. Broad age ranges occur on poor sites or where the species and regeneration mechanisms promote slow growth, where most older plants are continually retarded or killed by browsing or frost, or where few stems initiate each year so that the growing space is refilled slowly. Few stems initiate each year, but new stems can continue to initiate for many years where seed sources are distant; herbaceous or shrub competition is prevalent; or the first invading plants eliminate frost pockets, waterlogged depressions, or hot areas and more microsites become suitable for stem initiation.

An age range spanning 160 years occurred on a dry site in the Oregon Cascades after an extensive disturbance had removed the Douglas-fir and hemlock seed sources (Franklin and Waring, 1979). It apparently took many years for enough trees to become established and grow to exclude younger ones. A similarly wide age range occurred in a stand established after volcanic tephra deposits destroyed the previous stand in the Cascades of southern Washington (Yamaguchi, 1986). Although it took 180 years for younger stems to be excluded where the seed source was far away, parts of the same stand closer to living trees had an age range of only 70 years.

Trees which invade very early or very late may not be vigorous enough to compete successfully in the stem exclusion stage even after they have become established. Death of these trees narrows the age distribution. This narrowing is undetected in studies where the age range (such as many in Table 6.1) is reconstructed from living trees. The ranges reported, therefore, are conservative estimates.

Irregular distributions of age classes within a single cohort arise in several ways. Sometimes, most stems become established shortly after the disturbance, giving a skewed distribution of stem numbers and ages (Fig. 6.2A). This distribution is probably caused by favorable site conditions and abundant parent material (seed sources, roots or stumps for sprouting, or advance regeneration), so that most stems enter early and grow rapidly, leaving little growing space available for stems to become established later.

In other situations, a few stems become established at first, and many more invade later, giving the stand a skewed distribution in the other direction (Franklin and Waring,

1979; Fig. 6.2C). This distribution is most common where a harsh environment precludes establishment of many stems until a few trees have grown and modified the site. It can also occur where a few stems become established and mature enough to set seeds; these progeny then continue to recolonize the site.

Different species can invade a stand during distinct periods before the stem exclusion stage as a result of changing microsite conditions. More shade-tolerant species continue invading for a short time after conditions are no longer suitable for intolerant trees to invade. On extreme sites, a more sensitive species may not survive until other species have modified the microenvironment. Invasion periods correspond with good seed years for species which do not set seeds every year. Species which begin primarily from advance regeneration (e.g., northern red oak, Fig. 6.2A) or from buried seeds are nearly as old as the disturbance, while stems which seed into the same stand after the disturbance have a broader age range. Trees of a very fast growing species may be the last to invade a stand; they occupy the remaining available growing space and then exclude younger stems. Red alders exhibited this behavior following clearcutting and burning of large areas in the Pacific northwestern United States (Stubblefield and Oliver, 1978).

Age distributions in multicohort stands

Multicohort stands develop where disturbances do not completely remove the previous tree cover. The age-class distributions in multicohort stands follow disturbance, invasion, and mortality patterns (Hough and Forbes, 1943; Oliver and Stephens, 1977; Deal, 1987; Deal et al., 1989), just as they do in single-cohort stands. Multicohort stands can have a uniform distribution of cohorts if the disturbances occur regularly or an irregular distribution if the disturbances occur randomly. In some conditions, multicohort stands can have trees of every age (or age class) to quite an old age (Gates and Nichols, 1930; Morey, 1936; Maissurow, 1941; Hough and Forbes, 1943; Leak, 1975; Hett and Loucks, 1976; Oliver and Stephens, 1977; Tubbs, 1977; Lorimer, 1980; Larson, 1982; Deal, 1987).

Where disturbances occur at regular time intervals, trees of several cohorts (and age classes) can be found; however, the number of trees can vary greatly in each cohort. An uneven number of trees occurs among cohorts if one cohort is vigorous enough to exclude many stems of other cohorts if disturbances of unequal severity occur, or if a disturbance kills an inordinately large number of trees in older cohorts (Fig. 6.3A). For example, a hurricane in 1815 destroyed most trees in a mixed-species stand in central New England (Oliver and Stephens, 1977). The cohort which grew after this was vigorous enough in the stem exclusion stage to exclude most stems of new cohorts following subsequent, light disturbances (Fig. 6.3A). Similar age distributions have been found by Gates and Nichols (1930), Maissurow (1941), Hough and Forbes (1943), Hanley et al. (1975), Nakashizuka and Numata (1982a,b), Betters and Woods (1981), and Means (1982). The reverse-J-shaped distribution of trees by age classes (Fig. 6.3B) was found in Douglas-fir—grand fir stands in eastern Washington (Larson and Oliver, 1982) and in a selectively cut forest in Bavaria. Here, each disturbance allowed many new stems to

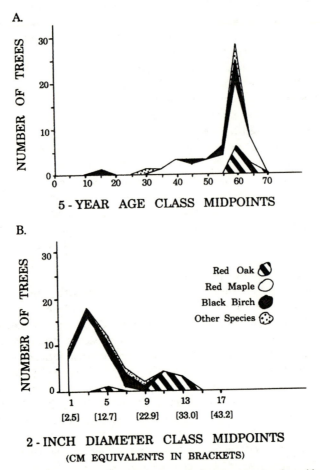

Figure 6.2 Age distributions can vary dramatically in single-cohort stands, often with no relation to diameter distributions. (A) Age and (B) diameter distributions of a hardwood stand beginning after clearcutting in central New England approximately 65 years earlier. Despite the reverse-J-shaped diameter distribution, most trees invaded soon after the disturbance, probably from advance regeneration and seedling sprouts (*Oliver, 1978a*). (See "source notes.")

invade. Progressively older cohorts declined in numbers both because of competition-related mortality (as occurs in single-cohort stands) and because of the death of a proportion of trees in each older cohort with each subsequent disturbance.

An approximately equal number of trees can be found within each cohort when light disturbances allow only a few stems to become established in each cohort. This pattern was found in a stand subjeced to many small disturbances in southeastern Alaska. The many small disturbances occurred because the unstable rooting medium of trees in the stad and surrounding hills caused large trees to fall through the stand periodically (Deal, 1987; Deal et al., 1991). Diameter distributions also showed roughly equal numbers in all diameter classes, although smaller trees were not necessarily younger. Trees of each age class were spaced throughout the stand. Few sites were suitable for germination and only a few stems apparently began after each disturbance. Mortality of these stems was also apparently low. Stands with relatively even distributions of tree ages can also be found elsewhere, reflecting the disturbance patterns (Hanley et al., 1975; Lorimer, 1980).

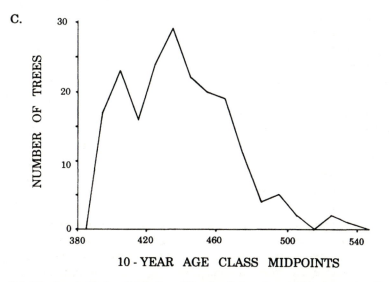

Figure 6.2 (*Continued*) (C) Age distribution of Douglas-fir trees in a stand in the western Oregon Cascade Range (*from Franklin and Waring, 1979*). Note that few trees invaded at first, creating an age distribution skewed in the opposite direction from the stand in (A). (See "source notes.")

Multicohort stands with trees in all age classes occur where disturbances are quite frequent or the period of stand initiation is quite long. Stems may still be initiating following one disturbance when a new disturbance occurs, allowing new stems to initiate in the released growing space. If the disturbances are regular enough, these stands can contain many small, young stems and fewer large, old stems—the reverse-J-shaped distribution (Fig. 6.3B). Very old stands (old growth stage) can be all-aged if the time since the last major or large minor disturbance has been so long that the older trees die on a regular basis, releasing growing space and allowing new stems to invade at a uniform rate.

Stands of short-lived species can also be all-aged if the older trees begin dying before new stems are restricted from invading following a disturbance. Some aspen stands in the Rocky Mountains appear in the all-aged condition where differential mortality occurs from shading and suppression at young ages (Fetherolf, 1917; Betters and Woods, 1981). The aspen stands initiate by root sprouts, and at first the root systems are interconnected. Upon death of a tree from suppression, the interconnection is lost. The root and above-ground growing space in the immediate vicinity are released, and new root sprouts begin before existing trees expand enough to reoccupy the growing space completely (Betters and Woods, 1981). Uneven-aged stands developing through this mechanism are rare in North America, since at least some species live longer than the time of stand initiation in most forests.

An aggregation of many small stands or "patches" of equal sizes through annual disturbances conceptually produces an all-aged forest. Theoretically, each patch acts as an isolated, single-cohort stand, and many concepts of all-aged forests are based on aggregates of such stands or "shifting mosaic steady states" (Pickett and White, 1985). If each patch is very small, the surrounding stands' edges can modify the patch's development, preventing initiation of a new cohort within it, with the result that a new cohort is not produced in each patch.

Species may possibly exist which utilize growing space so uniquely that they successfully invade the understory during the stem exclusion stage of other species. Evergreen magnolias and quaking aspens which regenerate from root sprouts may possibly behave this way. Basswood trees may also regenerate this way when new, younger stems grow up from the parent root collar. An asexual regeneration method may allow a stem to obtain energy from a parent tree while growing large enough to compete for growing space on its own. Western and eastern hemlocks and many other species which were thought to invade later actually initiate primarily following disturbances (Hough and Forbes, 1943; Oliver and Stephens, 1977; Stubblefield and Oliver, 1978; Wierman and Oliver, 1979; Oliver et al., 1985).

Cohorts can be irregularly distributed in time if disturbances strong enough to create a new age class occur at irregular intervals. Development patterns of these stands can resemble single-cohort stands during the long intervals between disturbances. Occasionally, isolated stems much younger or older than the waves of regeneration following disturbances are found within a stand (Fig. 6.2A; Hough and Forbes, 1943; Hanley et al., 1975; Oliver, 1978a). These trees apparently initiate while intense competition and exclusion of existing stems is occurring—during the stem exclusion stage. They are generally few in number and grow slowly unless released by subsequent disturbances. Such stems initiate where a very localized disturbance or other factor has made growing space available.

Spatial Distributions of Trees in Single- and Multiple-Cohort Stands

Natural stands can have less than 100 to over 100,000 stems per hectare (or acre; Maple, 1970; Mitchell and Goudie, 1980) at the end of the stand initiation stage. Disturbances affect the density of different species differently: those which expose mineral soil increase the number of trees germinating from wind-blown seeds, while a fire increases the number of sprouts from stumps and roots. The availability of different regeneration mechanisms for characteristic species—such as seed sources for light seeding species, advance regeneration for species which initiate this way—also determines initial stand density.

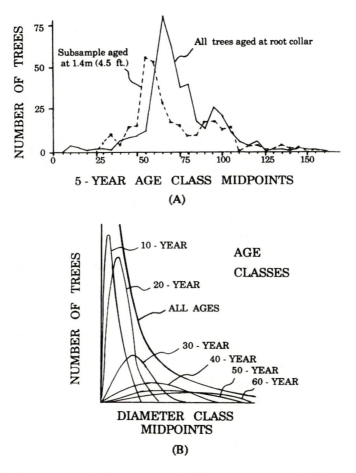

Figure 6.3 Tree ages in multicohort stands can be (A) irregularly distributed, (B) mathematically distributed in a reversed shape, or regularly distributed in all age classes (*Deal, 1987; Deal et al., 1991*). (A) Distribution of root collar ages of all trees and a sub-sample aged at 1.4 meters (DBH) in a mixed oak/maple/birch/hemlock stand in central New England. An irregular age-class distribution can correspond to intensity of disturbance and condition of a stand (*Oliver and Stephens, 1977*). (B) Idealized reverse-J-shaped distribution of tree ages. Some multicohort stands may approximate this tree age distribution. (See "source notes.")

Just as trees do not invade at the same time or at uniform time intervals, they do not begin growing at a uniform spacing unless planted that way. Most trees begin in aggregated, or clumped, patterns with some areas containing no individuals (Spring, 1922; Bray, 1956; Daniels, 1978; Rinehart, 1979; Yamamoto and Tutsomi, 1979; Nakashizuka and Numata, 1982; Peet and Christensen, 1987; Pukkala and Kolström, 1992). The initial, usually clumped, pattern is caused by the spatial distribution of advance regeneration and stumps and roots for sprouting, suitable seedbeds for germination, competition from other plants, behavior of the disturbance initiating the stand, and other factors. As a stand develops, a characteristic spatial pattern emerges as the result of inter- and intraspecific competition. More soil area is encompassed by the roots as the seedlings grow, and differences in very small-scale microsites become less important.

Mortality differs among species and sites and is very important in determining stand development patterns (Daniels, 1976; Peet and Christensen, 1987). Other conditions being equal, trees compete and die sooner at the narrower spacings within clumps; consequently, surviving trees change from a clumped to a random distribution and then approach a regular (evenly spaced) distribution (Curtis and McIntosh, 1951; Cooper, 1961; Laessle, 1965; Kitamoto and Shidei, 1972; Williamson, 1975; Akai et al., 1977; Christensen, 1977; Yeaton, 1978; Rinehart, 1979; Whipple, 1980; Good and Whipple, 1982). Dominant trees approach a regular distribution even more rapidly. Species in mixed stands also appear clumped at first but more evenly distributed with age (Grieg-Smith, 1952). Even though stands approach a more regular distribution, the spatial pattern continues to reflect the initial spacings and relative ages within the stand and the effects of stand edges. As stands become much older in the understory reinitiation and old growth stages, the older trees and all other trees return to more random, then aggregated conditions.

Trees of a stratum or cohort in multicohort stands sometimes are aggregated rather than well distributed throughout the stand (Greeley, 1907; Larsen and Woodbury, 1916; Show and Kotok, 1924, Baker, 1934; Viers, 1980a,b; Bonnicksen and Stone 1981; Nakashizuka and Numata, 1982). At other times, they are closely intermingled with other cohorts and strata. Aggregations, or *patches*, within stands can be classified as of one of two forms. One form is a discrete shorter stratum surrounded by taller ones, here referred to as a *gap* (Runkle, 1984). A tall stratum surrounded by shorter ones will be referred to as a *clump*. Edges of these patches are greatly influenced by their surroundings. If the patches are small, the relative amount of edge is large and most of the patch is strongly influenced by the environment beyond the patch (Nakashizuka, 1985). The relative influence of the surrounding forest declines as the patch becomes larger, until the patch can be considered an independent stand.

Each cohort can begin in a patch where germination (or other initiation) conditions are favorable (Akai et al., 1977; Collins et al., 1985), or trees of a new cohort can begin randomly. New cohorts do not necessarily appear beneath openings delineated by the disturbed area in multicohort stands (Oliver and Stephens, 1977). Rather, the new trees initiate in a broad area around a disturbance because the sunlight and root area released are not immediately beneath the canopy of the destroyed trees (Henry and Swan, 1974; Oliver and Stephens, 1977). In one study few stems appeared in each cohort of a mixed Sitka spruce-hemlock stand in coastal Alaska. These few stems appeared randomly distributed because only a few, randomly spaced sites were suitable for germination (Deal, 1987; Deal et al., 1989). Similarly, trees of all cohorts in a central New England mixed hardwood stand appeared randomly distributed, probably because many stems began as sprouts and advance regeneration and because suitable locations for germination were well distributed (Oliver and Stephens, 1977).

At lower latitudes new cohorts may initiate in more discrete gaps beneath an opening because the sun is more directly overhead for at least part of the day. In a multicohort, grand fir-Douglas-fir stand studied in the eastern Cascades of Washington, younger cohorts appeared in discrete patches—probably in areas devoid of grasses and, so, suitable for germination (Larson, 1982).

When a cohort begins in a clumped arrangement, component trees become more regularly spaced with age as the trees differentiate and as subsequent disturbances destroy some of them. By the time they reach the overstory, the number of trees has become so reduced that they are relatively randomly or evenly distributed (Williamson, 1975; Nakashizuka and Numata, 1982).

Development of Different Spatial and Temporal Patterns

The temporal and spatial patterns of trees in a stand are determined by the behaviors of component individuals and species and by environmental influences. Environmental influences can be fundamental to the site—for example, soil conditions which determine the growth rates of trees—or very temporary—such as places where a dead stem provides a shaded microsite for germinating seedlings to survive. The influences can result from conditions present before a disturbance initiates a cohort, from things which happen at the time of the disturbance, or from conditions after the disturbance.

Behaviors of the component individuals and species

Invasion patterns of different species are the result of their regeneration mechanisms, growth rates, mortality rates, and shade tolerances. Regeneration mechanisms allow different species to survive after different disturbances. High mortality often occurs during stand initiation (Fowells and Stark, 1971; Tappeiner and Helms, 1971; Glitzenstein et al., 1986). Each species and regeneration mechanism has a different mortality rate, which changes the abundance and distribution of stems as the stand develops (Oliver, 1978a; Peet and Christensen, 1987). Species such as western hemlocks, red maples, black birches, and bitter cherries have a high mortality rate compared with associated species. They are very abundant in very young stands, where they appear at first to dominate, but they quickly loose dominance to other species with lower mortality rates (Oliver, 1978a; Stubblefield and Oliver, 1978; Wierman and Oliver, 1979; Deal, 1987; Deal et al., 1989).

Certain regeneration mechanisms within a species are associated with a higher mortality rate than others. Red oaks initiating from seeds have a high mortality rate during the first few decades. Mortality of red oak stump sprouts is high until each stump is reduced to one or a few sprouts, after which sprout mortality is quite low. Red oaks from established advance regeneration have a low mortality rate. A species may appear to have a very high mortality rate as individuals begun from the less vigorous regeneration mechanisms rapidly die.

Growth rates among species and regeneration mechanisms also differ. Most true firs grow very slowly at first, while red alders (Stubblefield and Oliver, 1978) and most birches (Cline and Lockard, 1925; McKinnon et al., 1935) grow rapidly at first and reoccupy the growing space quickly. Sprouts and advance regeneration have preestablished root systems and generally allow trees to grow more rapidly than if the same species were to initiate from seeds.

Influence of events before a disturbance

Soils and tree patterns.   Uniform microsites allow many stems to become established in a randomor regular pattern.   Where seed sources are nearby, abandoned fields often grow to very dense stands because the soils and microsites are uniformly amenable to regeneration.   Forest soils are rarely uniform.   The ability to initiate and survive within each microsite determines a species' initial distribution.   Its growth rate in the different-microsites influences both its survival and how long it allows new stems to initiate—and hence the range of tree ages after the disturbance.

Fewer individuals are often found on very productive sites because the early invading trees and other plants expand rapidly and exclude younger stems.   Also, fewer germination sites are available for initiation on productive sites where depressed microsites become anaerobic after disturbances because evapotranspiration ceases and water tables rise.   Moderately poor sites often contain more species and individuals, since plants arriving first do not grow rapidly and exclude later arrivers.   Extremely poor sites also contain few stems, since a tree can become established only on rare microsites and during unusual weather conditions.

Soil patterns reflect their geomorphologic origin (Lyford, 1964*a,b*, 1974) and parent rock material, as well as rate of weathering—itself a reflection of microtopography (Lyford et al., 1963; Lyford, 1974).   Soil weathering patterns vary depending on the moisture, heat, and organic matter distribution, which in turn vary with microrelief.   Soil patterns are the result of weathering superimposed upon a pattern of chemical and textural variation, so that even within small areas there can be complex patterns which are difficult to decipher (Fig. 6.4A; Lyford, 1974).   The influence of short-range soil variations on stand development is very important because a species' ability to initiate, survive, and grow varies with microsites throughout a stand (Bratton, 1976).   Serpentine outcrops, for example, support vegetation which grows on soils derived from extremely basic material (Brooks, 1987).

Soils derived from transported soil material are often less homogeneous than soils formed by bedrock weathering in place.   For example, glacial outwash is the result of glacial and fluvial action.   Soils developed from outwash vary from fine silts to coarse boulders within only a few feet, with no distinguishable order to the variation.   Glacial till has a more uniform range of textures, but varies in depth, relative compaction, and availability of oxygen.   Up to eight conifer species and many shrubs were found within a small [0.004-ha (0.01-acre)] plot on rocky outwash where a glacier had retreated approximately 70 years before.   The trees ranged from newly germinating ones to over 60 years old; since the canopy had not yet closed, even more species could invade.   On nearby, more productive sites, only three conifers and few shrub species were present; the trees were much larger and were already excluding new individuals.   In still other areas, the bouldery soil was preventing all but a few species from becoming established (Oliver et al., 1985).

Alluvial soils vary from coarse sands near moving waters to fine textures farther away or in abandoned channels.   In floodplains, the growth potential can vary greatly with elevation or soil texture changes and strongly changes over even a few inches of topographic relief (Fig. 6.4B; Buchholz, 1981: Peterson and Rolfe, 1985).   Other soils such as those developing on severely eroded sites, avalanche tracks, or old beaches (Oliver, 1978b) also vary greatly over very small areas.

Isolated increases in soil productivity sometimes occur with no apparent explanation. Trees with significant stemflow of rainwater may increase nutrient concentrations in the soil beneath them (Gersper and Holowaychuck, 1971), or a dead animal may have decomposed and provided a nutrient-rich microsite there. Similarly, in earlier times flocks of passenger pigeons roosted in trees and deposited feces which affected local soil conditions for many years (Lawson, 1709; Savage, 1970).

Variations in soil texture, color, and litter cover affect soil temperatures, which affect the germination and survival of seedlings (Herman, 1963; Baker, 1968; Cochran 1969). Litter cover interferes with germination and therefore excludes new seedlings of some species (Davis and Hart, 1961; Krause et al., 1978; Harrington and Kelsey, 1979), although other species grow vigorously on organic matter (Thompson, 1980). Frost-heaving of rocks can expose mineral soil where wind- or bird-disseminated seeds such as eastern redcedars and junipers can germinate and grow (Livingston, 1972).

Most forest soils throughout the world possess *"mound and pit"* (or *hummocky*) microrelief caused by past windthrows (Figs. 6.4C and 6.5; Stephens, 1956). The exposed *mound* provides mineral seedbeds suitable for germination of light-seeded species, while the nearby *pit* often becomes saturated with water, making the soil anaerobic and unsuitable for root growth.

The elevation between mounds and pits may vary from less than 1 m (3 ft) to over 4 m (12 ft), and the moisture, nutrients, and organic matter conditions within the soil vary dramatically (Lutz, 1940; Mueller and Cline, 1959; Lyford and MacLean, 1966; Armson and Fessenden, 1973; Beatty and Stone, 1986; Bowers, 1987; Ugolini et al., 1989). Windthrows or other events which "plow" the soil alter the root environment and thus change the soils potential to support growth (Hutnik, 1952; Denny and Goodlett, 1956; Stephens, 1956; Lyford and MacLean, 1966; Armson and Fessenden, 1973). The windthrow events create raised, aerated soil conditions and disrupt adverse soil processes, thereby allowing forests to grow in some areas where they otherwise could not (Stephens, 1956; Ugolini et al., 1989). Even years after the obvious hummocks have subsided, the raised microrelief can still have a greater capacity to support roots (Lorio et al., 1972). The microtopography becomes less distinct but still retains some relief and variations in soil conditions for several hundred years. Where forestlands have been cleared and plowed for agricultural purposes, these hummocks disappear; however, the mound and pit microrelief returns after the land reverts to forests and trees are again blown over.

Trees and other plants differ in the location within the mound and pit topography on which they are most often found (Hutnik, 1952; Struick and Curtis, 1962; Bratton, 1976; Thompson, 1980; Beatty, 1984; Deal, 1987; Deal et al., 1989). Advance regeneration or stump sprouts survive on undisturbed soil between new mounds and pits. Trees germinating from seeds arriving after the disturbance often colonize windthrow mound tops, since these locations provide suitable mineral seedbeds (Lyford and MacLean, 1966). Yellow birch in northern New England (Hill, 1977) and Sitka spruce in coastal Alaska (Deal et al., 1991) regenerate from windblown seeds on mineral soils, and so are found frequently on windthrow mounds. When many trees are blown over, evapotranspiration is reduced, causing microdepressions to become saturated and new plants grow only on raised microrelief. Later, when plant cover is reestablished and evapotranspiration has again reduced the water table, shade may be too great for plants to initiate in the depressions. Greater accumulation of organic matter in the depressions further discourages growth (Hart et al., 1962). Once a mound has become established and new trees have grown on it, they too may eventually overturn, perpetuate a windthrow mound, and favor establishment of a new tree in that location. In this way, the spatial arrangement of trees can remain relatively constant for many generations.

Trees are often aggregated on raised microrelief in cool, wet regions or regions with heavy snow (Christy and Mack, 1984). These windthrow mounds and other protrusions contain more favorable soils than the colder, less fertile soils between. They also become exposed sooner during snowmelt and therefore provide a longer growing season. Dead logs or stumps in open areas also become exposed by the melting snow sooner than the general soil surface. Once exposed, these dark bodies reflect heat from the sun, melt the snow in a basin around them, and thus provide nearby plants with a longer growing season.

Depth of litter can be greater on ridges where the lack of moisture inhibits decomposition compared with the base of the slope (Nakane, 1975). These variations in organic matter can give different species competitive advantages during establishment. Some species become established in organic matter where rotting logs and other organic debris accumulate on the forest floor and provide favorable rooting media (Harmon et al., 1986). Plants become established on less decayed logs more often in mesic and bog conditions than in dry conditions (McCullough, 1948). In very swampy conditions, trees may become established only on logs lying above the general soil level (Hall and Penfound, 1943; Lemon, 1945; Dennis and Batson, 1974). A tree beginning on an organic substrate can become a stable tree supported by strong *prop* roots if there is enough light for the newly establishing tree's roots to grow rapidly to mineral soil before the organic matter rots (Figs. 6.5B and C); however, where overstory shade prevents rapid growth of prop roots, the organic matter decomposes first and the tree falls.

Plants sometimes grow primarily on the north or east sides of small hummocks in dry and hot environments in the northern hemisphere. There, shade offers enough relief from direct sunlight to allow plants to survive the initial few days of growth. In some places, this raised microrelief is mimicked by placing wooden stakes on the south sides of newly planted seedlings. Similarly, frost-sensitive species are eliminated from basins or drainages, since frost freely moves down hill. Microtopographic variations which prevent this cold air drainage allow these species to survive and thus change the plant distribution in the ensuing stand.

Infrequent droughts can kill trees in mosaic patterns, reflecting areas of shallow soils and sensitive species (Karnig and Lyford, 1968). Following such droughts, the stand appears as a mosaic of old trees which survived the drought on deeper soils and dead trees with younger trees and shrubs beneath them on shallow soils. The different responses of the young and older trees to subsequent disturbances will cause a distinct pattern in future forests as well.

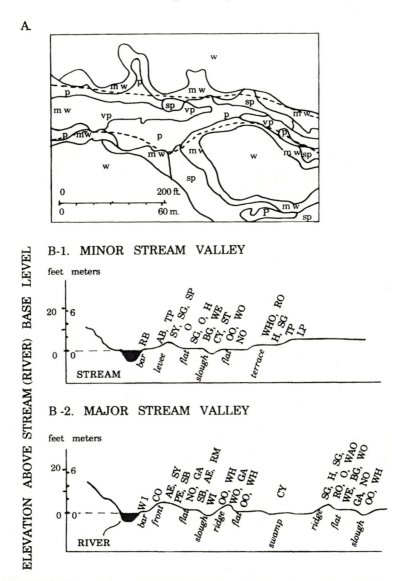

Figure 6.4 Microsite variations created by small-scale variations in soil properties. (A) Changes in soil conditions in a small area in Massachusetts. Dashed lines show soil boundaries given on a published soil map; solid lines delineate soil boundaries found by intensive study: w = well-drained, mw = moderately well drained, sp = somewhat poorly drained, p = poorly drained, vp = very poorly drained soils (*Lyford, 1974*). (B) Schematic cross section of a minor stream floodplain (B-1) and a major floodplain (B-2) in southeastern United States. Alluvial depositions strongly influence microsites, and subtle changes in microrelief and soil conditions strongly affect tree species distribution. (AB = American beech; AE = American elm; BG = blackgum; CY = cypress; CO = eastern cottonwood; GA = green ash; H = hickories; LP = loblolly pine; NO = nuttall oak; O = oaks; OO = overcup oak; PE = pecan; TP = tulip poplar; RB = river birch; RM = red maple; RO = red oak; SB = sugarberry; SG = sweetgum; SP = spruce pine; ST = swamp tupelo; SY = sycamore; WAO = water oak; WE = winged elm; WH = water hickory; WHO = white oak; WI = black willow; WO = willow oak.) (*From Dr. J. D. Hodges, Mississippi State University.*) (See "source notes.")

C.

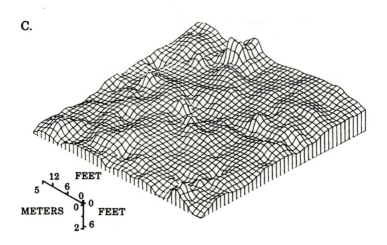

Figure 6.4 (*Continued*) (C) Windthrows create mound and pit microrelief and strongly influence microsites and soils in coastal Southeast Alaska (see Fig. 6.5; *Bowers, 1987*). *(See "source notes.")*

Previous plant and seed distributions. Spatial patterns, species compositions, and age ranges in subsequent cohorts are influenced by the distribution of plants before a disturbance. These plants in turn have been influenced by the previous overstory structure (Ffolliott and Clary, 1972; Veblen et al., 1979). Previous plants provide seeds, sprouting stock, and advance regeneration for the new stand; they exclude newly regenerating trees by acting as preestablished competition (Shirley, 1945; Smith, 1951; Wardle, 1959; Beck, 1970; Ruth, 1970; Webb et al., 1972; Marks, 1974; Daniels, 1978; Veblen, 1982; Maguire and Forman, 1983; Osawa, 1986); and they may leave allelopathic chemicals in the soil which inhibit germination and/or growth of certain species (Table 6.2).

Regardless of the form of vegetative reproduction—root or stump sprouts, rhizomes, bulbs, tubers, or advance regeneration—the spatial pattern and species which develop after a disturbance will first be based on the distribution pattern of parent material before the disturbance. The distribution of parent stock before the disturbance is itself the result of other factors. For example, a clumped pattern of newly sprouting stems would develop if a stand with sprouting species were destroyed while it was young enough to still retain a clumped distribution. If the stand was not destroyed until much older, after the surviving stems had become more regularly spaced, fewer, more regularly spaced, sprouting stems would grow.

The overstory-understory relation is a casual one, not a close, coevolved or obligatory relation between understory and overstory species (Carleton and Maycock, 1981). Both direct competition from existing plants and patchy germination caused by litter accumulation (McConnell and Smith, 1971) can influence the germination and growth of new trees and other vegetation and thus strongly influence subsequent tree distribution patterns for a long time (Shipment, 1955; Allen, 1956; Maguire and Forman, 1983). Understory herb cover fluctuates with time of year (Bormann and Buell, 1964); conse-

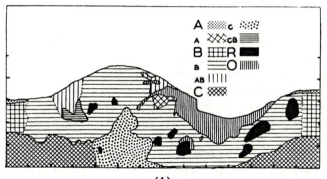

(A)

(B)

(C)

Figure 6.5 Windthrows and root mounds. (A) Schematic of microsites created by windthrow mounds resulting from overturning of trees. Capital letters designate undisturbed horizons: A = zone of eluviation; B = zone of illuviation; C = unweathered or parent material (corresponding small letters indicate disturbed material); O = organic debris. R = rocks. Scale at 30.5-cm (1-ft) intervals. (*From Lutz, 1940; Lutz and Griswold, 1939.*) (B) Yellow birch growing on top of tree blown over in hurricane of 1818 in central Massachusetts (Harvard Forest). The tree blew over to the back right; the person is standing in a pit. Mound and pit microrelief is still evident in the photograph taken 155 years later. The birch germinated shortly after 1818 and survived because enough trees blew over to allow the birch to grow vigorously and establish prop roots before the mound eroded away. (C) Sitka spruces in coastal southeastern Alaska. Such trees can be clumped on old mounds where windthrows create favorable microsites for germination and growth. (See "source notes.")

quently, herbs may compete aggressively and exclude tree seedlings during some times but appear to leave unoccupied areas of forest floor during other months. Certain species of polytrichum moss are favorable to eastern white pine germination, since they allow seeds to penetrate to the mineral soil; and they keep soil surface temperatures low, but they do not compete excessively for moisture (Smith, 1951).

Often one plant species is inhibited from growing near or beneath another by allelo-pathic chemicals produced by the repelling plant (Table 6.2; Stickney and Hoy, 1881; Muller, 1966; DeBell, 1971; Peterson, 1971; Tubbs, 1973; Rice, 1974; Gabriel, 1975; Gant and Clebson, 1975; Younger et al., 1980, Thibalut et al., 1982; Ward and McCormick, 1982). If a partial disturbance leaves a plant which produces an allelo-chemical, an area free of susceptible species develops beneath it in the next cohort. Similarly, advance regeneration of a susceptible species is confined to areas without an overstory tree with allelopathic properties. An inhibiting chemical may remain active in the soil even after a disturbance has removed both the susceptible and allelochemic trees of the previous stand. Some grasses and other understory species may have allelopathic effects on some tree species (Rietveld, 1975; Stewart, 1975; Walters and Gilmore, 1976, Horsley 1977a,b; Priester and Pennington, 1978; Fisher, 1980; Norby and Kozlowski, 1980).

The distribution of advance regeneration and buried seeds is also partly controlled by dispersal agents such as winds, animals, and floodwater patterns and by the proximity of seed sources.

Influence of the disturbance

The type of disturbance influences the subsequent tree patterns by determining which species and individuals survive. Residual trees affect invasion patterns, survival, and growth rates of trees of younger cohorts.

Disturbances by wind (Oshima et al., 1958; Iwaki and Totsuko, 1959; Sprugel, 1976), root rots (Cook, 1982), and some snow avalanches act as waves through a stand, leaving characteristic spatial-age distributions of progressively older trees in parts of the stand first impacted. The disturbance also changes the microtopography, which greatly affects the regeneration pattern.

Figure 4.8 implied that each disturbance type creates a uniform stand with different amounts of the forest floor removed. In fact, each type of disturbance creates a mosaic of small areas with different amounts of the forest floor and soil removed (Hofmann, 1924). After windstorms or logging operations there are areas where the organic matter and all above- and below-ground vegetation is removed. Wildfires can skip over forest floor vegetation, leaving patches of it alive (Bonnicksen and Stone, 1982). The amount of forest floor left intact after each disturbance would appear as in Fig. 6.6, with the more severe disturbances removing larger proportions of the forest floor.

Even where advance regeneration and sprouting stumps and roots are not present, the amount of forest litter left after the disturbance can inhibit the invasion of new stems and influence the pattern of invasion, especially if the disturbance reduces the amount of area with appropriate seedbed conditions. Harper et al. (1965) have found a limited number of microsites in mineral soil in which a seed of a given size can germinate. In other areas the seed may not come into close enough contact with the soil to germinate.

Table 6.2 Some Allelopathic Trees, Shrubs, and Other Plants and Some Plants They Affect

Allelopathic species	Some plants affected
Trees	
Black cherry	Red maple
Eucalyptus	Shrubs, herbs, grasses
Hackberry	Herbs, grasses
Juniper	Grasses
Oaks	Sweetgum, herbs, grasses
Sassafras	Elm, maple
Sugar maple	Yellow birch
Sycamore	Herbs, grasses
Walnut	Trees, shrubs, herbs
Shrubs	
Bearberry	Pine, spruce
Elderberry	Douglas-fir
Laurel	Black spruce
Manzanita	Herbs, grasses
Rhododendron	Douglas-fir
Sumac	Douglas-fir
Other Plants	
Aster	Sugar maple, black cherry
Bracken fern	Douglas-fir
Club moss	Black cherry
Fescue	Sweetgum
Goldenrod	Sugar maple, black cherry
New York fern	Black cherry
Reindeer lichen	Jack pine, white spruce
Short husk grass	Black cherry

SOURCE: Fisher, 1980.

The soil may contain too much pore space ("fluffed" soil) to allow sufficient capillary movement of water to the soil surface and to the germinant. Rocks and organic litter can prevent seeds from reaching moist soil.

A hot fire or alluvial deposition may reduce the effects of residual allelopathic inhibitions if the inhibiting chemical is either decomposed or buried by the disturbance. The reverse condition occurs following the draining of mangrove swamps. The decomposing organic matter becomes extremely acidic and inhibits plant growth (Pons et al., 1982). Disturbance characteristics may also stimulate certain regeneration mechanisms; for example, fires stimulate germination of seeds of pin cherries, fireweeds, and lodgepole pines and growth of bracken ferns from rhizomes; burial of vine maple and cottonwood stems stimulates growth of roots and shoots along the buried stem.

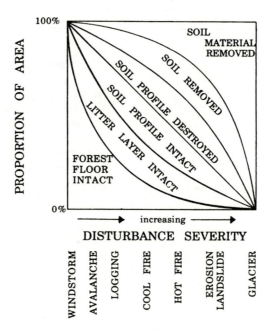

Figure 6.6 Specific disturbances often leave characteristic patterns of destruction (Fig. 4.8) with a variety of regeneration conditions, but proportionally different amounts of area in each condition.

Influence of events after the disturbance
The changes in temporal and spatial patterns of invading stems following a disturbance will be discussed in Chap. 7.

Applications to Management
The desirable structures in many single-cohort stands have been caused by the wide range of tree ages and uneven spacing of stems. Plantations of single species and ages, on uniform sites, may contain such uniform spacings and ages that growth patterns, stand structures, and wood properties develop which are dramatically different from the preceding, natural stands. Forest management goals of restocking stands within a short time (such as 5 years) after a timber harvest can create very unusual stands on many sites and may be unobtainable on some. The long time before tree closure in many natural stands has created much longer periods of herb and shrub growth than found in plantations. These herbs and shrubs may have had beneficial effects on soil development, tree pathogen eradication, and survival of browsing and grazing animals.

On the other hand, the uniform age and spacing of single-cohort plantations allows control of a stand's species composition, manipulation of the more uniform stands, growth of more uniform tree sizes and wood properties, and attainment of quicker financial returns. In general, stands of more uniform age offer more efficiency of management but may require more careful thinning to avoid stagnation (Chap. 8). Because of rapid site occupancy by fast-growing seedlings in many plantations, it may not be worthwhile to replace a few dead seedlings even one year after planting, since these trees will probably be excluded by the older trees before they become merchantable.

The patterns of initiation will have a great effect on the stem quality of individuals. Trees growing without trees of the same age next to them are more tapered, and the branches are larger. Stands which are excessively clumpy during the stand initiation stage have two categories of trees: (1) trees which grow in the centers of clumps and, consequently, are often small in diameter have small limbs, and are very cylindrical; and (2) trees which grow on the edges and, therefore, are very tapered and may have large limbs on one side. Those trees which initiate without close neighbors are branchy and in some instances may grow into "wolf" trees. These trees will always be of poor quality and take up much growing space; they generally should be removed during early thinning. Unless the initial spacing of natural stands is changed early in the life of the stand by precommercial thinning, the clumpy nature will affect growth and stem quality of the individuals for many decades.

The initial spacing of a stand can be manipulated to varying extents before, during, and after harvesting of the previous stand. Before harvest of the previous stand, the presence of advance regeneration—of future crop species or competitors—can be encouraged by thinning of the overstory (Loftis, 1983, 1985) or discouraged by cutting, browsing, underburning, or chemical treatments. Spatial patterns of trees seeding in after the harvest can be influenced by eliminating understory species which otherwise survive the harvest and exclude newly seeding plants. These competing weeds are generally less vigorous in the shade than in the opening after timber harvest. During harvest and subsequent site preparation, the disturbance patterns can be manipulated to eliminate or space advance regeneration and to create mounds or other favorable microsites and seedbeds at appropriate spacings. After harvest, spatial patterns can be manipulated by planting seedlings or seeds of desirable trees or by killing undesirable ones. Depending on the pattern of interaction among species, it may be unnecessary to eliminate certain noncrop species if it is known they will not compete vigorously with crop trees.

On droughty, cold, or otherwise harsh sites, it is sometimes desirable to keep some other vegetation where it provides shade and frost protection, making more microsites acceptable for tree germination and seedling growth. At other times, herbaceous vegetation competes for favorable microsites on harsh sites; the chances of tree germination and survival decline, and the stand remains very clumpy.

Many times, a desirable spatial pattern can be created by manipulating conditions in the previous stand before it is removed. Species such as oaks which grow well from advance regeneration can form dense regeneration when they begin in disturbed areas in the previous stand. Intentionally creating openings through partial cuttings in one stand may increase advance regeneration in the following stand. If loggers fell trees and leave the slash in clumps of advance regeneration, some of the advance regeneration in the clumps is killed and a more uniform spacing will be favored in the next stand.

CHAPTER 7

STAND INITIATION STAGE:
SINGLE-COHORT STANDS

Introduction

During the stand initiation stage, some plants die and others increase in size while others continue invading. The environment, growth pattern, and size of each plant change more dramatically during this early growth stage than during any other period—with many species becoming larger by an order of magnitude within a short time.

A stand's development pattern until the next major disturbance is largely determined during this stage. Which species invade, when, and in what spatial patterns they invade, and which ones survive are key determinants of a stand's future development. The invasion and survival pattern is also strongly influenced by the site, available species, type of initial disturbance, and other factors (Canham and Marks, 1985).

Factors Affecting Early Stand Development

Stand development patterns during stand initiation differ from later stand development when competition is primarily among woody species. During the stand initiation stage:

1. A wide range of herbaceous and woody plants grow together with more varied growth patterns than found in woody trees and shrubs.

2. Many more species, as well as individuals, interact than during later stages.

3. The relative range of ages of competing plants is greater than during the later stem exclusion stage.

4. The small sizes of all plants magnify the effects on growth caused by subtle variations in the microenvironment.

5. The stand environment changes rapidly during this stage compared with that of older stands, because the biogeochemical cycle is still reacting to the previous disturbance (Bormann and Likens, 1981) and because the growing plants are changing it.

6. The growing space is not completely occupied by existing plants as it is during later stages.

Many growth patterns of competing vegetation

Grasses, herbaceous plants, ferns, and other nontree species grow in more diverse patterns than those found among tree species. They all compete for growing space soon after a disturbance. At times nontree plants can permanently exclude trees from an area. In other cases they can temporarily inhibit them and/or alter the eventual tree species composition. At still other times, nontree species can aid establishment of trees and promote development of a forest stand. The herbaceous and woody plants vary dramatically in growth patterns, environments to which they are adapted, life spans, and subtle chemical and other mechanisms of competition. The mechanisms of competition among these various plants is beginning to be studied in depth (Harper 1977; Hara et al., 1993; Hara and Wakahara, 1994).

Competing plants can be of diverse evolutionary origins, although some with quite different growth patterns are closely related. Ferns, grasses, herbaceous angiosperms, woody shrubs, and trees have followed evolutionarily different pathways. Ferns have existed for 400 million years. Modern trees, shrubs, and many similar herbaceous plants became prominent later, with conifers becoming prominent 240 million years ago, and broadleaf plants becoming prominent about 135 million years ago (Scagel et al., 1969). Grasses became prominent only about 60 to 70 million years ago.

A single genus can contain both herbaceous and woody species, and thus include quite different growth forms. Plants of the birch (Betulaceae), oak (Fagaceae), elm (Ulmaceae), and some other dicot families are found only in woody forms. All conifers are woody. Monocots are technically nonwoody. Monocots such as orchids, lilies, rushes, and sedges are obviously nonwoody. Bamboos and palms are monocots and are technically nonwoody, but they develop hard stems and treelike forms. Many "herbaceous" and shrub species evolved through a genetic loss or change of certain anatomical features.

Lesser vegetation can roughly be classified as annual herbs, perennial herbs, and shrubs. Perennial herbaceous plants, including grasses and ferns, often grow rapidly from seeds, tubers, rhizomes, stolons, or bulbs near ground level each year. They grow upward and/or outward rapidly at first and occupy available growing space, competing vigorously with small trees and shrubs. Unlike woody plants, their height growth is not cumulative, and each year—or every several years for a few species—height growth must begin again from ground level. Growth patterns of the herbaceous stems can be similar to tree growth patterns (Halle et al., 1978). At other times, the terminal and lateral primary meristems of herbs produce flowers later in the season instead of branches and shoots in response to daylength or other stimuli (Harper, 1977). With no aerial vegetative buds to continue shoot growth in subsequent seasons, these shoots die. Aerial parts of grasses are generally the equivalent of leaves, which elongate from buds—or their anatomical equivalent—beneath or at ground level. Although considered herbaceous, stems of many such plants are quite woody. Bamboo species can offer extreme competition with trees (Tanaka, 1988; Cao, 1995). Perennial herbs can maintain a rhizome, stolon, bulb, tuber, or similar regenerating mechanism at or beneath ground level and are not dependent on germinating from seeds each year.

Woody shrubs accumulate growth as trees do and often grow more rapidly in height each year than many trees. Unlike trees, shrubs do not accumulate height growth near the top of the previous year's growth. Instead, shrub shoots tend to bend over each year; and the next year's shoots arise from the old shoot base, from along the old shoot, or from the bent-over tip, depending on the inherited morphology of the species.

Shrubs and perennial herbs sometimes regenerate from rhizomes—modified stems growing beneath the soil—or stolons—similarly modified stems growing on the soil surface. Rhizomes and stolons develop buds at intervals, which then form stems and roots and develop into new plants. As evolutionarily modified tree stems, rhizomes and stolons have an advantage over aerial stems in not being limited in size by gravity.

Annual and perennial herbs and shrubs which produce many seeds when young can genetically adapt more rapidly than trees. These herbs can pass through numerous generations during the same time period required for a single generation of trees. Herbaceous and shrub species often have competitive advantages over trees when the obvious height advantage of taller vegetation is not significant. Consequently, the smaller vegetation is commonly found after a disturbance but before trees grow tall, where trees have been excluded by the herbs and shrubs, when adequate growing space is available near the forest floor later in a stand's age (understory reinitiation stage), or where the seasonal lack of soil growing space on extremely droughty or waterlogged sites prevents vigorous tree growth. Annual herbs are generally adapted to rapid colonization of disturbed areas. Their ability to produce seeds during the year they germinate allows them to increase in numbers dramatically. Seeds of some annuals can remain dormant in the soil until another disturbances occurs—often not until many decades later—while seeds of other annuals cannot survive more than one or several years of dormancy. Some annuals and perennials grow and produce seeds during the favorable season and thus avoid extreme conditions. Herbs, shrubs, and trees which set seeds less frequently or maintain perennial, aboveground tissue survive where they can endure the year-round environment.

Grasses produce very numerous, surficial roots. These roots rapidly absorb the moisture of light summer rains before it can penetrate to the deeper roots of other species, thus giving grasses a competitive advantage on droughty sites. Polystrichum moss and moist mineral soil provided favorable sites for germination and early growth of white pine in New England, while grasses, pine litter, lichen, and dry mineral soil microsites were unfavorable. The moss and moist soil dissipate the soil surface temperatures, while the other substrates do not and so expose the germinants to lethal temperatures (Smith, 1951).

The effect of other plants on survival and growth of young trees depends on the environment and on the specific growth forms of the competitors. Annuals and perennials offer intense competition where soil moisture or aeration is limiting. In such conditions, woody plants often do not survive the first critical years and take advantage of their ability to accumulate growth. Annual and perennial plants can continue to dominate on these sites because each year they begin growth anew, or from a pre-existing root system, and exclude woody vegetation. Woody plants can dominate such sites if they become established and occupy soil and aboveground growing space in advance of other vegetation forms.

Woody plants survive root competition from herbaceous plants where soil moisture is less limiting. Shade and other competition from annuals and perennials slow woody plant growth and may kill some individuals; however, enough generally survive to occupy more growing space each year. Even when woody plants invade moist, well-drained sites later than perennials, they often survive and dominate, since the other plants relinquish some or all growing space each year. In fact, woody plants sometimes grow better if they invade later, after other plants have favorably altered the microenvironment. Of course, later invasion can adversely affect tree growth where other vegetation shades it, offers root competition, or produces inhibitory, allelopathic chemicals.

Many species and individuals

More species and individuals are generally found during the stand initiation stage than during later stages, since the unoccupied growing space allows many individuals to invade. Relatively few species and individuals survive as the plants increase in size, fill the growing space, and compete during the stem exclusion stage (Alaback, 1982a,b; Zamora, 1982; Oliver et al., 1985). Many individuals and species are interspersed during invasion and ensure diverse patterns of interaction. Subtle variations in ages, spacings, microsites, and genetic makeup create a variety of development patterns as the plants grow.

Species can be interspersed, or they can exist in distinct clumps of single species, depending on the initial disturbance patterns, innate site characteristics, and/or species regeneration mechanisms. Neighboring plants can be either closely or widely spaced for similar reasons. Favorable microsites for germination of a species may allow many individuals to become established, but the plants may not grow vigorously because of close competition or poor soils. Other microsites may allow only a few individuals to initiate, and these may grow well or poorly.

Some of the randomness surrounding which species will eventually dominate is eliminated by the large numbers of individuals and species present during stand initiation.

These numbers allow many combinations of species, spacings, and ages; therefore, species can survive and grow even if they are able to dominate only under a combination of factors whose probability of occurrence is quite small.

Broad relative range of plant ages

Because all plants are so young, plants may be competing with others many times older or younger than themselves during stand initiation, even in single-cohort stands. The relative differences between ages decrease as total ages increase.

Rapid morphological changes in young plants create a more fluctuating age pattern during stand initiation than occurs during later development stages. Older annuals and perennials may be dying while some trees shift from juvenile to mature growth habits and other plants invade. A truly all-aged stand can exist during this stage, with stems of every age since the last disturbance. The age and size distributions near the end of this stage can vary (Table 6.1 and Figs. 6.2 and 6.3).

Small sizes of plants compared with their physical surroundings

The small sizes of plants during stand initiation make them susceptible to factors that do not affect larger, older trees. In later development stages, trees are large and shade smaller plants or occupy larger areas of soil than herbaceous plants—usually giving the trees a competitive advantage. When first initiating, however, these trees are no larger than herbaceous plants and can be out-competed by the herbs. In fact, herbaceous plants often grow taller than trees during the first year. In this way, Douglas-firs—which can grow to over 65 m (200 ft) tall—can be killed by blackberries which rarely reach 2.5 m (8 ft) tall; balsam firs and red spruces are killed by raspberries; and grasses—which may not grow to over 1 m (3 ft) tall—can exclude eastern white pines (Smith, 1951) and prevent them from reaching their potential of over 65 m (200 ft).

A plant's roots and stem occupy only a very small area at first. Fluctuations in the environment of this small area dramatically affect growth and survival of the plant. Microsites can vary even in an area quite uniform in large scale (Fig. 7.1). This microsite variation can be caused by geomorphic, drainage, or disturbance factors (Lutz, 1940; Lyford and MacLean, 1966; Lyford, 1974). Relief of a few centimeters can dramatically change the germination and rooting environment of a seed (Hart et al., 1962; Lorio et al., 1972). The soil can be raised and subject to drying; "fluffed," or poor in water retention because of the loose structure; organic and unsuitable for certain plant roots; depressed and anaerobic; or baked by a severe fire and unsuitable for root penetration or nutrient release. Very hot fires can bake the soil where a dried, downed log has burned. Tree species may be unable to become established where herbaceous vegetation exists in an area (Shirley, 1945; Daniels, 1978). Seeds must land in places where their shape and size are compatible with the soil surface to germinate successfully (Harper et al., 1965). Before their roots penetrate deeper, newly germinating plants are subject to the extreme moisture fluctuations near the soil surface. Similarly, new plants are subjected to the temperature extremes found just above the soil surface. Small plants can be injured or killed by freezing if they are growing in depressed frost pockets or frost drains subject to convective cooling or exposed areas subject to advective cooling.

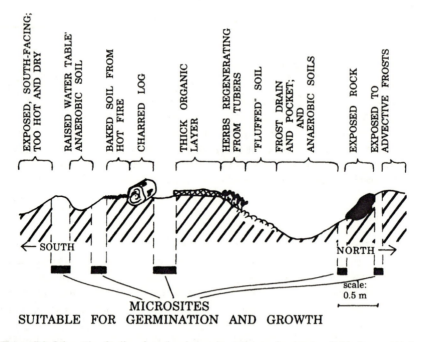

Figure 7.1 Schematic of soil surface showing various microenvironments suitable for germination and early growth of seedlings. Only limited places on a soil surface are conducive to successful regeneration of plants at a given time. Different microsites become more favorable and others become less favorable as plants grow. Other places are conducive to growth of sprouts or advance regeneration, depending on the location of former stumps, roots, and advance regeneration.

Extreme heat on a microsite with a southern exposure—especially on a dark soil—can reduce growth by increasing respiration or overheating (and killing) the stem at the ground line. Burned sites, mineral soils, and litter have dramatically different temperature fluctuations, which affect the germination and growth of various species differently (Herman, 1963; Baker, 1968; Cochran, 1969). Mineral soil and cracks exposed by frost-heaved rocks can provide suitable microsites for germination (Livingston, 1972). A microsite shaded by a large rock, a log, or slightly older vegetation can reduce growth by reducing photosynthesis or increase growth by improving plant moisture retention. Stumps, logs, or other raised areas can cause snow to melt faster and give nearby vegetation a competitive advantage. A plant can be crushed if growing where a dead plant of the same or an older cohort falls on it. Heavy rains can kill newly developing seedlings, especially on mineral soils (Yamamoto and Tsutsumi, 1985). Suitability for growth on these and other microsites can change dramatically during a season.

Small sizes also make young plants very susceptible to damage by animals. Large trees are not so severely browsed, since most of their succulent foliage and younger twigs are out of reach of browsing animals and the lower stem is often protected by thick bark. Both woody and herbaceous plants in the stand initiation stage are small and succulent enough to be killed or severely damaged by animals. Some herbaceous and woody plants contain chemicals in young, tender shoots which make them less palatable to animals. Animals generally seek a varied plant diet. Consequently, they will browse (and often kill) nearly all individuals of a rare species, but only a portion of the more numerous ones, reducing species diversity and inhibiting the chance of a locally rare species to dominate.

At first the plant is small, and very local conditions are influential on its growth. A tree's root and aboveground systems expand as it grows, covering larger areas and more microenvironments. Changes in light, moisture, heat, oxygen, or nutrients immediately surrounding its original point of initiation become less important as a tree grows, since the remaining shoot and root system can compensate for these changes. Stem growth upward first allows the terminal to avoid the extreme heat and cold just above the soil surface and later allows the trees to extend above other plants, thus receiving more sunlight. Outward growth of branches provides more photosynthetic area and modifies the soil surface environment. Trees also become more resistant to environmental extremes through development of thick bark and other features. Consequently, a tree is not so adversely affected by unfavorable conditions near its point of initiation as it grows larger. It is also not so positively affected by favorable conditions. The large size also reduces the proportion of plant destroyed with each bite of an animal and decreases the palatability and access animals have to the plants—thus changing the impact of browsing.

Rapidly changing environment

A stand's environment constantly changes; however, changes occur most dramatically during stand initiation. Previously unfavorable microsites may become favorable, and those previously favorable may become less so. The environment surrounding each plant is changing in response to three factors: the newly initiating plants, the disturbance which began the stand, and the weather fluctuations throughout the year and between years.

Plants moderate the soil surface environment as they invade a newly disturbed area. They keep the soil surface cooler during the day by casting shade and warmer and frost-free at night by reducing reradiation of heat and the flow of cold air. Species unable to survive immediately after a disturbance on extremely hot or cold sites often survive after other plants reduce the midday heat (Wahlenberg, 1930; Gaines, 1952). Similarly, species survive and grow in depressed areas after other plants reduce the cold air drainage. Older vegetation also reduces wind speed by blocking the wind. This blockage, with evapotranspiration from new plants, increases humidity as well. Sunlight reaching the soil surface is reduced; however, the shade is "low shade" (Fig. 3.6) and the light reduction may not be critical to growth.

The below-ground environment and growing space also change as the plants invade and grow. Reduction of sunlight on the forest floor keeps the soil from becoming as warm as it would be in unshaded circumstances. The soil becomes dry at greater depths through evapotranspiration as roots penetrate, even though shade may keep the surface soil moist. Increased evapotranspiration reduces the below-ground soil water volume, lowering the water table and zone of saturated, anaerobic soil and making the soil more suitable for root growth. Raised microsites previously favorable for plant growth because of available soil oxygen become droughty and less favorable with the decrease in soil moisture. Concave microsites whose soils were previously saturated and unfavorable become more mesic and suitable for growth. Growing plants also increase the nitrogen, organic matter, cation exchange capacity, and structure of the soil, making conditions favorable for growth of different species (Voigt, 1968). Where the disturbance removes all soil material and leaves only bedrock, primitive lichens and other plants grow for a long time before conditions are acceptable for other plants. At other times soil conditions are quite acceptable for immediate invasion of woody species. Insect as well as bacterial and fungal populations change with the plant communities. Immediately after a fire in the Cascades of western Washington, newly regenerating seedlings of Douglas-fir often become infected with teapot fungus (*Rhizina undulata*; Hepting, 1971); and in the eastern United States, unless replanting is delayed for several months, replanted pine stands become infested with the still-large populations of the Pales weevil (*Hylobius pales*) or the pitch-eating weevil (*Pachylobius picivorus*), which build up in the dead tops of previously harvested trees (Baker, 1972). Allelopathic chemicals from one initiating species can also inhibit other species.

Superimposed on the changing microenvironment created by the newly growing stems are the changes induced by the disturbance (Hofmann, 1924). A large disturbance stimulates changes in growing space which continue for many years. The organic matter killed by the disturbance slowly decomposes; the decomposition rate of existing organic matter changes; the nutrient status and exchange within the forest floor alters (Marks and Bormann, 1972); the soil structure, moisture, and oxygen conditions vary (Bormann and Likens, 1981; Ugolini, 1982); and the microenvironment continues to change.

Most disturbances do not remove all organic matter from an area. Dead root systems and foliage remain after timber harvest, fires, or tornadoes. Dead, fallen stems as well as foliage and upturned roots remain after large windstorms or avalanches. Dead, standing snags and possibly foliage are still present after fires or insect attacks. No organic matter is usually present after landslides, erosion, or glacial activity.

The dead organic matter decomposes over a few or many years, depending on its characteristics and the climate (Buchanan and Englerth, 1940; Kimney and Furniss, 1943; Shea, 1960a,b; Toole, 1965a, 1966; Wright and Harvey, 1967; Yoneda, 1975a,b; Grier, 1978a,b). Generally, the initial effect of rotting stems is to shade, insulate, and add organic matter to the forest floor, which increase the carbon/nitrogen ratio and reduce the rate at which nutrients become freely available in the soil. Decomposition of dead logs is closely related to their contact with moist ground, the availability of oxygen, and the environment (Harmon et al., 1986). On the other hand, a fire may release many nutrients previously tied up in organic matter.

Addition of more organic matter initially reduces the rate of release of nutrients from existing organic matter both by insulating the soil and by increasing the carbohydrates. This reduction can be overridden by other effects of a disturbance, and disturbances generally increase the decomposition rate of organic matter. Disturbances often mix organic matter with mineral soil, having an effect similar to plowing, and accelerate the decomposition rate. Increased sunlight on the forest floor warms it, creating a more rapid rate of microbial respiration and, consequently, decomposition. The increased warmth and other factors sometimes promote increased nitrogen fixation by microorganisms, thus reducing the carbon/nitrogen ratio and further increasing organic matter decomposition.

Nutrients are often more abundant in the soil water after a disturbance. At other times they are more tightly held within the living plants or dead organic matter. After disturbances, nutrients can be leached from the soil to the groundwater and to streams (Jordan et al., 1972; Vitousek et al., 1979; Bormann and Likens, 1981). As vegetation regrows, this release subsides. The environment created by a disturbance causes this release by changing the soil processes (Ugolini, 1982), by increasing the decomposition rate of organic matter, by eliminating the growing plants which otherwise absorb these nutrients, and by increasing nitrogen-fixing microbial activity in the soil.

Whether rapid decomposition of organic matter and release of nutrients increases or decreases growth of the new stand is largely dependent on climatic conditions. In warm, moist environments, decomposition rates can be so rapid that nutrients are released and leached beyond the rooting zone before roots of newly invading vegetation reabsorb them. Alternatively, in cool or dry environments, decomposition following a disturbance proceeds more slowly and invading plants reabsorb more nutrients before they are leached. In cool climates such as found in Alaska, accelerated decomposition after a major disturbance helps prevent the organic matter from accumulating enough to form bog *(muskeg)* conditions (Ugolini et al., 1989).

Disturbances such as erosion, plowing, and windthrows immediately increase or decrease growing space by altering the soil's physical properties. In addition, root systems of trees killed by a disturbance slowly decompose and the soil structure maintained by these roots collapses (Burroughs and Thomas, 1977; Ziemer and Swanston, 1977). The significance of this collapse of old roots depends in part on how rapidly roots of new trees reoccupy the soil. Disturbances also change the soil's physical environment by creating anaerobic conditions upon reduction of evapotranspiration. Unstable soil conditions can also lead to soil slumping, erosion, landslides, and other secondary disturbances.

The above changes in the environment of newly initiating stems must be superimposed upon normal daily, seasonal, and periodic fluctuations. For example, a newly exposed microsite with a southern exposure can be favorably warm during early spring or late autumn, but adversely hot during midsummer. Similarly, a depressed microsite can be prohibitively cool in early spring or late autumn, or too wet following the disturbance, but very favorable for growth at other times.

Presence of available growing space

A plant increases its growing space at the expense of other plants in all stages except during the stand initiation stage. During stand initiation, the growing space has not been fully occupied, and new plants invade the growing space without necessarily detracting from the growing space already occupied by others.

Pattern of Stand Development in the Stand Initiation Stage

Green plants beginning after a disturbance have certain commonalities despite wide differences in forms and rapidly changing environments. They all compete for approximately the same growing space. Competing plants have generally not coevolved any strong mutualistic associations, either by species "preying" on others or by species giving more than incidental "positive assistance" to others.

As a plant initiates through germination, sprouting, accelerated growth of advance regeneration, or other mechanisms, it first experiences a period of free growth (open growth) as it expands and fills unoccupied growing space without direct plant competition. During this time, it may die from heat, frost, drought, or other environmental limitations. If it survives, its growth rate depends on the environmental conditions, the genetic predisposition of the individual, and the vigor of the seed, stump, root, or advance regeneration from which it originated. At some point, it may be genetically predisposed to curtail expansion, set seeds, die, or alter growth in other ways. Barring this curtailment, the plant later enters a period of competition when other plants have filled growing space which would otherwise be directly available to the subject plant.

Free growth (open growth) period

The many forms of trees, grasses, herbs, ferns, and other plants create a variety of growth patterns for filling growing space (Dyrness, 1973; Zamora, 1982; Halpern, 1987). Annual plants compete vigorously for one season, or part of a season, and then die—relinquishing growing space which can be filled by other existing plants or by new ones. Herbaceous perennials, including grasses, can die back to their roots, similarly relinquishing aboveground growing space but maintaining that controlled by roots. Dead stems of these herbaceous plants can crush and shade other plants, further releasing growing space. On the other hand, woody trees and shrubs accumulate growing space by adding new growth to that gained in previous years. They do not need to expand each year to reoccupy the same growing space; they maintain this space and expand farther each year.

Energy for growth in young plants comes from the originating seed, stump, or root and can give some young trees more capacity to expand than others. As plants grow older, their energy for growth is more directly related to their growing space occupancy. Simultaneously, available growing space changes with time and as the plant grows, and only those plants capable of expanding to occupy other microsites survive. For example, a tree may germinate on a south-facing, raised microsite in the spring when only these areas are warm or aerobic, but the tree survives only if it expands its roots and branches to other microsites before the raised microsite becomes too hot and dry in summer.

Plants which become established first during the stand initiation stage may not ultimately survive. As the growing space becomes filled, some plants continue both to expand in the existing growing space and to compete for that of neighbors, while other plants die (Fowells and Stark, 1971; Tappeiner and Helms, 1971; Johnson 1981a,b; Glitzenstein et al., 1986). Death results from competition or the effects of the physical environment, or because the plants are genetically predisposed to dying (e.g., annuals). Dying plants relinquish growing space and allow others to invade and surviving ones to expand; however, the environment is different for those invading later. Later herbaceous invaders may die and relinquish the growing space again. Trees may continue to grow and expand. Invasion during this stage exhibits some elements of the "relay floristics" pattern (Fig. 5.1), since some plants established first modify the environment and allow later plants to survive. There is no evidence that the pattern is the result of coevolution among species, however.

Slight size, spatial, or microsite advantages in the occupation of growing space allow some individuals to survive while others die. Even within a species, older individuals may not gain enough competitive advantage to dominate. In one study a hemlock stand established by shelterwood cutting had an age range of 9 years (Jaeck et al., 1984; Table 6.1); however, trees which invaded during the first 2 years and the last 3 years were shorter than those invading during the middle period. Plants invading first may lose their vigor if they are subjected to harsher site conditions, residual pathogens, allelochemicals from nonwoody vegetation or the previous stand, or other influences. Once weakened, they may not regain the vigor of slightly younger trees. In another study, Douglas-firs first invading a xeric, windy meadow did not grow as rapidly as those invading in later years in their shelter (Agee, 1987). First invaders also grow more slowly if their wide spacing causes the same reduced height growth exhibited by some trees at wide spacings (Fig. 3.10). Late invaders grow between older trees and are effectively at narrower spacings.

Competition

Plants which survive during the free growth environment expand until they occupy all available growing space and compete with their neighbors. The free growth period is shorter where competing vegetation is closer (South and Barnett, 1986). This early competition is similar to the stem exclusion stage; however, the varied growth patterns of the nonwoody and woody plants and their immature ages create many differences. Unlike trees in the stem exclusion stage, many herbaceous plants die completely or die back to the root collar each year even when in a competitively advantageous position. Consequently, growing space is released and new plants initiate, or already existing ones expand.

Woody plants initiate and dominate after a disturbance in different ways. The relative frequency of each way is uncertain:

1. They initiate in an area first from seeds, sprouts, or advance regeneration and occupy suitable microsites.

2. They invade suitable sites previously occupied by annuals or perennials which die back to the root collar.

3. They invade older herb or shrub communities where existing plants have lost their vigor.

4. They invade areas where a few trees are reducing the ability of existing herbs and shrubs to occupy the growing space.

5. They invade after small disturbances release growing space previously occupied by herbaceous and shrub species.

6. Possibly, they invade areas where existing trees and shrubs are growing vigorously.

It is uncertain how common each invasion pattern is; however, most species probably begin in a combination of ways. Trees commonly dominate by invading before other plants have occupied all available growing space. Sometimes, the small size and few numbers of these trees cause them to be overlooked at first, and they are assumed to enter later (Drury and Nisbet, 1973; Oliver, 1978a; 1980; 1981). Growing space relinquished by short-lived plants can be invaded by trees, so they may dominate areas previously dominated by herbaceous communities without an intervening disturbance.

Herb and shrub communities can also develop through stages similar to those described for tree stands (Fig. 5.2). As perennial herbs and shrubs become old, they are less able to dominate the growing space and exclude new stems, in patterns similar to the understory reinitiation and old growth stages of forest stands (Watt, 1947). Trees can invade the deteriorating, old communities and eventually develop into forest stands.

In some cases, trees invade over many decades (Table 6.1), entering areas in which all the growing space was originally occupied by herbaceous and shrub species (Kamitani et al., 1995). A few trees can invade such areas, expand their crowns and root systems, shade and reduce the vigor of existing herbs or shrubs, and/or reduce adverse microsites for germination of other tree seeds (Agee and Dunwiddie 1984; Tanaka, 1988). Other trees, often progeny of the older trees, then invade between the older trees and compete vigorously with the lesser vegetation. Such patterns lead to subalpine tree clumps (Franklin and Mitchell, 1967; Brooke et al., 1970; Lowery, 1972) and occur in arid areas and other places where the environment or nontree competition is extreme.

Trees also become established in herbaceous and shrub communities when small disturbances kill previous plants and make the growing space available (Harper, 1977). These disturbances may be as small as the footprint of an animal, since the growing space necessary for germination is small. Each small disturbance creates a new cohort of plants, leading to a multicohort stand in a strict sense. For simplicity, however, the stand is considered a single cohort if the trees invaded during a single interval—albeit quite protracted.

A true relay pattern of invasion may occur where one species establishes conditions which encourage its replacement by another species. This condition is not as common as once believed; however, it may occur as a casual interaction among some species on some sites.

Later development of the stand initiation stage

Regardless of how a tree becomes established, it increases its ability to survive by expanding upward, outward, and downward. The expansion rate varies with species, site, and competition. With expansion, plants become more capable of enduring fluctuations in the environment. Growth also keeps competitors shaded, shorter, and less able to compete for growing space above and below ground.

Late invading trees can be at a disadvantage even before previously established ones occupy all the growing space. Younger plants are subjected to side, low, and high shade and root competition even before they are directly overtopped by slightly older plants of the same cohort. Late invading plants often grow more slowly and eventually become relegated to lower strata or crown classes. The disadvantages of late invasion are most pronounced in single species stands, since all trees have nearly identical growth patterns and growing space requirements. For example, replacing occasional, dead loblolly pine seedlings (interplanting) in fast-growing plantations even 1 year after initial planting has generally not been successful because the interplanted trees are at such a competitive disadvantage that they rarely grow to merchantable sizes.

As plants grow after a disturbance, they slow and then reverse many nutrient release, organic matter decomposition, and soil structure processes begun by the disturbance. At the same time, the height growth and buildup of organic matter increase the susceptibility of stands to windthrow or fire (Fig. 4.1). When and whether regrowth stops and reverses the losses of nutrients and soil structure before the general nutrient budget is diminished varies with climate, disturbance type, bedrock geology, disturbance interval, and invading species.

Not all trees which become overtopped during stand initiation are permanently relegated to less competitive positions. A tree can be overtopped by herbaceous vegetation or other tree species and eventually dominate the stand. The low shade of taller plants is seldom permanently deleterious if the lower tree stays within the zone of living foliage of the upper tree. On the other hand, when the tree is subjected to high shade, growth is reduced, even where there seems to be as much light reaching the lower tree (Figs. 3.6 and 3.7). For example, young tulip poplars survive dense low shade from honeysuckles and grow through it to dominate a stand. On the other hand, shade from taller trees in a shelterwood stand reduces the height growth of small tulip poplars and often kills them. Similarly, young northern red oaks can be overtopped by the low shade of competing maples and birches and still outgrow and dominate them (Oliver, 1978a; Chap. 9); however, they develop poorly beneath the high shade of multicohort stands.

Crowns close progressively earlier when more stems are present, and newly initiating trees and lesser vegetation are excluded earlier. Very few trees are actually needed to outgrow other vegetation and create a closed forest. Fully stocked stands of Douglas-firs on productive sites contain as few as 185 trees per hectare (75 trees per acre) when 100 years old (McArdle et al., 1949). When these few Douglas-firs initiated growth, millions of herbs and shrubs of many species and growth forms were probably growing among them.

Usually, nonwoody plants are eventually overtopped by trees and shrubs. They lapse beneath even the lower tree foliage into the high shade (Fig. 3.6) and often die (Fig. 7.2). Since trees rarely grow uniformly during stand initiation some places remain open where growing space is available or herbs and shrubs occupy it, even though trees have already occupied all growing space in nearby areas. As the available growing space diminishes, the herbs that survive—such as vernal and autumnal herbs—are those capable of tolerating low light conditions or developing in temporary growing space.

Woody shrubs behave like tree species at first, accumulating growth and occupying more growing space each year; however, shrubs do not sustain rapid height growth and later become overtopped and often killed by taller-growing tree species, in a pattern similar to the overtopping of one tree species by another (Chap. 9).

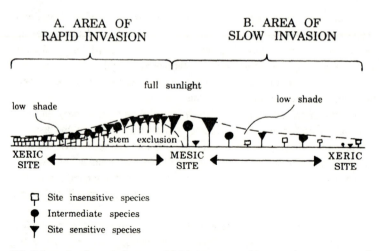

Figure 7.2 Schematic of area in later stand initiation stages. More species are present where the stem exclusion stage begins later (e.g., on xeric sites) because late-invading species still have available growing space for growth. The stand is approaching the bushy stage—the beginning of the stem exclusion stage in which trees have not begun to differentiate into crown classes and strata. (A) Some areas—especially those with rapid invasion and growth (high site)—reoccupy growing space and the local area reaches the stem exclusion stage sooner. (B) Adjacent areas, with slower invasion or less vigorous growth, are still in the stand initiation stage, with new stems able to invade in low shade.

Emergent Properties

The time a stand remains in the stand initiation stage varies from less than 5 years to over a century. A stand remaining in this stage for a long time appears as a stable community, rather than as one in the early stage of a dynamic process.

A stand is very attractive to grazing and browsing animals during stand initiation because most leaves, twigs, and buds are within reach. These animals dramatically influence which plant species gain the competitive advantage and dominate by giving an advantage to nonpalatable species and those which respond rapidly to grazing or browsing.

During stand initiation many observations of the cycle of species invasion, rise to dominance, death, and replacement by other species—especially where annuals and biennials complete their life cycles—have led to well-entrenched theories of plant succession following disturbances (Oosting, 1956; Odum, 1971). Plant species have appeared at various times to be mutually helpful ("facilitative") to other species becoming established, to be "tolerant" of other species, or to be "inhibitive" (Connell and Slatyer, 1977). The mutual "help" is currently believed to be *protocooperation*—a fortuitous occurrence, not a coevolved pattern.

The relay pattern (Fig. 5.1), in which one species appears to invade following another, partly exists because different species become dominant at different times and partly because some annuals and perennials germinate and grow during the first few years after a disturbance and then die, making growing space available for longer-lived species. Since all plants of a species have a similar maximum life span, the vegetation appears as waves—or *seres*—which become dominant, set seeds, and then disappear as longer-lived species emerge as dominants.

The many species and lack of dominance during stand initiation create a diversity of plant species. These many plant species, the ready availability of browse, and the horizontal and vertical spatial variation allow a large diversity of animal species.

During early stand initiation, many different herbaceous and woody species can dominate, depending on which had the initial competitive advantage. As the growing space is filled, fewer species emerge as dominants on many disturbed areas. This convergence of species is not because of a predetermined or coevolved direction to stand development; rather, fewer long-lived species exist in a region and therefore the potential range of species which live to dominate becomes reduced.

Analogies and extrapolations sometimes assume that all stands in a region will continue converging until there is a single dominant species given enough time. Such assumptions have not been substantiated by empirical evidence, however. Different, relatively stable forest structures can exist on a given site depending on which species was able to gain the competitive advantage during stand initiation. Because severe disturbances are more frequent than the life spans of trees in most regions, the forests are usually destroyed before any possible convergence of species can occur (Sprugel, 1991).

At the end of the stand initiation stage, all plants compete within a single stratum. It is difficult to know which species will dominate the stand without prior knowledge of the dynamics of the stand.

Applications to Management

The stand initiation stage is the only time when the forester can easily affect the species composition of the stand. Specific efforts can take advantage of subtleties which allow a preferred species to gain the advantage and dominate the site. Site preparation which mimics a disturbance giving the preferred species an advantage—or a competitor a disadvantage—can be beneficial. Most forest regeneration activities in North America do not attempt to eradicate all species except the targeted crop tree. Rather, the objective is usually to reduce the competitive ability of other plants enough to allow the crop species to gain and maintain a competitive advantage so it will eventually expand and take over the growing space occupied by other plants. Complete eradication is usually avoided because it is expensive and because soil structure and nutrients may be lost before the crop trees' roots occupy the soil growing space.

The stand initiation stage is the only time when there is much light on the forest floor. Stands which are artificially kept out of the full sunlight of the stand initiation stage by the suppression of disturbances or by allowing only minor disturbances (such as selective logging) undergo forest floor and soil changes. The high temperatures and light levels during stand initiation cause more rapid decomposition of the forest floor than at any other time—especially in cool wet areas such as the coastal Pacific northwestern United States and Alaska. Many herbaceous species prevalent during stand initiation fix nitrogen which is recycled in the stand during later stages. Shortening of the stand initiation stage through weed control and planting of vigorous seedlings at close spacings may reduce the nitrogen level and the growth of the stand. Heavy use of nonselective herbicides may also prevent this nitrogen fixation. Artificial fertilization can counteract this nitrogen reduction.

Site preparation can create microsites favorable for the desired species, and planting of seeds or seedlings can ensure that the desired species rapidly occupies the microsites. Favorable microsites can also be created by shading desired species with noncompetitive cover species or with artificial stakes in hot, dry areas, or by creating barriers to cold air movement in frost drains. Animal browsing can be minimized by applying repellents, by placing guards on desired species, or by judiciously using fertilizers at time of planting. Physical barriers such as dead, fallen treetops which prevent deer and elk from browsing on preferred seedlings may, alternatively, prevent hawks and bobcats from preying on rodents, which also eat seedlings; consequently, such barriers may aid or hinder growth, depending on the circumstances.

It is important to understand the causes if seedlings (especially planted seedlings) are dying, so that appropriate remedial action can be taken. Some causes (such as wet planting areas) create clumpy mortality, and others (such as bad planting techniques) create more random mortality. If mortality is clumpy, simply decreasing the planting spacing will not solve the problem, but can make it worse because seedlings will be too dense in areas where they survive. If the mortality is from reasons which make it somewhat random, increasing the number of seedlings can result in a better-stocked plantation.

Management activities are planned to coincide with growing space fluctuations on each site. Seedlings are usually planted when dormant in autumn, winter, and/or spring, depending on the accessibility of the site, the browse potential during winter, and the tendency of the seedlings to be pushed out of the ground by freeze-thaw cycles of the soil (*frost heaving*). The seedlings then take advantage of available growing space in early spring.

The stand initiation stage often becomes the "brushy stage" just after the beginnings of stem exclusion. It is important for the forester to know the growth patterns of each species at this time so that proper silvicultural activities can be planned. In the northeastern United States, pin cherries may dominate the stands during this stage but will not inhibit growth of the other hardwoods to the point that it is economically viable to combat the cherries. Sourwood trees can play a similar role in upland areas of the southeastern United States.

It is also important that foresters know all species and numbers of individuals in a stand during the stand initiation stage so that the eventual stand structure can be predicted. Some species cannot compete with either woody or herbaceous weed competition unless they are dominant in the brushy stage. Both loblolly pines and Douglas-firs must be dominant at this time to survive. Other species such as yellow birches can grow through the competition during the brushy stage. Two questions then face the silviculturist: Will the desired species die? And, will growth of the desired species be significantly slowed by the competition? Answers to these questions will then help make the decision for intervention, such as with herbicides. The forester must be able to predict the stand's development pattern both with and without investing money for a silvicultural operation. Prediction of stand structure at the end of the stand initiation stage needs to be good enough that investments can be compared between stands since it may not be feasible to treat each stand and these treatment activities have to be prioritized. The structure at the end of the stand initiation stage will determine the value of the stand for many decades into the future.

Too many trees of the crop species are more often the problem than too few where regeneration activities are practiced. The overabundance is partly because management efforts to give the crop species the competitive advantage grossly supplement seedlings of that species which would become established anyway, partly because silviculturists prefer to err conservatively on the side of too close spacing, and partly because little has been known about the desirable numbers to establish. Very few well-spaced trees are actually required to form a new stand.

Less intensive management efforts are being developed which identify or establish fewer crop trees and eliminate only those individuals of undesirable species which will not be outcompeted. These efforts—"enrichment" plantings and other mixed-species stand manipulations—require less energy but more knowledge than conventional forest regeneration operations do.

Trees are difficult to establish if they do not initiate before herbaceous and shrub species occupy the growing space. Early weeding operations which eliminate unwanted trees often do not require repeating, since existing vegetation reoccupies the growing space and excludes new invaders. Where animals eliminate desirable regeneration, new stems can become established only with difficulty. Management of power line, pipeline, or roadside rights-of-way where trees are not wanted can be done by ensuring that shrubs and herbs fully occupy the growing space (Niering and Goodwin, 1974). After this, trees become established very rarely and can be removed with little effort.

Encouraging rapid establishment and growth of tree species may be desirable for obtaining large trees; however, it may not be desirable for all management objectives. Rapid establishment may result in too many, uniformly growing trees with resulting problems of stagnation, small sizes, unstable or weakened trees, and possibly poor differentiation (Chaps. 8 and 9). Rapid early growth may also produce crooked stems and poor wood quality with some species and sites.

Later weeding mimics a partial disturbance in which existing stems reoccupy the growing space before a new cohort can become established. Selective killing or injuring by chemical or mechanical means can give the preferred species a competitive advantage even if it is partially overtopped by low shade or under severe competition from grasses. Selective killing of some of the crop species ("precommercial thinning," "thinning to waste," or "early stocking controls") can allow stems of desirable spacings, forms, and growth rates to have the competitive advantage; however, this operation can actually prolong the time before trees exclude understory vegetation. The increased spacing may allow undesirable species to grow rapidly and dominate the site before the crowns of the preferred species close; however, the preferred trees often have too much advantage for other species to initiate and outgrow them by the time these spacing operations occur.

CHAPTER 8

STEM EXCLUSION STAGE: SINGLE-COHORT STANDS, SINGLE-SPECIES STANDS

Introduction

A stand enters the stem exclusion stage when trees reoccupy all growing space and exclude new plants from becoming established. At first the variations in height within each species masks any differences in height growth between species, and the stand's structure is primarily the result of the time and position of each tree's initiation. The stand is considered in the *brushy stage*—at the beginning of the stem exclusion stage— before some species and individuals gain dominance over others and a pattern develops (Gingrich, 1971).

Trees increase or give up their occupied growing space and live and die by competing with each other during the following decades. Gradually, a distinguishable physiognomic pattern emerges in the stand, since the tree species behave in predictable ways in given environments. A stand's future changes in appearance, species composition, volume growth, resistance to disturbances, and other elements of stand structure are largely set by the spatial arrangements and initial heights of trees when the stand reaches the brushy

213

stage. Although predictable, the development patterns are not straightforward or simple. They vary between species, sites, spacings, stem age distributions, external influences such as animals and disturbances, and combinations of the above factors. The stand patterns are largely the result of the growths and responses to interactions of the component species with their characteristic individual tree growth patterns. Almost infinite variations of these stand development patterns can occur; however, the stand patterns can be anticipated and understood through a working knowledge of basic tree growth patterns, interactions, and responses to disturbances and competition.

This chapter examines development of the most simplistic stands—those with a single species. First, development patterns of very simplified stands which grow at uniform spacings and height growth rates will be discussed. Such uniform growth is only approached, to various extents, in real stands; however, their development is conceptually similar to Charles's and Boyle's ideal gas laws. Real gases only approximate the behavior of the "ideal" gases to varying degrees. Later in the chapter, the behavior of stands with less uniform growth patterns caused by variations in spacing, age, site, or genetic characteristics will be discussed. Finally, stand responses to removal of some trees by natural agents or silvicultural operations will be discussed.

Idealized Stand Development

Idealized stands established at progressively wider, uniform spacings are shown in Fig. 8.1. The trees are assumed to have conical crowns and strong epinastic control and to grow in the height-age pattern of tree 2, Fig. 8.2 (when its height growth is not reduced by a very small crown). The two dimensional "lollypop" figures are simplifications of interactions in a three-dimensional forest. A three-dimensional simulation of Douglas-fir stand growth based on height growth and similar crown interactions has been developed by Mitchell (1975).

At first, trees at all spacings are open-grown, and their crowns expand without external restrictions. Approaching crowns behave like trees in Fig. 3.14. When trees are short, the lower branches on adjacent trees overlap and then interlock (Cochrane and Ford, 1978), simultaneously slowing in lateral extension. With time, higher branches overlap and interlock; and lower branches become more shaded, grow little, become nonfunctional, and eventually die. As trees grow much taller, they become increasingly exposed to wind and sway more on their longer trunks. Approaching and intertwining branches from adjacent trees abrade each other (Tarbox and Reed, 1924; Spurr, 1964), causing branch tips to be knocked off and crowns to become separated.

After crown closure in the beginning of the stem exclusion stage, trees alter their shapes in response to the environment [called the "plastic phase" of growth by Harper (1977) and Hutchings and Budd (1981) and the "zone of suppression" by O'Conner (1935)]. Trees lose their lower, deeply shaded foliage and branches, and the stem grows more slowly and less tapered. Crown closure occurs later at wider spacings. At crown closure, widely spaced trees are taller, their crowns are larger (Bennett, 1960; Kramer, 1966; Stiell, 1966; Curtis and Reukema, 1970), and their limbs and main stems have larger diameters. There are, of course, fewer trees per area. The amount of foliage per area stays relatively constant for a given species after crown closure, regardless of spacing (Mar-Moller, 1947; Tadaki, 1966; Marks and Bormann, 1972; Long and Turner,

1975; Gholz and Fisher, 1982; Gholz, 1986). A slight "peak" of maximum foliage per area possibly occurs for the first 5 to 10 years after crown closure—perhaps when the crowns are intertwined—followed by a constant, slightly lower amount (Switzer et al., 1966; Tadaki, 1977).

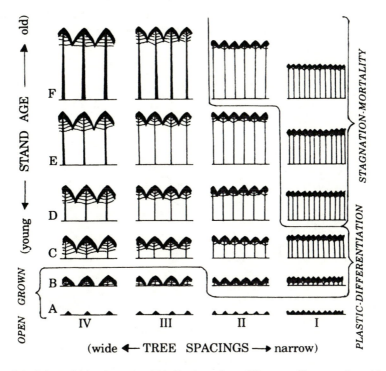

Figure 8.1 Schematic development of idealized stands at different, uniform spacings, identical potential height growths and tree architectures, and no differentiation. Trees change from the open-grown phase before crown closure, to the "plastic-differentiation phase" as the trees change shape in response to competition, to the "stagnation-mortality" phase, in which low vigor slows even height growth (*Oliver et al., 1986b*). (See "source notes.")

When all trees grow in height at an equal rate, they maintain the crown sizes they had at crown closure. The larger crowns of trees at wider spacings allow each tree to add more stemwood volume; however, stands at wider spacings contain less tree volume per hectare (acre) when young, since the larger tree sizes are more than offset by the fewer trees per hectare (acre).

Trees with small crowns at narrow spacings continue to slow in volume growth before trees at wide spacings do, since proportionately smaller fixed crowns have to support growing respiration surface areas (Chap. 3). Reduction in volume growth first appears as decreased diameter growth. The low priority of pest resistance also makes the stands more susceptible to diseases or insect attacks. If the trees do not die from pests or a general lack of vigor, they grow even taller but very little in diameter and become very unstable. The trees are so unstable that they often remain standing by mutually support-ing each other (Groome, 1988). If the closely spaced stand does not blow over in wind-storms or is not attacked by pests, the trees begin to slow in height growth (Fig. 3.15; Maisenhelder and Heavrin, 1957; Oliver, 1967; Reukema, 1970, 1979; Balmer and Williston, 1973; Balmer et al., 1975; Johnson and Burkhardt, 1976; Mitchell and Goudie, 1980; Lloyd and Jones, 1983; Omule, 1984; Liegel et al., 1985; Jones, 1986). When all stems slow in height uniformly, the stand begins the *stagnation/morality phase* (Fig. 8.1).

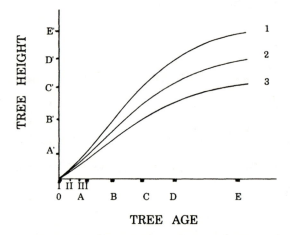

Figure 8.2 Height growth patterns of nonsuppressed trees in Figs. 8.1, 8.4, 8.5, and 8.6. All trees grow along height growth pattern 2 except in Fig. 8.5. Variations in height growth (1 and 3) are caused by microsite or genetic variations. Variations here are anamorphic (proportional), but can be polymorphic.

Trees at wider spacings have larger crowns, produce more total photosynthate (Gevorkiantz and Hosley, 1929; Holsoe, 1948; Hamilton, 1969), and so continue rapid diameter and height growth longer than closely spaced trees. Volume growth of a tree is proportional to its crown size (Ker, 1953), and so individual tree growth is greater at wider spacings (Bennett, 1960; MacKenzie, 1962; Hansbrough et al., 1964; Hansborough, 1968; Akai et al., 1968, 1970; Curtis and Reukema, 1970; Mohn and Krinard, 1971; Schlaegel, 1981; Omule, 1984; Jones, 1986; Oliver et al., 1986a,b; O'Hara, 1988). The relation of spacing to maximum diameter reached in uniformly growing trees is quite predictable within a species (barring differentiation) but quite var-ied between species (Fig. 3.17).

Eventually, trees at each successively wider spacing slow dramatically in volume growth (Fig. 8.3A), become unstable because of their small diameters relative to heights

(Fig. 8.3B and C), and finally become repressed in height growth (Fig. 8.3B; Eversole, 1955; Reukema, 1970, 1979; Balmer et al., 1975; Johnstone, 1976; Jones, 1977, 1986; Lloyd and Jones, 1983; Omule, 1984; Liegel et al., 1985). Tree falls, tree buckling, insect outbreaks, stagnation, and/or fires are common in crowded stands throughout North America and in Europe, Japan (Ishikawa et al., 1986), the Andean Highlands (*Pinus radiata* plantations; Galloway, 1991), and the Asian and South American tropics (Halle et al., 1978; Liegel et al., 1985; Bruenig and Y-w Huang, 1989).

Instability and height repression are less frequent at wide spacings. Since height repression occurs later at wide spacings and height growth slows naturally at older ages, the rate at which instability occurs decreases. Trees at wide spacings also have more opportunity to differentiate, which allows some of them to develop larger crowns and avoid height repression.

Trees with broad crowns or weak epinastic control behave like the hypothetical case described above, except the photosynthetic surface area is smaller at a given spacing in trees with broad crowns. Consequently, these trees slow in growth (Assmann, 1970), lose resistance to pests, become unstable, and possibly become repressed in height growth sooner at each spacing. Trees with broad crowns or weak epinastic control may also have less height growth at very wide spacings (Fig. 3.10).

Trees with more columnar crowns behave similarly, although they may maintain larger photosynthetic surface areas and the plastic responses occur later at each spacing.

Stand Development with Differentiation

The above discussion assumes that all stems develop at an idealistically uniform rate. Such uniformity does not occur, although research plantations often approach this condition (O'Conner, 1935; Kira et al., 1953; Reukema, 1970, 1979). Even in extreme conditions of very closely spaced stands, some trees slow in height growth more rapidly than others (Mitchell and Goudie, 1980).

Trees within a stand differ in growth rates and occupy different amounts of growing space (Daniels, 1976; Weiner, 1984). These variations give some trees competitive advantages and thus allow them to take over part of another tree's growing space. Growth patterns diverge as the redistribution of growing space occurs (Lutz, 1932; Deen, 1933; Kira et al., 1953, 1956; Hozumi et al., 1955; Kohyama and Kira, 1956; Yoda et al., 1957; Gilbert, 1965; McMurtrie, 1981; Britton, 1982; Cannell et al., 1984). This divergence is referred to as *differentiation* and is generally manifest first in diameter differences (Ker, 1953) and then in height differences, since diameter growth has lower priority for allocation of photosynthate (Kira et al., 1953; Mitchell, 1975; Hutchings and Budd, 1981). Differentiation leads to the assertion of dominance by some individuals and to changes in a stand's appearance (Lutz, 1932; Deen, 1933; Kira et al., 1956; Gilbert, 1965). Differentiation creates differences in crown sizes and positions, which allow the classification of trees by crown classes—dominant, codominant, intermediate, and overtopped (Chap. 5; D. M. Smith, 1986). Continued differentiation during which less vigorous trees reduce height growth but do not die—as in mixed-species stands—can lead to development of canopy strata (Chap. 5). As tree crowns expand and more dominant trees take over growing space from less dominant ones, the number of trees in the upper (dominant and codominant) position declines (Guillebaud and Hummel, 1949;

(A)

(B)   (C)

Figure 8.3 When trees grow tall without differentiating, (A) diameter growth slows (B) and (C) they become unstable and fall over, or (D) eventually slow dramatically in height. (A) Douglas-firs at regular, narrow spacing (on left) did not differentiate and so slowed dramatically in diameter. The edge tree of the same age on the right maintained a large living crown on the open (right) side and so continued rapid diameter growth (Wind River, Washington). (B) Spruce-fir stand in Maine grown at narrow spacing. The trees could not withstand the increased snow and wind from an adjacent opening because they were tall and thin. (C) Slash pines grown at close spacing in central South Carolina, to the north of their range, and off site did not differentiate well. They grew tall and thin and buckled under the weight of ice from freezing rain.

Warrack, 1952). Only rarely does a tree in the lower (intermediate or suppressed) crown classes latently "emerge" to the upper (dominant or codominant) crown classes in single-species, single-cohort stands. Trees increasing their growing space become large, while those reducing in growing space occupancy remain smaller and become less resistant to pests and more susceptible to buckling or tipping over (Chap. 3; Wiley, 1965). A differentiating stand appears stable and vigorous because the dominating trees grow large in diameter, although the less competitive trees become suppressed and weak and eventually die.

Many factors allow a tree to take over the growing space of another. Variations in species, spacings, microsites, ages, and genetic composition change the growth patterns among competing individuals and allow some trees to dominate others, as shown in the following examples. These examples assume that the species has the same crown shape, general height growth pattern, and response to decreasing crown size described above for the uniform stands in Fig. 8.1.

(D)

Figure 8.3 (Continued) (D) Very dense lodgepole pine stand in the Rocky Mountains of Canada. All trees are the same age, but the trees away from the edge had limited crown size and curtailed diameter and height growth. Trees adjacent to road had larger crowns and, so, grew taller. (See "source notes.")

Variations in spacings

Figure 8.4 illustrates the effects of variations in spacing on stand development. Trees generally grow at irregular spacings except in plantations with little mortality or in extremely dense, natural stands.

In Fig. 8.4, crown closure occurs sooner among more closely spaced trees. It occurs at different times on different sides of a tree (e.g., trees *g* through *k*, Fig. 8.4) where neighboring trees are at different distances, resulting in a tree with an uneven crown. The crown shape and surface area of each tree are defined by all the trees touching it (in this case, two trees, since only two dimensions are shown). Photosynthate allocation and resultant tree growth and shape consequently vary for each tree (Worthington, 1961).

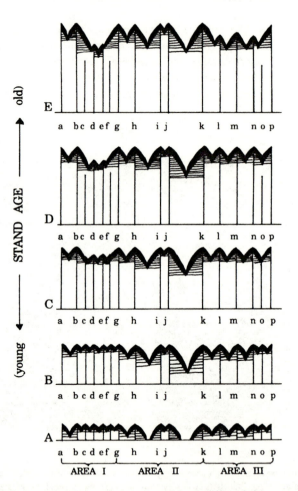

Figure 8.4 Schematic of irregularly spaced stand, resulting in differentiation among trees with uniform potential height growths, architectures, and soils (similar to Fig. 8.1). Trees at closer spacings have smaller crowns after crown closure and eventually become repressed in height growth, allowing more vigorous trees to expand their crowns and further suppress them.

The time of crown closure and stem exclusion also varies at irregular spacings. Crown closure generally describes a process between *two interacting trees* and occurs at different times within the stand. Stem exclusion generally describes development of a *whole stand* and refers to the time when the last available growing space becomes occupied and new stems excluded. Because of the looseness of the definition of "stand," it may be convenient to subdivide a stand at times, depending on the detail and use for which the stand is being described. For example, areas I, II, and III in Fig. 8.4 could be referred to as separate stands for some purposes. At time A, areas I and III would be in the stem exclusion stage, and area II in the stand initiation stage. For other purposes or at other times the whole area could be discussed as a single stand in the stand initiation stage (before time B) or the stem exclusion stage (after time B).

Trees spaced close together (*b* through *g*, Fig. 8.4) have early crown closure and crown recession, with correspondingly small diameter growth compared with those spaced farther apart (*h* through *m*). Trees with close neighbors on both sides maintain small live crowns and eventually slow in height growth (e.g., trees *c* through *f*, and *o*, beginning at time C). When an adjacent tree has wider spacing on the far side (trees *b*, *g*, and *n*), its live crown is larger, although irregular, and it maintains rapid height growth (Cannell et al., 1984). The adjacent trees with rapid height growth behave like trees growing above a wall (Fig. 3.14) and expand their crowns into and above their neighbors with repressed height growth.

When a repressed tree and its dominating neighbor are of the same species, the leaves of the repressed tree cannot photosynthesize at lower light intensities than its neighbor can and the overtopped foliage and limbs die. Repressed trees (e.g., *c*, *d*, *e*, *f* and *o*, Fig. 8.4) lose more crown and photosynthetic capacity and thus curtail their diameter growth, pest resistance, height growth, and, eventually, ability to add roots and leaves. Finally the repressed trees die from insects or diseases, from stem buckling or tipping, or simply from starvation.

This suppression and death is a natural result of stand development—an autogenic process. Such death is sometimes referred to as *"regular mortality."* The death of trees not being suppressed or otherwise weakened by the stand development process is referred to as *"irregular mortality"* and is an allogenic process. Irregular mortality is caused by fires, windstorms, acid rain, unusual insect or disease infestations, human activities, and similar agents. Unlike regular mortality, irregular mortality can commonly occur in any crown class.

The exact time of death fluctuates with environmental conditions. Insect defoliations or droughts weaken suppressed trees, but death may not occur until the following spring during shoot elongation or in midsummer during moisture stress (Staley, 1965; Karnig and Lyford, 1968). Insects and diseases also kill suppressed trees sooner than they would otherwise die. Heavy, wet snows or glaze causes thin, unstable suppressed trees to bend over and break or tip over from the base. A suppressed tree growing close to a more vigorous one lives inordinately long if root grafts from the dominant provide it with enough carbohydrates to survive (Bormann, 1965). Since the time of mortality is not fixed, estimates of mortality and of upper limits of stand volume are approximations which fluctuate with the environment.

Trees adjacent to a tree being suppressed expand their crowns over the suppressed trees, increasing their photosynthetic capacities and growth rates. The crowns of trees that were previously nonadjacent trees meet as intervening trees become suppressed (Fig.

8.4). Lower limbs of dominant trees facing suppressed trees are smaller than if the suppressed tree had not been there, since the limbs were shaded by the intervening tree before it became suppressed. After the surviving trees' crowns touch and shade each other, one eventually begins to dominate or both slow in growth. Suppression and dominance proceed more slowly when height growth slows as trees age (or as trees become very tall). Sunlight rarely comes from directly overhead even at midday except in tropical latitudes; at higher latitudes, it comes from more horizontal angles. Consequently, short trees in a stand probably photosynthesize less and become suppressed more rapidly at higher latitudes.

Differences in growth between trees as they differentiate are quite dramatic (Ford, 1975; West and Borough, 1983; Perry, 1985). Even the arbitrary classification of trees into crown classes has the effect of dividing the stand into tree populations with quite different growth rates (Oliver and Murray, 1983).

Differentiation caused by irregular spacing has been observed in many stands. When oaks grow in pure stands, those at close spacings die readily, whereas those at wider spacings live much longer. The widely spaced trees are larger in diameter—an asset for timber production and some wildlife and recreation uses—but have larger limbs—a drawback for timber quality production. Similarly, many single-cohort Douglas-fir stands in the Pacific northwestern United States grew at irregular, close spacings following logging of the late nineteenth and early twentieth centuries. Trees in these stands differentiated into crown classes, with the dominants becoming regularly spaced and containing relatively small knots from dead branches. These natural stands contrast with regularly spaced plantations established more recently, where most trees are either small and not well differentiated when planted at close spacing or quite large but limby (and contain much juvenile wood) when planted at wide spacings. As another example, a pure, single-cohort stand of western redcedars in western Washington began after a clearcut-and-fire episode 60 years before. It differentiated into crown classes; the suppressed redcedars were infected by *Armellaria mellea,* a disease which attacks weakened trees, while the upper crown classes remained vigorous and uninfected (Nystrom, 1981).

Trees growing next to a road or an opening are larger at a given spacing than trees within a stand, since the crowns of these edge trees expand into the opening (Fig. 8.3A and trees *j* and *k* of Fig. 8.4). Lazy foresters and others who observe stands only from the road, known as "windshield cruisers," often overestimate the size and numbers of trees within the stand.

Variations in microsites

Figure 8.5 shows the effects of variations in microsites on stand development. Microsite variations cause trees to grow along different height-age trajectories. Sometimes the height-age curves are proportional, or harmonized (anamorphic; Fig. 8.2). At other times, they are polymorphic, where one microsite favors rapid early growth and another favors rapid growth later (Weber, 1983). For the present example (Fig. 8.5), proportional (anamorphic) potential height growth patterns are assumed in the different microsites. Except for the microsite variation, all other factors—spacing, genetics, and age are identical. As before, trees have identical potential crown shapes and uniform responses to reductions in crown size.

As a stand develops, the time of crown closure varies but occurs first on most productive sites, since height growth is most rapid. Trees on poorer microsites (tree *d*, Fig. 8.5) soon become suppressed by adjacent, vigorous trees on better microsites, but suppression occurs slowly where all trees are on poor microsites (trees *h* through *j*). Trees on good microsites also become suppressed and die (tree *f*) if surrounded by other trees of equal vigor. Trees are most likely to become dominant if they grow on good microsites next to others on poor microsites or are on poor microsites but spaced far from good microsites. As the stand becomes older, trees are primarily found on the good microsites because trees on poorer microsites did not compete successfully with the faster-growing neighbors, not necessarily because trees could not grow on the poor microsites. In mixed western hemlock-sitka spruce stands in coastal southeastern Alaska, many hemlocks initiate on a variety of microsites—from convex windthrow mounds and rotten logs to concave depressions. After the first century, the dominants are those on convex slopes or on rotting wood, with the others becoming suppressed or dying (Deal, 1987; Deal et al., 1989).

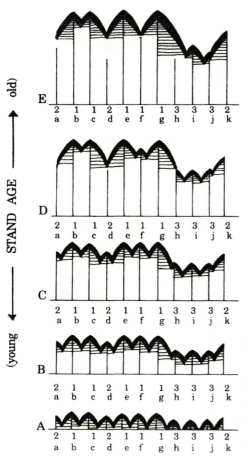

Figure 8.5 Microsite (or genetic) variations can also cause stand differentiation where trees grow differently in height. Sometimes, fast-growing trees can become suppressed and die (e.g., tree *f*) while slow-growing ones (tree *i*) dominate, depending on interactions with adjacent trees. (The numbers beneath the trees correspond to the height growth patterns of Fig. 8.2, caused by either microsite or genetic variations.)

More complicated patterns of dominance occur where site variations lead to polymorphic height-age curves. The processes causing these variations are similar to those in mixed stands growing at different spacings.

Variations in ages

Trees usually invade a stand over many years, rather than during a single year. Variation in ages also leads to stand differentiation. For example, only the ages of invading trees are varied in Fig. 8.6. All trees grow at the same height growth rate and pattern, at an equal spacing, on equal microsites, with identical genetic makeups, and with identical crown characteristics similar to those in Fig. 8.1.

As seen in Fig. 8.6, older trees generally have a growth advantage; however, trees with an age advantage do not always dominate (tree *b*), since growth of the surrounding

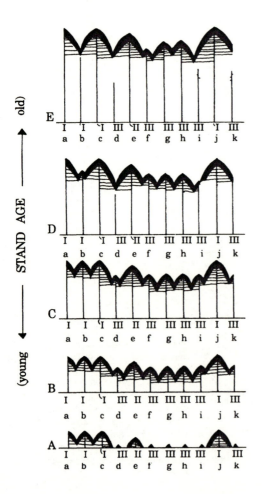

Figure 8.6 Stand differentiation can also result from trees established during different years even if potential height growth patterns, microsites, spacing, and tree architectures are held constant. Age variations and growth of adjacent trees influence growth, and older trees do not always dominate (e.g. tree *b*) and younger ones do not always become suppressed (e.g., tree *g*). [I = oldest trees; II = trees slightly younger (and so the height growth pattern #2 in Fig. 8.2 is shifted slightly to the right); III = youngest trees (but of same cohort).]

trees strongly influences a tree's development. Conversely, trees much younger than the others (*g* and *h*) can also dominate if surrounded by appropriate neighbors.

The advantage given to slightly older trees also varies with spacing, microsite, genetic makeup of the trees, and overall site productivity. Where early growth is slow, younger trees are at less disadvantage than where early growth is rapid.

Variations in genetic makeup

The causes of differentiation are difficult to detect or reconstruct and are often attributed to genetic variations, even though variations in microsite, initial spacing, or relative growth of neighbors may actually be responsible for the differentiation. The more dominant trees are often erroneously considered genetically superior.

Genetic variations exist to varying degrees in different species. Loblolly pine has a wide genetic variation. At the other extreme, red pine is genetically extremely homogeneous. Genetic variations can be expressed in height growth—both anamorphic and polymorphic variations—crown shape, growth efficiency, pest resistance, and other characteristics. Some variations are similar to the effects of different species, although variation between genotypes of the same species is generally less than between species.

Genetic variations in height growth lead to stand development patterns similar to those caused by microsite variations. Stands would appear as in Fig. 8.5 where anamorphic variation is assumed, with "productive microsites" representing trees with genetically more rapid height growth. With other factors equal, trees with genetically superior height growth potential generally become dominant. Even genetically superior trees, however, can become suppressed when surrounded by other superior trees, and inferior ones dominate if surrounded by appropriate neighbors.

Stands with little genetic variation would appear as in Figs. 8.1, 8.4, and 8.6. Narrow genotypes combined with uniformities of age, site, and spacing would lead toward stagnation (Fig. 8.1).

Genetic qualities of individuals are often overshadowed by variations in age, spacing, and microsite. Foresters selecting "genetically" superior trees for breeding programs do not always realize that a stand's natural dynamics can cause the largest tree in a stand to be genetically inferior.

Other factors

The exact factors which give a tree a competitive advantage and allow it to become dominant are often difficult to determine. Many other factors influence a stand's development pattern in addition to variations in spacing, microsites, ages, and genetics. Certain species differentiate more readily than others (Gilbert, 1965). Species with spreading crowns differentiate more readily than trees with narrow crowns (Hauch, 1910; Gilbert, 1965). Conifers generally have relatively narrow crowns and are more often found in single-species stands, are more frequently planted in homogeneous plantations, and are believed to approach stagnation more frequently than hardwoods. The regeneration method of each stem causes it to grow differently in height and thus affects its competitive position.

Emergent Properties of Developing Stands
Variations in species behavior

Species vary in their tendency to differentiate in pure stands. Southern pines—especially loblolly and slash pines—differentiated readily in one study (Pienaar and Shiver, 1984*a,b*) and less notably in others (Lloyd and Jones, 1983; Jones, 1986). Lodgepole (Horton, 1956; Johnstone, 1976; Mitchell and Goudie, 1980), red, and pitch pines are generally assumed to differentiate very little in pure stands. Douglas-firs seem intermediate between these. Red alders can approach stagnation (Warrack, 1949). Paper birches, yellow birches, and sugar maples also vary in their tendency to differentiate (Gilbert, 1965).

Differentiation is related to innate genetic variations within a species and to growth forms. Trees with flat crowns differentiate more readily (Hauch, 1910; Gilbert, 1965), since slight differences in heights lead to large differences in photosynthetic surface areas. Species which readily root-graft probably differentiate less, because trees share photosynthate. Intolerant species do not necessarily differentiate more readily than tolerant species; however, suppressed trees of tolerant species survive longer after differentiation has begun.

Site and differentiation

Stands on poor sites are generally believed to exhibit less differentiation than those on good sites. This poor differentiation can lead to growth repression of all stems and approach stagnation (Fig. 8.3). Growth repression can also occur on more productive sites (Oliver, 1967; J. H. G. Smith, 1986). The lack of differentiation on poor sites has been attributed to many factors. An inability to express genetic variation on poor sites may lead to less differentiation, although little research has investigated causes of stagnation. Poor differentiation may result from a situation in which many stems begin on poor sites, thus creating small crowns before any trees become large enough to dominate others. Poor differentiation may also be caused by less microsite variation on poor sites, which allows many stems to become established but does not create variations in growth rates.

A poor site does not allow the dominant and codominant trees to retain as much lower foliage, and more light reaches the lower trees. Consequently, intermediate and suppressed trees of a given species survive longer on a poor site, restrict crown expansion of the dominants and codominants, and thus slow differentiation.

Narrowly spaced stands

Trees at narrow spacings appear to stagnate more readily than at wide spacings, possibly because narrowly spaced trees compete before genetic variations in growth assert themselves. Also, narrow spacings are more common in homogeneous site and age conditions; and stagnation is more obvious where it occurs early, at narrow spacings. Lodgepole stands of over 50,000 stems per hectare (20,000/acre) slowed in height growth before they were 2 to 3 m (6 to 9 ft) tall (Mitchell and Goudie 1980), while stands of 42,000 stems per hectare (17,000/acre) of ponderosa pine slowed by 7 m (21 ft) tall (Oliver, 1967). Douglas-fir stands of 5400 trees per hectare (2200/acre) slowed in height by 15 m (50 ft) tall (Reukema, 1970); stands of wider spacings slowed at progressively later times (Reukema, 1970). Slash and loblolly pines can slow in height growth at even wider spacings and short heights (Lloyd and Jones, 1983).

Changes in spatial patterns

A stand can begin as a mosaic of clumps of trees at wide, medium, and narrow spacings. Trees at wide initial spacings reach crown closure later; their crowns, too, recede; and differentiation and suppression occur later. Suppression and mortality occur sooner at narrower spacings, so the surviving trees end up at spacings similar to the initially widely spaced parts of the stand. In this way, spacing between trees becomes more uniform with age. A stand's horizontal spatial pattern tends to shift from a clumped, or patchy, distribution to a more random, and then regular, distribution as it grows older (Cooper, 1961; Laessle, 1965; Daniels, 1976; Akai et al., 1977; Christensen, 1977; Yeaton, 1978; Rinehart, 1979; Whipple, 1980; Good and Whipple, 1982; Peet and Christensen, 1987). Dominating trees shift even more rapidly to the regular distribution (Curtis and McIntosh, 1951; Kitamoto and Shidei, 1972; Williamson, 1975; Rinehart, 1979; Whipple, 1980; Good and Whipple, 1982; Smith and Grant, 1986). The originally clumped arrangement of trees is caused by the irregular sites for germination or other regeneration mechanisms, while the later more regular distribution results from differentiation and regular mortality. The approach toward spatial regularity is unusual elsewhere in nature (Watt, 1947; Clark and Evans, 1954; Kershaw, 1963; Payandeh, 1974; Pielou, 1974) but has been observed in forest stands.

Waves of mortality

Large variations in stand development patterns can occur even within a species; however, the patterns can be generalized, since each species behaves only within certain limits. In general, trees become crowded as a stand grows, with crowding occurring sooner where stems are closer. The crowded stems either differentiate, die, fall over, or quit growing. More crowded parts of the stand usually differentiate first, and some trees die while remaining ones grow larger at a spacing more comparable with the rest of the stand. The resultant, more uniform stand then grows crowded and differentiates, and the spacing becomes even more uniform. More trees become suppressed and die (assuming the stand as a whole does not stagnate), and the remaining dominant trees again grow. Stands, therefore, may pass through waves of density-dependent mortality in the stem exclusion stage (Newnham, 1964; Yarranton and Yarranton, 1975, Harcombe,1986; Peet and Christensen, 1987; Harrison and Daniels,1988). The timing of these waves depends on the growth rates and spacings of the stands, on how rapidly differentiation occurs, and on how rapidly the suppressed trees die (Newnham, 1964). A first wave of mortality can be quite distinct. Additional waves can occur as the stand becomes older and again becomes crowded (Newnham, 1964), although irregular mortality and a slowdown in height growth can mask them. These waves affect both the time when understory reinitiation occurs and the vigor of the understory. These waves also determine when a stand reaches its maximum growth rate (Chap. 15; Peet, 1981; Peet and Christensen, 1987). Sometimes many overstory trees die at approximately the same time in tropical forests—

synchronous mortality (Mueller-Dombois et al., 1983)—as a different type of mortality wave.

Autogenic life span of stands with different initial spacings

The diameter growth of each tree first declines as a stand becomes crowded, creating thin, unstable trees (Long et al., 1981). The diameter growth of surviving trees increases as their crowns expand after differentiation and mortality; however, they do not attain the large diameters of trees initially grown at wide spacings. Trees which begin at dense spacings and dominate through differentiation become increasingly thinner and less stable than comparable trees initially grown at wide spacings. The instability is especially pronounced in species and sites where rapid height growth continues for many decades. Stands differentiating naturally from narrow spacings may have an autogenic life span which culminates in their buckling or tipping from instability, in spite of natural differentiation. Old growth stands containing very tall trees (e.g., Douglas-firs) may be special cases of stands which began with few trees at wide spacings which permitted them to maintain stability.

Responses to Thinning

External factors also cause trees to grow at different rates, thus changing the stand development patterns. Partial disturbances can remove or kill a tree. Growing space occupied by this tree suddenly becomes available and surrounding trees and newly initiating plants expand to refill it. How rapidly the surrounding overstory trees refill the growing space depends on the rate of crown and root expansion, which is dependent on the innate crown and root characteristics, tree age (Wiley and Murray, 1974), site characteristics, tree vigor (Krinard and Johnson, 1975), and amount of growing space released.

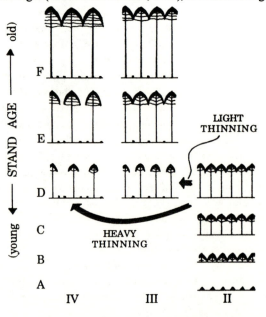

Figure 8.7 The same stand can be thinned to different spacings. Time of crown reclosure will be delayed at wider spacings, but crown sizes will eventually be larger.

(wide ← TREE SPACINGS → narrow)

The effect of thinning on stand development will first be considered in regularly spaced, undifferentiated stands (see Fig. 8.1). When trees are removed after crown closure (Figs. 8.7 through 8.9), the remaining trees behave like those in Fig. 3.14 in which a surrounding wall is removed. If the stems are extremely thin relative to their height, they fall over when the support of adjacent trees is removed (Groome, 1988). There may be a slight shock to the crown as the lower, previously shaded branches and stem first adjust to full sunlight. This shock often is not noticeable in stem diameter growth, and the trees expand into the growing space immediately. On extreme sites, however, a reduction in height growth for 3 to 10 years has been reported (Staebler, 1956a,b; Day and Rudolph, 1972; Hall et al., 1980; Harrington and Reukema, 1983; Maguire, 1983).

Trees surrounding the removed ones are "released" from competition. The increased light allows their lower branches to remain alive and continue growing outward as the tree grows taller. The crowns grow larger, since the live crown bases do not recede (Zahner and Whitmore, 1960). This unrestricted crown growth before reclosure with crowns of nearby residual trees is termed *free growth* (Hummel, 1951; Jobling and Pearce, 1977). Free growth differs from open growth, since the trees have experienced crown closure and the crowns do not necessarily extend to the ground in free growth. The spacing may be narrower than necessary for open-grown trees of a similar height. Surviving tree crowns reclose over the gap formed by the removed trees if residual trees are still growing rapidly in height, the crown shape is appropriate, and the opening is not too large (Fig. 8.7). The time of reclosure varies with tree height, growth rates, and spacing between trees (Figs. 8.7 and 8.8; Long and Smith, 1984). If and when crowns reclose, lower branches again behave as if they have reached a light-impermeable barrier, lose vigor, and begin to die. Afterward, the crown again maintains a stable size and grows upward as the stand grows taller. The result is first crown expansion and then a larger crown of constant size after crown reclosure (Briegleb, 1942; Kramer, 1966).

Thinning may also have the effect of inducing cone and seed production in some species (Croker, 1952; Allen, 1953; Wenger, 1954; Yocum, 1971).

Innate crown characteristics

Crowns of adjacent trees in thinned stands may not reclose as regularly as indicated above. Lateral branches of trees with weak epinastic control expand outward at a rapid but irregular rate. Alternatively, trees with strong epinasty but slow height growth take a long time to reclose over an adjacent opening. Stands on poor sites or very old stands with slow height growth do not respond rapidly to thinning or other allogenic disturbances. Trees with columnar crowns may not expand enough to refill an adjacent opening. Large openings, of course, reclose more slowly than small ones.

Even where adjacent crowns do not physically expand enough to cover an opening, they increase their photosynthetic surface area by *crown densification*. This densification occurs when newly sunlit crown parts increase the size and number of leaves and add numerous new buds (Maguire, 1983) and water sprouts within the crown and/or along the trunk. The available sunlight growing space is reoccupied by the existing crowns. Since sunlight passes diagonally through a stand during most of the day and year (especially in temperate latitudes), much of the light is intercepted by these

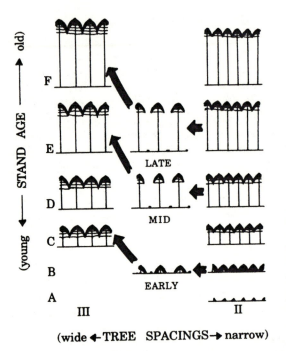

Figure 8.8 The same stand can be thinned at different times. Resulting crown sizes and tree heights will be similar after crown reclosure unless height growth repression occurred before thinning. Different timings of thinnings change both the volume of trees removed and the residual volumes.

overstory crowns as they become more dense. Growing space available to the understory within openings diminishes as the surrounding overstory crowns intercept more sunlight (Larson, 1982).

Released overstory trees photosynthesize more as their crowns and roots expand. Their growth rate first accelerates as they expand into the available growing space and then stabilizes as the growing space is reoccupied. Released trees add more diameter growth and maintain height growth longer than unthinned trees (MacKenzie, 1962). After they reoccupy the growing space, released trees resemble in crown size and respiration area trees which have grown from the start at wider spacing (Figs. 8.7 and 8.8); however, they have smaller diameters because the accumulated wood within their stems is less (Fig. 8.9).

Release of trees with small crowns
If trees are very crowded, they grow very slowly following release (Fig. 8.8; Staebler, 1956a,b; Krueger, 1959; Reukema, 1961; Krinard and Johnson, 1975) providing they do not fall over or break (Groome, 1988). Growth is especially slow if height growth has already been repressed. The released crown expands only as photosynthate is available for shoot growth (Mitchell and Goudie, 1980). As crown size gradually increases, more photosynthate is produced and height growth slowly increases, which allows the crown to expand. In species with strong epinastic control, the crown expands at a geometrically increasing rate until maximum potential height growth is regained. Then, it expands as described above, behaving both in gross structure and in growth rates similarly to trees of similar height (but younger ages) and spacings which have never been suppressed.

These released trees are always smaller in diameter than ones of similar height, age, and spacing which have not been suppressed (Fig. 8.9). They remain susceptible to stem buckling or tipping if they grow taller more rapidly than their diameter enlarges to support the extra weight.

Decades may pass between the release of trees with small crowns and their increase in diameter and height growth—if they increase at all. Release is first apparent in increased leaf size and number, followed by height growth; then, diameter growth increases at the base of the living crown and gradually increases downward. An increase in stem diameter growth can appear at the crown's base several years before it appears at breast height [1.4 m (4.5 ft)].

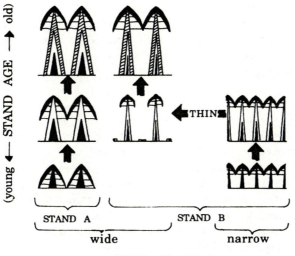

Figure 8.9 Effect of stand history on standing volume of stands with similar structures These stands at the same age, spacing, height, and species (top left) with closed crowns are currently growing (producing stem volume of striped area) at the same rate (per tree and per area) but contain different volumes (total stem volume per tree and per area) because of different past histories. Previously thinned stands have less standing volume because previous crowding allowed less early volume accumulation per tree before thinning (black stem area) and during lag time after thinning (white stem area). (*Oliver et al., 1986b*). (See "source notes.")

A previously crowded tree is physically susceptible to stem buckling or tipping immediately after release because of its small diameter relative to its height. After release, it initially becomes even more top-heavy as first the foliage, then the terminal, and then the upper stem increase in size. Severely repressed trees are not usually considered capable of responding to release because periodic strong winds, freezing rains, wet snows, or other factors can exacerbate their instability long before they grow stable.

Release of trees with large crowns

Trees with very large crowns maintain their height growth upon release. Their crown bases quit receding and their crowns expand and refill the newly available growing space. Before release, these trees contain almost as much photosynthetic surface area as open-grown trees, and their diameter growth is still rapid. These trees grow rapidly after release (Krueger, 1959), although sometimes the release is not noticeable since it simply delays growth slowdown. The trees continue to grow at rapid rates until well after crown reclosure.

Tree vigor, stand structure, and tree response to release

In thinned stands, trees which occupy their growing space most efficiently are generally those which are tall, with crowns of medium size (Kramer, 1966; Hamilton, 1969, 1981; Ford, 1975; O'Hara, 1988). Apparently, trees with larger crowns use growing space inefficiently because their lower limbs are inefficient (Woodman, 1971) and/or their large amounts of sapwood lead to high respiration rates (Sprugel, 1989). Small trees which do not occupy much growing space are not efficient, because their foliage is not in a good position to receive sunlight. The most productive thinned stands, therefore, are not necessarily characterized by a certain amount of foliage, but by an efficient vertical distribution of foliage in the canopy (O'Hara, 1989).

Release in well-differentiated stands

Response to release in well-differentiated stands depends on which trees are released (Meyer, 1931; Steele, 1955a,b). When suppressed trees are released, their limited photo-synthetic surface area expands very slowly, eventually causing resumption of normal height growth and crown expansion. This release of suppressed trees in single-species, single-cohort stands differs from release of certain lower-strata trees in multicohort or mixed-species stands. In single-species, single-cohort stands, suppressed trees generally do not grow rapidly soon after release (Ker, 1953; Staebler, 1956a,b; Yoda et al., 1957; Krueger, 1959; Reukema, 1961; Hamilton, 1969; Oliver and Murray, 1983; O'Hara, 1988). In mixed-species or multicohort stands, dominant trees of understory strata often grow rapidly after release (Oliver, 1976).

When dominant trees in differentiated stands are released, the large crowns and rapid height growths allow rapid crown expansion. The rate of reoccupation of growing space by surrounding, vigorous trees depends on such factors as stand age, height growth rate, and characteristics of the species. Volume growth per hectare, therefore, is generally greater when trees of lower crown classes are removed in thinnings, although in practice more dominant trees are sometimes removed to improve residual tree quality (McKnight and McWilliams, 1956) or to maximize cash flow (Emmingham and Hanley, 1986).

Responses to Limitations Other Than Light

The above discussions have implied light is always the most limiting factor for growth. On many droughty sites moisture is often the most limiting factor. In these cases or when some other factor limits growing space, trees respond similarly except that, instead of crowns enlarging, root systems expand until all growing space is refilled (McClurkin, 1958, 1961; Della-Bianca and Dils, 1960; Zahner and Whitmore, 1960; Barrett and Youngberg, 1965). Full occupancy of root growing space may occur before crowns actually touch (Hiley, 1948), so that stands on droughty sites may reach maximum stand density without crown closure. Roots of each tree do not occupy an exclusive area around its stem (Fig. 3.12B); therefore, where soil growing space is most limiting, trees may not space themselves as regularly as when sunlight and crown expansion limit growth and survival. Also, taller trees may not necessarily maintain a competitive advantage on dry sites, since the sparse canopies allow sufficient sunlight to reach short- er trees. Less is known about tree competition in droughty sites.

Applications to Management

Selections of superior trees for breeding purposes based on tree size relative to neighbor- ing trees may not result in any genetic improvement at all.

A stand's development pattern and possible ways it can vary with natural and human interferences are set by the initial distribution of trees. Narrow and/or regular spacings, uniform ages, and genetic similarities may lead to stands which eventually approach stagnation. Even where enough differentiation occurs that the stand does not stagnate, many trees may slow in diameter growth enough to become unstable. Other stands may develop so many dead trees that they burn up.

Most single-species, conifer stands in North America are growing at a narrow spacing. In some cases the stands need to be thinned to allow residual trees to grow to large sizes. In other cases, the trees are already too unstable to be thinned successfully. In both cases, markets and thinning and harvesting systems need to be found for the small-diam- eter trees to prevent insect outbreaks and fires and to recover some value from the stands. Markets need to be found for trees such as small red pines and lodgepole pines where stands have grown too unstable to be thinned. Southern pines also differentiate more slowly than many species, but a good pulpwood market has obscured this in the past. The reduction of bobtail truck operations and other systems for thinning small trees may create future problems in maintaining vigorous southern pine stands.

Although both natural stands and plantations go through the stem exclusion stage, the way that single-species stands appear during this stage has given plantations a bad repu- tation with many groups. Some people find single-species stands in this stage less aes- thetically pleasing than other types of stands. Many animal species cannot live in stands in the stem exclusion stage because there is little vegetation near the forest floor. Few plantations progress far into the understory reinitiation stage before harvest, so many people assume that the stem exclusion structure is unique to plantations.

The times of crown closure and slowing of diameter can be controlled by initial or early spacing of the trees. Trees at uniform spacings and of equal vigor are less likely to differentiate than those at initially irregular spacings. Regular spacings are more likely than irregular spacings to yield trees of uniform, predictable diameters when they begin to slow down. A narrow, regular spacing can allow the manager to gain more volume early and thin the trees at an early age and small size. Whereas this spacing gives the manager the most options for subsequent thinnings, there is also a commitment to thin to avoid approaching stagnation. There is a commitment to thinning or harvesting the stand before extreme diameter growth slowdown to avoid significant loss of diameter and vol- ume growth.

The stem exclusion stage of stand development is when most commercial thinning takes place and often final harvest as well. Three questions that must be asked when thinning are:

1. Is this tree merchantable?

2. What will the future value of this tree be (what is the growth rate)?

3. How much will this tree interfere with the growth of other trees in the stand and how will the remaining trees respond?

Answers to all three questions depend on the differentiation of the stand.

Removal of some trees—a silvicultural thinning—can be done for a variety of reasons:

To obtain wood or a positive cash flow

To obtain an early return on investment

To increase volume growth per area

To remove diseased, poorly formed trees, or undesirable species

To recover anticipated mortality

To keep a logging crew together and machinery operating (to recover at least some of the fixed costs)

To maintain the vigor of the remaining stand

To prepare the stand structure for nontimber uses (e.g., animal habitat)

Four factors are important to obtain maximum volume growth with thinning (Oliver et al., 1986b): (1) the stand must be thinned at the appropriate age for each spacing, species, and site, (2) the least vigorous—intermediate and suppressed—trees should be removed, (3) the remaining trees should be left at the appropriate spacing, and (4) sufficient time before rethinning or harvest should be given to allow the stand to recover from the thinning. More dominant trees can be removed for cash flow, stem quality, or other justifications; however, removal of these tends to reduce total stand volume growth.

Many thinnings are planned too late. Delaying thinning may add some additional diameter on the trees to be cut, but the suppressed and intermediate trees are usually growing at such a slow rate that this added growth is minimal. More importantly, the trees that will be left after a late thinning have already developed small crowns. After the base of the live crown has receded, the tree can respond to thinning only by increasing height or having branches in the upper part of the crown grow outward. Both eastern white pine and Douglas-fir stands maintain rapid height growth late in their lives; consequently, they will respond to thinning at advanced ages. Species such as southern pines slow in height growth at an earlier age and only early thinning will keep the crowns large enough for adequate growth. Also, very shade-intolerant species die before they become very suppressed, and thinning in these species must be done when the trees still have large enough crowns to respond to release.

CHAPTER 9

STEM EXCLUSION STAGE: SINGLE-COHORT STANDS, MIXED-SPECIES STANDS

Introduction

Variations in stand development caused by differences in tree growth patterns are greatly accentuated when competing trees are of different species. All species compete for approximately the same essential factors, with little variation on the physiological level in ability to grow under different amounts of shade, water stress, or nutrient availabilities. On the other hand, species show large differences in their abilities to survive limitations of essential factors and in their allocations of photosynthate. Different photosynthate allocations result in distinct differences in primary and secondary growth patterns and tree architectures between species. A species' architecture allows the species to gain competitive advantages over other species when grown under certain conditions. Stand development patterns often emerge as a result of these competitive advantages (Larson, 1992). The patterns are not obligatory or universal, and they vary with such conditions as site, spacing, regeneration mechanism, and age.

In most forests, stand conditions and species behaviors are uniform enough and competitive interactions are dramatic enough that emergent patterns are recognizable and predictable over extended areas. At the same time, however, subtle variations in stand conditions can change the competitive advantages of component species and the resultant forest structure. There are no obligatory patterns, and there will never be a universally predictable pattern to the development of certain species mixtures unless all stand conditions are known and kept constant. The inclusion of more species in a stand changes the interaction patterns and the resulting structure. Consequently, stands with increasingly diverse species compositions behave in increasingly complex, although not random, patterns.

General Patterns of Mixed-Species Stands

Development patterns

Many tree species often initiate in the growing space made available by a disturbance. Trees are quite short at first, and little differences in height appear between species. Any differences in height growth between species are masked by slight differences in ages, regeneration mechanisms and vigor, and microsites. As the trees grow taller, accumulating differences in height growth among species become apparent and a vertical *stratification* among species emerges. This stratification is a sorting of the vertical structures of a forest into superimposed horizontal layers (Wilson, 1953; D. M. Smith, 1962, 1973,1986; Marquis, 1967; Dawson and Sneddon, 1969; A. P. Smith, 1973; Roach, 1977; Oliver, 1978a, 1980; Stubblefield and Oliver, 1978; Webb, 1978; Wierman and Oliver, 1979; Skeen, 1981; Lincoln et al., 1982; Akai et al., 1983; Bowling and Kellison, 1983; Hibbs, 1983; Kelty, 1986; Larson, 1986; O'Hara, 1987; Clatterbuck et al., 1987; Deal, 1987; Yamakura,1987; Clatterbuck and Hodges, 1988; Cobb, 1988; Ishizuka, 1981; Guldin and Lorimer, 1985; Watanabe, 1985; Kushima, 1989; Deal et al., 1991; Long, 1991; Marquis, 1992; Osawa, 1992; Cobb et al., 1993; O'Hara, 1995; O'Hara et al., 1995). Faster growing species appear as the dominant trees at first, and slower ones occupy the lower crown classes. If dominant species sustain their height growth, other species can fall completely beneath the upper canopy.

Whether species which lag in height growth die or survive depends on the species' ability to survive where light and other factors reduce available growing space. If the species which lags behind cannot survive beneath the overstory, it dies as suppressed trees in single-species stands do, and the number of species in the stand is reduced. In this way, stands which begin with many species can be reduced to single-species stands.

On the other hand, the shorter species will survive if it can utilize growing space not usurped by the taller one. For example, if the shorter species can live at lower light intensities or retain its leaves and photosynthesize when the overstory is leafless, it may survive in the lower-canopy position. Trees which survive beneath the main canopy generally grow slowly because they receive very little sunlight (Sumida, 1993), and most of that sunlight is utilized for respiration or root and leaf formation. These shorter trees can shade the stems of the dominant trees, killing lower branches and keeping water sprouts from forming on them; such trees are sometimes silviculturally referred to as *"trainer"* trees because they improve the wood quality of more dominant ones (McKinnon et al., 1935).

Trees which lapse beneath the overstory slow in height growth because their terminals are battered by more dominant trees; they photosynthesize less in the shade and so generally grow less. The trees allocate more growth to branches in the high shade, and so the understory crowns become flat-topped.

Where closely spaced trees of similar tolerances lapse beneath the overstory, they compete with each other in a lower stratum and differentiate into dominant, codominant, intermediate, and suppressed crown classes (Fig. 5.3) within this stratum. Dominant and codominant trees in lower strata often have large crowns even though they are in high shade. These dominant trees do not grow much while in the understory; however, if they have large crowns with many living buds, their crowns can quickly expand and they grow rapidly in height if released (Fig. 3.11; Oliver, 1976; Hoyer, 1980)—provided they were not killed by the "shock" of release.

Canopy strata and crown classes

A fast-growing species forms an emergent A-stratum (Fig. 5.3) if only a few individuals are present. The same species forms a B-stratum above the rest of the stand if there are enough trees to form a closed canopy. As a mixed stand grows, trees of different species differentiate in height even more readily than in pure stands because of their different growth patterns. Trees which lag behind in height growth form lower strata if shade-tolerant enough. Within each stratum, trees further differentiate into crown classes (Kelty, 1986). It is unclear how many strata can exist in a stand. Including the A- and forest floor strata, four strata are very common in North American mixed stands.

A species generally does not live beneath itself in single-cohort stands except in the understory reinitiation and old growth stages. A species can be found in several strata if the lower-stratum individual had been overtopped by a species which did not offer competition with the individual in the upper stratum (e.g., cherrybark oak and sycamore, discussed below).

Stratification patterns can change as stands grow older. A stratum may become depleted as the component species die, depending on their mortality rates. Lower strata may grow upward as the overstory becomes less dense. In other cases, species in lower strata have slow but steady height growth, strong epinastic control, and stiff branches which enable them to grow into higher strata (Oliver, 1978a; Gara et al., 1980; Larson, 1986).

Emergent patterns

Certain emergent patterns are common to many mixed-species, single-cohort stands. One or a few species will often be so abundant and/or grow so rapidly soon after the disturbance that other species are not noticed (Drury and Nisbet, 1973). The dominant species can be mistakenly assumed to invade before the others and are referred to as *pioneer species*.

Certain species are commonly relegated to lower strata as the stand develops because of their slow early growth rates and tolerance to shade. These species survive in shade but grow very slowly compared to the A- and B-strata trees receiving full sunlight (Fig. 2.1). It is often mistakenly assumed that some of these species (e.g., western redcedar) grow slowly even if not overtopped (Nystrom et al., 1984). Other species commonly found in the understory, such as dogwood or sourwood, may truly not grow large even in open-grown conditions.

The size differences can lead to the mistaken conclusion that the lower-strata trees are younger, have initiated beneath the canopy of the higher strata, and will eventually replace the overstory stratum (Meyer and Stevenson, 1943; Jones, 1945; Braun, 1950; Phillips, 1959; Daubenmire, 1968; Minckler, 1974; Mueller-Dombois and Ellenberg, 1974). Where more trees are relegated to lower strata than to upper strata, the diameter distributions of all stems can show a reversed shape with few large stems and many small ones (Fig. 6.2A and B; Wilson, 1953; Blum, 1961; Gibbs, 1963; Marquis, 1967; Roach, 1977; Oliver, 1978a; Lorimer and Krug, 1983). This diameter distribution is sometimes assumed to indicate an all-age stand (De Liocourt, 1898; Meyer and Stevenson, 1943; Meyer, 1952; Hough, 1932; Worley and Meyer, 1951; Meyer et al., 1952; Eyre and Zillgitt, 1953; Phillips, 1959; Baskerville, 1960; Piussi, 1966; Leak, 1965; Daubenmire, 1968; Assmann, 1970; Minckler, 1974; West et al., 1981b) and can further lead to the assumption that understory trees are younger. Diameter distributions of individual species in mixed species stands can have characteristic shapes (Lorimer and Krug, 1983).

The forest floor stratum which invades during the understory reinitiation stage can consist of advance regeneration of the same tolerant species as those which occupy the C-, D-, and other lower strata but are contemporary with the A- and B-strata. Trees in the C- and D-strata are sometimes presumed to have invaded later, as did the forest floor stratum, but to have grown taller in an all-aged manner.

Patterns described above vary with differences in height growth patterns of component species, branch stiffnesses, shade tolerances, regeneration mechanisms and disturbance types, numbers of stems of competing species, sites, life spans and mortality rates between species, times of stem initiation, spacings among stems, and other factors. Each factor will be discussed below. Before these factors are discussed, however, several examples of mixed-species, single-cohort stands will be given.

Examples of Mixed-Species Development
Northern red oak, red maple, black birch, and hemlock (Figs. 9.1 and 9.2)

Northern red oaks, red maples, black birches, hemlocks, and other species grow together in single-cohort stands in central New England and elsewhere. The stands begin with so many maples and birches that the few oaks are easily overlooked and often assumed absent at first. All three species have similar height growth patterns for the first two decades on upland sites. After this, the oaks outcompete the others and form a B-stratum above them, causing the maples and birches to slow in height and diameter growth (Scholz, 1948; Oliver, 1978a; Hibbs, 1983; Kelty, 1986). The more numerous maples and birches at small diameters in the lower strata have mistakenly led scientists to assume they are younger than the oaks (same stands as Figs. 2.1 and 6.2A and B). The oaks which eventually dominate probably begin as advance regeneration beneath the previous stand.

Mixed stands of these species have certain silvicultural advantages compared to pure stands. The maples and birches can act as "trainers" (McKinnon et al., 1935) by keeping oak stems pruned but allowing oak crowns to expand above them when older without costly thinnings. The overstory oaks then grow as if they were in a single-cohort stand, with individual oak growth being more dependent on spacing of other overstory oaks than on numbers of lower-stratum trees (Kittredge, 1988).

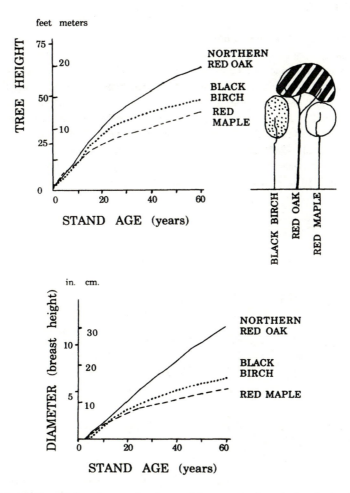

Figure 9.1 Height and diameter growth patterns of three species in a single-cohort, mixed-hardwood stand in central New England. All trees began growing together, but diameter and height growths of birches and maples slowed as the trees were overtopped by oaks. (*Oliver, 1978a.*) (See "source notes.")

Species such as sugar maples and beeches behave like red maples and black birches; other species such as paper birches and white ashes are stronger competitors and remain in the B-stratum with the oaks. Quaking aspens grow rapidly to an emergent position but are short-lived and die after about 5 decades. Hickories, on the other hand, first grow in lower strata but eventually can grow into the B-stratum, battering oak crowns with their stiff branches during windstorms.

Northern red oaks do not always outgrow the other species. Hemlocks and maples sometimes gain so much initial height advantage on very moist sites that the oaks can not outgrow them. The maples and hemlocks can also dominate in regions where the oak regeneration is not vigorous, where maples regenerate from very vigorous stump sprouts (Beck and Hooper, 1986), or where hemlocks grow from large advance regeneration (Kelty, 1986, 1989).

Cherrybark oak and sweetgum

Cherrybark oaks and other red oaks outgrow sweetgums, river birches, and hornbeams when they invade abandoned fields in river bottoms of the southeastern United States. They create stratified, mixed stands similar to the oak, maple, and birch stands described above. The few oaks and more numerous sweetgums maintain approximately equal heights for the first 2 to 3 decades; then the cherrybark oaks outgrow the sweetgums and form a pure B-stratum. The sweetgums here act as trainers for the cherrybark oaks (Clatterbuck et al., 1985; Clatterbuck and Hodges, 1988; Johnson and Krinard, 1988). Water oaks and swamp white oaks also outgrow blackgum and ironwood in a similar pattern in Mississippi bottomland forests (Bowling and Kellison, 1983).

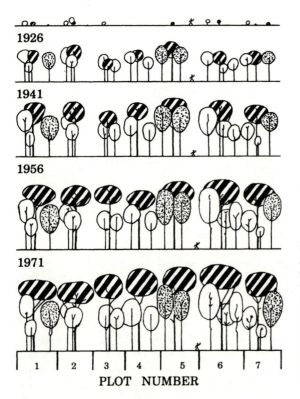

1911

1926

1941

1956

1971

PLOT   NUMBER

Figure 9.2 Development of single-cohort stand of red oaks (striped crowns), red maples (plain crowns), and black birches (dotted crowns) in central New England (same stands as in Figs. 2.1, 6.2A and B, and 9.1). Oaks eventually form a pure B-stratum. Person is 1.8 m (6 ft) tall. (*Oliver 1978a.*) (See "source notes.")

Cherrybark oak and loblolly pine

Loblolly pines overtop cherrybark oaks and relegate them to lower-canopy strata when they grow together in abandoned old fields on good sites. The oaks maintain upper-stratum positions for progressively longer periods when they are slightly older or are spaced farther from the pines. Since they compete vigorously with the oaks, the pines act more as competitors of than as trainers for the oaks (Clatterbuck et al., 1987).

Cherrybark oak and sycamore (Fig. 9.3)

Sycamores outgrow cherrybark oaks at first when planted together in well-drained river or stream bottoms. The oaks become overtopped when small and relegated to lower strata when spaced close to the sycamores. The sycamores act as competitors to the closely spaced oaks, or conversely, the oaks act as trainers to the sycamores. More widely spaced oaks are not overtopped by the sycamores and remain with them in the B-stratum. When not overtopped, the oaks maintain steady height growth after the first 2 decades while the sycamores slow down. As a result, oaks are found in lower strata when spaced close to the sycamores and in the B-stratum when spaced farther from the sycamores (Clatterbuck et al., 1987; Oliver et al., 1989).

Oak and tulip poplar (Fig. 9.4)

Tulip poplars have rapid and sustained height growth on all but the poorest sites. They generally overtop nearby oaks and other species. This overtopping commonly occurs with northern red oaks in southern New England, with white oaks in North Carolina (O'Hara, 1986), with northern red and black oaks on good upland sites in Tennessee (Smith, 1983; Beck and Hooper, 1986), with cherrybark oaks on riverbottom or loessial soils in Arkansas and Mississippi (Oliver et al., 1989), and with white pines in Maryland (Fenton and Pfeiffer, 1965). Tulip poplars are capable of overtopping even older oaks (O'Hara, 1986). By contrast, Virginia pines can outgrow and overtop tulip poplars on very poor sites, although on better sites poplars will predominate (Doolittle, 1958).

Douglas-fir and western hemlock (Fig. 9.5)

Douglas-firs and western hemlocks grow together in single-cohort stands at low elevations in coastal Washington (Wierman and Oliver, 1979). Douglas-firs overtop hemlocks after the first 2 or 3 decades, forming stratified stands with Douglas-firs in the B-stratum and hemlocks beneath or in the B-stratum where Douglas-firs are absent. Hemlocks can be more numerous at first, but they have a higher mortality rate than Douglas-firs. The stiffer branches of the preformed Douglas-firs protect their terminals while battering adjacent hemlocks with sustained growth and weaker branches. The hemlocks' upper stems eventually become crooked and flat-topped—probably from being battered and shaded.

These mixed stands are often characterized as western hemlock stands when young, since hemlocks predominate in both basal area and stem numbers. The continued growth of Douglas-firs and the higher mortality rates of hemlock make the Douglas-firs much more prominent with age. The stratification pattern can be used advantageously by planting few Douglas-firs and either interplanting hemlocks or allowing them to grow naturally from seeds. Hemlocks will grow large and utilize the growing space where

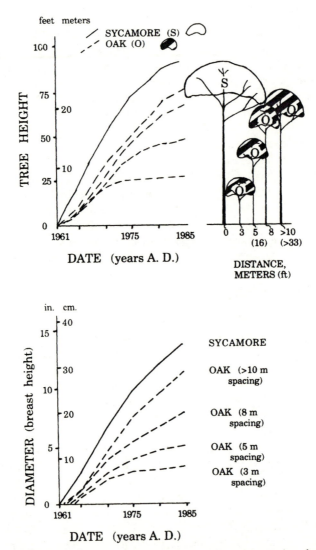

Figure 9.3 Development of single-cohort cherrybark oak and sycamore stand on abandoned field in Arkansas. Sycamores outgrow oaks when young, causing the oaks to slow in height and diameter growth sooner when closer to the sycamores. After 20 years, sycamores may slow in height growth while oaks not overtopped maintain height growth and widely spaced oaks remain in the upper stratum. (*Oliver et al., 1989.*) (See "source notes.")

there are no Douglas-firs. Where Douglas-firs survive, hemlocks will eventually lapse beneath them and act as trainers. Hemlocks stay alive beneath Douglas-firs longer than suppressed Douglas-firs do, so the suppressed hemlocks can be removed in thinnings or with the overstory at final harvest. The same stratification does not occur as readily at cooler, higher elevations where Douglas-firs do not grow as well as western hemlocks do.

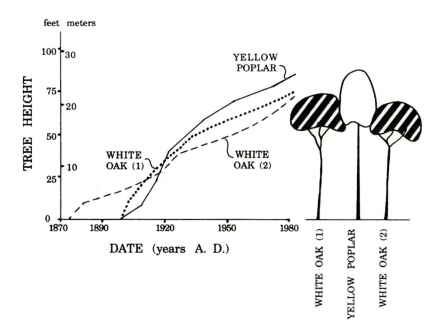

Figure 9.4 Height growth of yellow poplars and white oaks of the same and older cohorts following partial cutting in the North Carolina Piedmont. Yellow poplars outgrew the oaks even if the oaks were of a previous cohort. (*O'Hara, 1986.*) (See "source notes.")

Douglas-fir, red alder, western hemlock, western redcedar, and bitter cherry (Fig. 9.6)

Mixed-species, single-cohort stands of Douglas-firs, red alders, western hemlocks, western redcedars, and bitter cherries often appeared following clearcutting and burning in the late nineteenth and early twentieth century in coastal Washington. Bitter cherries predominated the new stand at first but died after several decades. The other species at first seemed to develop somewhat randomly, but by 50 years red alders generally occupied the B-stratum and western hemlocks and western redcedars were in lower strata (Stubblefield and Oliver, 1978). Douglas-firs were not found beneath red alders, but Douglas-firs and western hemlocks did grow as emergents above the red alder canopy after about five decades were not initially overtopped (Newton et al., 1968).

Stratification patterns of these species are largely dependent on height growth patterns and subtle variations in spacings and tree ages. Red alders grow rapidly at first but later slow dramatically compared to the conifers, resembling the patterns of sycamores with cherrybark oaks. The various species invaded during a period of about 20 years, and initial spacings were quite wide because few seeds invaded the large clearcut areas. Where Douglas-firs or western hemlocks were substantially older than the alders, they stayed taller than the alders and continued rapid height growth. Similarly, where the Douglas-firs or hemlocks were spaced far from the alders, the alder crowns did not overtop the conifers before the conifers outgrew the alders (Newton et al., 1968; Stubblefield and Oliver, 1978). Trees which escaped the alders' domination formed the emergent stratum.

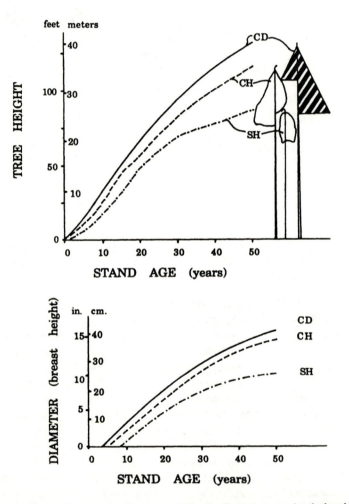

Figure 9.5 Height and diameter growth patterns of Douglas-firs and western hemlocks when growing together on productive sites in coastal Washington. Hemlocks (CH) not competing with Douglas-firs (CD) can grow as codominants in the B-stratum. Hemlocks (SH) close to Douglas-firs slow in diameter and height growth as they are overtopped by the faster-growing Douglas-firs. (*Wierman and Oliver, 1979.*) (See "source notes.")

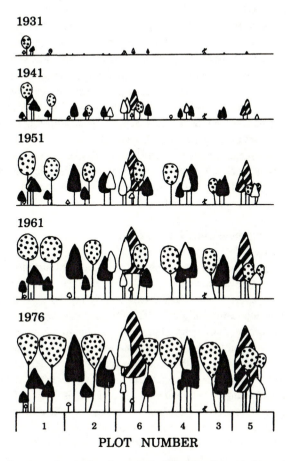

Figure 9.6 Development of a single-cohort stand of Douglas-firs, red alders, western redcedars, and western hemlocks in western Washington. (Dotted crowns = alder; diagonal bars = Douglas-fir; black = redcedar; plain = hemlock.) Hemlocks and redcedars can develop above or beneath alders, depending on initial spacing and slight age variations. Douglas-firs live only if they grow above the alders (same stand as in Fig. 2.1). (*Stubblefield, 1978.*) (See "source notes.")

Conifers became overtopped by alders when they were the same age or younger and were spaced close to the alders. Western hemlocks and redcedars can live beneath alder shade and therefore formed lower strata beneath the B-stratum of alders when over-topped. Douglas-firs cannot survive beneath alder shade and so were eliminated from the stand where overtopped by alders. Conversely, the alders died where enough conifers outgrew them to form a continuous canopy above them, since alders cannot live in the shade beneath any of the conifers.

Whether an area becomes dominated by Douglas-firs or red alders (and perhaps by hemlocks and redcedars as the alders die after many decades) is a chance occurrence of variations in tree ages and spacings during stand initiation.

Sitka spruce and red alder

Sitka spruces become relegated to lower strata and survive beneath red alders in coastal Washington, just as western hemlocks and redcedars become overtopped by alders. Sitka spruces, however, maintain strong epinastic control and have very stiff branches. Some spruces grow slowly upward and eventually emerge through the upper canopy of alders (Gara et al., 1980). Which spruces emerge through the alder B-stratum depends on subtle variations in heights, ages, and spacings of the spruces and alders.

The emergence pattern may have advantages for managing Sitka spruces. These spruces are attacked by Sitka spruce weevils (*Pissodes strobi*) when grown in full sunlight in many environments, and their stems become crooked. Spruces are less readily attacked when grown under alder shade. After they outgrow alders, their terminals may be too high to be attacked. Even if they are attacked, a long straight stem has already been produced beneath the point of deformation.

Western larch, lodgepole pine, Douglas-fir, and grand fir (Fig. 9.7)

Mixed-species, single-cohort stands in the northern Rocky Mountains contain various combinations of western larches, ponderosa pines, Douglas-firs, grand firs, western white pines, and lodgepole pines. Western larches grow rapidly to a pure upper stratum when mixed with Douglas-firs and grand firs. They act as emergents in the A-stratum where isolated or form a continuous, upper (B-) stratum where more numerous (Fig. 9.7A; Cobb, 1988). When lodgepole pines grow with larches, Douglas-firs, and grand firs, they occupy the B-stratum with the larches (Fig. 9.7B). They are larger than the larches for the first few decades. Eventually, the larches grow taller, and the lodgepoles die during the second century. Where larches and pines are isolated and cast little shade, they do not alter the other species' development patterns, and the other species stratify in a predictable pattern as well. Where larches and lodgepole pines are more crowded in the B-stratum, they grow less in diameter and reduce the height growth of lower-strata Douglas-firs and grand firs (Cobb, 1988; Cobb et al., 1993).

Douglas-firs first commonly overtop grand firs and relegate them to lower strata where the two species are the primary stand components (Larson, 1982, 1986). If a grand fir is not immediately beneath a dominating Douglas-fir, it continues height growth and eventually grows into the B-stratum with the Douglas-firs. Stiffer branches of grand firs abrade the dominant Douglas-fir crowns in winds, reducing the Douglas-fir crown sizes as the grand firs grow upward. Deteriorating crowns of the Douglas-firs can be noticed in the reduced Douglas-fir diameter growth even before the grand firs achieve an equal height (Fig. 9.7C).

Sitka spruce and western hemlock (Fig. 9.8)

Dense single-cohort stands of Sitka spruces and western hemlocks grow vigorously on upland, well-drained soils in southeastern Alaska after windstorms or similar disturbances destroy the previous stand (Deal, 1987; Deal et al., 1989). Hemlocks often regenerate much more numerously than the spruces. Height growths of the two species are similar for the first century, but the mortality rate of the hemlocks is higher. In addition, spruces generally stay in the dominant and codominant crown classes, possibly because their stiff branches abrade hemlock crowns. The spruces become more prominent as the stands grow older. Eventually the B-stratum can contain spruce dominants with hemlocks in the lower crown classes of the upper stratum and in all crown classes in lower strata.

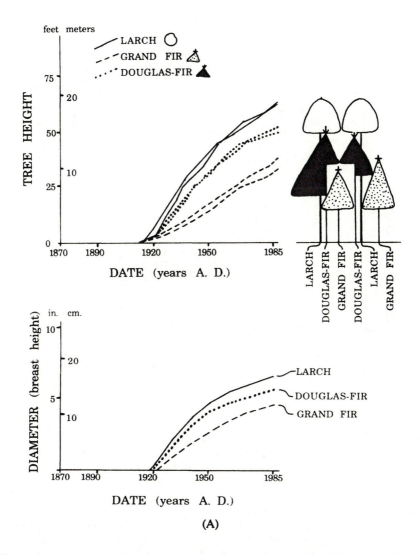

Figure 9.7 Development of single-cohort, mixed-conifer stands in eastern Washington Cascades. Stratification pattern depends on which species are present. (A) Larch, Douglas-fir, and grand fir stand. Western larches develop in the upper stratum, and Douglas-firs and grand firs in progressively lower strata. (*Cobb et al., 1993.*) (See "source notes.")

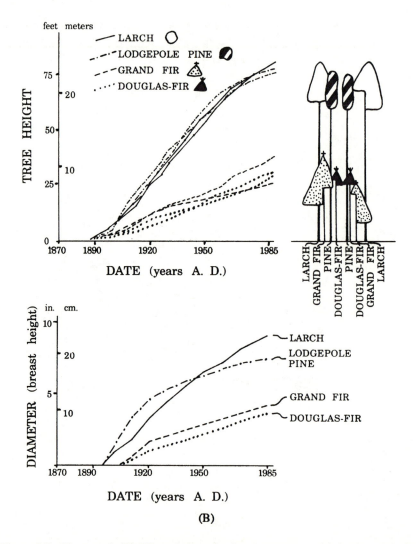

Figure 9.7 *(Continued)* (B) Where lodgepole pines are present with larches, Douglas-firs, and grand firs, pines first outgrow all species and then become slightly shorter than the larches. Douglas-firs and grand firs are even shorter because of added suppression by lodgepole pines. (*Cobb et al., 1993.*) (See "source notes.")

Pacific silver fir and western hemlock

Pacific silver firs and western hemlocks in the Western Cascade Range have very similar height growth patterns for the first 60 to 100 years, with western hemlocks growing only slightly taller (Herman, 1967; Grant, 1980). Since the species are quite similar in shade tolerance, hemlock-fir stands behave much like single-species stands. The firs seem to maintain rapid height growth longer than the hemlocks and eventually outgrow and dominate them—provided the firs have not already become dominated by the hemlocks. Height superiority of the firs seems to occur sooner on poorer sites (Grant, 1980).

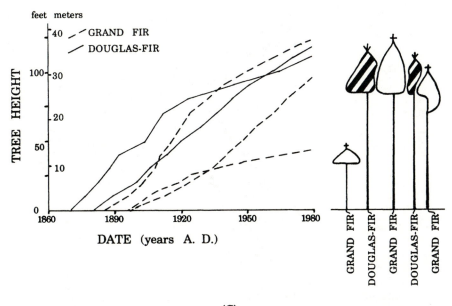

(C)

Figure 9.7 (Continued) (C) In stands where only Douglas-firs and grand firs are present, grand firs are initially shorter than Douglas-firs but often emerge through them, probably because of the grand firs' stronger branches. Diameter growths of Douglas-firs slow as grand firs emerge and abrade their crowns. (*Larson, 1986.*) (See "source notes.")

Because Pacific silver firs and hemlocks grow so similarly, the one dominating a stand often has gained the competitive advantage in ways other than height growth. Pacific silver firs survive as advance regeneration where the ranges of the two species overlap; western hemlocks cannot survive as advance regeneration (Thornburg, 1969; Kotar, 1972; Scott et al., 1976), but produces abundant, light seeds which travel far into disturbed areas. Stands dominated by firs are often found where disturbances such as snow avalanches or logging operations favored advance regeneration, and stands dominated by western hemlocks are found where fires or glaciers eliminated any advance regeneration and light-seeded species were favored (Oliver et al., 1985).

Western redcedar
Western redcedars are known for their slow growth and strongly tapered and fluted stem, even on very productive soils in the Pacific northwestern United States. Because of their slow early height growth compared to associated species, redcedars usually grow in subordinate canopy strata in single- or multicohort stands. They are common in the A- and

B-strata only when very old, probably after dominating trees of other species died and the redcedars were released.

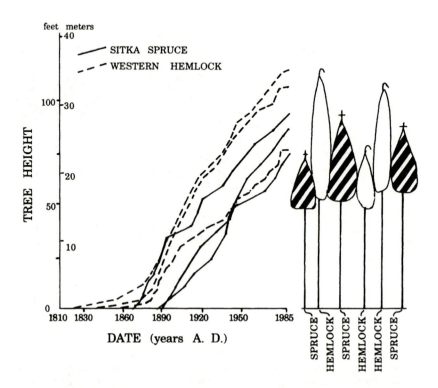

Figure 9.8 Development of single-cohort stand of Sitka spruces and western hemlocks in coastal, southeastern Alaska. Stand began in about 1865 after a large windstorm. Sitka spruces germinated after the windstorm but they were able to stay as dominants in the upper stratum with western hemlocks which began as advance regeneration. Hemlocks which germinated soon after the windstorm were relegated to lower strata. (*Deal et al., 1991.*) (See "source notes.")

Redcedars actually can grow rapidly in diameter and volume, producing stand volumes equivalent to those of Douglas-firs of the same age on similar sites by 50 years. Rapid growth occurs on the few occasions when redcedars grow in pure stands and so do not become overtopped and restricted in growth. In pure stands, the redcedars' very shade-tolerant lower limbs die, resulting in a much less tapered and fluted stem. The faster growth and cylindrical stem are seldom observed because redcedars become so readily overtopped that they seldom grow alone (Nystrom et al., 1984).

Factors Influencing Development
The preceding examples show variations in development patterns of mixed species stands, as well as patterns common to many stands of diverse species in different

regions. The patterns are not universal and are influenced by many characteristics of the competing trees and their environments. These characteristics are discussed below.

Height growth

Species with the most rapid early height growth generally form the upper strata where height growth patterns vary greatly. Rapid early height growths were the primary causes for alders dominating conifers in western Washington (Stubblefield and Oliver, 1978), for sycamores or loblolly pines dominating cherrybark oaks in the southeastern United States (Clatterbuck et al., 1987; Oliver et al., 1989), for western larches dominating mixed-conifer stands in the eastern Cascades of Washington (Cobb, 1988; Cobb et al., 1993), and for European birches dominating Norway spruces in Finland (Mielikainen, 1980, 1985), and for paper birches dominating white spruces in western and south central Alaska (Long, 1991).

Only slight differences in height growth exist among some species mixtures—such as northern red oaks, red maples, and black birches in New England, cherrybark oaks and sweetgums in the southeastern United States, and Douglas-firs and hemlocks in the Pacific Northwest. In these cases, other factors such as branch stiffness allow one species to stratify above another.

In mixtures such as Pacific silver firs and western hemlocks (Grant, 1980) or Sitka spruces and western hemlocks (Deal, 1987; Deal et al., 1991), little difference in height growth occurs. The species grow in the same stratum for many decades, differentiating into crown classes similar to those of single-species stands. In northern hardwood forests in the Great Lakes Region, U.S.A., stands with species of similar height growth rates and shade tolerances did not develop the stratification pattern as readily as stands with dramatic differences in these factors among species (Guldin and Lorimer, 1985).

Height growth patterns are quite different between interacting species in many cases, with one species growing rapidly at first while another grows slowly. The species with rapid early growth can overtop the other and, through shading of its crown and battering of its terminal, keep the subordinate one from growing taller (Oliver,1978a; Stubblefield and Oliver, 1978; Wierman and Oliver, 1979; Marquis, 1981b; Kelty, 1986). In cases such as hickories with northern red oaks (Oliver, 1975, 1978a), grand firs with Douglas-firs (Larson, 1982, 1986), and Sitka spruces with red alders (Gara et al., 1980), the overtopped trees maintain at least some height growth and eventually grow through the upper canopy.

Branch and limb stiffness

Physical abrasion by branches plays an important role in determining stratification patterns in many mixed-species stands. As trees grow taller and their limbs grow longer, they sway farther. Lateral branches of dominants batter against terminal shoots of other trees in windstorms, breaking the terminal shoots and reducing their height (Oliver, 1975, 1978a; Wierman and Oliver, 1979; Kelty, 1986).

The battering branches and terminals of adjacent trees are of similar strength and stiffness in single-species stands. Consequently, adjacent trees become equally battered. When one tree becomes taller than another, however, its lateral branches batter against the terminal of the shorter tree (Tarbox and Reed, 1924; Spurr, 1964). The taller tree does not suffer greatly if a lateral branch is broken in the process. However, height growth of the shorter tree is set back if the terminal is broken, and the tree may become overtopped by its neighbors before it can recover rapid height growth.

Battering branches and terminals of different species are of different strengths and stiffnesses in mixed stands. Usually the limbs and terminals of the tougher species remain intact while the others' branches or terminals are broken, giving the tougher trees the competitive advantage in maintaining height growth and dominating the upper canopy strata. Trees with preformed growth usually develop stiffer lateral branches (Chap. 3), which protect their own terminals while they batter against the weaker laterals and terminals of species with sustained growth. Stiffer laterals may partly explain how preformed northern red oaks stratify above sustained growth eastern hemlocks, red maples, and black birches (Oliver, 1975, 1978a; Kelty, 1986), how preformed Douglas-firs and Sitka spruces stratify above sustained growth western hemlocks (Wierman and Oliver, 1979; Deal, 1987; Deal et al., 1991), or how preformed cherrybark oaks stratify above sustained growth sweetgums (Clatterbuck et al., 1985; Clatterbuck and Hodges, 1988).

Where physical battering curtails a tree's height growth and relegates the tree to a lower stratum, the stem of the tree is often abraded and crooked. The crooks form when the terminal has been knocked out by a lower limb of a taller tree. After the lower limb has died and fallen, the abraded tree resumes height growth until again inhibited by another limb. This abrasion can be chronic, occurring during heavy winds. A similar *"sculpting"* of crowns has also been noted in tropical forests and has been referred to as *"crown shyness"* because the abraded crowns are separated when the wind is not blowing.

Branch stiffness may not be as important where large differences in height growth occur. For example, red alders have sustained growth with relatively limber branches but can stratify above preformed Douglas-firs (Stubblefield, 1978; Stubblefield and Oliver, 1978), and sustained growth tulip poplars and sycamores can stratify above preformed oaks (Beck and Hooper, 1986; O'Hara, 1986; Clatterbuck et al., 1987; Oliver et al., 1989) Shade-tolerant trees with stiff branches can also belatedly grow into the upper stratum, where their stiff branches abrade the crowns of taller trees as discussed previously.

Shade tolerance

The stratification pattern occurs only where species lapsing to the lower strata can live at lower light intensities than the trees shading them. For example, Douglas-firs will not form a stratification pattern with red alders, since the shade tolerances of the two species are very similar. The lower species dies when either species is overtopped. On the other hand, western hemlocks and western redcedars stratify beneath red alders, since they can live in the alders' shade (Stubblefield and Oliver, 1978; Stubblefield, 1978). Species of the same shade tolerances are often found occupying a stratum together (Nystrom, 1981; Nystrom et al., 1984; Guldin and Lorimer, 1985).

Tolerant trees living in the shade of another species survive on limited sunlight and generally grow very little. The shaded trees are generally dramatically smaller than their faster-growing contemporaries in the A- and B-strata, but certain conifers continue relatively rapid diameter growth even when overtopped by hardwoods. These conifers may grow beneath deciduous hardwoods by photosynthesizing during warm days in winter when the overstory is leafless (Kramer and Kozlowski, 1979). Eastern hemlocks, for example, can be larger in diameter than contemporary oaks overtopping them (Kelty, 1986), and western redcedars can continue moderate diameter growth even when over-topped by red alders (Stubblefield, 1978). It is unclear why the overtopped hemlocks and redcedars do not grow into the upper canopy; perhaps their terminals are in too much high shade to assume strong vertical growth or they become abraded on the lower hard-wood limbs above them (Kelty, 1986).

High shade, crown shapes, and height growth
Trees lapsing beneath the B-stratum move from a zone of low shade to the diffuse light of high shade (Fig. 3.6). Once below the zone of low shade, trees slow in height growth and change their terminal growth forms (Fig. 3.7; Busgen and Munch, 1929; Meng, 1986; Tucker et al., 1987). Tolerant trees survive in the high shade but grow very slowly in height. Species with strong epinastic control become flat-topped, with short terminal growth relative to lateral branch growth. Species with weaker epinastic control lose their terminals altogether. Even if trees beneath the B-stratum grow rapidly upward in high shade, their terminals enter a deeply shaded zone of little sunlight between the zones of high and low shade. The trees may reduce height growth even more in this zone. Consequently, the different shade zones often keep trees in distinct strata.

Spatial pattern
Spacing can determine which species dominate the upper strata when interacting species have markedly different height growth patterns. A species which grows most rapidly at first but later grows slowly dominates other species when grown at close spacings, since the species with rapid early growth overtops the others and relegates them to slow growth in lower strata. At wider spacings, a species with slower, continued height growth is not overtopped when small and eventually overtops species which later slow in height growth (Cline and Lockard, 1925). Examples of spatial pattern influence can be seen in the relation of cherrybark oaks to sycamores (Fig. 9.3; Clatterbuck et al., 1987; Oliver et al., 1989) and the relation of Douglas-firs and western hemlocks to red alders (Fig. 9.6; Newton et al., 1968; Stubblefield, 1978; Stubblefield and Oliver, 1978). Even when one species eventually overtops another, the sizes of the overtopped trees are greater and the ages when overtopped also are greater at increasingly wider spacings. Trees which grow at initially wider spacings also develop larger, more numerous limbs on their lower trunks (Clatterbuck and Hodges, 1988).

Slight differences in ages
A species invading a stand slightly earlier during the stand initiation stage sometimes gains enough height growth advantage that it dominates another species regenerating later, even though it would not have outgrown the later-arriving species if both had begun at the same time (Newton et al., 1968; Stubblefield and Oliver, 1978; Kelty, 1986).

Douglas-firs and western hemlocks can outgrow and overtop red alders when these conifers invaded a few years before the alders (Newton et al., 1968; Stubblefield, 1978; Stubblefield and Oliver, 1978). The conifers are generally overtopped by the alders when all trees begin during the same year.

On the other hand, cherrybark oaks which invaded old fields several years before loblolly pines remained dominant for more years than if they had been the same age as the pines (Clatterbuck et al., 1987; Oliver et al., 1989). Eventually, the pines' height growths were so much greater that the oaks did not maintain their dominance.

Regeneration mechanisms

Many species initiate by several regeneration mechanisms, and their early height growths vary depending on the regeneration mechanism. Advance regeneration northern red oaks and Pacific silver firs grow more rapidly than do seedlings of same species (Merz and Boyce, 1956; Carvell and Tryon, 1959; Aho, 1960; Bey, 1964; Sander, 1966, 1971, 1972; Johnson and Zingg, 1968; Clark, 1970; Utzig and Herring, 1974; Herring and Etheridge, 1976). Sprouts of many hardwoods grow much more rapidly than do seedlings, and the size and vigor of the parent stump greatly influence the sprout's early height growth.

The early height growth of a species regenerating by one mechanism can enable the species to overtop a competitive species, whereas another regeneration mechanism may not enable it to outgrow the same competition (Guldin and Lorimer, 1985; Deal et al., 1991). Red oaks beginning from advance regeneration and small sprouts were able to outgrow sweetgums, hornbeams, and river birches, but oaks beginning from seeds contributed little to the subsequent stand (Johnson and Krinard, 1988). Oaks, maples, and other southern hardwoods overtop (and kill) loblolly and slash pines when the hardwoods begin from stump sprouts, root sprouts, and advance regeneration after clearcutting. On the other hand, the pines dominate in abandoned old fields when all trees initiate from seeds.

Northern red oaks are much more competitive in mixed hardwood stands when they initiate from advance regeneration or seedling sprouts rather than from seeds. The stratification pattern found in central New England with the oaks dominating maples and birches probably occurs when the red oaks begin from advance regeneration (Oliver, 1975, 1978a). Red maples may not succumb to oak domination if they began from very vigorous stump sprouts farther south, in North Carolina (Beck and Hooper, 1986).

Pacific silver firs are more dominant in fir-hemlock stands in the North Cascades of Washington where the disturbance favors advance regeneration, since western hemlocks primarily regenerate from seeds after the disturbances at these elevations and Pacific silver firs can grow as advance regeneration (Thornburgh, 1969; Kotar, 1972; Long, 1976; Scott et al., 1976). Where all trees begin from seeds, western hemlocks predominate (Oliver et al., 1985). Western hemlock advance regeneration was able to grow rapidly enough in height to remain in the upper stratum with Sitka spruce for the first century following a windstorm in Southeast Alaska, while hemlocks regenerating from seeds after the windstorm were overtopped (Figure 9.8; Deal, 1987; Deal et al., 1991). On the other hand, advance regeneration oaks were unable to maintain their initial height advantage over tulip poplars which began as germinating seedlings at the time the oaks were released (O'Hara, 1986).

Site

Similar stratification patterns continue as a site varies slightly, since height growth rates of all species change. With dramatic variations in sites, however, site-sensitive species change growth much more than others do. At the extreme, one species may not be able to survive on a site where the other can; here the surviving species forms the dominant canopy even if it did not do so where both grew together.

Site-sensitive tulip poplars outgrow Virginia pines on productive sites but not on poor sites (Doolittle, 1958). On the other hand, tulip poplars outgrow oaks on both good and poor sites (Fenton and Pfeiffer, 1965; Smith, 1983; Beck and Hooper, 1986; O'Hara, 1986; Oliver et al., 1989)—although both species grow slower on poor sites.

At lower elevations, Douglas-firs outgrow western hemlocks on many productive sites in western Washington (Wierman and Oliver, 1979), but western hemlocks can live at much higher elevations than Douglas-firs (Franklin and Dyrness, 1973). Northern red oaks often outgrow red maples on mesic sites in central New England. Red maples, however, dominate very wet sites where the oaks cannot live.

Numbers of trees, life spans, and mortality rates

A species can appear to dominate a site when young simply because it has more stems than other species have. Sometimes this species maintains dominance and has many characteristics of a single-species stand; at other times it becomes overtopped by a species which, although fewer in number, grows larger and eventually dominates the upper stratum (Oliver, 1978a; Wierman and Oliver, 1979; Lorimer, 1981; Hibbs, 1983; Cobb, 1988; Cobb et al., 1993; Clatterbuck and Hodges, 1988; Deal et al., 1991). Where volume, number, or basal area of each species is measured, stands can appear to change species composition as the initially less numerous trees grow larger. Oak, maple, and birch stands in central New England change in species dominance as more numerous maples and birches are outgrown by the fewer, but eventually larger, oaks (Oliver, 1978a; Stephens and Waggoner, 1980; Lorimer, 1981; Hibbs, 1983). Cherrybark oaks, chestnut oaks, and swamp chestnut oaks similarly emerge over more numerous competitors and dominate stands on river bottom soils of the southeastern United States (Bowling and Kellison, 1983; Clatterbuck et al., 1985; Clatterbuck and Hodges, 1988). Mixed stands of Douglas-fir and hemlock in coastal Washington are often first characterized as hemlock stands because of the preponderance of that species but later are characterized as Douglas-fir stands when the fewer stems grow much larger than the hemlocks (Wierman and Oliver, 1979).

Species also have different life spans, or different mortality rates (Wilson, 1953; Connell and Slatyer, 1977; Keeley and Zedler, 1978; Peet and Christensen, 1980, 1987; Bowling and Kellison,1983; Harcombe and Marks, 1983). A species with a short life span often grows rapidly at first, sets seeds, and disappears from the stand. Bitter cherry behaves this way in many low-elevation stands in the Pacific northwestern United States (Stubblefield, 1978; Stubblefield and Oliver, 1978), as do fire and pin cherries in the northeastern United States (Marks, 1974; Hibbs, 1983). Some species do not necessarily have a short life span but have a high mortality rate compared to other species in the stand (Skeen, 1981). Western hemlocks can be much more numerous than Sitka spruces in young stands in coastal Alaska; however, many hemlocks but very few spruces die in the first few decades (Deal, 1987; Deal et al., 1991). These stands appear to change from being hemlock-dominated to being mixtures of spruces and hemlocks.

Even within a species, mortality rates can vary between regeneration mechanisms (Stephens and Waggoner, 1970, 1980; Beck and Hooper, 1986) or can be similar for all regeneration mechanisms (Stephens and Waggoner, 1970; Hibbs, 1983). Oak species regenerating from seeds after a disturbance generally have a much higher mortality rate than those regenerating from advance regeneration. Species regenerating from stump sprouts have a high mortality rate if all new shoots on each stump are counted. Generally, one to three shoots on each stump eventually survive (Wood, 1939); the mortality rate of stump sprouts is much lower if the number of sprouting stumps rather than the total number of stems is considered. Even for germinating seeds, survival is greatly influenced by seedbed conditions.

Compounding effects of many species

A third species in a stand can cause two species to interact differently than they would without the third species. Douglas-firs generally dominate hemlocks when only those two species grow together in western Washington (Wierman and Oliver, 1979). When red alders grow with the two species, alders generally overtop and kill Douglas-firs but hemlocks survive in the C-stratum beneath the alders (Fig. 9.6; Stubblefield and Oliver, 1978). Development patterns, therefore, become increasingly complex as the number of species in the stand increases (Guldin and Lorimer, 1985). Forests with many species— like Southeastern bottomland and cove hardwood forests and tropical forests—appear to be extremely complex because the large number of possible combinations of species, sites, spacings, and age ranges creates an extremely large—but not infinite or incomprehensible—variety of development patterns. Fortunately for forest ecologists and silviculturists trying to understand and manipulate these stands, disturbance types, previous stand histories, and other factors can be relatively similar over broad areas, and the number of patterns which commonly occur may be much less than the possible ones.

Other factors undoubtedly contribute to the development patterns of multi-species stands. Allelochemic actions by some species influence some of the patterns, as do insect and other animal behavior. Although it is helpful to describe and understand the various patterns which can occur, it is erroneous to assume that all species behave in similar ways or even that a single species will always behave in a single way.

Applications to Management

Mixed-species stands present many opportunities for manipulation. Mixtures of species also affect early stand treatment decisions. If the species which grows to the B-stratum is desirable, it is only necessary to assure that enough stems of this species exist to become crop trees (and enough extra ones for normal mortality) to ensure adequate stocking. Crop trees will lose lower limbs if trainer trees of species which will become relegated to lower strata are kept. Low-value trainer species can be useful by improving stem quality of crop species—especially if they lapse to lower strata than the crop species. Leaving these unmerchantable trees in a stand can cause the taller, high-value species to have a longer clear bole before large branches or the first major fork. The shorter trees also prevent water sprouts in the taller trees after a crown thinning. Sometimes two or more commercial species can be carried in different strata to the end of the rotation. In that case, species which will occupy both the B- and C-strata can be thinned to final, crop tree spacings.

Even with mixed stands, too many trees in any stratum will result in less yield per hectare (acre). As in single-species stands, if many trees die during the stem exclusion stage and are not utilized because they are too small or there is too little volume per hectare (acre) for a profitable timber sale, then all photosynthate that is contained in the wood of these logs will be lost. Decisions for management activities require an understanding of how the different species stratify when grown together at each spacing, on each site, and with different regeneration mechanisms. Thinning guidelines need to be based on both height growth patterns of the species and spatial patterns. Where an undesirable species will stratify above the desirable one, it is necessary to eliminate the undesirable species and space the desired species before stratification occurs.

The desirable spacing of upper- and lower-strata species may be dramatically different depending on the growth habits of each species and the site factors limiting growth. Red oaks and eastern hemlocks are very productive when oaks stratify above hemlocks in mixed stands in the northeastern United States. Because of the oaks' flat crowns, optimum spacing for B-stratum oaks may be wider than for C-stratum eastern hemlocks. There may be more hemlock stems per hectare even though the oaks are taller. Management of these stands can be confusing because of the differences in optimum spacing between species in different strata. Thinning hemlocks will not increase oak growth where water is not limiting, and thinning guidelines must treat each species separately. Where water is more limiting, wider spacing of either stratum allows remaining trees in both strata to grow more, as occurs in mixed larch, lodgepole pine, Douglas-fir, and grand fir stands.

Where a single species regenerates by several mechanisms in a stand, there is often an early decline in this species' numbers as stems regenerating by the less vigorous mechanisms disappear. The species will not necessarily disappear from the stand provided enough individuals of the more vigorous mechanisms are present.

When harvesting merchantable trees in mixed-species stands, it is important to realize that intermediate and suppressed trees in each stratum may not have died during the stem exclusion stage; they have simply been outgrown and suppressed by the other trees. The species in the lower crown classes in each stratum are generally deformed, unvigorous, and often contain rotten centers. They will not grow rapidly upon release, but they will still intercept sunlight and otherwise make establishment or growth of well-formed seedlings from advance regeneration or other mechanisms difficult or impossible. A commercial clearcut or shelterwood cut that does not remove the deformed, lower-crown-class trees in each stratum has many of the same attributes as a "high-grading" cut. Cutting in stratified, mixed stands to shift them from the stem exclusion stage to the understory reinitiation stage to obtain advance regeneration often includes removal of the lower-strata trees (Loftis, 1985).

CHAPTER 10

UNDERSTORY
REINITIATION STAGE

Introduction
Several decades after the stem exclusion stage begins, a low stratum of herbs, shrubs, and advance regeneration invades the forest floor, which until then had been relatively free of short living material. Forest floor vegetation develops as trees allow growing space to be available which the understory plants are physiologically capable of exploiting. Development of this forest floor stratum indicates the beginning of the understory reinitiation stage (Fig. 5.2).

Changes in the Overstory
Overstory trees grow very vigorously at the beginning of the stem exclusion stage. They actively occupy all growing space, quickly reoccupy any available growing space created by minor disturbances, and vigorously compete with neighbors for more space. Consequently, establishment of new stems is prevented. As overstory trees grow larger and older, however, they occupy the growing space less aggressively because of the nat-

ural development of the trees, changing effects of mortality, altered microenvironments beneath the stand, and minor disturbances.

Changing development and physiognomy of the trees

More light penetrates through the forest canopy, and roots less vigorously occupy the soil growing space as trees grow older and/or larger. Living branches in young stands often protrude into their neighbors' crowns, overlapping the foliage and effectively preventing any direct sunlight from reaching between crowns to the forest floor. As trees grow taller and the limbs grow longer, overlapping limbs abrade and break against each other until eventually the crowns no longer overlap (Tarbox and Reed, 1924; Spurr, 1964; Larson, 1982). Greater swaying causes adjacent crowns to stay farther apart as trees grow even taller, and increasingly more light penetrates between to the lower strata and forest floor (Fried et al., 1988). Overstory tree heights and crown sizes also become less uniform in older stands as differentiation proceeds (Chaps. 8 and 9). More light penetrates diagonally between crowns as well. Lower strata of mixed species stands absorb this sunlight and exclude the understory for a longer period than do single-species stands. Eventually, however, even lower-strata trees differentiate and their crowns separate, allowing light to penetrate to the forest floor.

Understory reinitiation occurs sooner on droughty sites than on mesic ones. Trees of most species retain less of their lower foliage on droughty sites, possibly as a physiological mechanism for retaining each tree's internal water balance (Grier and Running, 1977). With less foliage, more light penetrates through the canopy to the forest floor. Occasional rains can increase the soil moisture during seasons when the overstory's roots do not readily respond, further enabling forest floor vegetation to develop.

The fine, absorbing tree roots are constantly dying and regrowing on both mesic and droughty sites. The amount of fine root growth stays more constant on dry sites beginning several years after crown closure but decreases on good sites (Ovington, 1957; Santantonio et al., 1977; Keyes and Grier, 1981; Eis, 1986), possibly as internal recycling of nutrients increases. With the decline or lack of increase in fine roots, soil growing space is readily refilled upon death of old fine roots by the developing forest floor vegetation's roots. Insects and diseases specific to the roots of overstory species also build up in the soil, slowing growth of overstory trees while allowing new species to grow.

As trees mature, the number of primary growth centers (buds and root tips) and the amount of living, respiring tissue increase enormously. Because many primary growth centers compete for photosynthates, large trees cannot always expand limbs or roots as rapidly in one direction to refill growing space as they could when young. The greater respiratory demand also shifts photosynthate allocation away from expansion. Consequently, the rate at which trees refill temporarily available growing space can decline with age and size.

Changing effects of mortality

Death of a single tree through suppression does not release a large amount of growing space when trees are small at the beginning of stem exclusion. The tree had already relinquished most of its growing space before suppression and had only enough growing space at the time of death to maintain its respiration tissues. The crown and root area occupied by suppressed trees is so small that adjacent, living trees rapidly expand and reoccupy the growing space when the suppressed trees die. Even where many small trees die at once—as during a drought—the soil growing space and canopy area released by each one is small enough to be refilled rapidly by adjacent trees.

As a stand grows older, larger trees become suppressed and die (Figs. 8.4 to 8.6); consequently, more growing space is released by the death of each tree. Surrounding trees do not reoccupy the released growing space as vigorously as in young stands both because the surrounding trees grow less rapidly and because the area in which to expand is larger. Light and soil growing space, therefore, stay available on the forest floor long enough to be filled by newly initiating vegetation (Webb et al., 1972; Nakashizuka and Numata, 1982b). The area of released growing space and invasion of the forest floor is not necessarily directly beneath a gap created by the dead crown. Rather, forest floor vegetation initiates in a wide area around the dead tree; also, it does not grow uniformly within the gaps (Smith, 1963).

Changing understory environment

Light, temperature, and moisture continue to change in the forest floor and soil after crown closure, although the changes are not as dramatic as during stand initiation. These changes continue to alter growing space near the forest floor and so influence reinitiation of the understory.

The bottom of the living canopy is close to the forest floor soon after crown closure at the beginning of the stem exclusion stage. The forest floor receives little direct or diffuse sunlight, and conditions are not conducive to photosynthesis. As trees grow taller, the base of the living canopy rises and light conditions on the forest floor change to high shade, increasing the light intensity and quality so that plants can invade and survive, if not grow much. When a minimum light level is achieved in the understory, species with threshold requirements of light intensity necessary for germination begin growth as advance regeneration (Fried et al., 1988).

Soil conditions also change as the stand grows, shifting the physical location and amount of soil growing space. Understory plants may be able to invade as this shift occurs. In cool or dry regions, a thick organic soil layer can build up as the mineral soil becomes insulated and cooler. The area of active fine root growth rises into the organic layers (Vogt et al., 1981b). Nutrient availability and fine root development can also rise where podzolization is occurring. Following erosion or similar disturbances in most regions, roots increase the soil's water-holding capacity and hence soil growing space long after crown closure and stem exclusion as they penetrate clay soils of poor structure. Simultaneously, nutrients become more available as the A- and B-horizons develop following a disturbance which destroyed the previous soil horizons. Not only does more growing space become available, but angiosperm trees and shrubs more efficient in nutrient absorption from organic matter (Vogt, 1968) invade the forest floor where conifers dominate the overstory. Buildup of nitrogen also continues after crown closure as a soil ages, increasing the soil growing space and allowing new stems to invade.

The forest floor contains a cool, moist, carbon dioxide-rich environment which can allow plants to regenerate and survive at very low light levels. The amount of dead leaves, branches, and roots increases in the forest floor with stand age. Microorganisms decompose the organic matter and release carbon dioxide to the air (Hodges and Scott, 1968; Assmann, 1970; Woodwell and Botkin, 1970; Yoneda, 1975a,b). The higher carbon dioxide levels increase photosynthetic efficiency, so plants can survive at very low light intensities. Still air within the forest allows carbon dioxide to settle near the forest floor; and the moist, cool environment reduces drought stress and excess respiration during hot days (Geiger, 1965; Hodges, 1967). Consequently, forest floor plants live under marginal growing conditions in the modified microenvironment.

Changes caused by minor disturbances

Very minor disturbances regularly occur to forests and temporarily release growing space. Such disturbances during the stand initiation stage simply prolong the time of stem recruitment. Minor disturbances during the stem exclusion stage do not allow new stems to invade because established trees are so vigorous that they readily reoccupy any released growing space. As the overstory vigor declines later in the stem exclusion stage, small disturbances may release growing space which new plants invade first. Such minor disturbances include branch abrasion during windstorms, death of suppressed trees, uprooting of trees and exposure of mineral seedbeds, killing of roots by burrowing animals, and killing of fine surface roots and creation of seedbeds by ground fires or unusual droughts or frosts.

Development of the Forest Floor Vegetation

Seeds continue entering the stand via wind, water, and animals even during the stem exclusion stage. Now sexually mature, trees within the stand also produce seeds and add to those entering from outside. Fates of these seeds vary with species and local conditions. Soon after stem exclusion, the herbivore population which expanded when food was available during stand initiation is still large but lacks available food. The lack of green, succulent vegetation near the forest floor makes the seeds more visible and attractive to herbivores. Consequently, they consume many incoming seeds.

Seeds of some species germinate beneath the closed canopy of the stem exclusion stage, surviving on energy stored in the seeds. Life spans of these germinants varies with species, but none of the germinants grow very large. Eastern cottonwoods, for example, live only a few days without adequate sunlight whereas 15-year-old Pacific silver fir seedlings can be less than 5 cm (2 in) tall and have only 18 needles (Oliver et al., 1985). Seedlings which survive for very long either live on food stored in the seeds or are very shade-tolerant. Where seed sources of such species are nearby, the understory of the stem exclusion stage contains an almost invisible population of seedlings which germinate, grow very little, die, and are replaced by later germinants.

Some seeds can not germinate beneath closed canopies because the litter physically prevents them from reaching suitable germination sites or internal physiological mechanisms prevent their germination in cool or shaded conditions (Stanley and Butlers, 1961; Toole et al., 1961; Krugman et al., 1974; Osawa, 1986).

Some herbs and shrubs which began before the canopy closed continue growing slowly in the understory during the stem exclusion stage. Sword ferns survive beneath red alders in the Pacific northwestern United States, and mountain laurels live beneath eastern hardwoods this way. Vernal herbs in the forest floor utilize available growing space before deciduous overstory leaves develop each spring, although they do not grow vigorously during the early stem exclusion stage (Bormann and Buell, 1964; Jackson, 1966; Muller and Bormann, 1976; Muller, 1978). Some ferns can regenerate in the understory from broken pieces of rhizomes. In bright sunlight, the buds in the underground petiole apexes of rhizomatous ferns expand and form new rhizomes, and the ferns cover a large area rapidly. In deeper shade, few of the buds expand and the ferns expand slowly.

Light and soil growing space available to forest floor plants fluctuate but gradually increase during the transition from stem exclusion to understory reinitiation. The previously open forest floor gradually fills with new plants. Existing understory plants grow larger, and the seedlings which previously would have germinated and died now grow slightly larger and live longer. Seeds, rhizomes, and root sprouts of different species respond to the changed light levels, temperatures, or other measures of increasing growing space (Fried et al., 1988). Small disturbances create mineral soil or rotting logs as germination sites for some species. They also temporarily increase the growing space, allowing understory species to germinate and grow even though they later become suppressed. Unusual rains, overstory defoliations, or other factors increase growing space; and good seed years allow many germinants to exploit any available growing space. These events accelerate establishment of the forest floor vegetation. Root sprouts, rhizomes, stolons, and other regeneration mechanisms are not very vigorous during the stem exclusion stage; however, plants vigorously invade the forest floor by these mechanisms during understory reinitiation. Sugar maples, eastern hemlocks, and other species become established in large numbers during good seed years and gradually die; another wave later becomes established and dies even later (Hett and Loucks, 1968, 1971, 1976; Hett, 1971; Marquis, 1992). Oaks germinate, grow, and die back to the root collar (Merz and Boyce, 1956). Pacific silver firs and some Japanese firs (*Abies mariessi* Mast. and *Abies veitchii* Lindl.) exist as advance regeneration in two forms: (1) as very small seedlings (*seedling bank type*) and (2) as slightly larger saplings (*sapling bank type*) with well-developed lateral branches (Kohyama and Fujita, 1981; Kohyama, 1983; Oliver et al., 1985; Kanzaki, 1987). As the overstory canopy becomes more open, more seedlings survive the seedling bank and grow to the sapling bank regeneration which is easily visible on the forest floor (Schmidt-Vogt, 1986; Yamamoto and Tsutsumi, 1985)

Stands in the understory reinitiation stage can be considered multicohort stands, since new plants on the forest floor can be interpreted as a new cohort. Forest floor vegetation generally does not grow large and interact with the older cohort, however, and so does not alter the predominantly single-cohort character of the stand.

Understory reinitiation occurs earliest in stands of intolerant species or on droughty sites because much sunlight reaches the forest floor there. For example, understory reinitiation occurs in intolerant southern pine stands after only 4 to 6 decades (Fig. 10.1; Peet and Christensen, 1987). In more shade-tolerant Douglas-fir stands on good sites, however, it may not occur until 80 to 100 years (Fig. 10.2; Munger, 1930; Fried et al., 1988) or after 100 to 150 years in coastal Alaskan spruce and hemlock stands (Alaback, 1982a,b). In very shade-tolerant Pacific silver fir stands, understory reinitiation does not occur for over 120 years (Fig. 4.11A; Oliver et al., 1985). In mixed oak, maple, and birch stands in central New England, the forest floor reinitiates after about 80 or 90 years.

Shade-tolerant species usually predominate in the forest floor stratum, and can differ from an overstory of intolerant species. Shade-tolerant western hemlocks are generally found as advance regeneration in Douglas-fir stands in coastal Washington and Oregon (Franklin and Dyrness, 1973); shade-tolerant beeches and sugar maples are found beneath less shade-tolerant birches in northern hardwood forests; and maples, beeches, and other tolerant hardwoods are found beneath southern pine stands in the understory reinitiation stage (Glitzenstein et al., 1986). On the other hand, some relatively intolerant oaks live as advance regeneration (Liming and Johnston, 1944; Bey, 1964). These oaks grow for a few years, die back to the root collar, and resprout continually until released (Fig. 4.11C; Merz and Boyce, 1956).

Forest floor plants growing beneath the high shade of a dense overstory have different growth forms than when growing in the open. They grow little in height but expand outward with relatively long branches (Fig. 3.7), and their leaves are oriented perpendicular to the light. Shrubs and herbs similarly grow little in height but maximize foliage surface area exposed to light. Plants such as vaccinium, vine maple, and poison ivy have more erect, stronger stems in high light intensities but have weak central stems and many branches in the high shade on the forest floor. Overlap of leaves within or between shaded plants is rare at first. As the understory reinitiation stage progresses, more sunlight reaches the forest floor vegetation and both leaf area and leaf overlap increase.

Figure 10.1 Forest floor vegetation of brush and advance regeneration hardwoods beneath a 100-year-old longleaf pine stand in South Carolina.

Figure 10.2 Advance regeneration western hemlocks beneath a 100-year-old Douglas-fir stand in western Washington. The Douglas-fir stand began after a catastrophic fire. (Charred, old snag is from previous cohort.)

The spatial distribution of forest floor plants varies with soils and micro-topography (Struick and Curtis, 1962; Lyford and MacLean, 1966; Bratton, 1976; Gauch and Stone, 1979), rooting media, competition (Hough, 1937; Hough and Forbes, 1943; Wardle, 1959; Bormann and Buell, 1964; Bormann et al., 1970; Horsley, 1977a,b; Nakashizuka and Numata, 1982a), location of seeds or regeneration stock, season (Dansereau, 1946; Vezina and Grandtner, 1965; Bazzaz and Bliss, 1971; Bratton, 1976), and overstory conditions (Whitford, 1949; Bray, 1956; Moir, 1966; Smith and Cottam, 1967; Anderson et al., 1969; Thompson, 1980; Alaback, 1982a,b; Maguire and Forman, 1983).

Rotting logs provide growing sites for vaccinium shrubs, some conifers, and many other species when the other forest floor growing space is occupied (Christy and Mack, 1984). Some species cannot germinate on such sites (Jones, 1945). More competitive species exclude less competitive ones in ways similar to species invading disturbed areas, and the species and spatial distributions of forest floor vegetation are determined (Siccama et al., 1970; Fried et al., 1988). Plants germinating from seeds are not regularly distributed where the seeds are concentrated by animals or other vectors.

The overstory can also affect the spatial pattern of the understory. Oriental beech regeneration is missing where rhododendron is present beneath old beech forests in northern Turkey, perhaps because of allelopathy by the rhododendron or because small animals eat the beech nuts and root sprouts while living in the rhododendron. Western hemlocks grow as advance regeneration at low elevations in western Washington; however, they survive only in openings or on elevated microsites at upper elevations, whereas Pacific silver firs can grow beneath the dense forest (Thornburgh, 1969; Kotar, 1972; Scott et al., 1976). Apparently the lichens, branches, and other detritus from the overstory become compacted as the deep winter snows settle, and the flexible hemlocks become pressed to the ground and are killed. The small Pacific silver firs are much stiffer and remain upright and survive.

Advance regeneration trees generally live a few years or decades and then die, being replaced by younger trees which behave similarly (Hett and Loucks, 1968, 1971). Longevity of the advance regeneration varies by species, with cherrybark oaks able to live only 3 or 4 years, striped maple about 15 years (Hibbs, 1979), northern red oaks for several decades (Merz and Boyce, 1956), and some western conifers much longer. Understory Pacific silver firs less than 1 m (3.25 ft) tall can be over 100 years old (Fig. 4.11A; Utzig and Herring, 1974), although most die when much younger.

Eventual Fates of the Stands

The understory becomes more vigorous as more growing space becomes available, but forest floor trees do not grow rapidly into the overstory and create old growth stands. Sometimes, disturbances destroy the forest floor vegetation, the overstory, or both. If a disturbance does not occur, the forest floor vegetation becomes denser rather than noticeably taller as the overstory vigor declines. Understory trees in the understory reinitiation stage can be several meters (feet) tall (Fig. 10.2); they can also remain quite small even when quite old (Fig. 4.11A). The old age which advance regeneration achieves indicates understory trees remain quite small and distinct from the overstory for a very long time. Deaths of individual overstory trees have the effects of minor disturbances if the overstory is unvigorous, large, and old. Some understory trees respond to release, grow, and later become suppressed again by the overstory, behaving as a cohort in a multicohort stand. The gradual growing of advance regeneration to the overstory without significant disturbances creates an *old growth* forest with trees in all sizes (Fig. 5.2).

Applications to Management

The understory reinitiation stage is very important to management if advance regeneration is needed for the next generation as in many fir or hardwood stands or if advance regeneration hardwoods are a problem for the next generation as in many pine stands. Actions often need to be taken before this stage to increase or decrease the amount and type of vegetation initiating in the understory.

Early harvest of the stand can avoid advance regeneration where it is undesirable. Other methods of eliminating advance regeneration are less consistent. Fire can temporarily exclude the forest floor vegetation, especially for nonsprouting species; but the fire may also create mineral seedbeds and release growing space by killing fine roots of the overstory trees and thereby increase the forest floor vegetation. Fertilization can increase the overstory canopy density, thus reducing light to the understory and causing the forest floor vegetation to die. The forest floor vegetation may be easier to kill by manual or chemical means while living in the limited light beneath the overstory than after the overstory is removed and more growing space is available.

Advance regeneration is needed for the successful regeneration of many hardwood and conifer species in spruce-fir stands in Maine; in many Pacific silver fir stands in western Washington, Oregon, and British Columbia; and in hardwood stands where red oaks or black cherries are the desirable species. The optimal financial rotation for the overstory may occur before the stand is fully out of the stem exclusion stage. If desirable, the most dramatic way to shift a stand to the understory reinitiation stage is with a shelterwood cut, killing of lower-strata trees (Loftis, 1985), or heavy thinning. Lighter thinning of the overstory can stimulate the forest floor vegetation, but a vigorous overstory may eventually reclose and kill the younger vegetation. More shade-tolerant species will be favored in the understory if the thinnings are light, and the next generation will be more dominated by these species.

Establishment of advance regeneration can be inhibited by browsing animals. Deer cause this problem in much of the eastern United States. Where browsing animals are plentiful and advance regeneration is desired, preventive measures such as fencing, hunting, or putting browse protection devices on each seedling must be taken to ensure that stocking is adequate and that the understory does not become dominated by species that the deer find unpalatable. Many hardwood stands in the northeastern United States have become heavily dominated by undesirable beeches which grew as unpalatable advance regeneration before the removal of the overstory.

Stands in the understory reinitiation stage provide necessary habitat for many animal species. Although there is not as much browse as during the stand initiation stage, the added cover of tall trees is important for some species. Small mammals and turkeys and other bird species thrive in this stage.

Sometimes the problem is too much understory being established during this stage. Many species can initiate and become strong competitors with the desired species, and silvicultural activities may be needed to keep the initiating understory under control. In southern pine stands, for example, much silvicultural effort must be expended to suppress advance regeneration hardwoods if a new pine stand is to be regenerated from either pines in a shelterwood or pines planted after overstory removal.

The forester needs to monitor the understory carefully if there are plans to use the advance regeneration to regenerate the next stand. Some individuals will die if too shaded; others become flat-topped upon too much overstory (high) shade and crowded by other advance regeneration (side shade) and lose much of their live crowns. Most understory pines and some other species do not survive very long in the understory-reinitiation stage. Pacific silver firs, maples, and sugarberries can live long but are sometimes too weak to respond to release. Sometimes, advance regeneration individuals survive but do not grow if the overstory is removed; they take up valuable growing space at the expense of better-growing individuals. Advance regeneration hardwoods often develop poor forms in high shade. Upon removal of the overstory, these stems often die back to the root collar; and straight, vigorous stems grow in the full sunlight. Where the crooked stems do not die, it is desirable to cut them off at the root collar ("retune" them) to allow vigorous stems to grow as seedling sprouts. Most conifers, of course, will not sprout if severed at the root collar.

The growth response of each tree and shrub species, after it has changed growth form in the understory, varies with local soil and climate conditions. Local knowledge of the responses on each site is gained primarily through experience in a local area and is critical to the efficient promotion or control of advance regeneration and understory brush.

CHAPTER 11

OLD GROWTH STAGE

Introduction

Even after vegetation has reoccupied the forest floor, the forest floor vegetation grows very little for several decades or centuries. Eventually, however, individual, large overstory trees senesce and die (Franklin et al., 1987), releasing growing space. Other overstory trees do not reoccupy the released growing space rapidly, and newly germinating and sprouting trees and advance regeneration grow upward.

Patterns of younger trees growing upward vary with patterns of overstory death. The younger trees usually include a range of ages and sizes, since the overstory breaks up irregularly. The result is a stand containing trees of a wide range of ages and heights and with foliage well distributed vertically (Fig. 5.2). Stands with this large distribution of tree heights and ages can contain some relic trees which began immediately after a disturbance. Stands with younger trees growing upward but still containing some relic trees will be referred to as *transition old growth* stands. Barring a disturbance, all relic trees eventually die, and then the stand consists entirely of trees which grew upward through the deteriorating overstory. Stands in which the relic trees have died and which consist entirely of trees which grew from beneath will be referred to as *true old growth* stands.

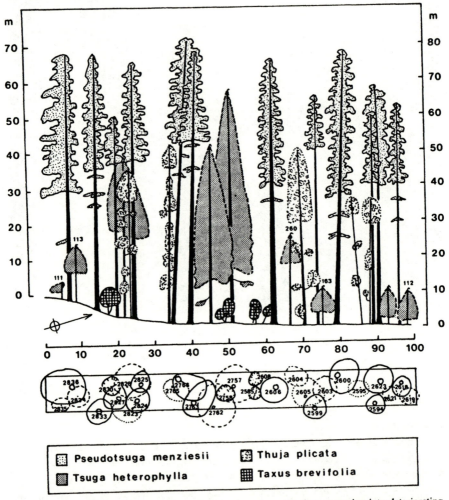

Figure 11.1 Shade-tolerant western hemlocks, and western redcedars emerging into deteriorating upper stratum of Douglas-firs in the transition old growth stage in western Washington Cascades. Douglas-firs probably began after a major disturbance, and other species began much later. (*Kuiper, 1988.*) (See "source notes.")

Process and Structural Definitions of "Old Growth"

In this book, "old growth" implies a uniformity of process. *True old growth* describes stands composed entirely of trees which have developed in the absence of allogenic processes. *Transition old growth* contains some trees which began after the initial disturbance and also large and numerous younger trees of allogenic origin. Using these definitions, not all old growth stands contain old, large trees or are dramatic in appearance.

Pure stands of short-lived black willows (Maisenhelder and Heavrin, 1957; Johnson and Burkhardt, 1976) or aspens (Jones, 1973; Mueggler, 1976; Betters and Woods, 1981) change to the old-growth stage rapidly while the trees are quite small, since these species do not grow large or live long (Table 11.1; Jones, 1945). Species such as coastal redwoods, Douglas-firs, eastern white pines (Nichols, 1913), and baldcypresses can achieve old ages and large sizes while still in the understory reinitiation (Fig.

TABLE 11.1. Upper Limit of Ages and Sizes of Selected Tree Species†

Tree species	Upper age limit, years	Upper size limit	
		Diameter, cm (ft)	Height, m (ft)
Baldcypress	1200	580 (19) [1]	55 (180) [2]
Douglas-fir	1375 *	550 (18) [1]	127 (415) [3]
Balsam fir	200 ±	215 (7) [1]	38 (125) [1]
Noble fir	600 ±	275 (9) [1]	100 (325) [4]
Pacific silver fir	590 *	245 (8) [1]	75 (245) [1]
Eastern hemlock	988 *	185 (6) [1]	49 (160) [1]
Western hemlock	500 ±	275 (9) [1]	79 (259) [1]
Eastern white pine	450 ±	365 (12) [5]	74 (240) [5]
Loblolly pine	300 ±	215 (7) [1]	56 (182) [1]
Longleaf pine	300 ±	125 (4) [1]	46 (150) [1]
Ponderosa pine	1047 * ‡	305 (10) [6]	80 (262) [1]
Coastal redwood	3000	640 (21) [1]	116 (380) [5]
Giant sequoia	1800	1130 (37) [1]	106 (347) [1]
Red spruce	400	155 (5) [1]	50 (162) [1]
Quaking aspen	200	140 (4.5) [1]	37 (120) [1]
Black birch	265	155 (5) [1]	25 (80) [1]
Eastern cottonwood	200 ±	365 (12) [1]	58 (190) [7]
Hackberry	200 ±	185 (6) [1]	41 (134) [1]
Red maple	150 ±	185 (6) [1]	55 (179) [7]
Sugar maple	400 ±	215 (7) [1]	42 (138) [8]
Northern red oak	300 ±	290 (9.5) [10]	49 (160) [1]
Sweetgum	300 ±	225 (7.4) [1]	61 (200) [1]
American sycamore	600 ±	490 (16) [9]	54 (176) [9]
Tulip poplar	300 ±	365 (12) [1]	61 (200) [1]
Black willow	70 ±	305 (10) [l]	43 (140) [1]
American beech	400 ±	185 (6) [1]	49 (161) [1]

†This table shows maximum possible sizes and ages of trees of various species in transition (and true) old-growth stands. * = maximum age reported; ages from Fowells (1965) and Harlow et al., (1979). Other information sources in brackets: [1] = Harlow et al., 1979; [2] = Hutchins, 1964; [3] = McWirtter, 1982; [4] = Mitchell and Wilkenson, 1988; [5] = Lang, 1952; [6] = Vines, 1977, [7] = Fowells, 1965; [8] = Wells et al., 1983; [9] = Deam, 1921; [10] = Gangloff, 1988.
‡Usually 500 ± years max

10.2) or transition old growth stages (Fig. 11.1). Stands of large old trees and dramatic appearances can also contain several cohorts of trees (Fig. 11.2).

Other definitions of "old growth" are common (Jones, 1945; Harris and Farr, 1974; Thomas, 1979; Runkle, 1981; Brady and Hanley, 1984; Franklin et al., 1981, 1986).

Climax stands are often considered to be identical with stands which developed without disturbances to the true old growth stage (Jones, 1945). The term "climax" has been used and redefined so many times (e.g., Cowles, 1911; Clements, 1916; Schenk, 1924; Jones, 1945; Braun, 1950; Whittaker, 1953; Churchill and Hanson, 1958; Scott, 1962; Budowski, 1970; Forcier, 1975; White, 1979) that it is avoided in this book. Climax is sometimes considered a stable "end" of relay floristics development patterns, so that senescing trees are replaced by other individuals of the same species rather than another species (Braun, 1950). Such stands are assumed to contain the same species in the understory and overstory and to develop in the absence of disturbances. True old growth stands in this book are defined as developing in the absence of allogenic disturbances; however, there is no reason to believe a single species or group of species continues to dominate the site.

Transition or true old growth on a site does not imply a certain array of species. Similarly, it does not imply massive trees or characteristic downed logs or organic material. Such features are often associated with transition or true old growth stands; but they are found in multicohort stands and single-cohort stands in the stem exclusion and understory reinitiation stages as well.

Figure 11.2 This stand of western hemlocks and Pacific silver firs in northwestern Washington has structural characteristics of old growth but is actually a multicohort stand.

Very old stands containing majestic trees in the southeastern United States or the tropics do not always contain large downed material because dead organic material decomposes rapidly (Olson, 1963; Toole, 1965a, 1966). As an exception, baldcypress and American chestnut stems are very resistant to decay and remain as downed logs for many decades (Cornaby and Waide, 1973).

Time when the old growth stage is reached

Barring subsequent disturbances, the time before a postdisturbance cohort is completely replaced by younger trees varies with species, site, stand structure, and other conditions. The natural life spans of trees vary from less than 100 years to over 1000 years (Table 11.1). A single-cohort stand of a single, short-lived species soon changes to the old growth stage as the short-lived trees senesce and younger ones dominate. If other species are not present, old growth aspen stands develop within a century in the Rocky Mountains (Jones, 1973; Mueggler, 1976; Betters and Woods, 1981). Black willow stands on river floodplains in the southeastern United States develop into mixed-species stands of sweet pecans, sycamores, red maples, boxelders, green ashes, elms, and sugarberries (Shelford, 1954; Maisenhelder and Heavrin, 1957; Johnson and Burkhardt, 1976) within 60 to 80 years as the short-lived willows die.

Sometimes a similar physiognomy exists in both stratified, single-cohort stands and transition old growth stands when the upper stratum species is very short-lived. Transition old growth stands with short-lived species in the overstory can appear as single-cohort, vertically stratified, mixed stands; but trees occupying lower strata are younger than the overstory trees. On the other hand, single-cohort mixed-species stands with several strata (e.g. Figs. 9.1 through 9.8) can resemble transition old growth stands (Oliver, 1978a, 1980; Stubblefield and Oliver, 1978) when still in the stem exclusion stage. Whether lower strata trees are in the same age cohort as upper ones or in younger cohorts is of more academic than practical importance in cases of short-lived overstory species.

From 100 to over 500 years is commonly required before most North American stands reach the transition old growth development stage. Douglas-firs which begin after a major disturbance in the Cascades Mountains and Pacific coast area can remain dominants for over 1,000 years. During this time, younger trees of many species grow beneath them and die in the transition old growth condition.

Poor sites decrease the time before overstory trees senesce and the transition old growth stage is reached, especially if the site reduces the vigor of the overstory species. Loblolly pines quickly lose vigor on droughty ridges in the Piedmont of the southeastern United States. Pine stands which grew on these ridges after old fields were abandoned can have younger hardwoods grow into the pine canopy in less than 100 years (Fig. 11.3). Alternatively, if the overstory species is well adapted to the poor site, the slower growth retards development of the old growth condition. Ponderosa pines on droughty sites in the Rocky Mountains, longleaf pines in the Sand Hills of the southeastern United States, and western hemlocks at upper elevations in the Cascade Range often stay alive much longer on poor sites than when found on better sites where they quickly grow large and are forced to compete with more species.

Figure 11.3 Hardwoods growing into pine overstory in a transition old growth condition on a poor Piedmont site in South Carolina. On this site, the pines are short-lived. They become weakened and no longer dominate the site before reaching majestic sizes .

Spacing and species composition of the overstory also affect how soon the older cohort dies and the old growth stage is reached. Closely spaced trees do not grow rapidly in diameter and so become subject to buckling or tipping, especially if they have rapid, prolonged height growth such as Douglas-firs or eastern white pines. Even if thinned to wider spacings, the increased diameter growth may not counterbalance the continuing height growth and the stand remains unstable. In fact, stands of very large, old Douglas-firs (or other species which continue to grow tall) may be special cases of stands which began at wide spacings or on poor sites, where height growth was not as rapid.

Species composition also affects how long the cohort initiating immediately after a disturbance dominates the stand. In pure stands, maximum age of the dominating species varies (Table 11.1). An initially dominating species in a single-cohort, mixed-species stand sometimes does not live as long as one in the lower strata. Consequently, instead of old growth being reached when the dominant species dies, trees of the initial cohort but previously in lower strata may emerge. An old growth condition does not occur when the gray birches, willows, or fire cherries in mixed-species stands die if the upper stratum is dominated by one of these species. Instead, previously subordinate maples, beeches, red alders, western hemlocks, or western redcedars (Oliver, 1978a; Stubblefield and Oliver, 1978) of the same cohort grow to the upper stratum and continue to exclude younger trees. Stands of very large western redcedars found on moist sites in western Washington probably began with an alder overstory (Fig. 9.6) which died after the first century and allowed contemporary redcedars to dominate for the next 1000 years.

Disturbance patterns, climatic fluctuations, and other factors external to the stand can also affect stand development into the old growth stage. Disturbances may destroy a stand or change it to a multicohort stand before it reaches the old growth stage. These disturbances are somewhat predictable by region, and their likelihood usually increases as the trees grow larger. Their affecting a particular stand, however, is still a matter of chance. Stands in certain locations are also more prone to destruction (e.g., stands in mountain passes are often destroyed by wind), and old growth stands are generally confined to protected areas. Similarly, climatic fluctuations during the life of many trees can be so drastic that older trees become weakened and die on some sites (McGee, 1984), increasing the rate at which the old growth condition is reached.

Development of transition and old growth stands

Increasingly larger intermediate and suppressed trees in the upper strata die as the understory reinitiation stage progresses. Eventually, dominant or codominant trees begin dying (Franklin et al., 1987). As some dominant overstory trees die, others weaken (Grier et al., 1974; Long and Smith, 1984) and die more readily. Consequently, the transition from a continuous overstory to a stand with an occasional, isolated overstory tree sometimes occurs quite rapidly, often within a few decades. Causes of the overstory's rapid decline have not been investigated in detail (McGee, 1984), but the overstory probably deteriorates because death of some trees allows more wind to penetrate beneath the canopy and blow over the remaining ones (Harcombe and Marks, 1983), because the falling trees scar the remaining ones, allowing rot to enter (Kirkland and Brandstrom, 1936; Isaac, 1940, 1956; Englerth, 1942; Englerth and Isaac, 1944; Shigo and Larson, 1969; Deal, 1987), because large dead trees concentrate pathogen and insect populations which then attack the living overstory trees, and/or because increased growth of the understory stresses overstory trees, making them die more easily (Johnson and Kovner, 1956; Zahner, 1958; Barrett, 1970, 1973, 1979; Oliver, 1984). More sunlight and wind hits previously protected and shaded parts of the residual trees; nutrients can become "tied up" in organic matter in the cooler understory and therefore unavailable; and the vigorous understory can reduce soil moisture and nutrients available to overstory trees.

Unlike younger, more vigorous overstory trees whose neighbors have been killed by a minor disturbance, old overstory trees generally do not respond rapidly to refill the available growing space when a dominant neighbor dies. Consequently, the understory trees which respond to the increased growing space are not later suppressed by a vigorously expanding overstory. These understory trees grow upward and form a new, often uneven overstory.

Death of each dominant tree does not necessarily create a distinctive "island" of regeneration in the hole left by the dead tree (Oliver and Stephens, 1977). Rather, trees initiate and grow in a large area surrounding each dead tree—probably because the root and sunlight influences of the dead overstory tree extended well beyond the projection of its crown. The occasional deaths of overstory trees probably create so much spatial overlapping of regenerating trees that the net result is a varied mosaic of young trees emerging into the overstory with a variety of ages and sizes (Peet and Christensen, 1987).

The understory can behave in several ways as overstory trees die, depending on the climate, species, and other conditions. In very cold climates where organic matter accumulates on the forest floor, nutrients become inaccessible through podzolization and organic matter accumulation. Overstory tree vigor then declines, and sphagnum moss and other plants capable of surviving the nutrient-poor conditions invade. Ultimately the trees die and the true old growth condition of such stands can be a sphagnum bog (Fig. 11.4; Zach, 1950; Heusser, 1952; Lawrence, 1958; Heilman, 1966, 1968; Strang, 1973; Viereck, 1973; Black and Bliss, 1978; Ugolini and Mann, 1979; Strang and Johnson, 1981). Bogs have replaced stands which developed for about 1200 years with no intervening disturbances since glacial retreat in southeastern Alaska (Vitousek and Reiners, 1975).

Figure 11.4 Forest floor litter builds up and soil conditions become unsuitable for tree growth without disturbances in cool climates such as southeastern Alaska shown here. Consequently, old growth stands can become sphagnum bogs *(muskegs)*, barring disturbances.

If less organic matter accumulates, understory trees remain more vigorous during the transition old growth stage. Trees which eventually form the overstory during the true old growth stage can be either tolerant or intolerant of shade. Shade-tolerant species sometimes become established in the understory reinitiation stage and slowly grow upward as the overstory releases growing space (Long, 1976). Not all suppressed understory trees remain vigorous enough to grow into the overstory. In stands with a deep organic layer on the forest floor, the primary germination sites for new seedlings are rotting logs or stumps, as discussed before. Trees on these microsites become very unstable if they are so shaded that prop roots do not develop before the logs or stumps deteriorate (Fig. 11.5). Where they exist only on unstable microsites, understory trees may not be stable enough to grow to the overstory unless a large minor disturbance releases enough sunlight for them to develop strong prop roots or creates new microsites where trees can initiate on stable substrates. Technically, stands with these minor disturbances would be multicohort stands and not in the old growth stage, although they would have old growth structures.

In other places, shade-tolerant trees become established on more stable microsites and grow slowly to the overstory (Connell, 1978). Such stable, shade-tolerant trees are found in the Pacific northwestern United States where Pacific silver firs and western hemlocks grow beneath overstories of the same species (Oliver et al., 1985) or where western hemlocks, Pacific silver firs, and grand firs grow beneath old Douglas-fir canopies. In the eastern United States, eastern hemlocks, sugar maples, and beeches grow into the overstory in the same way (Nichols, 1913; Stearns, 1951; Leak, 1961; Bormann and Buell, 1964; Brewer and Merritt, 1978; Runkle, 1981), and many hardwood species grow beneath shade-intolerant pines and hardwoods in mixed stands in the southeastern United States (Johnson and Kovr, 1956; Maisenhelder and Heavrin, 1957; Harcombe and Marks, 1978; Peet and Christensen, 1980).

Figure 11.5 Younger trees in old growth stands are often unstable where they germinate on unstable logs, stumps, and windthrow mounds. The low light intensities do not allow trees to develop large prop roots before the rooting medium deteriorates. (Straight young trees in background began after a later, larger disturbance. Compare with Fig. 6.5B and 6.5C.)

Shade-intolerant species also grow into the overstory in old growth stands. Newly germinating, intolerant species may grow rapidly in the increased growing space as the overstory deteriorates. These young intolerant trees sometimes outgrow the tolerant understory trees before they recover from previous suppression. Intolerant trees probably grow where a large amount of growing space is released quickly, as after the death of many overstory trees, a few short overstory trees, or overstory trees with broad crowns and large roots systems. For example, intolerant yellow poplars can develop as the overstory oaks, hickories, and yellow poplars die in eastern old growth stands (Buckner and McCracken, 1978; Barden, 1979; McGee, 1984, 1986a,b).

Intolerant species may also grow in old growth stands if the climate, site, or other factors prevent more shade-tolerant species from surviving or outcompeting them (Mueggler, 1976; Betters and Woods, 1981). On dry sites in Oregon and Montana, Douglas-firs grow in the understory beneath Douglas-firs while more shade-tolerant western hemlocks and other species cannot survive in the droughty soils (Tesch, 1981; Means, 1982). Similarly, intolerant ponderosa pines grow in the old growth understory on sites too dry for more tolerant Douglas-firs, hemlocks, or true firs (Franklin and Dyrness, 1973).

Younger trees replacing the overstory during the old-growth stage may or may not be the same species as the original overstory. In areas where few tree species grow, each species has a broad realized niche (Chap. 2) and consequently grows both as a dominant after a disturbance and as an emerging, younger tree in the old growth stage of the same stand. For example, Sitka spruces and western hemlocks predominate in most coastal forest stands of Southeast Alaska. Both are found initiating immediately after a major disturbance as well as growing into the overstory in very old stands (Deal, 1987; Deal et al., 1991). Tolerant Pacific silver firs behave similarly with western hemlocks at slightly higher elevations (Kotar, 1972; Thornburgh, 1969; Oliver et al., 1985). Shade-tolerant beeches and sugar maples sometimes replace themselves and sometimes alternate in replacement in eastern hardwood old growth stands (Runkle, 1981).

At other times, younger trees growing into an old growth overstory can be different species than the older overstory (Franklin and Dyrness, 1973; Veblen and Stewart, 1982a). Shade-tolerant western hemlocks grow beneath older Douglas-firs in old growth stands in coastal Washington and Oregon (DeBell and Franklin, 1987) and beneath Sitka spruces in coastal Alaska (Reiners et al., 1971; Alaback, 1984a). Similarly, shade-tolerant eastern hemlocks, sugar maples, and beeches grow beneath older eastern white pines and less shade-tolerant hardwoods in the old growth stage in eastern forests (Nichols, 1913; Henry and Swan, 1974).

Stems growing slowly in the overstory in transition old growth are distinctly younger than the overstory. At the earliest, the younger stems begin as forest floor advance regeneration during the understory reinitiation stage. As these stems grow larger after the understory reinitiation stage, they can occupy the growing space efficiently enough to exclude even younger stems. Consequently, the emerging understory in transition old growth stands can sometimes undergo a series of recruitment, exclusion, and then understory reinitiation stages as it develops (Bray, 1956; Whitmore, 1978; Nakashizuka, 1983, 1984a,b; Peet and Christensen, 1987). If this exclusion occurs, true old growth stands can contain a series of broad waves of age classes rather than trees of all ages.

Distinctions between old growth and multicohort stands with some old trees can be arbitrary. Transition old growth stands resemble multicohort stands in which younger cohorts are much younger than the oldest one. True old growth stands resemble multicohort stands with many cohorts at close time intervals. If waves of age classes occur in true old growth stands, these stands resemble multicohort stands with cohorts well distributed in time. Whether an individual tree's death is an allogenic or autogenic process is often academic, since autogenic factors generally predispose a tree to allogenic factors, and vice versa. Surviving trees do not behave differently whether the death of the overstory was allogenic or autogenic, and the degree to which outside factors cause deterioration and death of the overstory exists in a gradient anyway.

Fates of old growth stands

The fates of old growth stands vary. Eventually, a disturbance destroys all or part of the stand and a single- or multicohort stand develops. Before or after stands develop to the true old growth condition, they can be destroyed by a disturbance. True old growth stands have probably never been common because major disturbances occur quite frequently in most areas compared to the life spans of trees. Trees become more predisposed to disturbances as they grow larger; and so stands of progressively older, larger trees are increasingly rare (Henderson, 1994). The fates of stands as they approach the old growth condition vary with climate, species, and other factors. In some places the stands continue for a long time with a low mortality rate of the overstory and low replacement rate by the understory (Lorimer, 1980; Runkle, 1981, 1982). In other places, the overstory dies in a short time and understory trees emerge (McGee, 1984, 1986a, b). The true old growth condition can be reached quickly if trees are short-lived and disturbances are infrequent, but only slowly if the postdisturbance species are long-lived. In some places the stands can develop into upland sphagnum bogs; elsewhere they can remain as forests if the decomposition rate of dead organic matter is rapid enough.

Definitions of Old Growth Structures

In contrast to the above definition based on *processes*, "old growth" from an aesthetic or wildlife perspective usually implies massive trees, large downed logs, and an abundance of dead organic material (Franklin and Waring, 1979; Franklin et al., 1981, 1986; Parker, 1989). In these cases, old growth is viewed from a structural or physiognomic perspective rather than from a developmental one (Jones, 1945; Wallmo and Schoen, 1980; Franklin et al., 1986).

From a structural perspective, old growth has been assumed to have various characteristics:

- A reverse-J shape diameter distribution (Meyer and Stevenson, 1943; Minckler, 1971; Leak, 1973). This diameter distribution was assumed to indicate a reversed-J age distribution as well.

- A variety of tree species and other vegetation (Leak, 1973; Alaback, 1984a; Franklin et al., 1981, 1986).

- Many large, old trees, often at a wide spacing (Nichols, 1913; Franklin et al., 1981, 1986; Alaback, 1984a).

• A relatively continuous vertical distribution of foliage (Lutz, 1930; Hough, 1936; Morey, 1936; Franklin et al., 1986).

• Standing dead trees—snags (Franklin et al., 1981, 1986). Of course, the deterioration rate of these depends on their size and location (Engelhardt, 1957; Wright and Harvey, 1967).

• Large logs on the ground (Franklin et al., 1981,1986). Here, too, the time before these logs decompose depends on the climate (Toole, 1965a, 1966).

• A relatively equilibrium (or steady-state) volume, where growth equals mortality (Steele and Worthington, 1955; DeBell and Franklin, 1987).

• A relatively equilibrium (or steady-state) nutrient condition, where a large amount of internal recycling occurs (Sollins et al., 1980; Bormann and Likens, 1981).

Such conditions support complex interactions among animals, living plants, and dead organic material. Although intricate from biomass, productivity, and nutrient cycling perspectives and aesthetically fascinating, such conditions are unique to certain species and climates and are not necessarily the result of an absence of disturbances (Parker et al., 1985). Stands containing massive trees are often those of long-lived species which grow at wide spacings (Leopold et al., 1989). Stands of some species would never qualify as old growth by the structural definition, since the trees rarely grow large (Table 11.1). Trees reach their largest sizes when they begin after a major or large minor disturbance; therefore, very large trees are often found at the end of the understory reinitiation stage, during transition old growth, or in multicohort stands where old trees have grown very large. An old growth structure can be developed with management (Newton and Cole, 1987).

Large standing snags, downed logs, and abundant organic matter are not common in very old stands in some regions. Snags and logs exist in various degrees of decay (Harmon et al., 1986). Old stands contain large downed logs and organic matter where individual trees grow large, where a species' wood is resistant to decay and where organic decomposition is slow. Such stands occur in the cool, coastal conifer forests of the Pacific northwestern United States (Cole et al., 1967; Grier, 1978b; Grier and Logan, 1977; Waring and Franklin, 1979; Franklin and Waring, 1979; Franklin et al., 1986) and in the dry conifer forests of the Rocky Mountains (Thomas, 1979). Slow rates of decay in these forests make snags and downed logs important for timber production (Buchanan and Englerth, 1910; Kimney and Furniss, 1943; Engelhardt, 1957; Shea and Johnson, 1962; Wright and Harvey, 1967; Belluschi and Johnson, 1969), wildlife habitats (Thomas, 1979; Cline et al., 1980; Davis et al., 1983; Maser and Trappe, 1984), nutrient cycling (Cornaby and Waide, 1973; Balda, 1975; Sharp, 1975; Grier, 1978a,b; Breedy, 1981; Brown, 1985), and other geomorphologic processes (Swanson et al., 1976; Harmon et al., 1986).

Applications to Management

The development of old growth stands through the biological process of growing for very long in the absence of external disturbances does not necessarily lead to the popular conception of old growth stand structures of large, majestic trees. In fact, the large majestic trees can be the result of multicohort stands, mixed-species stands, or relatively young vigorous stands still in the stem exclusion or understory reinitiation stage. At the same time, the developmental (process) old growth stands may not contain large or majestic trees if the species involved are short-lived and do not grow large.

It is sometimes assumed that old growth stands can be created in a short time through silvicultural manipulation. Different aspects of old growth stand structures can for the most part be created in a relatively short time frame; but for stands to complete the process of growing without intervening disturbances takes more time and often requires careful planning of protected locations for the stands, intensive protection from fire, and luck to keep the stands from blowing over or being destroyed by insects or other disturbances.

For nontimber management objectives such as recreation and wildlife habitats, most concern is for an old growth structure, not the old growth process of stand development. Unlike older stands, the upper strata of young stands are vigorous and may suppress and exclude younger cohorts unless periodically thinned. Many different strata can be created with foliage extending horizontally at all heights. The trees will be in distinct age cohorts, however, because the stand will have developed from a series of partial disturbances (allogenic development) rather than from continual replacement (autogenic development).

It is also impossible to retain current old growth stands permanently. All stands are subject to disturbances. Even where stands are protected from most human intervention, there is a natural disturbance cycle typical of each stand type and location. The forester constantly faces the question whether to stop a disturbance (such as a fire) and retain the old growth structure or let the stand revert back to an early stage of development. To perpetuate stands with old growth structures for special plant and wildlife habitats and for recreation, foresters could plan and manage younger stands in protected locations which can be grown to old growth structures to replace the older stands which will eventually be destroyed.

The old growth structure has been defined in different ways in the Pacific Northwestern United States. Franklin et al. (1981, 1986) and the Wilderness Society (Morrison, 1990; Morrison et al., 1990) defined it as stands with large trees, multiple canopies, large dead snags and down logs and other attributes described earlier. The United States Forest Service defined it as stands over about 200 years old. Later, the Forest Ecosystem Management Assessment team expanded the concept to "late succession and old growth" stands, which were defined as stands over about 80 years old (FEMAT, 1993). The question of what definition of old growth to use for maintaining stands with "old growth" structures depends on the objectives of management. Any decision should be made by policymakers after a comparison of robust alternative management approaches for achieving the various objectives (Chapter 16). If the objective is aesthetic, the decision is based on beauty, accessibility, and other tradeoffs. If the purpose is to provide habitats for animals and plants depending on "old growth" as part of an effort to provide structures for all species across the landscape, the amount of old growth to be protected is limited by the need to provide stands with other structures as well—so a complete landscape of "old growth" would not provide maximum biodiversity. In this case, the stands with the greatest relatively "old growth-like" structures within the landscape at the appropriate spatial distributions could be protected, while other stands could be silviculturally manipulated to provide other structures. If there were insufficient amounts of old growth structures within the landscape, any existing "old growth" structures could be maintained and those stands which could most efficiently develop old growth structures through growth and silvicultural manipulations could be treated for that purpose.

CHAPTER 12

MULTICOHORT STANDS:
BEHAVIOR OF COMPONENT COHORTS

Introduction

Development patterns in multicohort stands are more complicated than in single cohort stands. In multicohort stands, each cohort is influenced by the presence of the other cohorts. Each cohort can also differ in species composition and spatial and temporal distribution of trees. As with single-cohort stands, species composition and spatial and temporal distribution of each cohort are strongly influenced by the disturbance type and other events immediately surrounding the cohort-initiating disturbance. Because each cohort can be so different, its influence on other cohorts in the same stand varies greatly in amount and quality of shade, microclimate, competition for soil moisture and nutrients, availability of seeds and root sprouts for regeneration, allelochemic influences, and other factors (Marquis, 1992). Trees in multicohort stands are influenced by all of the gradients of single-cohort stands and by the additional gradient in competition created from other cohorts.

This chapter will focus on the behavior of each cohort within a multicohort stand. The holistic development of these stands will then be discussed in Chap. 13.

Factors Influencing Development of Each Cohort

Influences common to single- and multicohort stands

Behaviors of trees initiating after minor disturbances are strongly influenced by the same factors as those that govern tree growth following major disturbances. Which species are present to invade the disturbed area limits the species composition of the new cohort. Root sprouts and seeds from trees surviving the minor disturbance; stump, seedling, and root sprouts from trees which did not survive; and advance regeneration potentially comprise the new cohort. The disturbance type acts as an "environmental sieve" to determine which species and regeneration mechanisms survive and dominate (Davis and Hart, 1961; Willis and Johnson, 1978; Viers 1980a,b, 1982). Soils, microsites, and spatial patterns also give one or another species a competitive advantage (Trimble and Hart, 1961).

Influences of associated cohorts

Trees of older cohorts which survive a disturbance influence regeneration and development patterns of younger cohorts in many ways. They provide a seed source for the new cohort. Surviving trees modify the wind, humidity, and temperature microclimates (Geiger, 1965). They also actively expand to refill the growing space released by the disturbance, as do newly invading individuals (Abrams and Nowacki, 1992; Mikan et al., 1994; Abrams et al., 1995). As the available growing space is refilled following a minor disturbance, the plants compete both within and between cohorts to obtain growing space held by others. Older cohorts are generally taller and so have an advantage in competing for sunlight, but they do not always have the same advantage in competing for soil moisture and nutrients.

Younger cohorts also affect the growth of older ones. At first, trees of the younger cohort are shorter and do not compete for sunlight, but in some instances younger trees grow as tall as the old ones and compete in the same stratum (Oliver and Stephens, 1977). Young trees modify the microclimate of older trees by reducing soil surface temperatures and wind speeds near the ground. The major impacts of younger cohorts— especially if they are shorter—are to compete with older trees for soil moisture and nutrients. Trees initiating in younger cohorts can also increase nutrient availability or insect and pathogen populations, depending on the species.

Low shade and high shade

Sunlight from a taller residual overstory casts high shade on the forest floor and regenerating trees (Fig. 3.6). The diffuse high shade provides enough sunlight for shade-tolerant trees to survive but not grow rapidly upward, and trees in cohorts living in high shade have distinctly flat-topped crown shapes.

Angle of sunlight

Direct sunlight practically never comes from immediately overhead. Sunlight comes from directly overhead in tropical forests for a few minutes around noon for a few days each year. In the United States on June 21, it reaches a maximum angle from the horizon of approximately 88° in Miami, Florida; 77 to 78° in Flagstaff, Arizona and Nashville, Tennessee; 75° in Pueblo, Colorado and Lexington, Kentucky; 70 to 71° in Lander,

Wyoming and Hartford, Connecticut; 65 to 66° in Caribou, Maine and Seattle, Washington; and 55° in Juneau, Alaska (Buffo et al., 1972). For the rest of the year, direct sunlight is from a lower angle. Even on June 21, direct sunlight comes from a low angle during most of the morning and afternoon and is at a high angle only for the few hours around noon.

Sunlight passing through a partial overstory left by a minor disturbance comes from an angle, and shadows are not cast directly beneath trees. Rather, shadows are cast in a direction opposite to the sun and change during the day and between days. The effect of a large tree in a multicohort stand is not necessarily to cast shade and inhibit growth directly beneath itself. Rather, it casts shade over large areas to its west, north, and east (in the northern hemisphere) during the day. Direct sunlight penetrates farther beneath a tree as the height to its crown base increases, allowing smaller trees to live beneath its canopy. At higher latitudes, the angle of sunlight is lower for more of each day and for more of the year; therefore, the importance of this side light increases.

Crown expansion and densification

Canopy reclosure after a disturbance generally occurs slowly but varies with species, site, age, and vigor of the residual trees (Chapman, 1942; Woods and Shanks, 1959; Zahner and Whitmore, 1960; Trimble and Tryon, 1966; Phares and Williams, 1971; Erdmann et al., 1975; Hibbs, 1982; Larson, 1982; Runkle, 1984). Very old trees or trees of lower crown classes are generally not vigorous enough to expand their crowns and roots rapidly compared to younger, vigorous dominant and codominant trees. Consequently, the less vigorous trees do not increase in competition with younger cohorts. Full reoccupation of growing space is sometimes assumed to be at crown reclosure—after the residual overstory crowns are touching. In fact, lower strata in multicohort stands become suppressed long before the overstory recloses. This suppression is partly related to reoccupancy of the soil growing space, but equally important is the increasing crown density of overstory trees (Chap. 3; Larson, 1982). Overstory trees· with weak epinastic control respond to removal of adjacent trees in the same stratum by accelerating growth of their lateral branches, reclosing the canopy over the gap, and often creating more flat-topped crowns. Trees with strong epinastic control often maintain their crown shapes; their lower limbs do not accelerate growth but the crowns become denser. The lower crowns stay alive instead of receding and grow outward at the normal rate controlled by the terminals. Because they are not being shaded, the lower crowns stay alive and, as the trees grow taller, the lengths of live crowns increase. The crown foliage stays alive, and water sprouts grow within the living crowns of many species. Consequently, the crowns cast more shade. The denser crowns effectively block direct sunlight to the forest floo · even though the crowns are not touching, since much of the shade is cast at an angle (Trimble and Lyon, 1966; Larson, 1982).

The relative importance of crown densification compared to canopy reclosure varies with latitude; canopy reclosure may be more important at low latitudes where sunlight is more frequently from high overhead. Species at high latitudes have more conical or columnar crowns and more epinastic control, whereas tree crowns at lower latitudes are generally more rounded, with weaker epinastic control (Fritschen et al., 1980; Kuuluvainen and Pukkala, 1989). At latitudes where sunlight from a low angle becomes more important than from directly overhead, trees of older strata increase their shade in proportion to their increased height growth—especially where the trees are so open that the bases of their live crowns are not receding. Consequently, where trees are do not grow rapidly in height because of age, site, or vigor, they do not rapidly increase their shade on younger cohorts.

Soil moisture and nutrients

Soil conditions are much different after a minor disturbance than after a major one. The soil surface is cooler after minor disturbances. In cool climates, organic matter may not decompose and release nutrients as readily. Residual trees continue to evapotranspire and so help keep moist sites from becoming anaerobic but continue to deplete soil moisture from dry sites. Rainfall passing through residual trees reaches the ground in a nonuniform pattern as stem flow and crown drip (Gersper and Holowaychuck, 1971), whereas rain is distributed more evenly over a site where trees are absent. In areas where significant soil moisture comes from fog condensing on overstory limbs and dripping to the ground, a partial overstory continues this moisture input.

Roots of different trees are intertwined, rather than being spatially distinct as tree crowns are (Fig. 3.12B; Stout, 1956). All soil growing space in a discrete area is not released when some overstory trees are destroyed. Rather, the released growing space is intermingled with that occupied by the intertwined roots of still living trees on a broad area.

Growing space and rooting patterns shift as a cohort becomes older. In cool environments, the trees' fine roots shift from the mineral soil to the organic matter as soil warmth and available nutrients shift upward. Older cohorts may have no advantage over younger trees in occupying soil growing space, since there is not a single, stable rooting zone which older trees can maintain.

Soil growing space previously occupied by the destroyed trees may not always be available after a partial disturbance. Destruction of some trees reduces evapotranspiration (Douglass, 1960). In cases of very moist soils, the soils can become anaerobic. Here, the partial disturbance decreases growing space and even kills trees not immediately destroyed by the original disturbance. Where soil growing space is made available by the disturbance, existing trees expand their roots into it, as do the initiating trees forming a new cohort. Root systems of residual trees already live throughout the soil, intertwined with those of recently killed trees. It is unclear how rapidly roots of older trees can expand into available growing space compared to roots of younger trees.

A partial disturbance can favor plant species different from those remaining in the residual stand. These new species' roots may expand into the available growing space more readily than the residual species can (Voigt, 1968). For example, where grasses become established beneath forest overstories, they can form a thick, shallow root mat—or sod—and exclude absorbing roots of trees, which then grow at deeper soil depths. During the growing season these grass roots absorb and evapotranspire the rains—the primary means of soil moisture recharge on droughty sites—and reduce tree growth. Similarly, where trees of more shade-tolerant species become established beneath a partial overstory of less-tolerant species, they expand their roots at light intensities too low for the overstory species to live had it become established in the understory. These understory plants of different species exploit soil growing space more efficiently under some conditions. Hardwood species, with their finer, more numerous roots, may exploit well-developed soils better; conifers may better exploit soils with poor structures and nutrients in mineralized forms (Voigt, 1968). These differences may partly explain the success of hardwoods growing beneath pines in the Piedmont of the southeastern United States as the degraded soils improve with pine growth (Zahner, 1958).

Some disturbances destroy or weaken roots of residual trees, allowing new plants to invade and grow rapidly. Ground fires burn soil litter and release growing space by killing fine roots even where trees are not killed. The weakened roots of residual trees do not expand rapidly, and newly initiating plants reoccupy the soil growing space. Consequently, ground fires often result in thick grass on the forest floor although they reduce tree competition by killing some trees. In droughty ponderosa pine stands in eastern Washington, residual trees increased growth following ground fires which killed intermediate and suppressed trees, but not as rapidly as when the forest was thinned by manual cutting. Unlike fires, manual thinning does not damage roots, but allows residual tree roots to reoccupy the soil growing space rapidly. Once the overstory tree roots reoccupy the soil growing space, grasses and other understory species do not readily invade.

Competition of different cohorts for soil growing space differs from competition for sunlight growing space. On sites where moisture is limiting, leaf canopies are often less dense (Grier and Running, 1977), more light filters to lower parts of the stand, and younger cohorts are more vigorous (Trimble and Hart, 1961). Unlike competition for sunlight, younger cohorts need not grow as large as the older ones before competing successfully for soil growing space. In parts of the southwestern and southeastern United States, removal of the understory trees and shrubs increases the water balance and growth potential of the overstory (Johnson and Kovner, 1956; Zahner, 1958; Barrett, 1970, 1973 1979; Oliver, 1984). Partial cuttings which leave dominant trees but create a new cohort in the understory sometimes decrease growth of the dominant trees on droughty sites, probably because the younger cohort occupies soil growing space previously available to the dominant overstory. On sites where soil moisture is not limiting, the understory's effect is much smaller (Stone, 1952).

The overstory, of course, decreases the understory's ability to exploit soil growing space by shading it and reducing its growth so that it cannot maintain an adequate root system. Very shaded fir seedlings, for example, had curtailed growth of fine roots compared to more vigorous seedlings (Kotar, 1972; Magnussen, 1980; Magnussen and Peschl, 1981). Grass sod beneath dense shade is similarly thinner than in open areas.

Both overstory and understory roots also live in the ephemeral, fluctuating soil growing space. This growing space exists during some seasons and years—such as periods of high rainfall—but not during less favorable times. Plants living primarily in the ephemeral growing space may not survive periods of stress, or they may expand into more permanent growing space before the ephemeral growing space disappears.

Less dominant trees often occupy the more ephemeral growing space and most readily die in times of drought. Alternatively, large dominant trees have large respiration and photosynthetic surface areas and therefore a constant, large requirement for moisture, nutrients, and photosynthates. Very large dominant trees may also be vulnerable to slight reductions in soil growing space.

Other influences

The understory and overstory influence each other in other ways. One species can contain chemicals in the foliage or elsewhere which inhibit other species (Table 6.2). Both overstories and understories can host pathogens or insects which invade other cohorts. Dwarf mistletoe (*Arceuthobium* species) in overstory pines, Douglas-firs, and true firs in the Rocky Mountains produce seeds which are dispersed beneath them and detract from growth of younger cohorts (Hepting, 1971). When grand firs grow beneath ponderosa pines and Douglas-firs on droughty sites in the eastern Cascades of Washington, they can become infected with *Armellaria obscura* root rot, which then develops enough vigor to invade pines and Douglas-firs. Insects also can be encouraged by one cohort and then detract from growth of another cohort in pine stands in the eastern United States (Baker, 1972).

Behavior of a Single-Cohort in Multicohort Stands
Initiation of a cohort

Minor disturbances release growing space by destroying trees or parts of trees. The growing space released can be primarily below ground if roots are mainly destroyed, or primarily above ground if only foliage and limbs are destroyed. Destruction of either above- or below-ground parts causes a shift in utilization of the other, since trees maintain a balance of roots and foliage.

The shade and moderate microclimate left after a partial disturbance may allow much more effective establishment of trees than the intense sunlight and extreme microclimatic variations of stand-replacing disturbances (Worthington, 1953; Franklin, 1963; Williamson, 1973).

Release of growing space allows new trees, shrubs, and herbs to invade by germination, sprouting, and release of advance regeneration (Nakashizuka, 1983). Both shade-intolerant and shade-tolerant species begin growing even after small disturbances (Brewer and Merritt, 1978; Hibbs et al., 1980; Barden, 1981; Runkle, 1981; Hibbs, 1982) when minimum sunlight thresholds for germination are reached or when germination and early sprout growth are triggered by available soil growing space (Toumey and Kienholz, 1931; Davis and Hart, 1961; Willis and Johnson, 1978). Many tolerant and intolerant species regenerated when roots were cut around a treeless plot in a closed red pine forest in New Hampshire. The cutting (trenching) released soil growing space but did not alter the available sunlight (Toumey and Kienholz, 1931).

Even though many species initiate after even small disturbances, available growing space strongly influences which species survive and how fast they grow (Clark, 1970; Franklin and DeBell, 1973). As the overstory and more shade-tolerant species rapidly refill the growing space, intolerant species are eliminated from the new cohort (Davis and Hart, 1961; Willis and Johnson, 1978). Twenty-one years after the many shade-intolerant and -tolerant species regenerated in the trenched plots in the red pine stand of Toumey and Kienholz (1931, described above), the very shade-tolerant eastern hemlocks were the dominants in the post-trenching cohort (Lutz, 1945). Less-tolerant gray birches, aspens, willows, and other species—which outgrow hemlocks in full sunlight—did not grow well in the shade and eventually died. Similarly, western hemlocks outgrow

Douglas-firs in coastal Washington if the overstory is retained long enough, even though Douglas-firs outgrow hemlocks in full sunlight (Franklin and DeBell, 1973). Shade-tolerant sugar maples, beeches, and red spruces similarly outgrow less shade-tolerant associates in partial shade (Hart, 1963; Leak et al., 1969; Runkle, 1981).

Spatial patterns of newly initiating cohorts are the result of suitable forest floor conditions for regeneration by the various mechanisms. Plants successfully become established on extremely unstable logs, stumps, and windthrow mounds only where a large disturbance allows new trees to grow rapidly and establish strong prop roots before the unstable substrate deteriorates (Fig. 11.5).

Younger cohorts appear in patches if suitable regeneration substrates are created only in the gaps created by the disturbances. When broader areas suitable for germination, sprouting, or release of a new cohort are created, the new cohort does not appear only in a small gap beneath the disturbed area, since both soil and light growing space are released in a broad area around and beyond a destroyed tree (Oliver and Stephens, 1977). Both the intertwining of roots from different trees and the varying patterns of light as the sun moves create the broad area of available growing space.

Growth beneath an overstory

The effect of available glowing space—especially sunlight—on height growth and survival of trees in a younger cohort is schematically shown in Fig. 12.1. For simplicity, soil, disturbance type, and climate are held constant, and amount of growing space occupied by older cohorts is viewed as a gradient. A stand developing after a major disturbance where no competition exists from older cohorts is one extreme of the gradient, and a cohort developing after essentially no disturbance to the older one is the other extreme.

The initiating cohort forms a single-cohort stand when it grows with no overstory. The trees develop in a predictable stratification pattern when other factors of spacing, site, climate, disturbance type, and regeneration vigor are constant (Chap. 9). At the other extreme of a very dense overstory, little stratification occurs because the newly initiating trees either die quickly or grow very little in height.

Between the extremes, emerging patterns vary (Jones, 1945; Monk, 1961; Auclair and Cottam, 1971; Forcier, 1975; Ricklefs, 1977). Species which dominate single-cohort stands dominate beneath light overstories (Marquis, 1977). More shaded overstories favor tolerant trees to dominate (Lutz and Cline, 1947; Gilbert and Jensen, 1958; Trimble and Hart, 1961; Stubbs, 1964; Hook and Stubbs, 1965; Minckler and Woerheide, 1965; Trimble, 1965, 1970, 1973; DeBell et al., 1968a,b; Tubbs, 1968, 1975, 1977; Metzger and Tubbs, 1971; Sander and Clark, 1971; Harkin, 1972; Franklin and DeBell, 1973; Leak and Solomon, 1975; Skeen, 1976; Franklin, 1977; Marquis, 1977; Leak and Filip, 1977; Brewer and Merritt, 1978; Connell, 1978; Hibbs et al., 1980; Runkle, 1981; Viers, 1980a,b, 1982; Parker et al., 1985; Parker and Sherwood, 1986). All species slow in height growth as overstory shade increases (Wahlenberg, 1948; Brender and Barber, 1956), but certain species—generally more shade-tolerant ones—maintain more height growth in deeper shade than others which may grow better in less overstory shade (Trimble and Hart, 1961; Stubbs, 1964; DeBell et al., 1968a,b; Willis and Johnson, 1978). Each species slows in height growth at a characteristic level of overstory competition (Brender and Barber, 1956; Tryon and Trimble, 1969; Akai,

1978). Eventually, species with weak epinastic control lose their excurrent form (Fig. 3.7; Busgen and Munch, 1929); those with stronger epinastic control also develop a flat-topped appearance but maintain a terminal (Kohyama, 1980; Meng, 1986; Tucker et al., 1987). At progressively greater overstory densities, the various species finally die, with more shade-tolerant ones dying later at greater overstory densities.

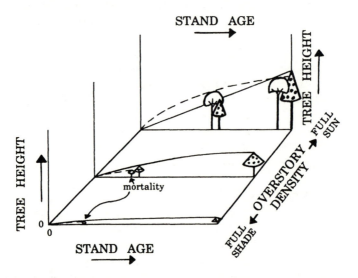

Figure 12.1 Schematic height growth of two species along gradient of overstory density. [Shade-tolerant species with strong epinastic control (triangular crown) and less shade-tolerant tree with weak epinastic control (round crown).] Height growth rates and abilities to survive are reduced for both species as the overstory density increases; however, the less tolerant species' vigor and height growth decline more rapidly, so that stratification pattern and survivorship change between species with increasing overstory density.

Trees of the same cohort in a multicohort stand first grow at approximately the same height, distinct from other cohorts. Each cohort excludes later arrivals and then begins differentiating and developing into crown classes and strata, as occurs in single-cohort stands. The process occurs more slowly in lower cohorts of multicohort stands because of the slower growth beneath high shade. Multicohort stands can develop strata and crown classes within each cohort, but the stratification patterns may differ from those in single-cohort stands. As differential tree growth continues in multicohort stands, each stratum can contain trees of more than one cohort.

Actually, understory cohorts behave slightly differently than outlined above. Figure 12.1 simplistically assumes that overstory density stays constant. Overstory density also changes, which affects the understory. If the overstory remaining after a disturbance is vigorous, it rapidly increases in density. If the residual overstory is not vigorous it declines in density after the disturbance.

Understory species beneath an overstory increasing in density vary in height growth according to the density of the overstory (Fig. 12.2). When several species interact beneath an overstory, their relative height growths shift according to both the overstory density and any differentiation and stratification among themselves. If overstory density increases, progressively more shade-tolerant species dominate the understory. The more shade tolerant species do not slow in growth as rapidly as shade intolerants. Consequently, increasing high shade causes the intolerant tree to become more crowded by tolerants of the same cohort. This crowding shortens the live crown of the intolerants and causes them to decline in height growth more than they would if the tolerant trees were not present (Wampler, 1993; Guldin, 1995). The increasing overstory also causes understory growth to decline, many species to die, and remaining trees to become increasingly flat-topped. Eventually, the overstory becomes so dense that younger cohorts die and the number of cohorts in the stand is reduced. At an extreme, a multicohort stand reverts to a single-cohort stand if one cohort dominates and excludes the others (Runkle, 1981; Hibbs, 1982).

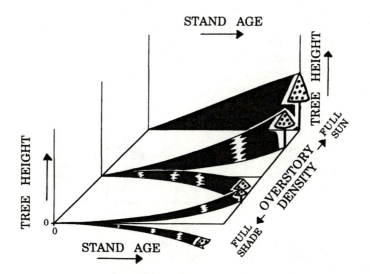

Figure 12.2 Overstory shade received by a tree changes as the height of tree or overstory density changes after a disturbance, so trees will not grow along straight lines of Fig. 12.1. As the understory tree grows taller, it sometimes receives more sunlight and increases in growth. Also, if the weak overstory is deteriorating, perhaps in the old growth stage, more sunlight will pass through with time, and understory tree growth will increase. If residual overstory trees are vigorous, they increase shade to the understory and cause understory height growth to decline.

On the other hand, understory trees accelerate growth beneath an overstory decreasing in density (Fig. 12.2). Here a species capable of growing fastest in more sunlight increasingly has the growth advantage where several species are together in the understory. It eventually dominates its contemporaries if not previously overtopped or killed by species which had the advantage in the earlier, shaded condition. The increasing sunlight available to the growing trees (Fig. 12.2) may explain why intolerant species can dominate gaps created in very old, shade-tolerant forests (Buckner and McCracken, 1978; Barden, 1979; McGee, 1984, 1986a,b).

Understory trees also become less shaded by the overstory as they grow taller (Yoda, 1974). They grow faster as they become less shaded along a trajectory of progressively less overstory density (Fig. 12.2). Where several species interact, those capable of growing fastest with more sunlight increasingly gain a height advantage and dominate the cohort, provided they were not initially overtopped and suppressed by species which had the advantage while shorter and in more shade.

Overstory density determines each understory tree's survival, height growth, and crown shape within a cohort. Trees within the cohort grow and differentiate slowly, since height growth is slow in partial shade (Figs. 12.1, 12.2). Crowns are flatter and wider than if they were not growing beneath high shade (Fig. 3.7; Busgen and Munch, 1929; Kohyama, 1980; Meng, 1986; Tucker et al., 1987), and side shade restricts crown sizes (photosynthetic surface area) much more readily than if the same species had grown under full sunlight and contained less flat crowns (Fig. 12.3). A large photosynthesis/respiration ratio is maintained in trees with flat crowns only when the trees are spaced far apart so the crowns can extend outward. This flatness, combined with slow differentiation, makes wide spacings very important for understory trees to maintain their vigor (Ferguson and Adams, 1980). Where trees of the same or similar species compete at close spacings in high shade and neither stratifies above the other, a disturbance which kills some trees allows the others to expand their crowns and therefore maintain an understory vigor which would otherwise be lost.

Figure 12.3 Crowding of lower-strata trees creates side shade and abrasion and so reduces crown size and vigor dramatically while high shade creates flat crown. (Compare with Figs. 8.1 and 8.4.) Resulting shallow, small crowns and slow height growths can reduce differentiation and growth potential after release.

Disturbances to an understory cohort

Disturbances to an understory cohort kill some or all trees. Ground fires, browsing, and some other disturbances selectively favor certain species. Disturbances which do not kill all trees of a cohort allow the remaining ones of that and other cohorts to expand into the newly available growing space. The available growing space is also invaded by a newly initiating cohort, whose trees then compete among themselves and with older cohorts. All trees then influence the development of this youngest cohort.

Release from overhead competition

When understory trees are released from varying degrees of overstory competition, they may die, neither grow nor die for many years, or begin growing depending on the severity of the environment and the vigor of the trees (Herring and Etheridge, 1976; Ferguson and Adams, 1980). After release, trees previously suppressed by high shade have branches, forks, and/or crooks characteristic of the species' epinastic control and the individual's degree of suppression (Fig. 3.7). Dominant and codominant trees in lower strata with large crowns can then grow rapidly after release (Ferguson and Adams, 1980; Marquis 1981a,b). Partial removal of the overstory also causes water sprouts to grow on the lower stems and crowns of certain species.

Many trees with large crowns grow slowly upon first exposure to full sunlight but soon begin rapid growth. The slow growth period is longer for taller trees (Oliver, 1976; Jaeck et al., 1984). Many released trees then resume height growth patterns characteristic of the site rather than their past history (Fig. 3.11). Very large trees of shade-intolerant species may even decline in growth while nearby, more shade-tolerant species increase growth (Isaac, 1956).

Applications to Management

Less is known about manipulating multicohort stands than about manipulating single-cohort stands of one or several species even though knowledge has increased dramatically during the past decade (Halkett, 1984; Kolstrom, 1992; O'Hara, 1995b). Each stratum of multicohort stands must be understood and then described and manipulated differently.

The most common type of uneven-aged stand is a multicohort stand (in contrast to an all-aged stand). To maintain the multicohort condition the forester must ensure that an adequate number of vigorous trees of desirable species is retained in each cohort and that the trees can and do respond to release. The slow differentiation of trees in lower strata can cause the trees to quit growing or not to respond to release. The forester must reduce high shade by cutting some trees in the upper strata and reduce side shade by cutting some trees in the lower strata to maintain the residual trees' vigor. A commercial harvest in the upper strata can be combined with a precommercial thinning in the lower strata (Goodman, 1930).

The success of managing multicohort stands is determined by both the skill of the forester and the species and structure of the stand. The condition of trees in lower strata often dictates the cutting operations for taller strata. If species in the lower strata characteristically lose their epinastic control, reduce their live crowns, and die readily, the overstory must be kept at a low density. Species are often not desirable for multicohort management if they are very intolerant and die in partial shade or if they readily become flat-topped in shade and form a fork when again released (Chap. 3). The competitive advantage in the younger cohorts can be shifted to the shade-intolerant or -tolerant species, whichever is desirable, by keeping the overstory thin or by allowing it to grow dense.

Species such as grand fir are more prone to disease problems when damaged by logging in the overstory than others, and they may be less desirable for multicohort management. The forester must determine how light is reaching these lower strata and how soon larger trees will grow and block this light. In hardwood stands in eastern North America the oldest stratum is generally highly differentiated. Most light in these stands comes from overhead or through openings on the sides of small gaps. In western mixed conifer stands, upper-strata trees often have long crowns, and much light enters the stand and travels at an angle below the crowns of upper-strata trees.

Harvesting the oldest cohort both releases and disturbs the younger cohorts. The equipment damages trees, soil, and forest floor (Trimble, 1963); slash is left on the forest floor; and the area becomes more susceptible to fires, insects, diseases, and other disturbances. The logging operation must be planned with these disturbances in mind.

CHAPTER 13

DEVELOPMENT OF MULTICOHORT STANDS

Introduction

Multicohort stands vary in stand histories and structures, leading to many interpretations of their development patterns. Despite these wide ranges of structures and histories, the development patterns are not random. This chapter assumes that trees of all cohorts are regularly distributed throughout the stand. Multicohort stands can also occur as mosaics of small single-cohort stands. In such conditions, most of the area in each small stand is influenced by the edges of adjacent stands. Stand edges and development of mosaics of stands will be discussed in Chap. 14.

Interpretations of Multicohort Stand Development
Relay floristics

Historically, multicohort stands were thought to develop according to the relay floristics pattern (Fig. 5.1; Egler, 1954). This pattern supported the concept that species had coevolved to a large degree of mutualism, whereby the species were dependent upon each other and species acted together as a "superorganism" (Chap. 2). Even where mutualism was not thought to be prevalent, stands were often assumed to develop by the relay floristics pattern, with one species invading and replacing another as it became more competitive in the changed environment of the developing stand (Clements, 1916; Oosting, 1956; MacArthur and Connell, 1966; Daubenmire, 1968).

The concept of relay floristics leading to all-aged stands has influenced many approaches to forest management, forest research, and environmental concerns about forests. Many harvesting patterns of mixed-species stands which have degraded forests were intended to mimic this perceived natural, uneven-aged development. Forestry research focusing on successional patterns, climax communities, and vegetation typing often contain an underlying bias that relay floristics patterns are occurring. Many environmental concerns about cutting of forests and establishment of plantations carry the underlying assumption that such activities interrupt a natural process of relay floristics.

Until recently, little direct work has examined whether all-aged stands actually exist. Meanwhile, the concept of all-aged stands with constant replacement of species has been reinforced by observations of the size distribution of trees in mixed-species stands, by apparently successful "selection" cutting patterns of mixed-species stands (Gilbert and Jensen, 1958; Leak et al., 1969; Filip, 1973; Leak and Filip, 1975), and by concepts that all-aged stands could persist by having small patches of regeneration grow after small disturbances (*gap phase development*). These observations and concepts will be discussed below.

Sizes and ages of trees

The number of trees in a single-cohort stand declines in a reverse-J shape when stand age (and therefore tree age) is graphed with tree number. All-aged stands have been thought to develop with a similar reverse-J shape distribution (Fig. 6.3B) of tree ages and numbers within a single stand (DeLiocourt, 1898; Raunkiaer, 1928; Hough, 1932; Meyer and Stevenson, 1943; Meyer, 1952; Meyer et al., 1952; Baskerville, 1960; Leak, 1965; Assmann, 1970; Minckler, 1971; Thimble and Smith, 1976; Schlaegel, 1978; West et al., 1981a,b); and since trees grow larger with age, it has been simplistically assumed that diameter can be substituted for age in assessing tree age distributions within stands (Hough, 1932; Meyer and Stevenson, 1943; Phillips, 1959; Piussi, 1966; Daubenmire, 1968; Minckler, 1974). It has then been erroneously presumed that stands with a reverse-J shape diameter distribution are all-aged or contain many age cohorts.

Multicohort stands can have a reverse-J diameter distribution (Hanley et al., 1975; Oliver and Stephens, 1977; Means, 1982), but the same distribution can develop in single-cohort (e.g., Fig. 6.2B; Wilson, 1953; Blum, 1961; Gibbs, 1963; Marquis, 1967, 1992; Roach, 1977; Oliver, 1978a, 1980). Small trees are not necessarily younger than large trees in stands of one or many species (Fig. 2.1; Blum, 1961; Oliver, 1978a, 1980; Stubblefield and Oliver, 1978). Diameter distributions in multicohort stands have a variety of patterns (Leak, 1964; Foiles, 1977; Betters and Woods, 1981; Lorimer and Frelich, 1984). Combinations of species or cohorts sometimes show the reverse-J diameter distribution even where individual species or cohorts have normal distributions (Assmann, 1970; Goff and West, 1975; Lorimer, 1980; Lorimer and Frelich, 1984). Extrapolating from tree size distributions to age distributions in stands can, therefore, be very misleading.

Selection harvesting of mixed-species stands

Timber harvest in mixed-species stands has been and is being done in similar patterns over the past century in much of the eastern and western United States and abroad (Mustian, 1977). Evidence in mixed tropical forests suggests the same patterns are also occurring there (Hartshorn, 1978). The similarities of cutting patterns are probably because of similar logistic and socioeconomic pressures as well as similar stand structures. The straighter stems and trees of economically valuable species (usually rapidly growing, relatively shade-intolerant species) in these mixed-species forests are harvested. The residual stands contain unvigorous, crooked, and otherwise deformed overstory trees which suppress any regeneration, so that neither overstory nor understory trees grow rapidly (Figs. 13.1, 13.2). This harvest pattern is known as *"high grading."*

Mixed stands of eastern hardwoods and western conifers have been entered repeatedly at intervals of 10 to 40 years. Merchantable stems were removed at each entry. The stands were successively reentered, *not* because the remaining trees grew vigorously, *but* because the species and sizes of trees considered merchantable changed. In the eastern United States valuable, often intolerant hardwood and conifer species were first removed; later, associated midtolerant species were removed; and still later, even smaller-diameter, tolerant species were removed (Fig. 13.2; Johnson, 1970; Johnson and Biesterfeldt, 1970; McLintock, 1979; Smith and Linnartz, 1980; McGee, 1982). In the northern Rocky Mountains, western white pines and ponderosa pines were first removed; later, Douglas-firs were removed—first of large diameters and later of small diameters; and, still later, grand firs and other true firs of the tolerant, lower strata were removed (Fig. 13.1; Daniel, 1980; Oliver and Kenady, 1982; Rickerd and Hanley, 1982). Even where they grew as multicohort stands, their development was quite different than historically thought. The result of these cuttings was to create unusual forests of crooked, often diseased trees with little future timber value.

The original stands existed either as single-cohort, mixed-species, stratified stands (Chap. 9) or as multicohort stands with many strata. The cutting patterns first removed the emergent trees and the dominant and codominant trees—those with the largest stems—in the upper stratum. Later, dominant trees in lower strata were removed. The cuttings released unvigorous, intermediate, and suppressed trees in each stratum and broke or scarred many of them (Hall, 1933).

The broken, scarred, and unvigorous trees did not grow rapidly after each logging episode (Trimble, 1963). Most photosynthate produced by their small live crowns relative to their respiration tissues was consumed by respiration and other uses. Little was left for stem volume growth. The photosynthate allocated to crown expansion allowed the taller, slowly growing trees to cast more shade and suppress lower strata. The few trees which did grow rapidly were commonly scarred and developed as hollow trees, or they had lost their epinastic control and so put much growth into large branches and crooked or forked stems. In addition to these cutting practices, exclusion of ground fires previously set by Native Americans and lightning allowed different species mixes to develop (Cooper, 1960; Heinselman, 1970a,b; Habeck, 1972; Cole, 1977).

The cutting practices and fire exclusion have allowed new cohorts of trees to enter the growing space and become suppressed by the overstory. New cohorts readily grew in dense patches from seeds of overstory trees and from sprouts and advance regeneration.

The overstory shade favored survival of more-tolerant species. Many eastern hardwood stands have changed in species composition from intolerant tulip poplars and oaks to tolerant hickories, maples, and beeches as the intolerant upper-strata trees were removed and tolerant species dominated the ensuing younger cohorts (Abrams and Scott, 1989; Abrams and Downs, 1990; Nowacki et al., 1990; Abrams and Nowacki, 1992; Nowacki and Abrams, 1992). Similarly, many western forests are changing from being predominated by intolerant pines and Douglas-firs to being dominated by tolerant true firs (Cole, 1977; Bickford, 1982; Cochran, 1982; Hall, 1982; Hopkins, 1982; Laacke, 1982; Rollins, 1982; S. T. Smith, 1982). Younger trees in these eastern hardwood stands become flat-topped and lose a central stem where high shade does not allow the terminal to maintain its epinastic control. Many western conifers maintain their terminal in the lower strata but become infected with diseases such as mistletoe, stem rots, and root rots (Aho, 1982; Oliver and Kenady, 1982).

Eventual fates of these partially cut stands are unknown. In the short run, overstory trees will grow little but will suppress the understory. The understory's competition for moisture and other growing space can weaken the overstory (Zahner, 1958; Barrett, 1970; Oliver, 1984; Smith and Grant, 1986), and insects and diseases can build up in these stands. There may also be genetic selection toward competitively inferior genotypes of tree species where this *"high grading"* is done (Howe, 1995).

Figure 13.1 Multicohort, mixed-species stand in central Idaho created by repeated partial cuttings of vigorous trees in each stratum as merchantability standards changed and different species and sizes became valuable. Result is a "high graded" stand where unvigorous overstory strata suppress growth of younger cohorts.

Figure 13.2 Multicohort, mixed-hardwood stand in central Tennessee caused by partial cuttings of most vigorous trees, as in Fig. 13.1. Hardwood trees with weak epinastic control create forks or crooks where released from suppression.

Despite the above examples, selection harvesting has successfully created stands with vigorous trees of many sizes and ages where it has been done with certain precautions (Fig. 13.3) to be discussed later.

Selection cutting developed in Germany in the late 1800's (Troup, 1926, 1952; Guldin, 1995). It was tried in the Pacific northwestern United States, where the objective was to remove the most mature trees and allow the smaller, immature trees to grow (Brandstrom, 1933; Kirkland and Brandstrom, 1936; Munger, 1950). The harvesting pattern was intended to mimic the all-aged, relay floristics patterns in which these forests were believed to grow by removing the larger trees and releasing the smaller (and presumed younger) trees (Munger, 1930; Isaac, 1943, 1956). In fact, the stands did not necessarily grow in the all-aged patterns. Under the name "Maturity selection," the result proved unsuccessful both because the remaining, smaller trees did not growth rapidly and because logging damage caused many of the remaining trees to rot (Isaac, 1956).

Selection cutting was first tried in the southwestern United States in ponderosa pine stands as "Improvement selection" (Pearson, 1950) shortly after 1900. Here, the objective was to leave trees of high quality and a variety of sizes at wide spacings to allow them to grow vigorously. This approach seems to be considered suitable even today for pine stands on droughty soils in the Inland Western United States (Fiedler, 1995; Guldin, 1995).

Selection cutting was established in mixed stands of loblolly and shortleaf pines and deciduous species. Entries were repeated on annual and periodic (5 to 7 years) cycles to "cut the worst and leave the best", removing both individual trees and groups of trees as appropriate (Reynolds, 1969; Guldin, 1995). After 40 years, the practice seems successful; however, aggressive control of more shade tolerant deciduous trees is considered necessary (Guldin, 1995).

Patch concept of stand development

Studies of stand development in the absence of major disturbances have suggested that new stems invade small openings in the forest after disturbances to groups, or patches, of overstory trees (Jensen 1943; Eyre and Zillgitt, 1953; Bray, 1956; Gilbert and Jensen, 1958; Woods and Shanks, 1959; Loucks, 1970; Forcier, 1975; Vankat et al., 1975; Williamson, 1975; Hartshorn, 1978; Skeen, 1976; Lorimer, 1977a, 1980; Pickett and Thompson, 1978; Whitmore, 1978; White, 1979; Ehrenfeld, 1980; Runkle, 1981, 1982, 1984; Nakashizuka and Numata, 1982a; Hibbs, 1982; Collins et al., 1985). The younger trees become suppressed and then released by later small disturbances until they finally reach the overstory. Each disturbance creates a gap of available growing space which is refilled by a new cohort of trees (Jones, 1945; Watt, 1947; Bray, 1956; Bormann and Buell, 1964; Johnson, 1972; Brewer and Merritt, 1978; Hartshorn, 1978; Whitmore, 1978; Hibbs et al., 1980; Lorimer, 1980; Pickett and White, 1985). Many small disturbances have been thought to create mosaics of small, single-cohort stands which, in aggregate, appear as a multicohort stand (Aubreville, 1938; Webb et al., 1972). A balanced, uneven-aged stand with an aggregated reverse-J shape diameter distribution would theoretically develop where the disturbances—and hence patches of single-cohort stands—were evenly distributed in size and time.

Actually, a gap is part of a continuum of disturbance sizes from very small disturbances which do not allow a new cohort to develop to very large ones which remove most or all preexisting stems. Where gaps in the existing forest develop, new cohorts do not necessarily grow within the discrete, small areas. In addition, very small stands do not behave like larger stands because the area near a stand's edge is greatly influenced by the adjacent stand (Smith, 1963; Nakashizuka, 1985).

Idealized Development of Multicohort Stands

Forests which develop by the relay floristics pattern are probably not prevalent if they exist at all—except for invasions in the stand initiation and understory reinitiation stages. Certain stand structures are often assumed to indicate the relay floristics pattern, but they can be found in both single- and multicohort stands (Oliver and Stephens, 1977; Oliver, 1978a, 1980).

Stands containing more than one age cohort have been found under natural conditions, where unplanned human activities have occurred, and where the management objective has been a multicohort stand. Stands with more than one cohort do not necessarily have a well-balanced distribution of tree ages, a reverse-J shape diameter distribution, or patches of single-cohort stands (Oliver and Stephens, 1977). In fact, multicohort stands have more varied structures than single-cohort stands have. Depending on tree ages, species, and spatial distributions, multicohort stands can exhibit characteristics very similar to those of single-cohort stands, or they can be quite different.

Species compositions of multicohort stands vary within and between cohorts depending on the disturbance type and size and on the component species' characteristics. Remaining trees expand and new stems initiate by germinating, sprouting, or accelerating from advance regeneration to refill growing space released by a partial disturbance. The general process is one of competition for a limited amount of growing space, and the behavior of the new cohort is similar to the stages in single-cohort stands (Fig. 5.2; Bray,

Figure 13.3 Multicohort, mixed Norway spruce and silver fir stand in Bavaria, Germany. Here, vigorous trees are released and are allowed to grow very large; unvigorous trees are removed in all strata; and all strata are thinned to maintain large live crowns. Both spruces and firs have strong epinastic control, so forked and crooked trees are uncommon (and are removed in thinnings). The result is a multicohort stand of vigorous trees, which has been managed continuously for over 200 years.

1956; Whitmore, 1978; Nakashizuka, 1983, 1984a,b; Peet and Christensen, 1987) except that there is also the influence of older cohorts. Which species have a competitive advantage depends on the type of disturbance and other factors described for single-cohort stands. New stems are excluded from initiating after the growing space is refilled. Then some existing trees expand while others relinquish growing space in the ensuing competition.

Each cohort in multicohort stands generally follows the initial floristics development pattern except that the disturbance initiating a new cohort does not necessarily kill all older trees (Fig. 5.5). Simplistic development of an idealized multicohort stand will first be described (Fig. 13.4). The stand contains a single species. All disturbances occur at similar time intervals, allow the same species to initiate, and create uniform spatial distributions of regeneration. Each disturbance removes the same number of trees in each cohort and leaves the residuals evenly spaced. A new cohort of trees initiates immediately after each disturbance, and the existing trees increase in size in response to the available growing space. All trees in each cohort grow at the same rate because of the idealized uniformity of the disturbance, spatial pattern, and overstory density.

Idealized height growth patterns

Uniform height growths within each cohort create a distinct stratum for each cohort. The tallest (oldest) cohort is removed by each disturbance, and the next cohort is released to full sunlight and grows most rapidly. When released, it forms the new upper stratum.

After release, height growths of the tallest trees resemble height growths of dominant trees in a single-cohort stand on the same site (Fig. 3.11; Oliver, 1976; Hoyer, 1980; Larson, 1982; Jaeck et al., 1984). Height growths in the other strata (cohorts) are most rapid soon after the disturbance—possibly after an initial shock induced by the changed environment—and slow as shade from taller trees (and possibly root competition) becomes more intense. Trees in progressively younger, subordinate cohorts grow progressively slower, since they are shaded by more tall trees. As trees in all strata grow, the shortest cohort slows in height growth first, being most shaded because of growth in the strata (cohorts) above it. Each progressively taller cohort slows later.

Tree shapes

Tree shapes in this idealized stand reflect their histories of suppression and release from high shade. Trees which maintain a central stem when shaded (Fig. 3.7) increase height of this stem after each disturbance. Height growth slows again as overstory density increases, creating a distinct branching pattern (Fig. 13.4). Understory trees with weak epinastic control become flat-topped (Fig. 3.7) if the overstory is dense enough for them to lose their excurrent form. Upon release by the next disturbance, one or more branches grow upward and create a crook or fork at the height of release. The trees again become suppressed as overstory density increases and again lose their excurrent form, and another fork or sweep appears when they are again released.

Variations in Development Patterns and Emergent Properties

The above example is useful for illustrative purposes, but these idealized conditions would never exist. Even if all other conditions are closely controlled, removal of some overstory trees leaves gradients of light and soil conditions from the new openings to nearby residual trees (Nakashizuka, 1985). Times and intensities of disturbances do not occur extremely regularly in nature (Raup, 1964; Sprugel, 1976). Species composition, tree numbers, and spatial patterns of each cohort also vary, creating different patterns of tree invasions, heights, growth rates, and other variables in multicohort stands. The patterns which emerge from multicohort stands are much more varied than the originally assumed characteristics of stands developing in the relay floristics pattern (Fig. 5.1A).

The tree invasion patterns and subsequent distributions of age cohorts and tree numbers of multicohort stands reflect the disturbance and mortality patterns and site characteristics (Marquis, 1992). Disturbances can occur regularly enough in time and intensity to create several, discrete cohorts and appearances very different from those of single-cohort stands (Figs. 5.4A, 13.5). Alternatively, they can occur irregularly and leave stands with one dominant cohort and structures very similar to those of single-cohort stands (Figs. 13.6, 13.7).

Diameter distributions

Many attempts have been made to quantify the structure and growth of "uneven-aged" stands and forests. Balanced, uneven-aged forests have been assumed to have a reverse-J shape diameter distribution (Meyer, 1943, 1952) which can be described by a "q-Ratio" (DeLiocourt, 1898). This ratio has been used repeatedly as a guide for management

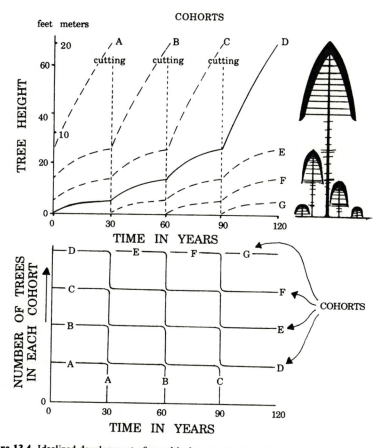

Figure 13.4 Idealized development of a multicohort stand with cohorts evenly distributed in time. Individual cohorts are identified A through G. Removal of oldest cohort and thinning or others every 30 years releases growing space, promotes new cohorts, and causes increased growth in existing trees, but all trees except those in the oldest cohort eventually become suppressed again. (*Larson, 1982.*) (See "source notes.")

(Reynolds, 1969; Mosher, 1976; Alexander and Edminister, 1977; Fiedler, 1995). The q-Ratio expresses the ratio of trees in one diameter class to trees in the adjacent class of larger diameters. The value is based on the assumed rate of decline in tree numbers between subsequent diameter classes and the size of the largest tree. For example, a stand with a q-Ratio of 2.0 would have twice as many trees in the 15-cm (6-in) class as in the 20-cm (8-in) class, assuming 5-cm (2-in) diameter classes. It was then assumed that the q-Ratio should remain constant in time, that it could be used as a guide for management, and that the diameter distribution was identical to the age distribution and represented the relation between tree aging and mortality in an uneven-aged stand. The q-Ratio value has been assumed to range from 1.3 to 2.0 for 5-cm (2-in) classes (Daniel et al., 1979). Declining distributions which are not exponential can be described by a few different q-Ratios or by an increasing ratio between adjacent diameter classes (Leak, 1964).

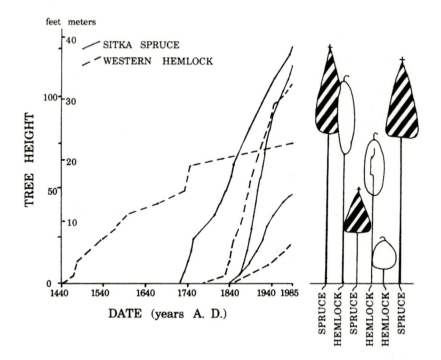

Figure 13.5 Development of selected trees in mixed Sitka spruce and western hemlock stand in coastal South Alaska. Some trees grew rapidly to the overstory after initiation; others became suppressed and may reach the overstory after a series of suppressions and releases. (*Deal et al., 1991.*) (See "source notes.")

The *q*-Ratio is useful for *describing* the diameter distribution of a stand before and after treatment. It can also roughly indicate size classes where there may be so many trees that smaller trees will become suppressed. Other factors should also be considered when studying or managing multicohort stands:

—the diameter distribution at the present time may not predict future changes in diameter distributions;

—diameter distributions should not be considered good indicators of individual tree growth, vigor, and form--especially in stands which have been "high graded" or which contain species which lose their epinastic control upon suppression from high shade (Figure 3.7B).

—for many management objectives, it may be desirable to have other size distributions (other than a smooth distribution of diameters) in multicohort stands (O'Hara and Valappil, 1995).

Problems have developed in trying to apply the q-Ratio as a forest management guide (Adams and Ek, 1974). Uneven-aged stands rarely have a uniform regeneration pattern, so a given diameter distribution—even if it were reflective of age—is rarely constant over time.

Variations in height growth

Trees of each cohort can grow uniformly in height and form a separate stratum (e.g., Fig. 13.4) where relatively light, uniform disturbances occur in stands with species of similar shade tolerances and height growth rates. Infrequent, discrete disturbances in stands of tolerant species with relatively similar height growths can also create stands where diameters reflect tree ages.

Figure 13.6 The hurricane of 1938 blew over a tree in this single-cohort hardwood stand in central Massachusetts, which began in 1870. (Person is standing in pit behind windthrow mound created in 1938.) The small trees (on the mound in front of the person) began immediately after the 1938 hurricane but they grew little because overstory trees reoccupied most of the growing space. (Photograph taken in 1974.)

Trees of the same cohort can also grow at dramatically different rates so that trees of several cohorts are found in each stratum (Figs. 13.5, 13.8; Oliver and Stephens, 1977; Kelty, 1986). Some trees grow directly to the overstory when initiating after a minor disturbance (Oliver and Stephens, 1977; Whitmore, 1978; Hibbs et al., 1980; Orians, 1982; Lorimer, 1983a; Deal, 1987; Deal et al., 1991); others become suppressed in the understory and eventually grow to the overstory following later disturbances (Oliver and Stephens, 1977; Hibbs, 1982; Deal, 1987; Deal et al., 1991; Donnelly and Murphy, 1989). The variations in height growth are caused by genetic differences in growth between species and because sunlight and soil growing space beneath a partial overstory is not uniformly distributed. Sitka spruces sometimes grow rapidly to the overstory in multicohort stands in coastal Alaska, whereas contemporary or even older hemlocks become suppressed or achieve the overstory through intermittent periods of height growth and suppression (Fig. 13.5). Spruces which do not grow straight to the overstory probably do not survive in the understory shade (Deal, 1987; Deal et al., 1991). Yellow

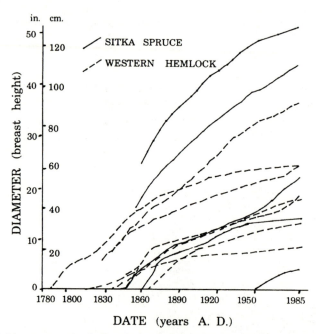

Figure 13.7 Diameter growth of selected stems in multicohort stand of Sitka spruces and western hemlocks in coastal Alaska, with cohorts unequally distributed in time. Absence of trees beginning between 1860 and 1945 caused the stand to have some properties of a single-cohort stand. *Deal et al., 1991.*) (See "source notes.")

poplars similarly can outgrow older oaks in multicohort, upland hardwood stands in the southeastern United States (O'Hara, 1986). Young oaks either grow rapidly to the overstory stratum in mixed stands in New England and Wisconsin or become suppressed in the understory and are released by subsequent disturbances (Oliver and Stephens, 1977; Lorimer, 1983a). Eastern hemlocks, birches, and maples less frequently grow straight to the overstory but instead reach it through a series of periods of suppression and release (Fig. 13.8; Marshall, 1927; Oliver and Stephens, 1977; and Marquis, 1981a,b). Contemporary birches and maples growing near fast-growing oaks and released birches and maples of an older cohort can be overtopped and relegated to the lower strata, in a pattern similar to that in single-cohort, mixed-species stands (Fig. 9.1).

Species composition

Species compositions of different cohorts can vary or remain constant, depending on which species are present for recolonizing the disturbed area, relative shade tolerances of the species, overstory density and vigor, species favored by each disturbance, and mortality patterns of the developing cohorts.

Multicohort stands with few species vary little in species compositions among cohorts simply because other species are not present as competition under any circumstances. Spruce-hemlock stands in Alaska (Figs. 9.8, 13.5, 13.7) maintain this composition in all cohorts for this reason (Deal, 1987; Deal et al., 1991). Similarly, loblolly pine—a very

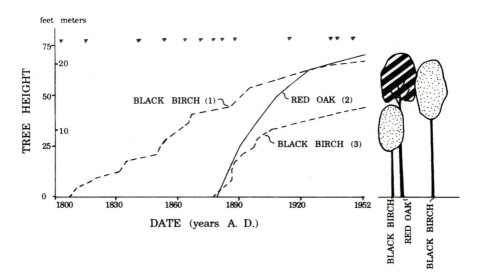

Figure 13.8 Growth of selected trees with steady diameter and height growth (red oak, 2) and step-like diameter growth (black birch, 1) in the same stand as in Fig. 13.9. Both trees reached the dominant canopy—the red oak by steady growth, the black birch, (1) by responding to a series of releases. Tree 3 (black birch) fell behind its contemporary red oak to form a lower stratum. Triangles note times of disturbances. (*Oliver and Stephens, 1977.*) (See "source notes.")

shade-intolerant species—grows in lower strata of multicohort stands where other species are not present to outcompete it and the overstory is not dense enough to kill it.

Different cohorts also maintain the same species composition where conditions giving rise to each cohort are similar. For example, silver firs and Norway spruces are maintained in all cohorts of a selectively cut forest in Germany (Fig. 13.3), since all cohorts begin after similar partial cuttings.

Different species are favored by different types of disturbances within a stand where many species are present (Willis and Johnson, 1978). Where deer browsing of hemlock seedlings reduces the number of these advance regeneration stems, future cohorts released after minor disturbances consist primarily of sugar maples (Frelich and Lorimer, 1985). Ponderosa and lodgepole pines and Douglas-firs invade and predominate cohorts after ground fires in the northern Rocky Mountains; but true firs, which are not favored by fires, grow well after partial cuttings. Firs increase in younger cohorts where cuttings replace fires as partial disturbances (Bickford, 1982; Cochran, 1982; Hall, 1982; Hopkins, 1982; Laacke, 1982, Rollins, 1982; Smith, 1982*b*). Both fires and xeric sites cause Douglas-firs to be favored on many dry sites in the Oregon Cascades (Means, 1982).

Where many species are present, a cohort becomes dominated by shade-tolerant species if it grows beneath a dense overstory (Marquis, 1992). Another cohort will become dominated by shade-intolerant species if it grows beneath a light residual overstory (Minckler and Woerheide, 1965; Gilbert and Jensen, 1958; Skeen, 1976; Leak and

Filip, 1977; Oliver and Stephens, 1977; Hibbs et al., 1980; Runkle, 1982; Wampler, 1993). In a mixed-species, multicohort stand in central New England, relatively intolerant birches and oaks dominated those cohorts which arose after large disturbances, whereas shade-tolerant hemlocks dominated cohorts which arose after very light disturbances (Fig. 13.9; Oliver and Stephens, 1977). A ground fire killed some of the overstory in mixed Douglas-fir and lodgepole stands in northern Washington 55 years ago. Where the residual overstory was dense, the relatively shade-tolerant Douglas-firs dominated the following cohort and the intolerant lodgepole pines dominated open areas (Finney, 1986).

Different mortality patterns among species may cause older cohorts to have different species compositions than younger ones even if the cohorts were nearly alike at similar ages. The oldest cohort in a multicohort stand in coastal Alaska was dominated by very large spruces (Deal, 1987; Deal et al., 1991). This cohort probably contained both hemlocks and spruces originally, as did subsequent ones. Hemlocks of the older cohort apparently died, leaving only spruces.

Overstory and understory vigor

Vigor of trees in each cohort depends on which trees survive the disturbances, growth of the other cohorts, and site conditions. Differentiation occurs in each cohort; some trees expand and become more vigorous at the expense of others. A disturbance sometimes makes a preexisting or new cohort grow poorly. Both the overstory and understory can lose their vigor, or one or the other can be vigorous while the other is weakened. It is rare, however, for both to remain vigorous for extended periods following a disturbance.

Vigorous overstory. Surviving trees in cohorts older than a particular disturbance usually maintain their vigor if the disturbance leaves dominant trees in each stratum, the trees are not wounded by the disturbance, site conditions are suitable, and the residual trees are not extremely old. Residual trees in all strata in selectively managed stands in Bavaria contained large crowns which had not become very flat-topped (Fig. 13.3). These trees grew vigorously after release.

Weakened overstory. Surviving trees lose vigor after a disturbance when site conditions and the developing understory stress them, the disturbance injures them, only less vigorous intermediate and suppressed trees are left, or the residual trees are very old. Sudden release can weaken trees which were vigorous before release where the site is very hot and xeric (Staebler, 1956a,b; Bradley, 1963; Harrington and Reukema, 1983). On some sites, younger cohorts can compete enough for soil growing space that older trees lose their vigor. The understory retards growth of overstory southern pines (Zahner, 1958) and ponderosa pines (Barrett, 1970, 1973, 1979; Oliver, 1984) dry sites. Although degree of retardation has often been considered minimal, vigorous younger cohorts can have a fatal impact on the overstory. A younger cohort of grand firs in the lower strata of a ponderosa pine, Douglas-fir, western larch, and fir stand on droughty soils in eastern Washington provide a vector for Armellaria root rot (*Armellaria obscura*) to infect and kill all trees. Dominant oaks sometimes decline dramatically when

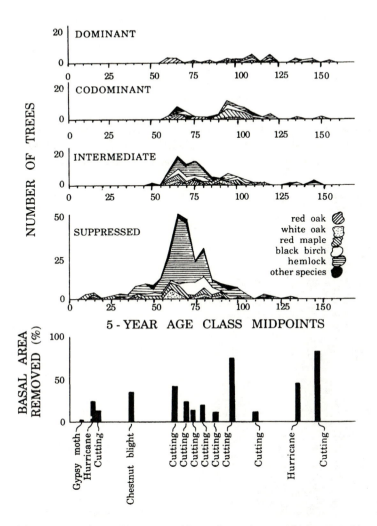

Figure 13.9 Ages, species, and canopy positions of trees in a multicohort stand in central Massachusetts. Both young and old trees occupy each canopy position ("dominant," "codominant," "intermediate," and "suppressed" correspond to a mixture of crown classes and strata). (*Oliver and Stephens, 1977.*) (See "source notes.")

released from surrounding competition in the Midwest because of increased competition from younger hardwoods beneath them.

Isolated seed trees of southern pines frequently die of wind, lightning, logging injury, and other factors (Trousdell, 1955). Bark beetle infestations (*Dendroctonus frontalis*) in southern pines on Piedmont soils may partly be caused by understory hardwoods which grow vigorously and compete for moisture and nutrients with the overstory pines.

Previously intermediate and suppressed trees generally do not grow vigorously when left after a disturbance (Fig. 13.10; Roach and Gingrich, 1968; Oliver and Murray, 1983;

McGee and Bivens, 1984). Both mixed-hardwood stands in the eastern United States and mixed-conifer stands in the West have been left in this condition by repeated cutting practices (Figs. 13.1, 13.2). At other times, intermediate or suppressed understory trees in mixed stands eventually grow vigorously if enough time elapses after release (Marquis, 1981a,b; McGee, 1981; McGee and Bivens, 1984). In the old growth stage of single-cohort stands, the overstory can also lose its vigor.

Vigorous understory. Shorter trees are often most vigorous soon after a disturbance, before taller trees shade them too much. Sometimes, however, a very vigorous understory causes the overstory to decline, further increasing the understory's vigor.

Figure 13.10 This unvigorous Douglas-fir was released 25 years ago. Although the crown has become full, the tree has grown less than 1.25 cm (0.5 in) in diameter since release.

Understory trees stay vigorous where they are not extremely crowded by their contemporaries or overshadowed by a dense overstory. Vigorous understory trees in microenvironments of relatively little competition can grow rapidly to the overstory, or they can maintain large, somewhat flat-topped crowns capable of responding to release if growing beneath high shade.

Weakened understory. After a disturbance, understory vigor generally declines as residual overstory trees grow, except where the overstory trees grow weaker. Residual overstory trees often increase growth very rapidly and prevent the understory trees from growing very rapidly, especially where the overstory is not seriously disturbed (Fig. 13.6; Minckler et al., 1961; Arend and Scholz, 1969).

Where unvigorous trees are left in the overstory—such as after "high grade" partial cuttings in mixed-species conifer and hardwood forests (Figs. 13.1, 13.2)—overstory trees increase their foliage and root area long before their stem growth increases markedly (Fig. 13.10). This increased foliage and root area suppresses the understory strata, and both the overstory and understory remain stressed.

Understory trees can be weak if they are very crowded and little differentiation occurs even if the overstory is sparse enough to allow vigorous growth of understory strata. The understory tree crowns mutually shade each other when close together, reducing their size and growth even more than trees in full sunlight (Fig. 12.3). Such strata approach a stagnated condition similar to that in single-cohort stands (Figs. 8.1, 8.3D). Silviculturists managing multicohort stands in Bavaria thin crowded understory strata to allow room for crown expansion of remaining trees.

Regeneration medium also influences vigor of understory trees. A favorable seedbed created by a disturbance combined with abundant seeds from the overstory can lead to a very densely regenerating cohort, but the component trees can lose their vigor if they grow too uniformly. Sometimes a disturbance does not create many favorable microsites for germination, sprouting, or release of advance regeneration. Then older cohorts continue to occupy most soil growing space, and newly initiating trees grow opportunistically on the unstable, raised soil of windthrow mounds or on rotting organic debris.

Spatial patterns

Each cohort does not necessarily develop as a clump within the area in which an older cohort was disturbed (Chap. 6). Even where a young cohort is clumpily distributed, the spacing between stems becomes more random and then regular as differentiation and mortality (or artificial thinning) occur (Williamson, 1975; Nakashizuka and Numata, 1982a).

Commonness of Multicohort Stands

It is uncertain how common multicohort stands are or were before European colonization of North America. Old land survey records in northern Maine suggest that extensive, stand-replacing (major) disturbances did not occur there before extensive logging. Rather, the forests were affected by partial disturbances and grew as multicohort stands (Lorimer, 1977b). Old timber cruise records in western Washington also suggest that trees of many cohorts were distributed throughout the forests (Leopold et al., 1989), although very old trees were often so isolated that a single-cohort, younger stand could be identified between them. Similar multicohort stands were common elsewhere (Gates and Nichols, 1930; Maissurow, 1935; Morey, 1936; Hough and Forbes, 1943; Hanley et al., 1975; Hartshorn, 1978; A. S. White, 1985). Other forests formerly existed more often in single-cohort stands (Spaulding and Fernow, 1899; Hough, 1932; Hough and Forbes, 1943; Lee, 1971; Henry and Swan, 1974; Franklin and Waring, 1979; Mitchell and Goudie, 1980; Larson and Oliver, 1982; Oliver et al., 1985; Deal, 1987; Deal et al., 1991; Lorimer and Frelich, 1994). Partial cutting patterns in mixed-species eastern hardwoods and western conifers have created many multicohort stands, where clearcuttings in the southeastern and Pacific northwestern United States have led to single-cohort stands. Veblen (1985, 1986) found the species composition of temperate rain forests in

Chile could not have been maintained without large disturbances. Most natural stands probably exist in a gradient between absolutely single-cohort stands and well-balanced multicohort stands. A multicohort stand with only two cohorts, with most trees in only one cohort, or with all cohorts nearly the same age has many attributes of a single-cohort stand.

There appears to be a tendency for the single-cohort condition to be maintained in stands. As the upper stratum grows, it reduces the vigor of understory trees—especially of younger cohorts containing species which are not more shade-tolerant. The stand approaches a single-cohort condition as trees of younger cohorts die. As the overstory density increases, trees of younger cohorts grow more slowly. They remain short enough to be browsed by animals or over-topped by annual plants for many years. They become susceptible to being crushed by falling overstory limbs, leaves, and other detritus. Where they germinate on an unstable rooting medium, they tip over and die. The result is exclusion of understory cohorts and reversion to a single-cohort stand.

The remaining overstory becomes more vulnerable to disturbances after a relatively dense overstory becomes partly destroyed by wind. The overstory becomes less wind-firm; dead trees increase the fuel buildup and potential for a large fire; and dead and weakened trees increase the insect population and thus the chance of insects spreading to healthy trees. Consequently, the overstay cohort or whole stand can eventually be destroyed, to be followed by a single-cohort stand.

Applications to Management

Selection cutting, leaving trees distributed in all age classes, has often been advocated because this cutting pattern has been assumed to mimic natural forest development most closely (Zon and Garver, 1930; Carver and Miller; 1933; Davis, 1935). Actually, stands develop naturally in single cohort patterns and a gradation of multicohort patterns, and there is no reason to believe one pattern is more innately "natural" or will lead to more innately "stable" forests than another.

Various types of multicohort stands have been advocated for a variety of reasons. Most recently, they have been advocated as an alternative (or a supplement in places) to clearcutting. Retaining trees of a variety of canopy positions during harvest is expected to provide features of old growth structures while still allowing trees to be provided for commodity values (Franklin et al., 1986; Hopwood, 1991; DeBell, 1989). Recent studies in both the Pacific Northwestern and Inland Western United States suggest that relatively few trees of older cohorts can be left without dramatically suppressing growth and killing trees of younger cohorts. Retaining about 37 vigorous, dominant overstory trees per ha (15 per acre) reduced growth of the younger cohort of Douglas-firs by about 50 percent over the subsequent 60 years in coastal Washington Douglas-fir stands (Wampler, 1993); retaining more than 50 overstory Douglas-firs per ha (20 per acre) killed the Douglas-firs in the younger cohort. Studies in Montana (Fiedler, 1995) suggest that moderate numbers of trees of intermediate cohorts can suppress or exclude regeneration of younger cohorts; consequently, vigorous thinning of these cohorts is suggested. Difficulties with residual trees blowing over (Adler, 1994) suggests that soils, topography, and stands with trees suitable for partial retention may need to be selected carefully.

Multicohort stands containing several evenly distributed age classes with vigorous trees in each cohort are quite artificial and difficult to perpetuate. Such management has been done successfully in relatively few places—such as some areas of Germany and Switzerland—and is most successful with species with certain characteristics (Fig. 13.3). Shade-tolerant species with strong epinastic controls, narrow crowns, and the tendency not to produce water sprouts are most desirable for this silvicultural regime. The shade tolerance allows more and denser strata without the lower trees dying. Similarly, strong epinastic control allows trees in the lower strata to maintain a central stem and conservative crown after release. If a species does not contain a narrow crown or maintain it after release, the lateral branches spread, shade, and kill lower trees. Species which produce water sprouts may be grown in multicohort stands, but their production of sprouts at each release results in stems with many branches which are deleterious for wood quality. Species of little shade tolerance and weak epinastic control can be grown in balanced, multicoholt stands, but the density and number of strata and cohorts need to be low to maintain the vigor of younger cohorts.

Certain diseases can make multicohort management difficult or impossible. An upper stratum infected with dwarf-mistletoe will constantly infect lower strata and reduce the value of these trees. Root diseases also cause problems where the only cost-effective treatments are species conversions.

When managing multicohort stands, a primary concern is to ensure the younger trees do not lose their vigor from either high shade or side shade before the overstory density is reduced. On drought-prone sites the older trees may also become weakened by dense, younger cohorts. The forester must maintain the vigor of lower strata by activities such as precommercial thinning that do not reduce the rotation length of the overstory (Goodman, 1930).

Managing the spatial distributions of cohorts is very important in maintaining multicohort stands. Younger cohorts are often clumpy, whereas trees in progressively older cohorts are more evenly spaced throughout the stand. This clumping has caused differences of opinion among silviculturists over the desirability of single-tree selection cutting versus group selection cutting. Most management of multicohort stands involves some of each cutting type. Upper-stratum trees can be harvested, and progressively lower strata can be commercially or precommercially thinned at the same time. The forester must remove trees in a variety of ways to achieve these goals. Multicohort management must have flexible cutting guidelines which can be implemented by a forester on the ground who assimilates all considerations necessary for both the present and the future.

Most growth per tree and usually per hectare (acre) occurs in the oldest, tallest cohort at any time. Manipulating this cohort has the most direct effect on yield. Where economic efficiency is desired, reducing the cutting cycle by a certain proportion (e.g., 3 years in a 30-year cycle in a stand with 3 age cohorts) would have less effect on mean annual increment than reducing the rotation of a single-cohort stand by a similar proportion (e.g., 9 years in a 90-year rotation).

It is very costly in both biological yield and economic return to convert a single-cohort stand to a multicohort stand, or vice versa (Larson, 1982; Haight, 1987; Guldin,

1995). In both situations growing space is left unoccupied for a long time, and trees are harvested either before or after their time of optimal volume or value growth.

Techniques, equipment, measuring criteria, and accounting procedures are being developed for managing multicohort stands (Duerr and Bond, 1952; Hann and Bare, 1979; Becker, 1995).

A "q-Ratio" (Fiedler, 1995), modified applications of the density management diagram (Long, 1995), and other techniques (O'Hara and Valappil, 1995) have been developed to describe and determine numbers of trees in each cohort (or size class) in multicohort stands. Recent techniques (Long, 1995; and O'Hara and Valappil, 1995) give the manager flexibility to manage stands with varying numbers of trees in each cohort or stratum—so that a wide variety of structures can be attained (e.g., Figure 6.1). The ability to manage and calibrate management of multicohort stands can proceed rapidly based on these techniques using an adaptive management approach (see "Applications for Management," Chapter 15).

Volume growth in multicohort stands will be discussed in Chapter 15.

Many structural components of "old growth" stands can be managed through variations of selection silvicultural regimes (McComb et al., 1993; Oliver et al., 1994a; Guldin, 1995). Selection regimes to create and maintain multicohort stands is often advocated as the only "natural" method to manage stands by those still assuming a "natural" forest consists entirely of stands with "old growth" structures (United States Congress, 1994). It must be remembered, however, that the "old growth" structure is only one of the stand structures found in forests; and many species can not live in this structure. Consequently, a forest where all stands are managed by selection regimes to maintain only old growth structures may prove no more favorable for protecting all species than a forest where all stands are managed by regimes which maintain only stand initiation and stem exclusion structures (Guldin, 1995).

CHAPTER 14

STAND EDGES, GAPS, AND CLUMPS

Introduction

A *stand edge* is a disjunction in a forest where the populations of trees on either side have distinctly different structures—numbers, ages, species, growth rates, or spatial distributions (di Castri et al., 1988). The edge is theoretically a single plane between adjacent stands, but an area on each side of this plane is modified by the presence of both stands, creating a broad zone which could be considered the *edge*. This edge can span from one to over four dominant tree heights in width, depending on the stands' characteristics and the criteria for delineation (Ranney et al., 1981). Although the edge is the result of the influences of both stands, it is not an average of the two stands' structures. In fact, the edge often contains more and larger stems and more species of different shade tolerances than are found within the stand on either side (Wales, 1972; Ranney et al., 1981).

The sharpness of the boundary between stands varies. Sometimes, the change is gradual and the transition from one stand to another occurs over a large area. Frequently, however, a rather sharp edge or boundary will exist.

Marked gradations in the environment and stand structures can occur over short distances. In multicohort stands, overstory trees can be spaced far enough apart to create a gradation of understory environments from full shade to essentially full sunlight. The distances between overstory trees can be so large that a small stand or patch exists, and multicohort stands can be considered mosaics of small stands. As a compromise between splitting an area into many small stands or lumping the stands, small areas of uniform structure within a stand but containing shorter trees—generally of a younger cohort than surrounding trees—are referred to as *gaps* within a stand (Watt, 1947; Bray, 1956; Williamson, 1975; Ehrenfeld, 1980; Runkle, 1984; Brokow, 1985). Similarly, a small area of uniform structure within a stand which contains taller trees generally of an older cohort than the surrounding trees—is referred to as a *clump*.

Causes of Stand Edges

Stand edges are caused by changes in any of the environmental gradients which alter the relative sizes, growths, or competitive advantages of the component species on the two sides. Consequently, an edge is found where soil conditions change and different species or growth patterns are found on the two sides, where climatic conditions change, where different types of disturbances initiated the stand on either side, where the times since the stands began are different, or where the residual densities of the overstories after partial disturbances differ.

Soil variations

Changes in soil conditions occur on geomorphologic boundaries or where variations in drainage, aeration, chemical properties, or other factors change the abilities of species to survive, grow, and compete successfully. Slight changes in soil properties where many species grow, as in the river floodplains of the southeastern United States, can give different species competitive advantages and so create different stands and an edge between them (Fig. 6.4B). Which species dominates each site depends on the regeneration mechanisms favored by the soil, each species' ability to survive flooding, soil deposition, and other disturbances, and the ability of each species to grow rapidly in the particular soil conditions. Where alluvial soil conditions govern species composition and stand boundaries, stand areas often take the geomorphologic shape of the soil created by the river; consequently, stands follow the flood deposits, cutoffs, and bands of soil formed along stream and river channels, which may be hundreds of meters (or feet) long but less than 3 meters (10 ft) wide. Similarly, changes in soil parent material, such as along serpentine outcrops, allow different species to survive and dominate, often creating distinct edges where different parent materials meet (Kruckeberg, 1954).

Previous treatments of the soil by different landowners also give different species competitive advantages in adjacent stands. Where the soil structure has been damaged by poor agricultural practices, abandoned fields often contain more pines, which have a competitive advantage under the less mesic conditions than in nearby, undisturbed forest soils.

Extremely localized changes in soil properties (Fig. 6.4A) generally will not create a noticeable edge if enough microsites allow trees of the eventually dominating species to become established and close their crowns in the B-stratum. For example, to maintain a uniform oak- or alder-dominated stand without an edge in Figs. 9.2 or 9.6, only enough microsites are needed for the oaks or alders to become established and grow to dominant positions, even if much of the intervening area is not suitable for these species.

Different soil conditions can also cause the same species to grow so differently in adjacent areas that two stands and an edge between them can be identified. An abrupt change from mesic to xeric soils can cause these differences. On the mesic site, trees often grow more rapidly and are widely spaced, since the stem exclusion stage is reached sooner. Adjacent stands can show dramatic difference in spacings, tree heights and diameters, and understory vegetation even if both stands begin after the same disturbance. Such adjacent stands of the same species occur in Douglas-fir forests in western Washington at the juncture of a glacial till and outwash. Similarly, they can be found in longleaf pine stands in the Carolina Sand Hills because of the soil variations (Oliver, 1978b). Different rooting depths of adjacent soils cause differential survival of trees during drought years (Karnig and Lyford, 1968); new stems can invade the areas of shallow soils and dead trees during moist years, creating a multicohort stand on the shallow soil adjacent to a single-cohort stand on the mesic soil.

Differences in growth patterns on different sites can also cause stands to respond differently to disturbances. Denser stands are more susceptible to stem breakage or tipping in winds, for example. Stands on less favorable sites are more susceptible to insect outbreaks (Mott, 1963; Fauss and Pierce, 1969; Osawa et al., 1986). Where different species occupy different sites, one may be more susceptible to fire. Consequently, it is not unusual for disturbance patterns, and cohorts of trees, to follow soil patterns (Hemstrom and Franklin, 1982), thus further defining the edge along a soil boundary.

Microclimate variations

Changes in microenvironments also create stand edges. In dry climates in the northern hemisphere, certain species cannot germinate and survive on southerly and westerly aspects but grow well on northerly and easterly ones. Similarly, stand establishment rates on southerly and westerly aspects may be so much slower that a stand with fewer stems and a more open canopy results. Near the northern part of its range, a species may grow only on the warmer southerly or westerly slopes; again, a different species composition would be found on northerly and easterly slopes. Frost pockets or changes in amount and directions of freezing winds similarly cause different stands to develop.

The stand edge along a change in the microenvironment is further accentuated where disturbance patterns follow microenvironment patterns, just as they follow soil patterns (Hemstrom and Franklin, 1982).

Ages of adjacent stands

Adjacent areas of similar soils and microenvironments can be disturbed at different times, creating markedly different stands and an edge between them. One side of the edge will be temporarily open after a disturbance to one stand. Different species grow along the edge than farther within the open stand because of the different microclimate (Geiger, 1965; Ranney et al., 1981) and animal populations there (Odum, 1959; States, 1976; Johnson et al., 1979; Thomas, 1979). Eastern redcedars, cherries, and other species whose seeds are disseminated by animals often grow along stand edges where animals concentrate—seeking perching sites on fences and limbs, shelter of the taller forest, and food opportunities. Hardwoods are often found in hedgerows between fields because animals bury seeds there and because farmers do not destroy the regeneration by plowing either because the trees are too close to the fence or wall or because the farmer knows field edges are less productive than centers. Eastern white pines and southern pines often grow along field and pasture edges. Where rocks were piled along field edges, farmers refrained from cutting edge trees for fear of damaging their axes. Around pastures, pines were not browsed by cows as readily as hardwoods. After the pastures were abandoned, the limby pines could still be found in rows indicating old field edges. Very large trees of older cohorts are found along old property lines in many regions. These trees have not been cut because the exact location of the property line was unclear, and it was easier for each landowner not to harvest the trees than to resolve their owner-ship.

When an edge is between tall (older) trees and short (younger) trees, shade-tolerant species often survive in the high shade at the edge, and shade-tolerant species can domi-nate the edge long after the shorter stand has grown tall. Similarly, shade-intolerant species can grow along a stand edge next to an adjacent opening even when shade-toler-ant species dominate the stand's interior. In both ways, species composition of the stand edge can be different from that in the interior (Wales, 1972; Ranney et al., 1981; Whitney and Runkle, 1981).

Trees along edges of an older stand often grow at closer spacings because of the near-by seed sources of the adjacent stand and because of the ameliorating microclimate (Gysel, 1951; Trimble and Tryon, 1966; Wales, 1972; Ranney et al., 1981). On the other hand, edges can contain fewer trees if older trees from the adjacent stand keep trees from growing nearby in the younger stand or if animals living in the older stand eat the seeds or browse the newly initiating stems (Smith, 1975; Bullock and Primeck, 1977; Howe, 1977; Vander Wall and Balda, 1977; Thompson and Wilson, 1978).

Different disturbance types
Adjacent areas of similar soils and microclimates can contain different species if differ-ent disturbance types occurred on them (Chap. 4). The boundary of the disturbances would develop as an edge with distinct characteristics even if both disturbances occurred at the same time.

Eastern white pine or southern pine stands often dominate abandoned fields or pas-tures, while hardwoods dominate adjacent areas which previously contained hardwood advance regeneration or stems from which stump sprouts grew. If a very young southern pine or white pine stand is destroyed after crown closure but before advance regeneration hardwoods invade, pines can again seed in and dominate the stand. An adjacent pine stand which contained hardwood advance regeneration—because it was older (in the understory reinitiation stage) or contained a different spacing—would be replaced by a hardwood stand if destroyed by the same disturbance (Cline and Lockard, 1925; McKinnon et al., 1935).

Adjacent stands of mountain hemlocks and Pacific silver firs grow on similar soils in upper elevations of the Washington Cascades, depending on whether glacial retreat or an avalanche began the stand (Oliver et al., 1985). Glacial retreat favors light-seeded hemlocks, and avalanches favor Pacific silver fir advance regeneration. If one forest landowner burns a cutover stand at upper elevations in the Cascades of Washington, the advance regeneration Pacific silver firs are destroyed and the new forest grows slowly as hemlocks, some Pacific silver firs, and other species slowly invade (Schmidt, 1957; Thornburgh, 1969). An adjacent stand can be dominated by advance regeneration Pacific silver firs growing at close spacings where this landowner does not burn after clearcutting (Deer, 1982; Emmingham and Halverson, 1982; Packee et al., 1982).

Different residual overstory densities

Adjacent stands can contain different stand structures and species compositions if the disturbance creating one stand removed all of the previously standing trees but only a minor disturbance occurred to the other stand. For example, 50 years following a fire in northern Washington, areas where all overstory trees were destroyed are predominated by lodgepole pines. Adjacent areas where the fire had not burned are still occupied by trees preceding the fire—originally a mixed lodgepole and Douglas-fir stand but becoming increasingly dominated by Douglas-fir as the lodgepoles die. At the borders between these two conditions and in areas where only some overstory trees were killed, a post-fire cohort dominated by the more shade-tolerant Douglas-firs is developing (Finney, 1986).

A stand of many hardwood species in the southeastern United States contains straight stems and a characteristic stratification by species where a single, major disturbance created it. An adjacent stand contains many crooked stems and a less regular arrangement of species by strata where minor disturbances created a multicohort stand. Similar variations occur in adjacent stands in other regions. Adjacent hardwood stands can both be single-cohort stands and contain similar species, but they appear distinctly different if one began from sprout origin after cuttings or burning of the previous stand and the other began without large stump sprouts.

Differences in more than one factor

A change in one factor affecting a stand's structure often causes other factors to change. Soil or microclimate variations create different growth patterns, so disturbances act differently on each area and create different stands (Hemstrom and Franklin, 1982). A disturbance may destroy one stand but not affect the other if it is of a different age, structure, or species composition or is on a different aspect. Stands of different ages result from such disturbances. A disturbance may similarly create stands of different species compositions. If only some trees were destroyed in one stand, the area develops as a multicohort stand while an adjacent area where all trees were destroyed develops as a single-cohort stand. Different species and different impacts of disturbances can further cause soil or microclimate conditions to diverge between adjacent areas.

Once an edge and two different stands are established for any reason, the edge and differences between the vegetation on the two sides can stay for a very long time—often much longer than the stands themselves. Experienced land surveyors familiar with forest development can often identify an old stand edge—property line—after several generations of forests have grown and been cut, burned, or otherwise destroyed. Subtle differences in species, tree shapes, and tree numbers indicate where a stand edge was—where trees on adjacent areas grew differently in the past.

Development Patterns of Stand Edges

Stand edges have been described as "cantilevered" if the edge is delineated at the stems of large edge trees, "canopy dripline" if the edge is maintained at the projection of the canopy of edge trees, and "advancing" if trees are continuously extending farther into an opening (Gysel, 1951). Stand edges change over time with vegetation development and disturbances on either side; these changes confound attempts to delineate edges (Lieberman et al., 1989; Belsky and Canham, 1994). Edges can be distinctly different from the adjacent stand on either side, containing either more or fewer stems, species of greater or lesser shade tolerance, more or fewer species (Gysel, 1951; Ranney et al., 1981), and different limb and stem sizes. These characteristics vary with the conditions under which an edge develops. Stand edges can roughly be divided into three types:

1. Those in which the bordering stands begin after the same disturbance and are therefore of the same age.

2. Those in which a stand begins next to an opening such as a field, stream, or different soil type which does not support a forest.

3. Those in which a stand begins next to an older one.

Edge between two stands of the same age

Edges developing between two stands which begin at the same time vary in appearance, depending on characteristics of the two stands. Where both stands are of the same species, the edge can be caused by differences in growth rates, spacings, and/or invasion times. For example, if one stand grows at wide spacing and another at a narrow spacing, the edge appears as the area between trees j and k of Fig. 8.4. Similarly, if the two stands vary in site, the edge appears as the area between trees g and h of Fig. 8.5.

Characteristics of an edge developing between stands with different species compositions vary with the species. A stratification pattern could occur on the edge (Chap. 9), with one or the other species forming large crowns extending (cantilevered) over the edge and overlapping (and possibly killing) the other.

Edge of a stand developing adjacent to an opening

Development of a stand adjacent to an opening is shown in Fig. 14.1. This opening could be a stream, road, field, or grassland. Trees growing immediately along the edge are not shaded on one side and retain a long live crown on that side. These trees therefore have large limbs on the open side. Many edge trees lean into the opening, either because of the heavier crowns on that side or because the trees turn toward the sunlight (Young and Hubbell, 1991; Young and Perkocha, 1994). Edge trees are often larger in diameter than interior trees because the side sunlight allows these trees to maintain larger live crowns (Fig. 8.3A). Where interior trees are very dense and repressed in height growth, edge trees can also be taller (Fig. 8.3D). Edges of stands adjacent to openings do not necessarily begin with trees at closer spacings than the interiors, but the side sunlight allows these trees to survive and grow at closer spacings.

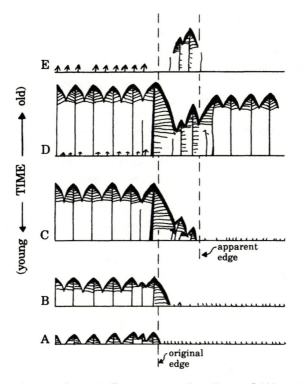

Figure 14.1 Development of a stand adjacent to an opening. (A grass field is assumed to exist on right at times A, B, and C.) (A) Trees grow and eventually close canopy. (B) Edge trees keep full crown to open side, growing larger and suppressing immediately inward, contemporary trees. Edge trees produce shade, litter, and root competition which reduce adjacent grass growth in opening. (C) Weed and tree species germinate in fluctuating growing space created by edge trees. Newly growing trees have full crowns to the outside and create a new apparent edge. (D) Old trees to the left develop to the understory reinitiation stage and develop understory advance regeneration. Trees have grown in the old field at the right. Original and apparent edge trees keep the old form. New, apparent edge trees are perhaps not growing vigorously in height because of competition from original edge trees. New, stand-grown trees to the right grow rapidly. (E) After blowdown or clearcut, small trees along the apparent edge were left because they were too small to be blown over or cut. Advance regeneration to the left was released. Residual edge trees, as well as differences in understory vegetation and regeneration mechanisms of resulting trees, cause the edge to be visible for one or more generations after it was created.

Lateral branches extend into the adjacent opening, as do roots if the opening is a field or grassland. This extension gives edge trees a further growth advantage. When extending into a field or grassland, the branches shade shorter vegetation, and the roots occupy soil growing space, reducing the vigor of the grasses, herbs, and shrubs (Williams-Linera, 1990; Matlack, 1994). Tree species from the adjacent stand sometimes invade the area of reduced herb vigor along the edge, thus advancing the stand edge farther into the field or grassland.

Where a stand grows adjacent to an opening, the edge trees can be quite resistant to windthrow (Geiger, 1965). Crowns of the edge trees extend closer to the ground, thus keeping the wind from blowing into the stand beneath the B-stratum. In other conditions, the assymetric crowns created by trees growing on an edge (e.g., Figure 14.1) can make them unstable and subject to falling over (Young and Hubbell, 1991; Young and Perkocha, 1994). The stems and roots of edge trees also grow strong and resistant to winds (Mergen and Winer, 1952; Mergen, 1954; Young and Perkocha, 1994).

Foliage and limbs extending to the ground also offer protection for animals against wind and sun as well as hiding cover from predators. Animals remain near this shelter but browse on the shorter vegetation of adjacent fields and grasslands. Such edges promote many species of wildlife (Thomas, 1979).

Stand edges adjacent to openings are also created when part of an existing stand is removed, as when clearing part of a forest to create an agricultural field. The newly created edge is first invaded by many trees, shrubs, and herbs. Later, preexisting and newly established trees dominate. The newly established trees often are species which compete in full sunlight and may be different from those in the stand's interior. Species differences between the stand's edge and interior occur when the sunlight and disturbance regimes which create the edge are different from those which create the rest of the stand (Gysel, 1951; Ranney et al., 1981).

Edge of a stand beginning next to an older stand

A stand can begin next to an older one where a field is abandoned and a new stand grows up next to an established edge (Fig. 14.1; Ranney et al., 1981) or where part of a stand is destroyed and an edge is created within the old forest (Fig. 14.2). Where an edge is created in an existing forest, changes occur both within the existing forest and in the newly established opening. The overstory trees of the remaining stand can react to the opening in several ways. They may become subjected to sunscald or produce water sprouts (Blum, 1963; Kormanik and Brown, 1964; Kormanik, 1968). They may slow in growth temporarily, similarly to thinning shock (Staebler, 1956a,b; Bradley, 1963; Harrington and Reukema, 1983). The older trees may extend their branches into the opening (Trimble and Tryon, 1966; Hibbs, 1982; Runkle, 1982), and their foliage may increase in density (Trimble and Tryon, 1966; Larson, 1982). Because they had not grown under such exposed conditions, the existing edge trees may also break or be blown over by the wind (Alexander and Buell, 1955; Gratkowski, 1956; Steinbrenner and Gessel, 1956).

Some light from the adjacent opening penetrates beneath the overstock edge trees, and death of trees in the nearby opening releases soil growing space. Consequently, a new cohort of trees and shrubs begins growing beneath the canopy near the edge. This cohort begins from advance regeneration, root sprouts, or newly germinating seeds. The germinants will be closely spaced because of the seed source immediately overhead. As overstory foliage density increases, understory trees become more shaded and slow in height growth and become flat-topped (Fig. 3.7). At progressive depths within the stand, there is more shade and the understory is more suppressed and consists of more shade-tolerant species.

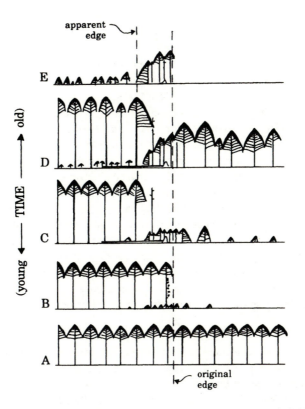

Figure 14.2 Development of a forest along an edge created in an existing stand. (A) Before edge is created, as by wind or cutting, the stand is in the stem exclusion stage. (B) After a disturbance, edge trees can develop sunscald and water sprouts. Regeneration often develops here very rapidly because of favorable microsites and proximity to seed source. (C) Edge trees begin dying and/or blowing over. More trees invade the previously opened area to right. Regenerating trees farther into the stand do not develop as well. (D) Previously disturbed area to the right has developed a new stand in the stem exclusion stage. Understory trees along the edge have become suppressed. Stand to the left has developed to the understory reinitiation stage with advance regeneration. Edge trees which withstood initial change in edge have developed large crowns. (E) After clearcut or windthrow to overstory, small trees along edge were left and advance regeneration develops as forest to the left. Apparent edge has shifted to the left of the original edge. Differences created by the edge can exist for one or more generations after the edge was created.

Within the opening next to the overstory trees, these overstory trees cast shade and occupy soil growing space. The shade and soil growing space occupancy decline at progressive distances from the stand edge, creating a gradient of environmental conditions. In the greater shade near the edge, height growth is slower than at farther distances, trees can have the flat-topped appearance of high shade, and more shade-tolerant species can be favored (Akai, 1978). The effects of an adjacent, taller stand can affect tree growth for 25 to nearly 70 m (80 to 230 ft) into the stand, depending on the growth measure and latitude (Hansen et al., 1993; Zens, 1995; both measures done in Northwestern North America). Shade tolerant species and young trees on the north edges (in the northern hemisphere) of taller stands showed more growth reduction than on south edges in studies of Douglas-firs and western hemlocks in mixed stands in Vancouver Island, British Columbia, Canada (Zens, 1995).

Tree spacing of the young cohort is often narrower near the edge because of the favorable microclimate and because of the abundant seeds from the nearby stand (Gysel 1951; Bray, 1956; Trimble and Tryon, 1966; Wales, 1972; Ranney et al., 1981). On the other hand, animals hiding in the taller stand and browsing in the younger one can reduce the number of young trees along the edge. The overall effect of the edge is a band of relatively closely spaced trees on both sides of the edge, which grades into more shade-tolerant species and advance regeneration and then disappears within the older stand and grades into less shade-tolerant, larger trees within the newly created stand.

An adjacent field or grassland (Fig. 14.1) can be invaded by trees after agricultural abandonment or climatic change. Depending on the microclimate and factors limiting growing space, the invading trees may be more numerous near the stand edge if an extremely hot or cold climate limits the invasion and is ameliorated by the edge's effect, or they may be more sparse if existing edge trees already occupy the soil growing space or if animals browse and kill invading trees. Even after a stand grows on the adjacent area, the edge can be identified by large living or dead branches on former edge trees and by differences in tree spacing and species between the edge and other parts of the stands.

Characteristics of Edges

Edges differ from the stands on either side in numbers and species of trees, tree heights and diameters, limb patterns, animal populations, and microclimates. The width of this "different" area influenced by an edge varies both with properties of a given edge and with the characteristic being considered. Distances of influence away from an edge are not the same for root occupancy, sunlight and shade influences, seed dispersal, wind reduction, microclimate amelioration, and animal browsing. The edge was found to influence vegetation structure as far as 15 m (50 ft) into a stands in Minnesota, U.S.A. (Ranney, et al. 1981) and as far as 65 m (215 ft) into stands in Vancouver Island, Canada (Zens, 1995). Edge influences on temperature have been found for as much as 240 m (790 ft) into a forest from an edge in Washington, U.S.A. (Chen and Franklin, 1992; Chen et al., 1993). The edge effect has been suggested to extend as much as 120 m (400 ft) elsewhere (Franklin and Forman, 1987).

Root occupancy by edge trees

Roots can extend 40 m (125 ft) or more from a tree (Lyr and Hoffmann, 1967). Surviving edge trees have roots which extend into the newly created stand, although the density of roots diminishes at greater distances from the edge. Consequently, competition of edge trees for soil growing space diminishes at greater distances into the newly created stand up to 40 m (125 ft). Similarly, newly destroyed trees release growing space inside the older stand in decreasing amounts to a maximum of 40 m (125 ft) from the stand edge. Tree roots may grow even longer, and root grafts may maintain occupancy of soil growing space. In these cases root influences can extend even farther beyond the stand boundary.

Shade influences

Light for tree growth comes directly from the sun and as reflected light from elsewhere in the sky (Geiger, 1965; Wales, 1972; Hutchison and Matt, 1977; Canham et al., 1990; Chen et al., 1993; Matlack 1993, 1994). Direction of reflected light is more difficult to calculate than that of direct sunlight, and reflected light is not as strong and significant for plant growth. In general, reflected sunlight increases with movement from beneath a canopy to beyond the forest edge.

Far more important for plant growth is direct sunlight. The depth to which it penetrates within and beyond the edge is dependent on aspect of the stand edge (Wales, 1972; Swift and Knoerr, 1973; Matlack, 1994), tree height, crown depth, topographic slope and aspect, and latitude (Buffo et al., 1972; DeWalle and McGuire, 1973). The sun is at different angles throughout the day. The highest angle is at noon when the sun is directly south (in the northern hemisphere). Lower angles occur in morning when the sun is to the east and in afternoon when the sun is to the west. Sun angle is also affected by latitude and day of year. Everywhere in the northern hemisphere the highest angle is on the first day of summer (usually June 21). North of the Tropic of Cancer, this angle is always less than 90° (directly overhead). At noon in mid-June (when the sun is at its highest), the sun is 87° high in Miami, Florida; and an edge tree 37 m (120 ft) tall casts a shadow 2 m (6 ft) long. At the same time on the same day, the sun is 55° high in Juneau, Alaska; and an edge tree of the same height casts a shadow 26 m (84 ft) long.

Light penetrates beneath the taller stand's edge tree crowns and into the understory to a distance dependent on the sun's angle and the heights of the crown bases. Areas closer to the stand's edge receive more sunlight; the maximum distance the sunlight penetrates varies with time of day, season, latitude, and height to crown base (Buffo et al., 1972). Lower sun angles in the morning and evening allow the sun to penetrate farther into the stand, so east- and west-facing edges have more edge influences within the stand. Similarly, the sun is at a lower angle at more northerly latitudes (in the northern hemisphere) and can penetrate farther beneath the southern overstory edge trees.

The maximum distance which sunlight enters into a stand below the canopy is described by the formula:

$$\text{Distance} = \frac{\text{height to crown base of edge trees}}{\text{tangent of sun angle}}$$

The distance which shade extends from a stand edge into an adjacent opening is described by the similar formula:

$$\text{Distance} = \frac{\text{height of edge trees}}{\text{tangent of sun angle}}$$

In the northern hemisphere direct sunlight comes from the east, south, and west in morning, noon, and afternoon, respectively. Stand edges facing east, with the taller stand to the west, receive direct sunlight beneath the canopy in the morning. The adjacent opening is fully in the sun at this time. As the sun moves during the day, more area beneath the older stand and then in the opening becomes shaded. In late afternoon all of the opening is in shade. Stand edges facing west create an opposite effect of shade in the morning and sunlight in the afternoon. In both stands, sunlight changes during the day in a regular pattern beneath the taller stand and in the opening at greater distances from the edge. Stand edges facing east are generally cooler than west-facing edges, since much of the morning sun's energy is consumed warming the night air and evaporating dew in the east-facing edge.

Stand edges facing south in the northern hemisphere, with the taller stand to the north, receive direct sunlight throughout the opening for the entire day. Direct sunlight and high evapotranspiration can make these edges quite hot and dry (Salisbury and Ross, 1969; Hutchinson and Matt, 1977). On the other hand, stand edges facing directly north, with the taller stand to the south, never receive sunlight beneath their canopy (except perhaps in very early morning and late afternoon at northerly latitudes when the sunlight is from the northeast and northwest; this sunlight is probably too weak to stimulate much photosynthesis). The adjacent opening receives direct sunlight for longer periods at progressively greater distances from the edge.

Relatively shade intolerant and xeric species generally become established in the direct sunlight of edges opening to the south or west. Conversely, edges with openings to the north or east contain more shade-tolerant or mesic species. Edges facing directions intermediate between the four cardinal directions described above have intermediate characteristics of shade, sunlight, and species growth. In the southern hemisphere, conditions of edge direction and latitude are somewhat reversed.

Seed dispersal

Where an opening is regenerated from seeds entering after the disturbance, edge trees of the adjacent stand strongly influence the species and numbers of regeneration. The number of seeds reaching an area decreases geometrically with distance from the parent tree (Fig. 4.2B). How rapidly the number decreases depends, of course, on seed weight and flight characteristics, parent tree height, and wind conditions. Pacific silver firs have heavy seeds and often dominate edges of stands begun after burns in the upper elevations of Washington, while hemlocks, with lighter seeds, are more dominant farther from the edges (Schmidt, 1957; Thornburgh, 1969). Seedling survival for some tree species is favored under the microclimate and disturbance regimes of forest edges (Yamamoto and Tsutsumi, 1984).

Wind and temperature amelioration

A "wall" of edge trees can modify the wind speed, wind direction, and temperature both inside and outside the older stand (Geiger, 1965; Moen, 1974). Progressively farther inside the stand conditions become more like those in the stand's interior; progressively beyond the edge into the opening, conditions become more like those in the opening. An edge may have higher soil moisture but lower litter moisture than inside an adjacent stand (Chen et al., 1993; Matlack, 1993). Air temperature can be higher and relative humidity can be lower at an edge than inside a stand (Chen and Franklin, 1992; Chen et al., 1993; Matlack, 1993).

Animal influences

Animals utilize edges for protection provided by taller stands and food available from shorter ones (Leopold, 1932). In doing so, they also modify conditions within these edges. They change the numbers and species of trees along edges by distributing seeds in their feces and from their fur. Eastern redcedars and cherries are common along edges for that reason. Through their browsing, animals reduce the numbers of all species or preferred browse species (Bullock and Primack, 1977; Howe, 1977; Vander Wall and Balda, 1977; Thompson and Wilson, 1978).

The distance which animal influences extend within and beyond a stand edge varies with animal species and other conditions (McGinnes, 1969; Patton, 1975). Birds deposit seeds of selected species in their feces generally within the limit of extension of branches on which they perch. Browsing animals extend their influence as far as they wander into the adjacent opening. Although variable, elks and other large animals wander as much as 200 m (650 ft) into an opening (Hanley, 1983), whereas smaller animals generally wander less than 50 m (150 ft) beyond a protective edge.

Patches—Gaps and Clumps—and Independent Stands

Patches differ in the amount of their area which is under the influence of the environment outside the patch. Differences are caused by different shapes and sizes of patches (Kelker, 1964; Patton, 1975).

Four zones of influence can be found between an open area and a single clump of trees. In the center of the clump, there is no effect of the open area; well away from the clump there is no effect of the clump. In the open area next to the clump is a zone influenced by the clump, and next to the edge within the clump is a zone influenced by the open area. Within the two zones adjacent to the edge, both below-ground edge processes (based on root extension) and above-ground edge processes (based on branch extension and shading) are occurring. Four similar zones are associated with a gap in a stand.

A large gap has a center with a very different climate than the edges or under surrounding stands. There are differences in wind movement and cooling processes. An opening of 0.04 ha (0.1 acre) in a hardwood stand on the Cumberland Plateau of Tennessee had little difference in microclimate between the edges and the center, but a gap of 0.20 ha (0.5 acres) or more showed a major difference in wind, temperature, and moisture between the center and the surrounding stand. The difference created a pronounced edge effect in vegetation in the larger gaps (Smith, 1963).

In small gaps, roots from the surrounding stand occupy the site and, except in the tropics, direct sunlight never reaches the forest floor because the sun is never directly overhead. Species which survive the intense root competition and low light levels are found there. In progressively larger gaps, an area appears on the north end that does receive direct sunlight. This lighted area is crescent-shaped in circular gaps because of the sun's daily movement. The zone of direct sunlight is widest at the north because at noon the sun is to the south and at its highest. Species of increasing shade tolerance dominate as one moves from north to south in the gap. Root competition is similar at a uniform distance from the edge all around the gap and will not differ from north to south. Most gaps have species which are shade-tolerant and can survive the root competition. As gap size increases, zones where shade-intolerant species can survive increase. It is inappropriate to consider that certain species grow only in small gaps. Species which survive in small gaps will be found in shaded portions of larger gaps. In increasingly larger gaps, the proportion of area dominated by shade-tolerant species decreases and the proportion and absolute area dominated by shade-intolerant species increases.

Patches (gaps and clumps) with more edge length for a given area are more influenced by the surrounding environments (McGinnes, 1969; Patton, 1975; Collins et al., 1985; Nakashizuka, 1985; Forman and Godron, 1986; Franklin and Forman, 1987). Conversely, with less edge length for a given area, more area is influenced primarily by the environment within the patch. The proportion of a patch's area influenced by the edge is expressed as the *"interior-to-edge ratio"* (Forman and Godron, 1986). Since circles contain the least circumference-to-area ratio, patches of this shape contain the greatest area influenced by the patch's own environment for a given area.

Progressively larger areas contain proportionally smaller circumference-to-area ratios for a given shape of patch. Larger gaps, therefore, contain proportionally less area influenced by the surrounding stand, and larger clumps influence proportionally less area beyond the clumps (Fig. 14.3). For circular stands, the proportion P of total area of the patch and edge *not* influenced by the edge is relative to the diameter of D of the unaffected patch and the depth of edge E by the formula:

$$P = \frac{(D/2)^2 \times 3.14}{(D/2+E)^2 \times 3.14}$$

The depth of the edge's influence varies depending on which influence is considered, latitude, and height of edge trees. It can be less than 3 m (10 ft) or more than 30 m (100 ft). Where it is as deep as 30 m (100 ft), only 0.30 ha (0.72 acres) is required to have the center of a gap under no influence from the edge. Even the most shade-intolerant species, therefore, need only a relatively small gap to grow vigorously (Franklin and DeBell, 1973).

If the patch becomes larger and the influence of outside stands becomes less important, the patch assumes more characteristics of independent, single-cohort stands. Any definition of what constitutes a new stand and what is simply a minor variation in structure within a given stand is not biologically based; it is simply arbitrary and can vary with the objectives of the classification. There is, however, a minimum size below which all parts of the stand become strongly influenced by the environment outside the stand's borders (Wales, 1972; Ranney et al., 1981). Depending on the depth of the edge's influence this size could be several hectares (acres) or more. Only at those sizes—when the patch begins to behave as an independent stand—would each patch behave as part of the mosaic which in aggregate can take on the attributes of a balanced, all-aged forest (provided equal areas of patches were created each year). Each patch would be quite large, and the entire area would appear more as an all-aged forest than as an all-aged stand.

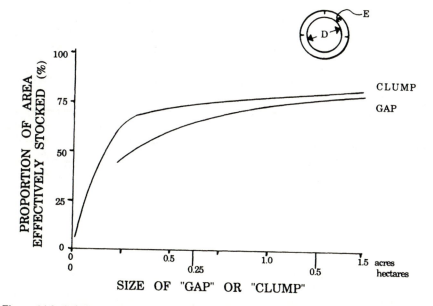

Figure 14.3 Relation of size of gap or clump which is 100 percent stocked (excluding edge) to effective stocking of patch (gap or clump) including edge. Gap or clump also encompasses a 4.6-m (15-ft), nonstocked edge. (The gap or clump is here assumed to be circular. E = width of edge with reduced growth inside a gap or outside a clump. D = diameter of center of gap or clump unaffected by edge.)

Applications to Management

Stand edges affect many management activities. Stands are generally managed as single units and edges tend to be perpetuated for many generations of trees. Recognizing edges and causes of edges can benefit land surveying, soil mapping, and similar activities. Managing edge trees with the rest of the stand means that compromises must be made on the edges, since these trees will seldom reach financial maturity at exactly the same time as the rest of the stand.

Some risk of windthrow can be reduced by designing the edge to be protected by the landscape rather than creating an edge perpendicular to the storm wind direction. When new edges are created, the forester should avoid leaving tall, thin trees on the edges or placing the edges in unstable or shallow soils.

Small gaps in the forest will probably have roots of the edge trees throughout them. These edge tree roots especially reduce available soil growing space when gaps are created by group selection or where treefalls create natural gaps.

Where immature, vigorous trees are left in clumps after a harvest or fire, there is always a question of the minimum size of clump which should be retained and the maximum size which should be destroyed and regenerated to a single cohort for greatest timber production. The size of the clump's edge will usually be the determining factor. The clump's edge trees will generally develop poor form and slow growth of regeneration in the edge just outside of the clump. Figure 14.3 suggests that clumps less than 0.12 ha (0.33 acre) in size have so much edge influence that there may be relatively little benefit to timber production in saving them.

Managing edges is also a means of managing many wildlife species. Aesthetically, edges are especially important in the near viewshed of heavily travelled roads. Most people object to abrupt edges which do not follow natural contours of the landscape. Some managers purposely make small clearings along roads to give travellers better views of distant scenery.

Many studies and management concerns have been directed at newly created edges with the stand initiation structure on one side. As stands regrow, edges create areas with deep crowns, wide variations in tree sizes, and other features described in the old growth structure (Figure 14.1D and 14.2D; Chapter 11). It is possible that "regrown" edges may provide habitats for many species requiring these structures, and that such edges can be managed as part of a landscape to provide these habitats.

CHAPTER 15

QUANTIFICATION OF STAND DEVELOPMENT

Introduction

Stand structures vary because individual trees accumulate organic compounds in different ways and at different rates. These stand variations have been pictorially and verbally explained previously in this book. This chapter will mathematically describe the changes in tree and stand volumes. For clarity, graphs will generally be used instead of mathematical formulas to describe relations.

Measures of Growth and Yield

Total tree volume and sizes of component fine roots, coarse roots, phloem, living xylem, buds and root tips, and living foliage are difficult to measure. Sometimes, surrogate variables are used as indirect estimates of tree or component volume, where one variable is correlated to a more easily measured one. The imprecision of these correlations and the lack of more accurate data on growth of all parts of trees limits the understanding and quantification of tree growth and yield.

Tree parts can be expressed in volume or weight. Volumes are usually calculated in cubic measures (e.g., cubic meters or feet) or measures of tree utility for a specific purpose (e.g., board feet, cords, steres, and others). Tree or component weight can be expressed as green (or fresh) weight or dry weight. The proportion of water in living organic matter varies by plant part, species, time of year, and environmental conditions but can easily be over 50 percent of total weight. Variations in fresh weights between trees and tree components may reflect local moisture conditions as well as more profound growth differences (Hinckley and Bruckerhoff, 1975). Fresh weights are often used where their values can be converted to dry weights for the particular species and component, where variations caused by water are not critical to the results, or where there is an economic use in expressing the fresh weight. Dry weight is often more closely correlated to the amount of carbon fixed—and hence the amount of energy consumed.

Tree and stand growth

Standing tree or stand volume is referred to as *yield*. Change in tree and stand volume with time is referred to as *growth*. These terms will be described separately. For each tree, the total organic matter added each year is referred to as *gross growth*. If any loss in tree size from branch, root, or leaf loss is subtracted, the change is referred to as *net growth*. Net growth is not really a measure of the change in living tissue; some organic matter which dies stays inside the tree, such as inner xylem, and does not get subtracted. In this book, the real accumulation of living material with the subtraction of all mortality of living material, even that which remains inside the tree, is referred to as *true net growth*. If a tree increases in size but as much tissue inside the tree dies in a year as is added, net growth shows an increase even though true net growth is zero.

True net growth of a tree is very hard to measure. Wood dies each year and is enclosed in the stem. Vessel cells and tracheids are vital to survival of the rest of the tree only when they are devoid of cytoplasm, and it is uncertain whether these cells should be considered alive or as dead as other completely dead, nonfunctional cells. Net growth is much easier to measure and is used in most quantitative studies of stand growth. Even when using net growth, some components are difficult to measure, such as fine roots and foliage. These components are estimated from other measures such as diameter at breast height. When the changing size of trees is important for forest management—as for timber or wildlife management—net growth is an appropriate and useful measure. Net growth is often used as an approximation of true net growth when calculating biologically meaningful measures within a stand. Net growth is an imprecise surrogate of true net growth, and the abilities to predict biologically meaningful measures of changes in living tissues from net (or gross) growth are not very accurate. Because of this lack of accuracy, the precision of growth and yield relations is not extremely high, and attempts to be very precise are inappropriate.

The growth terms used for individual trees are used for stands also. *Gross growth* of a stand is the addition of all material to the stand. If the stand is measured twice, the difference between the measures of living trees is *net growth*.

$$\text{Gross growth} = \text{net growth} + \text{mortality}$$

True net growth of a stand is the actual change in living biomass of a stand, which subtracts unmeasured mortality of cells within each tree. Net growth of a stand is not biologically valid for many applications because most of the biomass that is subtracted when a tree dies is actually dead wood which should have been subtracted previously.

Ecosystems studies refer to *net primary productivity* (Whittaker, 1975; Waring and Schlesinger, 1985). This term is meant to describe the change in biomass plus all biomass that was produced but lost to litterfall and insect grazing. In the actual calculation of change in biomass, the net growth of each tree is summed and mortality is added to this figure (the same as net growth + mortality). More often, the stand is measured twice and the mortality is added to the difference between the measures (gross growth). Since loss of organic matter to litterfall and grazing is added to the change in biomass, the measure is intended to include all organic matter produced but lost during the measurement period. The measure is inaccurate when a large tree dies between measurement times, since the entire biomass of the dead tree is added to the measurement period even though much of the biomass was created before the measurement period. Consequently, growth during this measurement period is overestimated. If net growth were used, the growth increment of the dead tree between the two measurements would be lost and production would be underestimated.

Tree and stand yield

Gross tree yield refers to the accumulated gross growth of the tree since the beginning of the period of measure. In single-cohort stands, gross tree yield is the sum of all living and dead organic material produced since the tree regenerated. *True net tree yield* would be the sum of all living tissue volume and weights. As with growth, the calculation of true net tree yield would subtract the dead center of the living tree when living tissue is considered. As with growth, true net tree yield is very difficult to measure and is not used. Instead, the dead xylem interior of living trees is included when estimating *net tree yield*. *True net stand yield* would similarly be the sum of only the living tissue of all living trees; however, *net stand yield* is usually considered as the living and dead tissue contained within all living trees in the stand. Because the dead xylem at the trees' centers is calculated as living until the entire tree dies, net stand yield fluctuates more with death of a tree than is biologically correct.

Yield differs from growth. Growth reflects the recent physiological condition of a tree or, in aggregate, a stand. On the other hand, yield reflects the past history as well as the current condition of a tree and stand. Two trees or stands have similar gross or net growths if they are now developing at similar rates, but they can have entirely different gross or net yields if their past histories differ.

The rest of this chapter will be primarily concerned with gross and net growth and yield instead of true net and gross growth and yield. Very little work has been done with true growth and yield. The limitations in biological meaning of the other measures of tree changes must always be kept in mind, however.

Single-Species, Single-Cohort Stands

Growth and yield will first be discussed in connection with stands of one species and cohort. These stands are relatively simple because they usually form a single stratum.

Growth is relatively easily calculated because the single species allows relatively uniform growth conditions and because the single stratum means all interactions occur in full sunlight or side or low shade. Growth and yield of single-species, single-cohort stands have been most thoroughly studied and form the basis for understanding growth and yield of more complex stands with many species, strata, and cohorts. Stands with little differentiation (Fig. 8.1) will first be discussed; then the effects of differentiation, mortality, and thinnings will be considered. After this, the relation of growth to standing volume (growing stock) will be considered.

Illustrations in this chapter simplistically assume the growing space occupied by each tree is directly proportional to crown size, implying growing space is first limited when sunlight becomes limiting. Of course, other factors can also limit growing space; but similar general growth relations occur when other factors limit growing space, with specific variations depending on the limiting factor.

Tree size in this chapter generally refers to stem volume for several reasons. Much information is available on stem volume and growth, since studies of growth and yield have generally measured this. Stem volume has been correlated to growth of other tree components and to total tree volume—with varying accuracies for different species and growth conditions. Stem growth is very sensitive to changes in growing space occupancy, since diameter and volume growth results largely from secondary (cambial) growth, which has relatively low priority in allocation of photosynthate (Chap. 3).

Undifferentiated stands

Tree and stand growth. Changes in sizes of components of an open-grown tree with strong epinastic control are shown in Figs. 3.13 and 3.15. The resultant growth is the difference between the source of photosynthate and the sinks of respiration and growth. Volume growth first increases during open growth as the photosynthetic area increases faster than the respiration area and then declines as the photosynthetic area declines relative to respiration area.

When trees grow in limited growing space (in this case limited by a relative shortage of sunlight; Figs. 3.14 through 3.16), the respiring surface area and other photosynthate sinks continue to expand until little photosynthate remains for stem growth. Growth declines sooner where the growing space is less. Trees with weak epinastic control behave similarly, except that, at close spacings, they have more photosynthetic surface area relative to their spacing than at wider spacings (Fig. 3.18; Assmann, 1970).

Stand growth is the combined growth of all individual trees. A stand grows very inefficiently when measured as growth per stand area until all growing space is occupied at crown closure (or full root occupancy where soil moisture or nutrients are limiting). The trees continue expanding until all available growing space becomes occupied. Each tree then continues to grow in the limited growing space it initially occupied or, if differentiation occurs, expands or contracts its growing space to the advantage or expense of others. Where trees are uniformly spaced and differentiation does not readily occur (e.g., Fig. 8.1), growing space is partitioned equally among all trees and the resulting growth of all trees is similar. Growth per tree in closed stands is less than in contemporary open grown trees, but a closed stand's growth can be greater than in a stand containing open-grown trees because the trees occupy the growing space more efficiently (Fig. 3.18).

After all growing space is occupied, growth efficiency in uniform stands is the same whether it is expressed as growth of all trees per area (acre or hectare) or growth of each tree per area of growing space occupancy (O'Hara, 1988), since the former is the summation of the latter.

Tree growth efficiency—tree growth per growing space allocated (O'Hara, 1986) or stand growth per area in uniform stands—varies with spacing and age as well as by species. Growth efficiency is higher soon after crown closure, indicating all growing space is occupied. Stands of closely spaced trees increase growth efficiency sooner than wide spacings do, since crown closure occurs sooner at close spacings. Narrower spacings eliminate lower branches and decrease the foliage area per tree, but growing space per tree is also less and efficiency of the tree with the reduced crown may first be greater than that of wider spacings. As spacing becomes narrower or trees grow taller at a given spacing, some of the most productive foliage in the central crown is shaded and eliminated and tree growth efficiency declines markedly. Alternatively, trees with weak epinastic control (Fig. 3.18) are more efficient at narrower spacings than are trees with strong epinastic control.

Older trees become efficient at wider spacings as all growing space becomes filled (Fig. 3.18; Sjolte-Jorgensen, 1967). Stands of widely spaced, old trees however, do not grow as efficiently as stands with narrower spacings because the lower foliage has to support long, respiring branches and so supports less growth per unit of growing space occupied than foliage in the upper crown. At progressively narrower spacings, trees with strong epinastic control curtail growth efficiency sooner as they become older. In trees with weak epinastic control, growth efficiency has the opposite effect, and tree growth efficiency declines at only extremely wide spacings.

Generally, growth efficiency reaches a maximum at later times for progressively wider spacings in all trees but at earlier times in trees with weak epinastic control than in trees with strong epinastic control. Growth efficiency reaches a maximum at some spacing and time for each species, after which growth efficiency is less at any spacing (Fig. 3.18; Bolt, 1971). Time of maximum growth efficiency is not known, but it probably varies by species and sites. It appears to occur long after crown closure.

Tree and stand yield.
Tree yield. Individual tree net yield is the accumulated growth of the tree minus the needles, fine roots, bark, and branches which have fallen off. In undifferentiating stands where growing space per tree does not change after crown closure, yield of a species depends on initial spacing, site, and tree age.

Growing space occupancy can also be expressed as ground area occupied by each tree, spacing between trees, or numbers of trees per unit area (hectare or acre). Tree age expresses the accumulation of growth. For some species and sites, tree height can be substituted for a combination of site and age (Chap. 3). For the present discussion, the same volume per tree or per area is assumed to accumulate when trees of a given spacing reach a certain height, regardless of how many years passed before the tree reached that height. The substitution of height for age is probably most accurate in trees such as coastal Douglas-firs which maintain constant height growth for long periods (Assmann, 1970; Williamson and Curtis, 1984; Mitchell and Cameron, 1985; Oliver et al., 1986a).

Even with coastal Douglas-firs, volume yield seems to be slightly greater at a given tree height on poor sites than on productive ones. Tree height cannot be used to equate yields on different sites for time-related measures such as annual ring widths, numbers of branches, or annual growth. Trees and stands which curtail height growth as they approach suppression (Figs. 8.1 and 8.3D) appear in a "suspended" state if tree height is substituted for age (Mitchell and Goudie, 1980), since they appear not to be advancing in age because they are not increasing in height.

Tree size can be related to time and growing space, or spacing, from several perspectives (Figs. 2.8, 3.17, and 15.1). Figures 2.8 and 15.1B are based on competition-density relations hypothesized by O'Conner (1935) and measured by Kira et al. (1953), Yoda et al. (1963), White and Harper (1970), Drew and Flewelling (1977, 1979), and Peet and Christensen, (1987). Figures 2.8, 3.17, and 15.1A are based on the correlated curve trends studies of O'Conner (1935), Pienaar (1965), and Oliver et al. (1986a). To maintain consistency with previous studies, the spacing and volume per tree axes are expressed in logarithmic form. The two graphs can be combined in a three-dimensional relation shown in Fig. 15.1D (Oliver et al., 1986b; McFadden and Oliver, 1988; O'Hara and Oliver, 1988). This relation is similar to the three-dimensional graphs of Sato (1983), Chang (1984), and Lloyd and Harms (1986).

As shown in Figs. 2.8, 3.17, 8.1, and 15.1, trees first undergo a period of "free growth" in the open-grown condition before they occupy all growing space (in this case, before crown closure). At this time, growth is identical for trees at all spacings (Marsh, 1957; Pienaar, 1965). Trees at closer spacings fully occupy all growing space sooner and slow in growth (Eversole, 1955; Bennett, 1960; Sjolte-Jorgensen, 1967; Jack, 1971; Stiell, 1976; Reukema, 1979; Oliver et al., 1986b; J. H. G. Smith, 1986). After crown closure, growth proceeds on limited growing space but tree shape changes (Figs. 3.14 and 8.1). This time of continued growth but changing tree shape is considered the "plastic phase" of tree response (Harper, 1977; Hutchings and Budd, 1981). Different plant components (foliage, branches, stem) grow at different rates at different spacings [trees/ha (acre)] and ages during this plastic phase. These changes in relative sizes affect the growth-spacing-time relations where only part of the tree (e.g., the stem or above-ground components) is measured (Hozumi et al., 1956; Kira et al., 1956; Westoby, 1977, 1984; Mohler et al., 1978; White, 1981). At later times, trees at progressively wider spacings curtail even height growth (Chap. 3). If differentiation, stem buckling, or other mortality does not occur first, the trees eventually approach stagnation.

Variations in the shape and values of the relation of tree volume to spacing and height (a substitution for age) occurs between species and (possibly) between sites. Species with weak epinastic control show more converging tree volumes at different spacings because close spacing increases relative growth efficiency. Crowding may increase height growth in some plants (Hozumi et al., 1955). In general, however, crowding is believed to reduce total volume growth of trees (Hozumi et al., 1955, 1956; Yoda et al., 1957). Relative allocations of photosynthate to height and diameter also vary between species even if both species have similar epinastic control (Fig. 3.17).

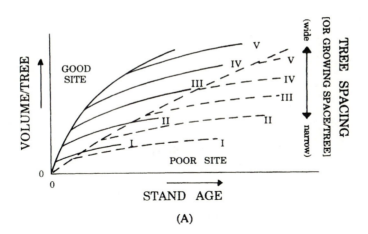

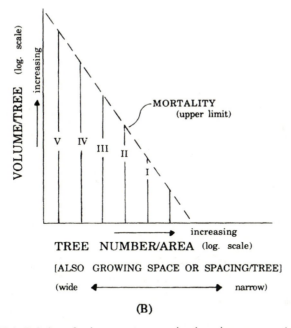

Figure 15.1 Relation of volume per tree, spacing (growing space per tree), and age in single-cohort, single-species stands. Spacing and volume per tree are expressed on logarithmic axes. (Figures are similar to Figs. 2.8 and 3.17.) (A) Correlated curve trends perspective (O'Conner, 1935) showing differences between sites. Trees achieve same maximum volumes at a given spacing on low and high sites but at a latter time on the low site. (B) Competition-density perspective of volume per tree and spacing (Kira et al., 1953; and others). Mortality is often believed to occur at a slope of $-3/2$. When mean tree diameter is used as a surrogate for mean tree volume, relation is analogous to Reineke (1933), McCarter and Long (1986), and Long et al. (1988; Fig. 15.13).

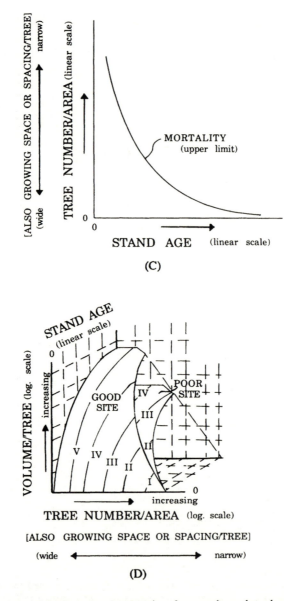

Figure 15.1 (*Continued*) (C) Reverse-J perspective of tree number and stand age. This relation has often been extrapolated to assume that all-aged stands contain the same relation (Fig. 6.3B). When the tree number axis is logarithmic, the relation becomes more linear. (D) Three-dimensional view of volume per tree spacing (or tree number), and stand age. Lower sites cause the same relation to occur at slower rates. When viewed from side, D appears as A; from the front, D appears as B; from the top (with axes in linear scale), D appears as C (*Oliver et al., 1986b.*) (See "source notes.")

Stand yield. Yield per stand (per unit area) is the sum of the yields of all individual trees. It can be related to spacing and age in uniform, even-aged stands by multiplying the vertical axis (volume per tree) by the horizontal axis (number of trees per area) of the three-dimensional figure (Fig. 15.1D; O'Hara and Oliver, 1987, 1988). The resulting yield surface is shown in Fig. 15.2A; two-dimensional views are shown in Figs.15.2B to D.

From the two-dimensional perspective of number of trees per area and stand volume (Fig.15.2B), the relation is similar to that of Gingrich (1967), Reukema (1975), Reukema and Bruce (1977), and Seymour and Smith (1987). Basal area is often used as a surrogate measure of stand volume, discussed later.

Unlike volume per tree, volume per area is much greater when there are more trees per area (narrower spacings) before the crowns close, as can be seen most readily from the volume per area/stand age perspective (Figs. 15.2C and D). Before crown closure, greater tree numbers cause greater total volumes at narrow spacings even though all trees are the same size at all spacings. After all growing space is occupied, the changes in tree shapes (plastic responses) cause trees at narrow spacings to grow less than those at wide spacings (Sjolte-Jorgensen, 1967; Liegel et al., 1985).

Two patterns of growth have been hypothesized for nondifferentiating stands reaching crowded conditions. These are the *constant-yield effect* (Fig. 15.2C) and the *crossover effect* (Fig. 15.2D). Both patterns are discussed below, but it is probable that the crossover effect occurs.

Constant-yield effect. Experiments with annual and woody plants (Kira et al., 1953, 1954; Ikusima et al., 1955; Hozumi et al., 1956; Shinozaki and Kira, 1956; West and Borough, 1983) have suggested that stands at all spacings eventually grow with the same volume and at the same rate per area at the same time in the absence of self-thinning, but progressively wider spacings begin growing at this volume later (Fig. 15.2C). Once this volume is achieved, all stands maintain the same volume, which increases with age. The constant-yield effect is philosophically supported by the early ecological concepts that stands converge in properties toward an ultimate, uniform condition (Chap. 2; Hozumi, 1977).

Crossover effect. Some evidence suggests that constant yield may not always occur (Kira et al., 1956; White, 1980). Alternatively, the crossover effect (Fig. 15.2D) suggests that the earlier reduction in tree size at narrow spacings diminishes the effect of more stems until the total volume per area at narrow spacings becomes surpassed by stands at progressively wider spacings. This crossover effect has been observed in spacing trials of Douglas-fir (Reukema, 1970, 1979) and loblolly pines (Peet and Christensen, 1987), although some mortality was occurring in both studies. The crossover effect has also been observed elsewhere when only parts of the plant were considered (Kira et al., 1954). Trees in undifferentiating stands eventually slow in height growth, approach stagnation, and add no more stand volume growth with time (Hutchings, 1979; Hutchings and Budd, 1981). Where tree height is substituted for age on the horizontal axis of Fig. 15.2C and D, the undifferentiating stands appear static since they do not increase in height. In fact, mortality and differentiation usually occur, creating a slightly different growth pattern, as will be discussed.

Regardless of which pattern occurs, time of maximum stand yield is related to spacing, with progressively wider spacings reaching maximum yield later (Pienaar and Shiver, 1984a,b; J. H. G. Smith, 1986). Beginning at narrow spacings, stands at progressively wider spacings reach maximum volumes later until physical limits of tree size restrict total volume achieved at wide spacings. Consequently, stands at very wide spacings (V, Fig. 15.2C and D) never achieve the volume of moderately spaced stands (III). Because the physical limits of tree size vary with species (Table 11.1), the spacing which achieves maximum stand volume varies by species. One tree per hectare would obviously never achieve the volume of more closely spaced stands of any species, but species with long lives and continued height growth, such as coastal Douglas-firs and Sequoias, may achieve maximum volumes at much wider spacings than short-lived species which do not grow very tall—such as gray birches.

Annual stand growth rate at each spacing and time is referred to as *current annual increment;* annual growth average over a short interval (5 or 10 years) is referred to as *periodic annual increment.* Current annual increment is represented by slopes of the lines in Fig. 15.2C and D. A stand's average yearly growth since stand initiation is referred to as the *mean annual increment.* Mean annual increment can be represented as the slope of a straight line from the graph origin to the point of age and spacing being considered. Maximum mean annual increment is obtained at an intermediate spacing and age, which varies by species. Lines showing the slopes of mean annual increment at its culmination for each spacing are shown in Fig. 15.3.

Growth and mortality. When stands become very crowded, some mortality usually occurs instead of a prolonged period during which all trees stagnate. Death occurs from suppression where stands are differentiating, from insects and diseases where trees are weakening, and/or from buckling where tree stems become very tall and thin (Chap. 3). Death from suppression occurs when a tree does not occupy enough growing space to provide photosynthate both to maintain its respiring tissues and to provide renewal of roots and other tissues necessary to sustain itself. Time of death depends on amount of growing space available to the tree and tree size—the amounts of photosynthesizing and respiring tissues. Very large Sequoia trees apparently live long when they have large crowns and root systems, which allow rapid photosynthesis; very slowly growing bristlecone pines can grow very old because they have little living tissue to maintain through respiration.

Very little of the tree is generally still alive when actual death occurs from suppression, even though calculations of net growth assume the entire standing tree is alive until exact time of death, as discussed earlier. A tree undergoing suppression or stagnation decreases its growing space occupancy and decreases its photosynthetic tissue relative to its respiration tissue. Fluctuations in growing space caused by seasonal changes or extreme weather conditions reduce the growing space available to each plant. Plants producing barely enough photosynthesis to survive under normal conditions die because they are not able to sustain themselves under more limited conditions. Mortality in suppressed or stagnated trees does not generally occur regularly. Rather, it often occurs during or shortly after extreme weather or other adverse conditions, even though the differentiation or stagnation which creates suppressed trees and predisposes them to mortality occurs at somewhat predictable intervals during stand development (Chap. 8).

Yield and mortality. Mortality generally occurs at the stage when stands were previously described in this book as stagnating (Fig. 8.1)—when photosynthesis can not satisfy respiration demands. Mortality occurs at progressively earlier times and to smaller trees at narrower spacings. The result of mortality is an upper limit to the mean size of a

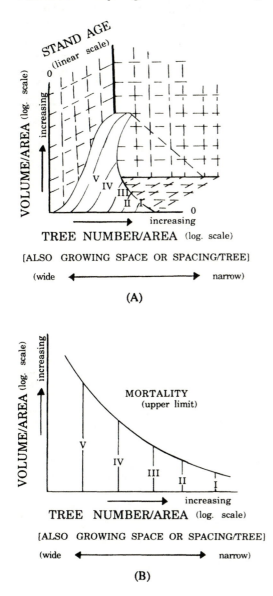

(A)

VOLUME/AREA (log. scale)

increasing

TREE NUMBER/AREA (log. scale)

[ALSO GROWING SPACE OR SPACING/TREE]

(wide ← → narrow)

(B)

Figure 15.2 Relation of volume per area, spacing, and age for single-cohort, single-species stands. (A) Three-dimensional figure created by multiplying vertical axis of Fig. 15.1D by tree number/area axis (Sato, 1983; Oliver et al., 1986; O'Hara and Oliver, 1987, 1988). (B) Relation of volume per area to spacing (Fig. 15.2A viewed from the front). When basal area per area is used as a surrogate for volume per area, the relation is analogous to that of Gingrich (1967) and Seymour and Smith (1987; Fig. 15.15).

given number of plants which occupy an area (Fig. 2.8 and 15.1B; Reineke, 1933; Ando, 1962; Yoda et al., 1963; Curtis, 1970; Westoby, 1981; Long and Smith, 1984; Weller, 1987). This relation of average growing space (or trees per area) to volume growth per tree has been empirically observed to have an upper limit with a slope of approximately -3/2 when both axes are logarithmically transposed (Yoda et al., 1963; White and Harper, 1970; Drew and Flewelling, 1977, 1979; Osawa and Sugita, 1989), although values between -1.2 and -1.8 have been reported, depending on site, species, weather conditions, and other factors (Mohler et al., 1978; White 1980; Weller, 1987; Norberg, 1988).

The intercept also varies between species but has been assumed not to vary with site, climate, or environmental conditions. The invariability is known as the "Sukatschew effect" (Harper, 1977). Other studies suggest that varying site conditions (Westoby, 1984)—especially extremes of nutrient conditions (Hozumi, 1983)—and shade (White and Harper, 1970; Ford, 1975; Westoby and Howell, 1981; Lonsdale and Watkinson, 1983) can alter the position.

This *"mortality line"* is also known as the *"self-thinning line," "-3/2 power thinning line," "upper limit of stand density," "maximum stand density line,"* or the *"full density curve."* It can be seen in the three-dimensional relations of volume per tree, spacing, and age (Fig. 15.1D) and volume per area, spacing, and age (Fig. 15.2A; Oliver et al., 1986b; O'Hara and Oliver, 1987, 1988). Various interpretations and explanations of the shape

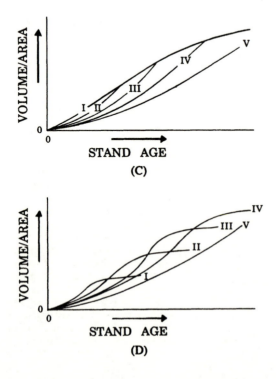

Figure 15.2 (*Continued*) (C) and (D) Relation of volume per area to age (Fig. 15.2A viewed from side). Volume growth (slope of lines) is greater at later times for wider spacings. Two possible relations have been hypothesized: the *constant yield effect* (C) and the crossover effect (D). (C) In the constant yield effect, stands at all spacings reach a constant yield but at later times at wider spacings; they then grow along the constant yield until mortality causes them to fall beneath the yield. (D) The *crossover effect.* Stands at narrow spacings reach a high yield sooner but then are surpassed by stands at wider spacings, even in the absence of mortality.

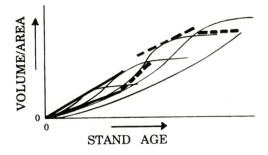

Figure 15.3 *Mean annual increment* (MAI) for each spacing and time is slope of line between that point and origin (solid straight lines) of Fig. 15.2D. *Current annual increment* (CAI) is slope of line at given spacing and time (dashed, straight lines).

of the upper-limit line have been given (Westoby, 1977; Miyanishi et al., 1979; White, 1981; Lonsdale and Watkinson, 1983; Perry, 1984).

From the volume per area/stand age perspective, mortality either prohibits stands from proceeding along the line of constant yield of Fig. 15.2C for very long if this growth pattern is correct or causes stands to be reduced in stand volume and tree number after they are surpassed in stand volume by stands at wider spacings if the crossover effect pattern (Fig. 15.2D) is correct.

Mortality from suppression gives a reverse-J shape to the relation of number of trees and time, since the number of trees which can occupy a site decreases exponentially with time and with an increase in average size. This relation (Fig. 15.1C) can be seen as one perspective of the three-dimensional age, tree size, and tree number relations of Figs. 15.1D and 15.2A. The exact shape of this tree relation varies with both site and species. Trees increase in respiratory surface area in proportion to their age and crown size rather than their diameter, since they grow taller with age and the living xylem is related to their canopy size (Waring et al., 1977). In three dimensions, mortality is probably a plane as shown in Figs. 15.1D and 15.2A. In spite of the many studies relating mortality to tree size and number, the upper limit of stand density before mortality occurs is probably not extremely closely related to tree size, since tree size is a result of the tree's past history (Fig. 8.9). It is probably more closely related to the tree's present photosynthetic and respiration surface areas, which are a product of its growing space and age (or height).

Stands undergoing differentiation

Tree growth and yield. Most stands undergo differentiation to varying extents and for various reasons (Chap. 8). Differentiation has a profound effect on tree and stand growth and yield; it occurs at different times and rates depending on the species, spacing, site, and other factors (Lutz, 1932; Deen, 1933; Gilbert, 1965). It occurs as some trees usurp that growing space formerly occupied by other trees (Hutchings and Budd, 1981). As trees become more dominant and increase their growing space, they increase in ability to grow, whereas the ones becoming suppressed are less able to sustain their respiring tissue (Yoda et al., 1957; Oliver and Murray, 1983; O'Hara, 1988). The dominating tree develops as a tree growing above confining walls (tree 3, Fig. 3.14); the tree becoming suppressed does not simply grow as a tree in a constant, narrow space (tree 2, Fig. 3.14), however. The growing space available to the suppressed tree actually decreases; and since the tree has increasingly less photosynthate to contribute to its respiration and growth demands, diameter growth slows very rapidly and the lower stem eventually does not even add annual diameter growth rings. Then height growth begins slowing and the tree lapses beneath the other trees and probably dies.

It is unclear how rapidly the volumes of living xylem and roots equilibrate to the expanding or contracting foliage as a tree becomes dominant or suppressed. If living stem and root tissues which require photosynthates rapidly contract as a tree becomes suppressed, the tree does not rapidly expend its photosynthate on excess tissue. Alternatively, if equilibration is slow, then a tree becoming suppressed expends its photosynthate on more living tissue than would a tree which had always occupied little growing space, and further suppression and death will be very rapid. Similarly, if a dominating tree's living xylem and root tissue expands as rapidly as its crown expands, then the tree's respiration rate also increases rapidly, leaving less photosynthate for allocation to growth than if the respiring tissue expands more slowly.

Yields of trees undergoing differentiation can also be characterized by tree size, individual tree growing space, and age (Fig. 15.4). These figures are similar to those in Fig. 15.1; however, stand average values such as average spacing or number of stems per area cannot be substituted for growing space per tree as they could in undifferentiating stands because each tree occupies different, changing amounts of growing space during differentiation.

Trees which differentiate can begin at equal (e.g., Figs. 8.5, 8.6, 15.4) or unequal spacings (e.g., Fig. 8.4). Each tree increases in size according to the amount of growing space it occupies (Fig. 15.4). Unlike stands not undergoing differentiation, growing space for each tree changes with time in differentiating stands even after full occupancy of growing space. Trees becoming suppressed occupy decreasing amounts of growing space, causing them to move to the right in Fig. 15.4B. Trees dominating others gain more growing space and move to the left. The times of crown closure and full occupancy of growing space are much more varied in these stands than in regularly spaced, undifferentiating stands where all trees become limited in growing space at the same time. In fact, a tree may become limited in growing space on one side before another in an irregularly spaced, differentiating stand.

When occupying a given amount of growing space, trees becoming suppressed are larger in diameter than contemporary trees which had always grown in the smaller amount of growing space, because part of the suppressed tree's size was attained when the tree occupied more growing space. Trees becoming more dominant are smaller in diameter than trees of the same age which had always lived in the larger amount of growing space for similar reasons (Fig. 8.9).

Projected sizes of trees undergoing suppression or expressing dominance do not follow the three-dimensional surface shown in Fig. 15.1D. Instead, trees undergoing suppression rise above this surface and those becoming more dominant fall beneath it (Fig. 15.4C). Trees becoming suppressed quickly reach the area of mortality, although they are larger than trees occupying the same growing space in undifferentiated stands when they die.

Growing space attained by dominating trees equals that relinquished by trees undergoing suppression, since total growing space stays about the same in a stand. A differentiating stand has some trees becoming much larger while others change little because of the dramatically different growths of dominating and suppressed trees. Single-cohort stands have often been assumed to have a normal distribution of tree sizes (Meyer, 1930; Hough;, 1932; Baker, 1934; Schnur, 1934; Lee, 1971; West et al., 1981a,b). A well-

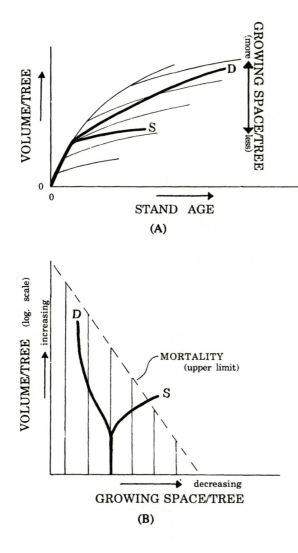

Figure 15.4 Schematic growth patterns of dominating (D) and suppressed (S) trees in differentiating stands which began at regular spacings in a single-cohort, single-species stand. Figures are similar to Fig. 15-1, except "growing space per tree" replaces "tree number" or "spacing" on one axis. (A) and (B) The dominating tree grew larger than the undifferentiated tree at the same spacing as it occupied more growing space. The suppressed tree accumulated less volume as it lost growing space.

differentiated stand can have a bimodal or skewed distribution of tree sizes (Meyer, 1930; Hozumi et al., 1955; Kohyama and Kira, 1956; Obeid et al., 1967; White and Harper, 1970; Ford, 1975; Westoby, 1977; Gates, 1978, 1982; Mohler et al., 1978; Aikman and Watkinson, 1980; Hutchings and Budd, 1981; Kohyama and Fujita, 1981; Britton, 1982; Gates et al., 1983; Cannell et al., 1984; Weiner and Thomas, 1986). The bimodal or skewed distribution is most recognizable in diameter distributions, since diameter growth is very sensitive to changes in growing space.

A skewed distribution of diameters can occur when the intermediate and suppressed trees die readily, as in a stand of a single, intolerant species or in mixed stands where the overtopped trees die. The diameter distribution may be skewed to the left (negative) when the stand is young and skewed to the right (positive) when the stand is older (Gates et al., 1983; Perry, 1985). In mixed-species stands where the trees which become suppressed and relegated to lower strata are tolerant and survive (Oliver, 1978a), a more bimodal distribution occurs.

The impact of differentiation on the relation of *average* tree size to tree numbers and ages is shown in Fig. 15.5. As the suppressed trees die, average tree size increases (Hamilton and Christie, 1974). In three dimensions, average tree size probably follows a trajectory beneath the surface described for nondifferentiating stands (Fig. 15.6; McFadden and Oliver, 1988). Stands of different species or different rates of differentiation follow slightly different trajectories.

Stand growth and yield. Yields per area in stands undergoing differentiation cannot be estimated as readily as in the hypothetical stands described above which undergo no differentiation. Stand yield in differentiated stands is obtained by summing sizes of individual trees (Reiners, 1983; Sprugel, 1985) instead of expanding average tree sizes to stand values, as was done in nondifferentiating stands. Consequently, yield patterns of differentiating stands vary with the amount of differentiation occurring within the stand (Fig. 15.7; Newnham, 1964).

Growth rates in differentiating stands approximate those of undifferentiated stands of different spacings (Fig. 15.2) to varying degrees. Stands which began at regular spacings (Fig. 15.7A) first have different yields depending on the spacings. They first show yields similar to those of undifferentiated stands which began at the same spacings (Fig. 15.2C and D; Pienaar and Shiver, 1984a,b). The stand which differentiates poorly continues to grow similarly to an undifferentiated stand until it becomes extremely crowded. As each spacing becomes very crowded, some trees die, and net stand yield actually decreases slightly (Peet and Christensen, 1987; Harrison and Daniels, 1988). Then the remaining trees grow more and stand volume again increases. Mortality often leads to regularly distributed trees at spacings predetermined by the initial spacings, and stands which undergo mortality in waves may actually increase in stand volume in ascending waves of growth and mortality (Newnham, 1964).

Stands at more irregular spacings or ages (Fig. 15.7B and C) show less initial variation in yield associated with different tree numbers in regularly spaced, undifferentiated stands (Fig. 15.2C and D). Irregular spacing creates unequal growing space for each tree, resulting in some trees having a little growing space at crown closure and others having much growing space at (later) crown closure. Differentiation and mortality occur throughout the development of these stands, and there are no dramatic waves of mortality and accompanying, temporary declines in stand yield (Newnham, 1964). Stand yields at irregular initial spacings vary even more independently of tree numbers than do stands of regular spacings which readily differentiated. Mortality generally depresses growth of differentiating stands, so that yield occurs within a zone (Fig. 15.7C) bounded on the upper side by the growth of nondifferentiating stands at different spacings (Fig. 15.2C and D; Hamilton and Christie, 1974) and fluctuates within the zone depending on variations in each stand's structure (Briegleb, 1942).

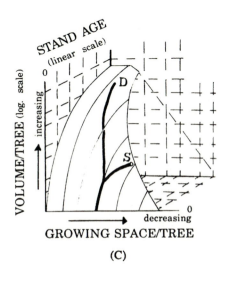

GROWING SPACE/TREE

(C)

Figure 15.4 (*Continued*) (C) In three dimensions, the dominant tree growth would project beneath and behind the plane of trees maintaining equal growing space (Fig. 15.1D), whereas the suppressed tree growth would project in front of the plane. Differentiating trees which began at irregular spacings would behave similarly, except that not all trees would have the same initial growing space.

Growth and yield with thinning

Undifferentiated stands. Growth and yield following thinning will first be considered in an undifferentiated stand where all trees occupy equal growing space. If a thinning is done while the trees are still vigorous enough to respond (Krinard and Johnson, 1975) and it uniformly removes half of the trees, for example, the growing space available to each remaining tree immediately doubles and the trees expand to refill it (Fig. 15.8; Griffith, 1959; Malmberg, 1965; Beck and Della-Bianca, 1972; Pienaar and Turnbull, 1973; Omule, 1985). This change causes each residual tree's growth trajectory to move behind the three-dimensional surface (Fig. 15.8C) if all growing space is reoccupied (and the trees were competing prior to the thinning). Release to wider spacings proportionately increases the growing space available to each tree. Tree growth then increases in proportion to the increase in growing space (MacKenzie, 1962). When the available growing space is refilled, the trees grow at the same rate as do trees originally established at the post-thinning spacing; however, because tree size is partly based on its past history, the released trees approach, but never reach, the size of those which grew at the post-thinning spacing all along (Fig. 8.9; Pienaar, 1965). Once the new growing space is refilled, growth is in a constant growing space and has a new three-dimensional response surface parallel to but behind the unthinned one (Fig. 15.8C). The distance (time) between the original response surface and the new surface depends on how rapidly the trees refill the growing space released by the thinning. In undifferentiated stands, this time depends on the size of growing space released and the vigor of residual trees. Thinning at younger ages, when crowns are relatively large, increases rate of growth response to thinning (Wiley, 1968; Tappeiner et al., 1982; Omule, 1984).

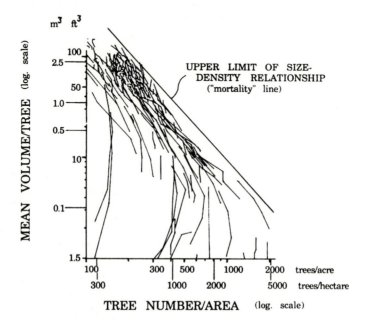

Figure 15.5 Changes in average tree size over time in Douglas-fir stands. (Each line represents a stand.) As the suppressed trees die, the average tree size increases but the number of stems decreases creating a growth pattern which moves up and to the left below the mortality line. (*Drew and Flewelling, 1979.*) (See "source notes.")

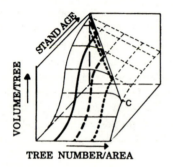

Figure 15.6 Trajectories of three differentiating stands (heavy dashed and solid lines) from the three-dimensional perspective. Volume per tree falls below the upper limit because naturally thinned stands have smaller stems than similar trees which began at final, wide spacings (Fig. 8.9). (*McFadden and Oliver, 1988.*) (See "source notes.")

A.   MODERATE DIFFERENTIATION

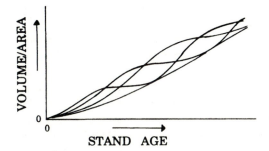

B.   STRONG DIFFERENTIATION

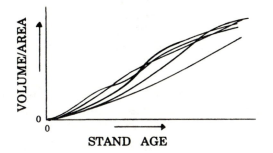

C.   GENERAL RANGE OF STAND VOLUMES

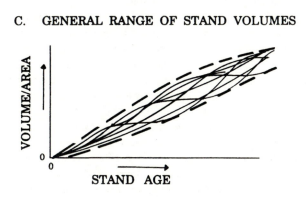

Figure 15.7 As (A) moderate or (B) strong differentiation and mortality occur in stands, volume per area for different initial numbers follows less distinct lines than in undifferentiated stands (Fig. 15.2D), since trees change amount of growing space each occupies. Tree growth would increase as some trees die and the remaining trees fill the growing space. (C) Average result for many stands (or for a single stand with irregular initial spacings) would be a band (within dashed lines) of cumulative volume with time. Stand growth (slope of individual lines) would have little relation to stand volume, but average growth for many stands (or irregularly spaced stands) would be proportional to stand volume.

Undifferentiated stands of tall trees with small crowns, relatively stagnated trees (Krinard and Johnson, 1975; Mitchell and Goudie, 1980), or trees on poor sites do not refill the growing space and grow at the maximum rate very quickly after thinning (Fig. 15.8), since trees grow into the growing space very slowly. Thinning to a wide spacing also causes growing space to be refilled more slowly (Johnson, 1968; Hamilton, 1981).

Stand yield after thinning in an undifferentiated stand is also described by a line which moves behind the three-dimensional surface (Fig. 15.9A). Stand volume is reduced to half where half of the trees are removed, and volume increases as the growing space is refilled (Fig. 15.9B; Pienaar and Turnbull, 1973; Pienaar and Shiver, 1984; Seymour and Smith, 1987). Volume growth per area is first slower in the thinned stand than in the unthinned one (Pienaar, 1979) until residual trees refill the growing space. When refilled, stand volume increases parallel to but behind a stand originally planted at the wider spacings. Consequently, thinning may eventually increase volume growth and standing volume per area provided the residual trees can refill the growing space rapidly (Fig. 15.9C; Steele, 1955a,b; Griffith, 1959; Worthington et al., 1962; Malmberg, 1965). Often the volume increase is not fully realized until one or two decades (or even longer) following thinning (Hamilton, 1976; Pienaar, 1979). Frequent thinnings, therefore, may not allow a stand to achieve high volumes (Heiberg and Haddock, 1955; Oliver and Murray, 1983).

Differentiated stands. Thinning differs from natural death by suppression to varying extents, depending on age of the stand thinned and the trees removed in the thinning. In young stands, trees which die from suppression usually occupy little growing space by the time they die; consequently, surrounding trees do not expand extensively before refilling the growing space (Ford, 1975). In older stands suppressed trees can occupy more growing space at the time of death, since the minimum amount of growing space necessary to sustain the larger amount of respiring tissue is greater.

The growing space released in a silvicultural thinning can mimic natural death if suppressed trees are removed shortly before they die. If the stand is young or suppressed trees are small for other reasons, stand volume is not reduced dramatically (Perry, 1985). Consequently, little growing space is released by removal of these trees, and more dominant trees do not grow dramatically faster. The primary effect of thinning small, suppressed trees is the economic advantage of recovering wood which otherwise dies and rots and attracts insects to the stand. In older stands where suppressed trees are larger, increased growth of residual trees upon death or thinning of suppressed trees is more dramatic because more growing space is released. Removal of suppressed trees is similar to the reduction in stand volume and increase in residual tree growth when the suppressed trees die, except that the volume could be economically recovered (MacKenzie, 1962). Trees removed in thinning are sometimes included when calculating net yield instead of being considered as lost. Such yield recovered from thinnings is not considered as *standing yield*—that volume remaining in living trees within the stand.

Removal of progressively more dominant trees by thinning releases more growing space, creating a longer time before its reoccupancy (Warrack, 1959; Long and Smith, 1984). Dominant residual trees refill the growing space created by thinning more rapidly, allowing the stand to increase yield faster than if less vigorous, more suppressed trees are left (Meyer, 1931; Steele, 1955a,b; Staebler, 1956a,b; Yoda et al., 1957; Hamilton, 1969; Reukema and Bruce, 1977; Oliver and Murray, 1983).

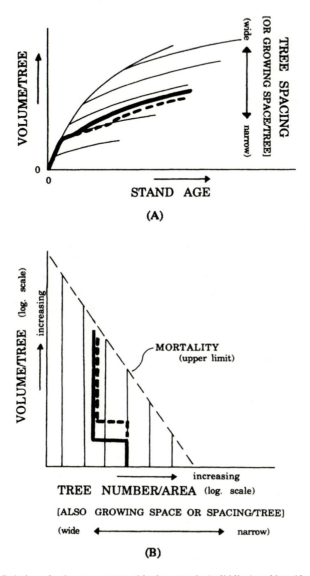

Figure 15.8 Relation of volume per tree to thinning at early (solid line) and late (dashed line) times in single-cohort, single-species stand. (Compare with Fig. 8.8.) Early thinning would allow greater individual tree volumes (A), but trees would not achieve the same volume as they would have if they had begun at wider spacings.

Stand growth efficiency may be greater after reoccupancy of growing space if codominant trees are left after thinning instead of dominant trees, although the stand probably recloses most rapidly where dominant trees are left. The large crowns of dominant trees may not be growing as efficiently as smaller crowns of more codominant or intermediate trees (O'Hara, 1988). In addition, crowns do not always reclose following thinnings; and growing space may be reoccupied by crown densification (Chap. 3), as occurs in multi-cohort stands. In this case, tree height and position of the foliage become more important than crown size in determining each tree's growth efficiency (Ford, 1975; Waring, 1983; O'Hara, 1988). Stand growth efficiency (stand growth per unit of foliage) is also affected by the arrangement of foliage or the stand structure (Dean et al., 1988; O'Hara, 1989).

An optimal thinning regime may include heavier thinnings when stands are younger and most capable of quickly reoccupying available growing space. Light thinnings in older stands would minimize unoccupied growing space where large trees are less capable of expanding into unoccupied growing space (Staebler, 1960; Kramer, 1966; Wiley and Murray, 1974; Hamilton, 1976, 1981; O'Hara, 1987). Lighter thinnings in older stands may also prevent residual trees from developing excessively large crowns and becoming inefficient users of growing space.

Growth, growing stock, and stand density

As trees increase in size, their growth rates at first increase, since they occupy more growing space. Similarly, stand growth at first increases as the trees become larger. There is, therefore, a rough relation between total stand volume, which is a surrogate measure of degree of competition, and stand growth (Langsaeter, 1941; Mar-Moller, 1947, 1954; Davis, 1955, 1956; Gross, 1955; Bickford et al., 1957; Staebler, 1958, 1959;

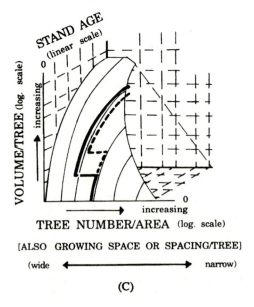

Figure 15.8 *(Continued)* Growth projections of thinned stands (C) would be beneath and behind plane of undifferentiated stand (Fig. 15.1D).

D. M. Smith, 1962, 1986). Stand growth is less in extremely crowded or widely spaced stands but is similar over a range of intermediate degrees of competition. This growth relation can be applied either to unthinned stands at each spacing at a constant age or to stands kept at different approximate degrees of competition by frequent thinnings. Some measures of degree of competition are referred to as *stand density*. Various studies have used stand volume and basal area as measures of stand density and have attempted to relate them to stand volume or basal area growth.

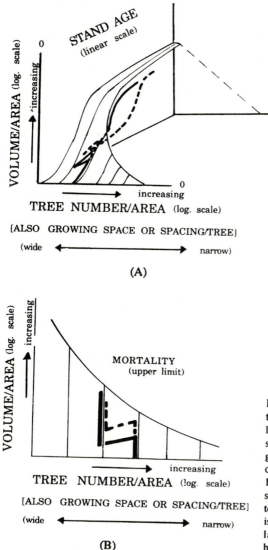

(A)

(B)

Figure 15.9 Relation of stand volume to early (solid line) and late (dashed line) thinning in single-cohort, single-species stands (Fig. 8.8). (A) Volume growth would project behind the plane of the undifferentiated stand (Fig. 15.2A). (B) The volume of thinned stands would appear to grow parallel to that of unthinned stands when time is not considered, but it would in fact lag in time and might not reach as high a value.

A refinement of the stand density measure is the combination of stand volume (or an approximate surrogate such as basal area) and number of trees, or average tree size (Reineke, 1933; Hummell, 1954; Curtis, 1970, 1971, 1982; King, 1970; MacLean, 1979; Ernst and Knapp, 1985). The various measures of stand density—stand basal area and stand volume—are somewhat related to growth, since they are related to amounts of foliage and other physiological measures (Waring et al., 1981; Gholz, 1982; Long and Smith, 1984). These stand density measures can be only approximately related to growth, since measures of density generally consider both living tissue which is closely related to leaf, root surface, and respiration area—and a variable amount of dead heartwood (Bickford et al., 1957).

Stand yield in the near future is closely related to present stand volume (a.k.a. *growing stock*) because changes in stand yield occur slowly. On the other hand, the relation of stand growth—the change in stand yield—to growing stock is more complicated (Assmann, 1954; Mar-Moller, 1954; Holmsgaard, 1958). The relation is sometimes referred to as the "Langsaeter relation" (Langsaeter, 1941; Smith, 1962). Growth can be correlated to growing stock from the relations shown in Figs. 15.2C and D and 15.7. The slope of each line at each point in Fig. 15.2C and D is the growth rate, and the vertical axis indicates the growing stock at that time. The relation of growth rate (slope) to volume (growing stock) for Fig. 15.2D is shown in Fig. 15.10. Stand growth is not closely related to growing stock where trees grow uniformly (Fig. 15.10). At any given growing stock (stand volume the vertical axis of Fig. 15.2D), stand growth (slope of the line) can vary dramatically between stands (Pienaar and Turnbull, 1973). Consequently, stand density does not define a unique growth condition and can give misleading results when used as a refined correlate of growth (Bickford et al., 1957; Assmann, 1970; Oliver and Murray, 1983).

There are an optimum age and an optimum spacing of maximum stand volume growth which vary by species; at wider spacings, stand volume growth continues to increase with age but does not reach as high levels as the optimum spacing and age.

Where stands are irregularly spaced and/or undergoing differentiation, stand growth is an aggregate of individual tree growths (Fig. 15.7C). Stand volume then increases in a general "band" with time, and the average relation of volume growth (slope of the band) to standing volume (vertical axis) is shown in the dashed line of Fig. 15.10. Growth is

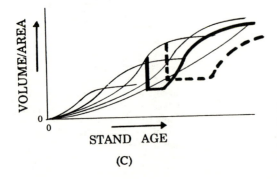

Figure 15.9 (*Continued*) (C) The stand volume after thinning can exceed the unthinned stand volume if there is enough time after thinning and residual trees are vigorous.

also roughly related to stand density when considering many stands of different spacings, a few stands with irregular spacings, or stands which otherwise are actively differentiating. For unthinned stands, maximum growth occurs at increasingly wide spacings and greater densities as the stands grow older (Fig. 15-2D; maximum slopes of lines in Figs. 15.2D and 15.9C indicate maximum growth for a stand at each spacing). Stand density is even less closely related to growth following thinning regimes (Fig. 15.9C). Stands of similar stand densities (standing volumes) have dramatically different growth rates (slopes in Fig. 15.9C) following thinning, depending on timing of thinning, how long since thinning occurred, and which trees were left after thinning (Holmsgaard, 1958; Gruschow and Evans, 1959; Brender, 1960; Buckman, 1962; Clutter, 1963; Baskerville, 1965; Assmann, 1970; Beck and Della-Bianca, 1972; Pienaar and Turnbull, 1973; Oliver and Murray, 1983; Oliver et al., 1986b; Curtis and Marshall, 1986).

Stand density may be a useful correlation to stand growth in coarse estimates (Bickford et al., 1957), but stand density may be a kludge for more refined estimates of growth and yield. A *kludge* is a "once satisfactory solution based on old technology which has been patched and repatched to meet demands for increased precision or detail and which should be abandoned for a new and simpler approach made possible by new technology" (Furnival, 1987). Basal area, a very common measure of stand density, was initially used to estimate standing volume and later to estimate gross growth over large areas. Only later, perhaps as a kludge, has it been applied to estimating growth of individual stands (Spurr, 1952; Bickford et al., 1957).

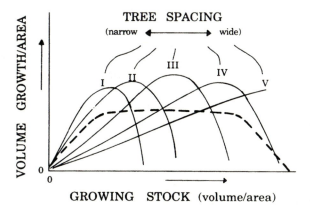

Figure 15.10 The relation of volume growth to growing stock (here expressed as standing volume) for unthinned stands derived from Figs. 15.2 and 15.7, since growth is the slope (derivative) of the relations in these figures. Undifferentiated stands would show a distinct relation for each spacing (lines I through V) as well as for site and species. Differentiating stands or aggregations of many stands would show the average relation of the dashed line (taken from Fig. 15.7B and C).

Constant final yield

Figures 15.2D and 15.7 suggest that an upper limit of stand yield exists for each age, species, and site. This upper limit at first becomes greater with stand age and then appears to approach a constant at very old ages. Variations in volumes among stands with equal initial spacings increase as differentiation and mortality occur. Consequently, most stands appear to approach a single range of standing volume, regardless of initial spacing (Baker, 1953). A "constant final yield," which all stands of a species and site approach asymptotically regardless of age, has been suggested for each species and site —not to be confused with "constant yield" (Kira et al., 1953, and others) discussed earlier in this chapter. Process or empirical evidence for stands approaching this hypothesized asymptote are limited. More probably, an upper limit of stand volume exists for each site, spacing, and species which a stand may or may not reach depending on its particular structure and degree of differentiation.

Summary of quantitative relations

The various relations between measurable characteristics have been developed into a matrix (Fig. 15.11; Leary, 1987).

Mixed-Species, Single-Cohort Stands

Less is known about volume growth of mixed-species stands than about pure-species stands. Growth and yield projections sometimes attribute relative growth and yield of each species to the proportional numbers, volume, or basal area of the component species. It is more appropriate to relate stand volume growth and yield to the stand development patterns and structures. Mixed stands can be more or less efficient at growth and/or yield than pure stands, depending on the species involved, the stage of development of the stand, the stratification pattern, and the site.

The overall relation of tree number to volume per tree in mixed stands is similar to that in pure stands (Fig. 15.1B; J. White, 1985). Where crown closure has occurred within a lower stratum of mixed stands, the upper limit of the tree volume/number relation of this shaded stratum may be -1 (White and Harper 1970; Ford, 1975; Hutchings and Budd, 1981), whereas the upper continuous (B-) stratum has a relation of $-^3/_2$. The more flattened appearance of shaded trees (Fig. 12.3) may account for a relation of 1:1 for tree volume to spacing (number), since little height growth occurs and both spacing and volume relations are primarily two-dimensional.

Species characteristics

Mixed stands can be more productive than pure stands where the species' niches are dramatically different (Kelty, 1992). If a stand has stratified with a very shade-tolerant species beneath a less tolerant one, the growth of the two species can be greater than for either species alone (Assmann, 1970; Boardman, 1977, Kramer and Kozlowski, 1979), as has been found in mixed northern red oak and eastern hemlock stands in Massachusetts (Kelty, 1989). Comparisons of many studies of mixed and pure species stands, however, shows a variety of patterns of yield depending on characteristics of the individual species (Kelty, 1992).

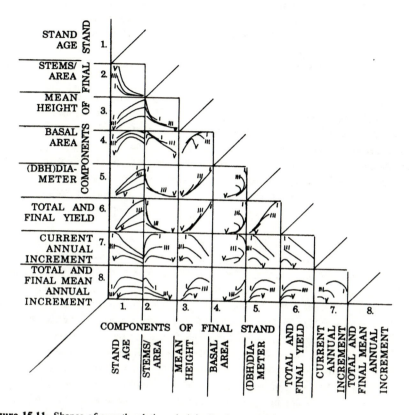

Figure 15.11 Shapes of growth relations (originally from Scotts pine; *Leary, 1987*). V, III, and I = progressively better sites. (See "source notes.")

Volume growth can be increased where one species makes more growing space available for another. The increased nitrogen from red or sitka alders can give greater volumes to mixed Douglas-fir and alder stands on poor sites than can be found in pure stands of either species (Tarrant and Trappe, 1971; Miller and Murray, 1978; Haines and DeBell, 1979; Harrington and Deal, 1982; Binkley, 1984; Binkley et al., 1984). Douglas-fir stands produce greater volumes than alder stands on most sites, and nitrogen from the alders enhances growth of the Douglas-firs. During early growth, the alders may outgrow and kill the Douglas-firs, however, unless careful measures are taken.

Stage of development

At first, all species generally grow in the same stratum following a disturbance, and each tree expands in an open-grown pattern. Each crown size (and probably soil growing space) then becomes restricted by its neighbors of any species. The growth/growing space relation is different for each species, and volume growth per area is the sum of the growth of each tree. Growth per area is, therefore, first roughly related to both the relative number and spacing of trees of all species.

As some species outgrow others, the effect is similar to differentiation in single-species stands, except that volume growth per area largely reflects the volume/area relations of the species achieving the A- and B-strata. The effect is also similar to that of a stand being thinned (Fig. 15.8), with the A- and B-strata species changing in volume from that determined by the narrow spacing of stems of all species to a wider spacing of only A- and B-strata trees.

Accumulated volume of each tree is determined by growing space and depends on both free growth before restriction and growth rate after restriction. Once stratification has occurred, most growth is done by the A- and B-strata trees, and volume growth reflects growth characteristics of species in this strata. The lower strata then grow very little and primarily influence total standing volume and net growth by surviving or dying. If the species being relegated to the C- and lower strata cannot live beneath the more dominating trees, the volume per area is wholly dependent on the upper-strata trees, as in single-species, single-cohort stands. Mortality of these overtopped trees causes low net growth per area.

Where the lower-strata species live beneath the more dominating species, they grow relatively little and gross growth per area depends on the upper-strata trees; but if the lower strata trees do not readily die, both standing volume and net growth are generally higher than in a single species stand of the A- or B-stratum species (Braathe, 1957; Assmann, 1970; Perry, 1985). Where pure stands of the shade-tolerant, lower-strata species are more productive than pure stands of the more dominating species, the standing volume and net growth per area of the mixed stand can be (1) intermediate between that of pure stands of either species, as in the case of red alders and western hemlocks (Stubblefield and Oliver, 1978), *Eucalyptus saligna* and *Albizia falcataria* in Hawaii (DeBell et al., 1989), and in some herbaceous communities (Trenbath, 1974; Harper, 1977), or (2) greater than either species, as in the case of mixed Douglas-fir and western hemlock stands (Wierman and Oliver, 1979). Spatial patterns also can cause variations in the relative volumes of pure and mixed stands (Mielikainen, 1980, 1985).

Site characteristics

Where moisture limits growing space for much of the year, lower strata in mixed stands may actually limit growth of overstory trees (Dale, 1975) and even reduce stand yield if overstory trees die. On more mesic sites, the presence of the understory seems to have little effect on the overstory, and the growth of the understory is additive to the overstory (Kelty et al., 1987; Kelty, 1989; Kittredge, 1988).

Other factors

Mixed stands may allow greater yield than pure stands where they slow or prevent spread of pathogens or insects (Driver and Wood, 1968). They may also be beneficial for wildlife (Thomas, 1979) and for wood quality where they allow differentiation of a few trees to the overstory but trainer trees in lower strata keep the overstory trees pruned.

Multicohort Stands

Volume growth of multicohort stands varies greatly with stand structure, site, and species as in single-cohort stands. Most of the stand's volume growth occurs in trees exposed to direct sunlight in the A- and B-strata. If these trees are vigorous, volume growth will be quite rapid. Trees in lower strata primarily grow after a disturbance and before the overstory becomes very dense; their presence makes them capable of rapid growth when the overstory is removed.

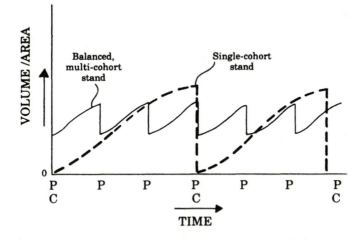

Figure 15.12 Schematic comparisons of cumulative stand volumes of single and multicohort stands. Partial cuttings of multicohort stands occur at times P; harvests of single-cohort stands occur at C. Average standing volumes and yields would probably be about the same in both single- and multicohort stands, since the volume of multicohort stands never becomes as low as that of young, single-cohort stands or as high as the volumes of stands about to be harvested. (*Larson, 1982.*) (See "source notes.")

When growing vigorously, multicohort stands with balanced sizes and ages between cohorts (similar to the ideal stand of Fig. 13.4) have different sizes and flows of timber volumes than single-cohort stands. The range of growth (mean annual increment) over extended periods is within the same order of magnitude as single-cohort stands of the same species and on the same site (Figure 15.12, Larson, 1982). Studies in the southeastern United States showed cubic volume yields in multicohort stands of loblolly and shortleaf pines to be only two thirds of the yields expected from single-cohort plantations (Guldin and Baker, 1988; Guldin, 1995). Tree size and quality can be higher in multicohort stands, however (Guldin and Fitzpatrick, 1991). Exact comparison is difficult because there are few examples of vigorous multicohort stands, measuring growth in these stands is difficult, and it is difficult to determine if comparable stands are on similar sites. The overstory of a multicohort stand is never as closed as that of single-cohort stands in the stem exclusion stage, or the shorter trees (younger cohorts) would die. Consequently, a multicohort stand does not produce the high volume growth of a single-cohort stand late in the rotation. When the taller stratum of a multicohort stand is removed, however, lower-strata trees reoccupy the site more quickly than in single-cohort strands, and the volume growth is never as low as during the early stand initiation stage of a single-cohort stand.

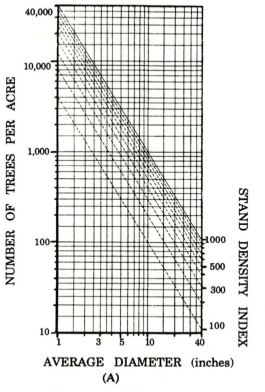

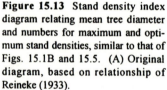

Figure 15.13 Stand density index diagram relating mean tree diameter and numbers for maximum and optimum stand densities, similar to that of Figs. 15.1B and 15.5. (A) Original diagram, based on relationship of Reineke (1933).

Applications to Management
Stand management guides

Many thinning and spacing guidelines and diagrams are available for the forester. These help foresters determine when the stand is or will become overly crowded and subject to mortality, instability, or infestation by insects or diseases as well as when maximum growth per hectare (acre) will occur. Guides are also being developed which help determine optimum wildlife habitats, such as optimum hiding cover for elks (Smith and Long, 1987). Final thinning decisions are made on the ground on a tree-by-tree basis, and the guidelines assist in achieving target spacings to the extent possible. The guidelines are interrelated, since they use the same basic characteristics of tree volume (or diameter or basal area, which are allometrically related), spacing (or trees per hectare), volume or basal area per hectare (a product of volume per tree and trees per hectare), and time (or its surrogate:dominant height). Taper and stem form differ throughout the rotation and are affected by the spacing and vigor of each tree, but they are usually not considered in the guidelines. Commonly used spacing guidelines are:

• Reineke's stand density index (Fig. 15.13A; Reineke, 1933), which has been modified by McCarter and Long (1986) and Long et al. (1988; Fig. 15.13B). This diagram is similar to Figs. 2.8 and 15.1B.

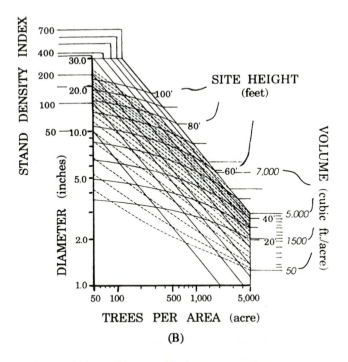

Figure 15.13 (*Continued*) (B) Modification of Reineke's (1933) diagram to conform to format of $-{}^3/_2$ density management diagram (Fig. 15.14; *McCarter and Long, 1986*). (See "source notes.')

• $-{}^3/_2$ density management diagram (Fig. 15.14; Drew and Flewelling, 1979).   This diagram is also similar to Figs. 2.8, 15.1B, and 15.13B.

• Basal area diagram (Fig. 15.15; Gingrich, 1967). This diagram is similar to Fig. 15.2B.

Guidelines not developed into diagrams include the spacing/top height guide (Wilson, 1946) and the live crown ratio (D. M. Smith, 1986). The correlated curve trends diagram (Figs. 2.8, 3.17, and 15.1A; O'Conner, 1935) is useful for determining minimum spacings which will allow trees to reach a given diameter. This diagram has not been calibrated for many species. The various guides are interrelated, but the values do not match each other because of variations in the allometric relations used to compare them. Thinning can especially change the relationships among tree measures (Larson and Cameron, 1986).

The best guideline is the one with most relevant variables for a given area and stand type, which has been calibrated for the stand types, and which the forester understands. In a stand to be spaced and thinned many times, the forester might use the spacing/top height ratio when the stand is young and change to a $-{}^3/_2$ diagram after the trees have enough volume for it to be accurate.

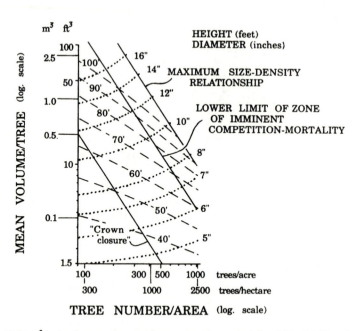

Figure 15.14 $-3/2$ density management diagram-based on principle of Fig. 15.1B and related to stand age through dominant heights. (Stand was calibrated for coastal Washington Douglas-firs; *Drew and Flewelling, 1979*). (See "source notes.")

The dominant-height values on the $-3/2$ diagram (Fig. 15.14) and the stand density index diagram (Fig. 15.13B) allow estimates of future stand growth. The basal area diagram also can be calibrated for tree height by using the type of relations in Seymour and

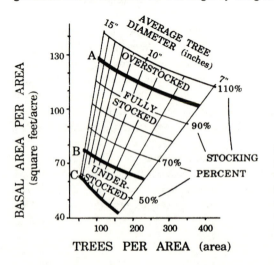

Figure 15.15 Basal area diagram relating basal area (a substitute for volume per area), diameter, and tree number to determine if desired stand density is achieved, based on principle of Fig. 15.2B. (*Gingrich, 1967.*) (See "source notes."

Smith (1987). The forester can plan different thinning regimes based on the current tree spacing and anticipate when the stand will again become overcrowded (Oliver et al., 1986b). These times of overcrowding are *decision times* when a stand can be harvested with maximum volume or thinned again. If thinned again, the forester must wait until another decision time before harvesting it for maximum volume. The estimated decision times allow the forester to plan stand structures, silvicultural operations, and timber flows. The times are not exact, but they permit general planning.

Foresters are usually unwilling to cut enough trees in a thinning to allow the remaining trees to grow vigorously. The diagrams model stand growth after thinning and give the foresters an objective way of assessing appropriate spacings. They assume the spacing is quite uniform in a stand. One cannot leave trees in clumps after thinning and assume they will grow as if they were at the average spacing per hectare. The clumps develop as if the trees are at a high density within the clumps and at a low density on the edges.

Various techniques for describing and determining appropriate stocking for each cohort in multicohort stands are being developed. A q-Ratio approach is based the assumption that the diameter distribution in a multi-cohort stand follows some form of "reverse-J"shape (Figure 6.3B, Meyer and Stevenson, 1943; Meyer, 1952); different values of "q" indicate different levels of increase in tree numbers in progressively smaller diameter classes (Chap. 13). A technique has been developed using modifications of the density-management diagrams (of Figures 15.13 and 15.14) in which the trees are allowed to be partitioned among cohorts (Long and Daniel, 1990; Sterba and Monserud, 1993; Long, 1995). A technique is being developed based on allocation of growing space, as measured by diameter and correlated to leaf area; this technique allows different numbers of trees to be allocated to different cohorts, providing the total amount of trees does not exceed the available growing space (O'Hara and Valappil, 1995).

Objective, analytical systems to classify stand structures are beginning to be developed (McNicoll, 1994; O'Hara and Milner, 1994). These systems rely on decision systems based on quantifiable measures of stand structure components. Such systems will become increasingly important as analytical approaches to landscape management become more in demand.

Adaptive management

Being aware that similar development processes occur in diverse forests can be used to advance forest management at the stand and landscape levels. Research and management in relatively unstudied forests can be done by comparing stand structure patterns and local species' physiological and morphological characteristics to those elsewhere in the world. Processes in the unstudied area can be hypothesized from analogous structural patterns and species characteristics in stands whose development processes have been well studied. The "null hypotheses" would be: there is no reason to believe the stand development processes are different (Oliver, 1992d).

The hypotheses can be tested using usual research techniques, or silvicultural manipulations can be used to test the hypotheses. Changes in structures over time following a silvicultural manipulation can be predicted according to the hypothesized stand development processes. Components of the stand can be monitored, and the accuracy of predictions will determine the validity of the hypothesized processes. Comparisons of actual to predicted structures not only will test hypotheses, but will allow processes to be understood better and new processes hypothesized.

This approach (Walters, 1986) has been proposed for forest management (Baskerville, 1985; Oliver et al., 1992) as "adaptive management." The adaptive management approach is similar to other iterative management approaches such as "quality control (Ishikawa, 1982) and "time series analysis" (Bowerman and O'Connell, 1987). The basis of the approach is to make a projection (hypothesis) of how the system will develop under a given set of conditions; to implement the conditions; to monitor, analyze, and learn from the results; and to adjust the conditions and knowledge base.

Many stand development processes are known well enough to be compared to newly studied areas using adaptive management. The development processes form the bases for silvicultural keys, guides, and practices, which can be treated as hypotheses. Individual conditions and insights into possible processes in new areas will help determine exactly which hypotheses are appropriate. Elements of adaptive management are being incorporated into silvicultural techniques in tropical forests (Jordon, 1982; Larson and Zaman, 1985; Ashton et al., 1989; Hartshorn, 1989a; Anderson, 1990; Galloway, 1991). A process-based stand development model which can be calibrated and recalibrated to different species will facilitate adaptive management (Larsen, 1994).

CHAPTER 16

FOREST PATTERNS OVER LONG TIMES AND LARGE AREAS

Introduction

The appearance of stability of forests in one place and the similar patterns of forests across landscapes have been widely observed by naturalists for many centuries. The observations have led to the assumption that forests are stable in species composition and structure and develop according to preset patterns across large areas. The assumptions have generally been based on observations of short duration and small areas. Times of most observations were less than the life spans of one generation of trees, and the areas observed have usually been within a single or limited number of geologic or climatic provinces. Only within the past few decades have communication, transportation, and information storage and processing allowed events to be understood which occur over long times and broad and distant areas.

In fact, forests constantly and dramatically change in both time and space. Forest species compositions have changed over the past few thousand years in such diverse places as the Amazon basin (Gentry 1989, Irion 1989, Uhl et al. 1990), temperate and boreal North America (Davis, 1981; Cwynar, 1987; Ruddiman, 1987; Pielou, 1991;

Brubaker 1988, 1991; Mack et al., 1978*a,b*; Karsian, 1995), and Asia Minor (Aytug et al., 1975).

The earth's climate has also changed dramatically; and plant and animal species are constantly migrating in response to climate and soil changes (Tallis, 1991; Davis, 1986; Brubaker, 1991; Pielou, 1991; Gates, 1993). Because of disturbances, species migrations, climatic variations, and other factors, a forest of a given structure and species composition exists at only one time and in one place. The patterns often appear unchanging because the rate of change is slow relative to the memories and life spans of humans. The patterns appear repeated across a landscape because the limited number of species and physiological responses and similarities of soils, climates, and disturbance patterns over large areas allow only a limited number of patterns to occur (Oliver, 1992*d*).

Formerly, the patterns were thought to be obligatory—the result of coevolved, mutualistic interactions. The patterns are now believed to be the result of competition among relatively new assemblages of species. The mutualistic perspective, however, still influences many studies of forest development.

Vegetation patterns across a landscape are neither completely random nor completely predictable. Biotic and abiotic influences on the trees can be anticipated, and physiological processes of the trees are partly understood. Consequently, the resulting development patterns can generally be anticipated—even if not completely understood. The probability of a pattern occurring at a given time can be estimated and simulated (Mitchell, 1975; Shugart and West, 1980; Shugart et al., 1981); however, there are enough variations in climate, soils, species behaviors, and other factors to make the occurrence of a particular pattern far from certain.

The influences of climate, soil, disturbances, species behaviors, and time have been dealt with in preceding chapters of this book. The competitive processes and mechanisms through which these influences interact also have been covered. This chapter will discuss development of forests on a single area over long times and patterns of forests across a landscape.

Changes on a Single Area over Time

Each forested area is initially defined by the species present. As a general pattern, a forest of any species composition is characterized by an extended time of slow, subtle, but continuing and irreversible change—forest development—punctuated by occasional, sudden, dramatic, and destructive disturbances (Botkin, 1990).

The component species

Forest patterns change only as the component species are capable of changing. Consequently, forest development depends on the species present and their capacity for change. Each species has a different growth pattern. The number of growth patterns within a forest increases dramatically as the number of species increases. Growth patterns of a single species can vary in different parts of the species' range or when interacting with different species, further increasing the complexity of forest patterns.

Even with these complexities, a limited number of tree species are found in a forested area. Most species exhibit similar behaviors over large areas and under similar circumstances; consequently, only a finite number of patterns occur within a forest. North

American forests contain few species in arid or cool areas or where other species have not yet migrated. As few as five tree species are found in many forested areas of northern Canada (Raup and Argus, 1982) and as many as 70 tree species in warm, moist river bottoms of the southeastern United States (Putnam et al., 1960). Some tropical forests contain even more tree species. As a general rule, warm, mesic areas contain more tree species than cool or arid regions, although past extinction and migration patterns may have left some places with inordinately few trees. Europe north of the Alps, for example, contains few species compared to nearby Asia Minor possibly because the Alps blocked north-south migration of species in northern Europe during glacial periods.

Different plant species are found over a range of environments because they are versatile, resilient organisms that evolved to survive in a variety of situations. Most species are adapted to regenerate in several forms, allowing them to respond to a variety of disturbances. Many species can survive in both full sunlight and deep shade; others can survive both very droughty and very saturated soil conditions. This resiliency allows a species to occupy a range of environments and further elucidates the complexity of forests in a region.

Some species are less versatile and can survive only under quite limited conditions. Consequently, they are found only in a small range of temperature, nutrient, or moisture conditions. Such species are susceptible to extinction if the limited conditions under which they grow are destroyed. Threatened plants, such as many subalpine and alpine species, are often very tolerant of certain environmental conditions—temperature and moisture extremes but not others—competition or certain types of disturbances. When the environment changes in a way in which the plants are not tolerant, they quickly become excluded from the community. Many rare and endangered species are often found together in small areas and are locally abundant where the environment is suitable for them to survive.

The forested landscape is composed of constantly migrating and evolving plants which happen to be together at a particular place and time. The plants generally do not remain together in a stable environment long enough to coevolve. Consequently, they compete with each other for the same basic moisture, sunlight, nutrients, and other growing space factors. Their versatility allows many species to survive in the same geographic range, climate, and soil conditions. Species survive in the same stand where they occupy different niches. At other times, two species can live in the same geographic range but generally do not live together within the same stand. For example, western hemlocks can live in the same stand with either Douglas-firs (Fig. 9.5) or red alders (Fig. 9.6) because hemlocks become overtopped by both species and can tolerate the shade of either. Red alders and Douglas-firs, however, may not live together, since either one will overtop and kill the other depending on the initial stand conditions.

The competitors in a given area are constantly changing through evolution, migration, and extinction. Plant evolution is the slowest process because the times when new genetic combinations become established are infrequent—generally following disturbances. Relatively rapid (but still quite slow) genetic changes—*hybridization*—can occur where previously isolated populations meet. These junctures—*suture zones*—occur in several places, including central New England, the Rocky Mountains, and central Texas (Remington, 1968, 1985) and possibly the Kenai Peninsula, Alaska. Suture zones occur when species that might hybridize are kept separate by a barrier. In northern New England, species that grow on stand edges and open fields were probably kept separate from similar species farther south by a usually continuous forest. When the forests grew back after agricultural land clearing and abandonment of the past few centuries, the species were probably able to come together physically and hybridize.

Plant species migrations from one geographic area to another can occur in very few generations—quite rapid movement in plant terms, but quite slow in human terms. Tree, shrub, and herbaceous plants can invade deglaciated areas very soon after the ice has melted (Cooper 1923, 1931; Oliver et al., 1985). Many tree species have occupied their present distribution in the eastern United States for only 2000 to 6000 years—about 20 generations (Mack et al., 1978a,b; Davis, 1981; Brubaker, 1991; Karsian, 1995). Plants have the greatest opportunity to migrate by invading a new area immediately after a disturbance, before the growing space is refilled by other species. Consequently, plants may migrate faster along frequently disturbed corridors such as stream-side areas, ridgetops, and—more recently—roadways. Species do not seem to migrate in concert (Spaulding and Martin, 1979; Van Devender and Spaulding, 1979; Van Devender et al., 1979), so that a group of species apparently does not coevolve as it moves across a landscape, and a single community does not last for an extended period. Tundra and alpine communities can be rapidly invaded by forests with climatic changes or changes in disturbance patterns (Franklin et al., 1971; Heikkinen, 1984).

A plant population can change from abundant to extinct either rapidly or slowly. An introduced predator can extinguish a species within several decades as almost occurred with the American chestnut, which now lives only asexually as a sprouting bush (Hawes, 1968; Stephens, 1976). Rapid extinction seems to be occurring with the American elm because of the Dutch elm disease (Clinton and McCormick, 1936; Hepting, 1971; Stephens, 1976). Introduced competitor plants can slowly reduce the range of a native plant, so the original plant becomes relegated to a smaller niche (Elias, 1977). Such reductions in realized niche size through introductions do not seem to have occurred dramatically in tree species in North America, although some native grasses and subalpine herbs have become rare through the spread of nonnative plants. Extinctions also occur through changes in climates, disturbance patterns, or soil conditions. Elimination of wildfires reduced the numbers of jack pine stands in the midwestern United States, concomitantly nearly eliminating the Kirtland's warbler dependent on these stands (Walkinshaw, 1983). The Franklinia, a bush native to eastern Georgia, became extinct (except for a cultivated plant) during the active land clearing of the early 1800s (Ayensu and DeFilipps, 1978). Herbaceous plant species have become endangered or extinct where the habitat has been destroyed by land clearing, draining, and conversion to domestic uses, commercial sale of endangered plants, loss of insect pollinators (Elias, 1977) and prevention of fires and similar disturbances.

Changes between disturbances

The rate of change between disturbances—stand development—is so slow in most forests that casual observers do not realize the changes. It is only when measures, photographs, or other objective observations are made at widely different times that the development is noted. For many years in the early twentieth century, the southern New

England landscape had the reputation of being cutover hardwood scrubland originating after clearcutting of the white pine stands (Fig. 16.1; Harvard Forest, 1941; Stephens, 1976). In fact many of the stands were in the early stage of developing into stratified oak, maple, and birch stands (Figs. 9.1 and 9.2), and the forests have grown into large, tall stands of valuable oaks (Oliver, 1978a). Similarly, in the southeastern United States much of the abandoned farmland was planted or seeded naturally to southern pines in the 1940s, 1950s, and 1960s (USDA Forest Service, 1982). Ironically, because of the presence of hardwood advance regeneration, many cutover southern pine stands presently being harvested will probably regrow to hardwoods as the New England stands did after harvest of white pines (McKinnon et al., 1935). Many open ponderosa pine and lodgepole pine stands of the Rocky Mountains commonly remembered in cowboy movies of the 1930s and 1940s have since been filled with a dense cohort of conifers—a natural result of stand growth following fire and grazing exclusion of the late 1800s and early 1900s. By the 1960s, the dense young conifer stands were admired by recreationists, and cutting and/or thinning them was opposed; by the 1980s, these stands had grown enough to become overcrowded, and insects, mortality, and fires threatened or destroyed them (Kilgore, 1973; Agee, 1974; Parsons and DeBenedetti, 1979; Butts, 1985; Agee and Huff, 1986; O'Laughlin et al., 1993; Sampson and Adams, 1994). A similar pattern is happening in northern Arizona, where a large and dense age class of ponderosa pines began about 75 years ago (Pearson, 1950; Daniel, 1980); the area has slowly changed from open grassland to dense forest, and it will probably grow so dense that insect outbreaks, mortality, and fires will occur unless measures are taken (Covington and Moore, 1994).

The rates of change in forests vary between sites and regions and with species. The changes were not appreciated where logging companies often abandoned land after harvesting the merchantable timber. More enlightened companies retained the land and have been able to harvest a second tree crop. Presently, forest growth is appreciated enough that many forest product and other companies are viewing forest land as an investment. The furniture industry generally bases its product lines on which species will develop to merchantable sizes in large quantities over the next planning period.

The slow changes are reflected in animal populations as well. Cutover lands in many parts of the eastern United States led to high carrying capacities of deer and other, small animals when many areas were in the stand initiation stage. As the stands developed into the stem exclusion stage, less browse was available, only a smaller population could be supported, and many animals starved (Marquis, 1975b). There is concern that regrowth of forest openings following farming and grazing abandonment will lead to extirpation or extinction of butterflies in parts of England. Even if openings are maintained, extirpation and extinction could result from inbreeding if there are no corridors of open structures to allow the animals to migrate from one opening to another (Young, 1992). The regrowth of forests after abandonment of agriculture in many parts of the eastern United States has increased habitats for beavers, bears, and alligators—all of whose numbers are increasing after diminished populations for nearly a century. In fact, the return of beavers and beaver ponds may increase the population of cypress trees, which germinate well along pond edges. The change from open land to forests through fire control and regrowth after timber harvest in parts of North America has eliminated browsing lands

(A)

(B)

Figure 16.1 Changes in vegetation composition strongly reflect history, since different social patterns create different disturbance regimes. The following are the reconstructed changes of a single area of central Massachusetts from 1740 to 1930: (A) Early clearing of original forest by colonists from coastal New England and abroad (ca. 1740). (B) Abandonment of farmland, allowing lig seeded white pines to enter as people moved to the more fertile Ohio Valley and farther west (ca. 1860).

(C)

(D)

Figure 16.1 *(Continued)* (C) Development of hardwoods from advance growth as white pines were harvested (ca. 1909). (D) Young hardwood stand in "brushy stage," having developed despite efforts to grow another rotation of white pines (ca. 1930). Hardwood stands have now developed to stratified stands dominated by oaks (Figs. 9.1 and 9.2). Similar patterns are occurring in southeastern United States, although farm abandonment occurred more recently. (*The Harvard Forest, 1941*.) (See "source notes.")

and populations of moose (Brassard et al., 1974; Krefting, 1974) and has hidden migration routes for mountain sheep between summer and winter ranges in parts of the Rocky Mountains (Geist, 1971), and has probably caused at least one butterfly species to become extinct (Fry and Money, 1994).

A general pattern of change in a stand has been described (Fig. 5.2). During stand initiation, native, exotic, and migrating species can become established on the area. Many species are adapted to short life cycles which are completed during the stand initiation stage. These species disappear as a stand becomes older. The remaining plants expand and exclude new ones from entering—the stem exclusion stage. The reduction in species as the stand develops leads to an apparent convergence of the stand composition from many, varied species to a few, predictable ones. In cases of mutually exclusive species such as alders and Douglas-firs—any of several types can exist on an area; and a mixture of species can exist where the species can coexist.

A stand becomes susceptible to insect and disease attacks, fires, and windthrows as it develops further, depending on its particular structure. Greater differentiation occurs where trees grow at irregular spacings and ages. Some become dominant and stable while others become suppressed and die. The time when stands become unstable depends on their growth rate and spacing and the characteristics of the particular stands, species, and disturbance agents.

Formerly, conifer stands of Douglas-firs and other species generally grew at wider age ranges than artificial plantations, allowing a few trees to become more dominant before the stands became unstable. Many plantations of uniform spacings and ages may become unstable, relatively stagnant, or susceptible to insects and diseases in the future; more likely, most will experience a growth slowdown compared to more widely spaced stands. Similarly, the uniformly aged, dense stands of Pacific silver firs and hemlocks which began from advance regeneration after clearcutting in the past 3 decades in the upper Cascades of Washington may become unstable, as did the spruce-fir stands in Maine which began in a similar fashion many years earlier (Fig. 8.3B). As a general rule, forest plantations in the northern hemisphere have been planted at quite narrow spacings, and a problem among silviculturists in many countries is how to manipulate these stands of small, closely spaced trees. Mixed-species stands often develop stably where trees in the upper strata are widely spaced.

Minor disturbances allow residual overstory trees to expand into the growing space while new trees begin growing. Only shade-tolerant trees of the new cohort survive if the residual overstory is dense. Partial cuttings and the exclusion of fire (allowing shade-tolerant, fire-susceptible species to invade) have led to a shift of species compositions from pines and Douglas-firs to true firs in many parts of California, Oregon, Washington, and Idaho (Bickford 1982; Cochran, 1982; Hall, 1982; Hopkins, 1982; Laacke, 1982; Rollins, 1982; J. H. G. Smith, 1982b). Similar practices have led to a shift to maples and other shade-tolerant species elsewhere (Lutz and Cline, 1947; Trimble and Hart, 1961; Stubbs, 1964; DeBell et al., 1968b; Oliver and Stephens, 1977).

Future species compositions and structures of stands on a given site are predictable and have been modeled. Most models have been for timber production estimates, and they project future volumes, numbers, and sizes of stems. Some models, however, project future stand structures forward to a more sophisticated degree (e.g., Mitchell, 1975; Shugart et al., 1981; Larsen, 1994).

Disturbances

The slow changes in stands described above are punctuated by disturbances which destroy some or all vegetation (Botkin, 1990). These disturbances can vary from small and frequent to large and infrequent. The types and timing of disturbances and the resulting forests in each region are somewhat anticipatable by studying natural processes and social trends (Cotton, 1942; Veblen et al., 1981; Godron and Forman, 1983). Natural disturbance patterns and responses have been discussed in earlier chapters. The type of human-caused disturbance in an area is the result of both the natural environment and the socioeconomic conditions. The natural environment often dictates the life style of the people—grazing in arid areas, farming in more fertile mesic areas, harvesting timber in heavily wooded areas, recreation in mountainous areas, and others. The socioeconomic condition of the people will determine whether the area is heavily impacted by firewood cutting, overgrazing, frequent underburning (or controlling fires), subsistence farming (or farm abandonment), recreation activities, and timber-harvesting practices. Where mixed-species stands occur, a "high grade" harvesting approach has left relatively unvigorous stands in many diverse regions (Figs. 13.1 and 13.2, Chap. 13).

Disturbances can be so common and subtle that they are not noticed or, if noticed, are perceived as not altering the landscape; but instead being an integral part of it. Such disturbances include grazing, light and frequent fires, and some forest cutting—especially selective cutting, thinning, coppice harvesting, or other harvesting of small trees. The short natural fire cycles in pine forests in the southeastern United States and Rocky Mountains (Table 4.1) resulted in such common surface fires that they were hardly noted as unusual events (Houston, 1973). The dense growth of hardwoods in the southeastern United States and the dense conifer regeneration in the Rockies following fire control indicate the importance of these fires in determining the forest type. Grazing by the high elk population in the Hoh Rain forest of Washington State is so subtle that the openness beneath the large overstory trees appears to be "natural." Grazing by cattle in the Rockies also contributed to the openness of the forests before grazing and fire were controlled (Daniel, 1980). Around the Mediterranean, grazing and browsing by sheep, goats, and cattle has been common for thousands of years, and a plant community of graze-resistant shrubs—maquis—has developed in many places (Shamida, 1978). This vegetation is considered so natural that conservationists are worried that it may disappear if traditional grazing practices are changed (Tomaselli, 1977). Grasses in Mongolia and the North American great plains were well adapted to the regular disturbance of grazing—by cattle, horses, and sheep in Mongolia and by buffalo in the great plains. The grasslands of eastern Washington and Oregon, however, were not naturally subjected to heavy grazing, and the introduction of cattle and sheep reduced the grasses dramatically. Graze-resistant grasses were imported from Mongolia in the late 1940s to withstand this new disturbance.

At the other extreme, the time between disturbances can be long and the effects can be dramatic. Where the disturbance is observed or remembered, it can seem like an isolated, unique, or unnatural event. Before present communication, transportation, and record-keeping abilities, the magnitude of large disturbances was not appreciated, and the disturbances were often forgotten or assumed not to occur. As late as 1938, the hurricane that blew over 243,000 ha (580,000 acres) of forests in New England (Baldwin,

1942; Smith, 1946; Spurr, 1956) was assumed by many residents to be a local storm. Few people were aware that storms of similar magnitude occurred in 1815, 1788, and 1635 (Perley, 1891; Brooks, 1939, 1945; Tannehill, 1944). In Washington state, many people remember the storm of 1962, which blew over 7.4 billion board feet of timber, but are unaware of a similar storm in 1921 which blew over 8 billion board feet (Washington State Library, 1975) or one that occurred on northern Vancouver Island in about 1900. Many majestic, "old growth" stands in the Olympic Peninsula of Washington state began after large fires between 400 and 700 years ago (Agee and Huff, 1987), but these forests are currently assumed to be indestructible by many conservationists. Old settlers in the Rocky Mountains are often not aware that fires originated the age classes of the forests in which they grew up. As more research is done in forests in all parts of the world, the ubiquity and importance of natural disturbances has become more appreciated.

A disturbance—large or small, frequent or infrequent—both alters the existing stand and establishes the pattern for development of the future stand. The disturbance dictates whether a single- or multicohort stand is created, depending on how much of the previous stand is removed (Fig. 6.1).

The time and type of disturbance (Fig. 4.8) determines which regeneration mechanisms are present and are favored by the disturbance. Figure 16.2 shows a major, light disturbance which kills all overstory trees but does not destroy the forest floor (e.g., an avalanche). It favors sprouting and rhizomatous shrub and herb species if it occurs during the stand initiation stage but favors species with wind-blown seeds if it occurs during the stem exclusion stage after the sprouting and herbaceous species disappear. If the same disturbance occurs during the understory reinitiation stage, another group of species—those which arise from advance regeneration—predominate on the site (Oliver et al., 1985).

Precolonial disturbances were probably less "clean" than modern ones; disturbance edges were often irregular; and disturbed areas contained patches where all trees were destroyed and other patches where few were destroyed. Some precolonial areas in the eastern hardwood region (Hough, 1932; Hough and Forbes, 1943; Henry and Swan, 1974), the Rocky Mountains (Lee, 1971; Mitchell and Goudie, 1980; Larson and Oliver, 1982), the Pacific Northwest (Franklin and Waring, 1979; Oliver et al., 1985), and coastal Alaska (Deal, 1987; Deal et al., 1991) commonly contained single-cohort stands which grew after catastrophic fires, windstorms, glacial retreats, and avalanches even before European colonists changed the disturbance patterns. Other parts of the northeastern conifer (Lorimer, 1977b) and eastern hardwood regions, (Hough and Forbes, 1943), the Rocky Mountains (Hanley et al., 1975; White, 1985), the Pacific Northwest (Leopold et al., 1989), and coastal Alaska (Deal, 1987; Deal et al., 1991) contained multicohort stands where natural disturbances were light.

Different species can occupy the same area at different times if the type and timing of the stand initiating disturbances are different. The shift in New England forests from hardwoods before agricultural clearing to pines after agricultural abandonment (Fig. 16.1) was because the light pine seeds have the competitive advantage when invading fields and pastures. These "old field" white pine stands were so prevalent in central New England that the region was described as a white pine forest type by the botanist Charles

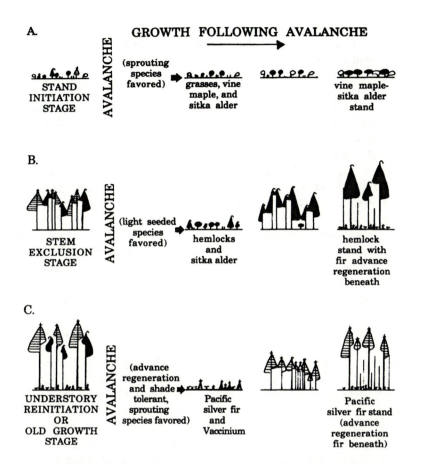

Figure 16.2 Which species dominates a stand depends on time of impact as well as type of disturbance. When the avalanche runout zone impacts Pacific silver fir and western hemlock stands of the Cascade Range of Washington at different development stages (shown down left side), different vegetation dominates. (A) When impact occurs in the stand initiation stage (Fig. 5.2) sprouting maples and sitka alders are favored, creating an alder/maple thicket. (B) When impact occurs in the stem exclusion stage after alders had been overtopped and killed by conifers, light-seeded hemlocks and annual plants have a competitive advantage and form a hemlock stand. (C) When impact occurs in understory reinitiation or old growth stages, Pacific silver firs from advance regeneration have a competitive advantage and dominate the site for subsequent centuries. (*Oliver et al., 1980*). (See "source notes.")

S. Sargent in the census of 1884 (Sargent, 1884). The pine stands shaded the grasses in the stem exclusion stage, and advance regeneration hardwoods developed when the stands reached the understory reinitiation stage. Upon clearcutting of the pines, the hardwoods had the competitive advantage and grew rapidly to form a hardwood stand. This hardwood advance regeneration was so vigorous that attempts to reestablish pines by planting them and cutting the hardwood stems was not successful (Lutz and Cline, 1947). The same shift from conifers to hardwoods with changes in disturbance types occurred naturally as well. A pine-dominated "old growth" forest of central New

England was found to have originated from a very hot fire in about 1665 which had killed any hardwood sprouts and advance regeneration; the 1938 windstorm blew over this pine stand and gave the competitive advantage to hardwoods, which will probably dominate the stand until a future disturbance many decades or centuries from now (Henry and Swan, 1974).

A very similar pattern of disturbances governing species composition is occurring in many parts of the southeastern United States (Orwig and Abrams, 1994). Clearing of the original forest occurred at the same time as in New England, but abandonment of the agricultural land occurred approximately 100 years later—from about 1900 to 1965. Seeding of southern pines into the old fields was supplemented by federal programs promoting the planting of pines. The rapid growth of southern pines in the warm climate provided harvestable timber which was cut rapidly, sustaining the southern pulp, plywood, and lumber industries. As in the white pine stands in New England, the southern pines developed hardwood advance regeneration; these hardwoods grow vigorously after the pines are harvested and form hardwood stands unless vigorous silvicultural efforts give the southern pines a competitive advantage. Small landowners own many of these cutover stands and often are not undertaking such efforts; consequently, the forest composition in many areas is changing to hardwoods. This trend, while predictable, has been dramatic enough by the early 1980s to cause concern about a lack of future pine forests among southern pine industries (Boyce and Knight, 1980; Haynes and Adams, 1983; Brooks, 1985).

Forests of even more diverse geography and species composition can also follow a uniformity of process when responding to the disturbances. The black cherry-dominated forests of the Allegheny Plateau which began in the late 1800s (Marquis, 1975b), the spruce-fir forests of Maine which were extensively harvested (or killed by the spruce budworm) in the late 1970s (Ghent et al., 1957; Ghent, 1958), and the young Pacific silver fir stands of the Washington Cascades which began after harvestings of 1945-1985 (Scott, 1980; Oliver and Kenady, 1982) all developed from similar disturbance patterns. All three forest types had advance regeneration beneath them at the time of overstory removal, and this advance regeneration grew rapidly to dominate the new stand. Partial cuttings in the decades before harvest in spruce-fir forests in Maine and in mixed hardwood forests in the Allegheny Plateau had stimulated advance regeneration which later responded to clearcutting. The Pacific silver fir stands contained advance regeneration because they were in the understory reinitiation stage when harvested. The same species will not necessarily dominate each stand following the next harvest unless the disturbance timings and types are mimicked to give the same species a competitive advantage. In fact, clearcutting of the Allegheny hardwoods without the presence of advance regeneration black cherries has led to difficulties in regenerating any tree species—especially the black cherries (Marquis 1992; Marquis et al., 1984). Clearcutting and insect killing of the spruce-fir forests of Maine before advance regeneration had been established has led to stands dominated by hardwoods and raspberries (Osawa et al., 1986; Seymour, 1992). The Pacific silver fir stands are not yet large enough to be harvested again, but light-seeded brush species will probably dominate the site following harvest unless steps are taken to ensure that advance regeneration is present before harvest (or blowdown) of the dense stands.

Fire control in forests which had frequently burned has led to a shift of tree species. Eastern oak uplands were often maintained by fires, which killed less fire-resistant species; these stands now contain more maples, tulip poplars, and other fire-susceptible species (Abrams and Scott, 1989; Abrams and Downs, 1990; Nowacki et al., 1990; Abrams and Nowacki, 1992; Nowacki and Abrams, 1992; Abrams et al., 1995). Many areas previously dominated by fire-resistant southern pines have become replaced by fire-susceptible hardwoods after fire exclusion. In the Rocky Mountains and Cascade and Sierra Nevada Ranges, previously open forests of resistant pine species and Douglas-firs are now becoming densely stocked with both these same species and fire-susceptible true firs (Day, 1972; Houston, 1973). Fire control is similarly causing a shift from pine- to hardwood-dominated stands in Minnesota (Frissell, 1973). Very large fires would probably have occurred in the southeastern United States following Hurricane Hugo of 1989 (Marsinko et al., 1993), in the northeastern United States following the 1938 hurricane (Henry and Swan, 1974), and in the Pacific Northwestern United States following the 1921 windstorm (Pierce, 1921; The Timberman, 1921; Boyce, 1929) were it not for active human efforts.

Recent human social developments have changed the disturbance patterns dramatically, with concomitant changes in their influences on forests. These anthropogenic changes in disturbance patterns may threaten extinction to some species by exclusion of fires (Walkinshaw, 1983), by timber harvesting (Thompson and Baker, 1971; Gutierrez and Carey, 1985), and by extensive land clearing (Elias, 1977).

A common tendency is to assume forests in North America existed in a stable, equilibrium condition ("steady state") before colonization from Europe—beginning in 1492. Some assumptions of previous stability arise from writings of Rousseau and Emerson and the early works of Thoreau in the eighteenth and nineteenth centuries (Dickerson, 1995). These philosophers attached a romantic sentimentality to an idyllic, past, primitive life; and they have instilled the erroneous idea that a stable forest community would exist if it were not for the activities of industrialized societies. This early assumption of a stable equilibrium is being increasingly discounted by scientific evidence (Raup, 1964; Botkin and Sobel, 1975; White, 1979; Oliver, 1981; West et al., 1981a; Pickett and White, 1985; Botkin, 1990, 1994, 1995; Stevens, 1990; Pielou, 1991; Sprugel, 1991; MacCleery, 1992; Sampson and Adams, 1994; Zybach, 1994).

Forests of the eastern and western United States, Canada, and elsewhere were commonly disturbed by Indian fires (Bromley, 1935; Day, 1953; Thompson and Smith, 1970; Frissell, 1973); by grazing of woodland bison and buffaloes, elks, and deer; by roosting of passenger pigeons; by floods; and by Indian agriculture (Lawson, 1709; Savage, 1970). In 1709, a naturalist traveling through South and North Carolina reported numerous agricultural fields recently abandoned by the Indians after European diseases had dramatically reduced their population (Lawson, 1709; Moodie and Kaye, 1969; Savage, 1970; McAndrews, 1976). Also, fire was used by Indians as a tool to drive game in eastern forests (Lawson, 1709; Savage, 1970), as it was in western forests and rangelands and among nomadic tribes in Asia (Pyne, 1982). Until only 14,000 years ago, glaciers covered most of Canada, the Great Lakes region, northern Washington, Idaho, and Montana, New York, New England, and parts of California (Fig. 4.7). Many species migrated both within and south of the glaciated regions. The sea level was lower during the glacial maximum because water was trapped in the glacial ice (Shepard, 1963; Emery, 1967); forests inhabited areas presently beneath the oceans. Dramatic changes—flooding, exposing of glaciated areas, frost heaving, and migrations of species—occurred while the glacier was retreating (Goodlett, 1954; Wright, 1968; Whitehead, 1972). In fact, central Canada is still undergoing an uplift in elevation in response to removal of the ice's weight. Similar glacial events occurred elsewhere in the world, with similar or even more dramatic results (Chap. 4). Soon after the glacier's retreat, many large animal species—the mammoth, the giant ground sloth, the native horse, the camel (Janzen, 1982)—became extinct quite rapidly. Both the cause of this extinction—sometimes attributed to overkill by humans (Martin, 1967; Haynes, 1976)—and the results of the changed animal populations dramatically shifted the disturbance patterns in North America.

Similar shifts in disturbance patterns have been occurring for thousands of years elsewhere in the world. Many forests of central Europe were maintained in a sprout hardwood (coppice) condition by the continual cutting for fuelwood. After the shift to alternative fuel sources, these areas grew to forests of large trees. Wars, plagues, and famines have changed populations dramatically, causing shifting uses and abandonment of lands. Pollen records in Asia Minor show a shift from trees toward shrubby and herbaceous vegetation with the increase in population and land use of the past 500 years (Aytug et al., 1975). The Spessart Oak Forest of central Germany is renowned for growing oaks on a 350 year rotation (Fig. 16.3). In fact, the forest first became established

Figure 16.3 This 350-year-old stand is part of Germany's famous Spessart Oak Forest, which has been managed for many decades for high-quality oak logs. The stands origin, however, was a sociological event—the changing land use at the end of the Thirty Years War, 1618 to 1648.

when the land was abandoned during population movements at the end of the Thirty Years War (1618-1648). During the Crusades, Europeans burned the forests along roads in Asia Minor to eliminate hiding cover from which enemies could ambush travelers; a similar use was made of chemical defoliants in Vietnam in the 1960s and 1970s.

Disturbance patterns and forest structures and compositions have not been stable since 1492 (Smith, 1976; Nowacki et al., 1990; Abrams and Nowacki, 1992; Orwig and Abrams, 1994; Abrams et al., 1995), and there is no reason to believe they will be stable in the future. Expansion of human influences for the past few centuries have changed the type, frequency, and extent of disturbances—leading to more changes in forest structures and species compositions (Goodlett, 1954). Population shifts, agricultural practices, and coppice cuttings have caused lands to be farmed, used for fuelwood lots, abandoned, selectively harvested for timber, grazed, and clearcut-harvested at various times in the past. Rapid changes in silvicultural practices and policies have left forests with a wide variety of species compositions, age distributions, stand structures, and stand sizes. Often, government policies cause disturbance patterns which are reflected in stands for many years or centuries afterward. Just as the Spessart Oak Forests are a reflection of the end of the Thirty Years War, many pine stands of the southern United States can be dated from the government subsidy programs which peaked in the mid-1950s (Fig. 16.4; USDA Forest Service, 1988).

Attempts to control natural disturbances cause shifts in disturbances. Attempts to maintain the Mississippi River flowing past New Orleans (Chap. 4) are keeping

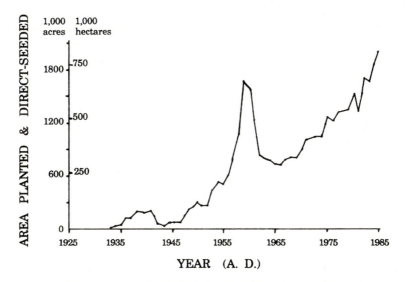

Figure 16.4 Area planted and direct-seeded in the southeastern United States to pines (USDA Forest Service, 1988). When farmland was abandoned in the southeastern United States (similar to Fig 16.1B) during the mid-twentieth century, pines were actively planted. Greatest planting occurred when government programs offered greatest incentives—during the Soil Bank Program of the late 1950s and early 1960s. The southern pine industry is currently harvesting these stands and planting more pines.

sediment from being deposited elsewhere in Louisiana (McPhee, 1990). As the previous sediment settles and new sediment is not added, saltwater is encroaching upon freshwater areas at a rapid rate (Baumann and Adams, 1981; Salinas et al., 1986; DeLaune et al., 1987). Changes in disturbance patterns—and forest development patterns—will occur in the future, although the directions of these trends is not certain. If the world population continues to shift from rural to urban living (Wallenstein, 1983), rural land will be usedmore efficiently for mass production of agricultural and forest products. The shift will allow food to be produced on less land area, and more land will revert to forests. If the cost of alternative energy sources stays low, the forests will be used less for fuelwood—again allowing more forests to grow. Finding uses for the wood to help pay for their manipulations will become a problem. Silvicultural manipulations will be wanted to help reduce the susceptibility of forests to fires, insects, and other disasters.

Other, unanticipated changes, such as problems with air pollution, will certainly arise as well. Changes in the earth's atmosphere alter the competitive status of species. Natural disturbances or human activities (e.g., harvesting) which destroy old trees poorly adapted to the changed environment accelerate the transition from maladapted individuals to those better adapted to new environmental conditions (Binkley and Larson, 1987).

Forest Patterns over Large Areas

Forests can have uniform structures or uniform patterns of stands of several different structures across a landscape, within a region, or even in different parts of the world. On a gross scale, the similarities of most forests is because trees generally have similar physiological processes and appearances—trunks and green foliage, and they interact to give a limited number of basic appearances (Oliver, 1992d). On a finer scale, related species are often found in different parts of the world; these related species behave similarly and create forests with structures which appear to be similar. Within a region, forests appear to be uniform where the region contains the same species, similar soils or soil variations, a similar climate, and similar disturbance patterns and histories.

Only a few basic structures appear in forests throughout the world on a very coarse scale. Trees generally grow into closed forests where moisture is not very limiting. In areas where moisture is very limiting, tree crowns may not be touching when the growing space is completely occupied, and stands remain indefinitely as open savannas. Savanna-like conditions can also appear during the stand initiation stage, after partial disturbances in stands on more mesic sites, and in bogs where nutrients are limiting (Fig. 11.4). Most mesic stands develop as closed forests after the first few years or decades and spend the majority of their existence in a closed condition. On a very coarse scale, therefore, the pattern of closed or open forests appears throughout the world (Riley and Young, 1966). A limited number of slightly more subtle refinements appear within the closed forest structures. Forests can develop with one or many species, single- or multiple-canopy strata, and single or multiple cohorts.

Within the northern hemisphere, forests on different continents often appear to be similar because the same genera—of the Holarctic group (Walter, 1973a)—are generally found in all places. Different species of the same genus often behave similarly.

Consequently, isolated forests of completely different species can have very similar appearances. A stratified, single-cohort oak-beech stand in Turkey appears to be very similar to single-cohort oak-beech stands in the eastern United States, with the tolerant beech able to live under shaded conditions. Pines throughout the hemisphere are relatively shade-intolerant and are generally found on droughty sites—often in single-cohort stands following agricultural or grazing abandonment or in multicohort stands following ground fires. *Populus* species are found along rivers in arid areas of western North America and Mongolia. Mixed birch-fir stands can be found in northern North America, in northern Scandinavian countries, in Russia, and in northern and mountainous parts of Asia. Subtle variations also exist between species within a genus which allow some genera to survive in different niches, giving the forests different structures. Firs vary from very shade-tolerant—such as grand fir, European and Pacific silver fir, and balsam fir— to relatively shade-intolerant—such as Noble fir. Japanese chestnut and Chinese chestnut trees are resistant to the chestnut blight (*Endothia parasitica*), which essentially eliminated the American chestnut (Jaynes et al., 1976). Some species of oaks and firs can exist as advance regeneration; other species cannot. Oriental and American beeches regenerate vigorously from root sprouts; European beech does not.

Similar forest patterns are often found within a single region. Here the same species are found; and soil, climate, and disturbance patterns can be similar. Some species have large ranges which span several regions. At these times the species may behave differently enough between regions to create different forest patterns. Douglas-firs grow as dominants on mesic sites in the rainy Pacific coast region and often slightly subordinate to pines or larches on much drier sites in the Rocky Mountains. Western hemlocks are found as advance regeneration at low elevations but not at upper elevations, where lichens and other debris embedded in the snow press it to the forest floor (Kotar, 1972; Scott et al., 1976; Thornburgh, 1969).

Subtle differences in behavior of a species also occur within a region. In arid regions, many species grow on cool aspects but not on hot ones, creating patterns of different species on different sides of the same hill. The ability of species to grow after release also can differ between sites within a region—with growth being more probable on cooler, moister sites.

Even with these variations, a finite number of patterns occur in a forest. The number of species which invaded in the stand initiation stage becomes reduced during later competition, leaving a limited number of species to dominate. Which species dominates depends on disturbance patterns, soil conditions, and climate conditions—all of which are not random within a region. Where more species occur, subtle differences in climates, soils, and disturbances can cause different species to dominate. Consequently, more variation in vegetation patterns is found in mesic and more tropical regions where more species occur. At high latitudes, very few species are found to dominate large areas of all soil conditions simply because they are the only species found within the area. On the other extreme, so many species exist in river bottom forests of the southeastern United States that subtle differences in soils, disturbance types, and spatial patterns can cause different species to dominate in mosaics of small stands (Putnam, 1951). Tropical forests can contain even more species, and the result is even more intricate patterns, although the processes of stand development appear to be the same.

A group of forest stands can form a pattern across a landscape. Each stand represents a contiguous area of uniform disturbance history, climate and soil growth potential, species composition, and stand structure. Stands of similar structures, species, growth rates, and disturbance histories are often found repeatedly throughout the forest, interspersed by other stands in different conditions. Similar stands are not always randomly distributed but can be found in various arrangements. These arrangements can be described by sizes, shapes, and spatial distributions.

Repeatable patterns of stands across a landscape can appear uniform or vary in a fine scale or on a coarse scale. The sizes of repeatable, uniform areas are sometimes referred to as *grain* (Forman and Godron, 1986). Mosaics of small stands with similar patterns interspersed with other, repeating stand patterns over a small area have much variation across the fine scale. If the same mosaics of small stands repeated themselves over a large area, the variation over the coarser scale would be relatively small. Alluvial floodplain forests have large variations in the small scale, since soil variations create many small stands; but the large-scale variation is small because similar stands appear repeatedly over large areas. Alternatively, uniform, narrow aspen stands along a series of streams dissecting uniform ponderosa pine stands have little variation within each stand on the small scale but large variation on the coarse scale between pine and aspen stands.

Shapes of the repeatable patterns are generally most important where they form, or do not form, connections between similar stands. These connections—*corridors*—are important for maintaining migration routes for animals and plants. Even large forested areas can behave as isolated "islands" devoid of certain animals if they are not large enough to support the complete range of these animals and the surrounding area is hostile to migration (Harris, 1984).

Stands of similar appearances can be spatially distributed in clusters where similar stands are found in groups separated from other groups of similar stands. The distribution can be random where similar stands are randomly interspersed with other stands. The distributions can be uniform where similar stands are regular distributed. These patterns are conceptually similar to tree spatial patterns (Chap. 6). One group of similar stands can be regularly distributed across a landscape; another group in the same landscape can be clumped; and other groups can have other distributions.

Sizes, shapes, and spatial distributions of similar stands can often be predicted, since they are based on the spatial variation of factors influencing stand growth. The spatial distributions of stands are based on the region's geomorphology, climate, disturbance patterns, time since various disturbances, and species living in an area. In fact, the sizes, shapes, and spatial distributions often can be used to understand the nature of these influences on an area

Geomorphology and stand spatial patterns

A region's geomorphology largely determines the soil, slope, aspect, and elevation patterns. These in turn influence which species have a growth advantage and so dominate each site. On the small scale, microsite variations can be determined by soil drainage and deposition of alluvial sediments and sands (Fig. 6.4A, B. and C; Lyford, 1974; Oliver, 1978b). Stands along streams often have corridor shapes, since the soils and drainage patterns develop parallel to the stream. Glacial and wind-blown depositions can cause stand patterns to be clumped, random, or regular. Topographic or soil features often cause disturbances to impact different slopes, aspects, soil conditions, or elevations differently. Consequently, the resulting stand development patterns can be quite characteristic of the integration of disturbances and the topography (Camp, 1995).

From an intermediate-scale perspective, areas of about five to several hundred hectares (or acres) often have uniform stand conditions based on more gross geomorphologic features. In glaciated areas, the vegetation pattern closely reflects the depositional pattern following the glacier (Stout, 1952; Goodlett, 1954; Lyford et al., 1963). Soils—and plants having competitive advantages—on nonglaciated areas also reflect geomorphologic origins such as differences in chemical composition of the parent material and differences in depth and texture of alluvial, aeolian, fluvial, and colluvial deposits.

Geomorphology also influences streamflow patterns and valley shapes (Fig. 16.B). Natural weathering, soil conditions, and soil movement create a gradient in productivity from low to high between the ridge and valley bottom in nonglaciated areas (Hack and Goodlett, 1960; Yamamoto, 1992). Different species dominate at different positions along this gradient. Consequently, stands can have shapes, sizes, and patterns characteristic of the drainage system (Fig. 16.6). Stands often have a corridor shape at this grain size as well, since both valleys and ridges are interconnected for long distances (Fig. 16.6). Vegetation patterns follow drainages in glaciated areas as well, except that more productive areas are often on ridgetops where till was deposited, whereas side slopes of coarse outwash terraces and valley bottoms of outwash gravels and sands can be more arid and nutrient-poor.

On a regional perspective, vegetation types follow geomorphologic formations. Here too, features such as drainage, glaciers, elevation, slope, bedrock, and sediment patterns form distinct soil conditions (Boyd and Penland, 1988) and resultant vegetation patterns. Similar vegetation often appears clumped on the similar geomorphologic formations. On a continental scale, vegetation can be partly explained by geomorphologic provinces. Similar soil and climate conditions are often confined to a single geomorphologic province or to the mountains or valleys at the junction of provinces (Fig. 16.7). Corridors of similar vegetation can also follow both mountain and river systems. Since geologic provinces are unique and support characteristic vegetation, areas of similar vegetation are generally clumped at this level of resolution.

Climate

Climate and geomorphology are, of course, interrelated and influence each other. Within a small area, microclimates influence vegetation patterns in frost pockets and extremely hot, dry, or water-saturated areas. Each condition gives a different species combination the competitive advantage, creating vegetation patterns which may be isolated or in corridors and clumped, random, or regular.

Similar stands of five to several hundred hectares (or acres) can be distributed according to climatic influences—especially in arid areas or areas of extreme or greatly varying elevations or temperatures. A given forest type is often found at lower elevations on the cooler slopes near the warmer, drier end of its range (Daubenmire, 1968; Bozkus, 1988; Leopold et al., 1989).

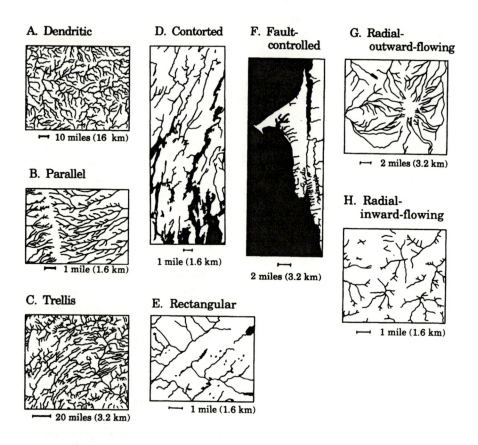

Figure 16.5 Vegetation patterns often are characteristic of bedrock geology patterns, since vegetation patterns follow soil productivity patterns which reflect stream, river, and valley patterns. Characteristic river and valley patterns and their geomorphologic causes are (A) Dendritic pattern: characteristic of plains and plateaus where horizontal rock layers influence stream locations. (B) Parallel pattern: characteristic of coastal plains, uplifted and drained lake floors, and tilted mesas where the land slope causes major streams to be roughly parallel. (Individual stream systems may appear to be dendritic.) (C) Trellis pattern: characteristic of folded rocks. Valleys follow weak rock formations. (D) Contorted pattern: characteristic of metamorphic bedrock where alternating weak and resistant bands create valleys and ridges. (E) Rectangular pattern: characteristic of areas where a series of faults or a well-developed joint system determines pattern of valley cutting. (F) Fault-controlled pattern: characteristic of area where fault forms line of weakness thereby creating a valley. (G) Radial pattern, outward flowing: characteristic of volcanic cones or other areas where streams radiate outward from a single point. (H) Radial pattern, inward flowing: characteristic of resistant surface formation overtopping porous lower layer, where streams converge into sinkholes. (*Shimer, 1972.*) (See "source notes.")

Within a region, climatic variations further influence vegetation types. From the continental perspective, broad vegetation types often follow general temperature and precipitation patterns (Fig. 16.7C) and fluctuations of patterns (Lamb, 1988; Brubaker, 1991;

Gates, 1993). On the global scale, broad vegetation patterns can be understood from basic weather patterns (MacArthur, 1972).

Disturbance patterns

Disturbances affecting broad areas have occurred in many tropical (Uhl et al., 1990; Oliver, 1992; Nelson et al., 1994), temperate (Pickett and White, 1985; Lorimer, 1977; White, 1979; Oliver, 1981; Osawa, 1992; Marsinko et al., 1993; Oliver et al., 1995), and boreal forests (Lutz, 1956; Heinselman, 1981; Osawa, 1986, 1989, 1994; Osawa et al., 1994). Disturbance patterns can be superimposed on geomorphologic and climatic influences in further explaining the arrangements of stands and forests. Disturbances, also, are not completely independent of the other two. Within a small area, minor disturbances create patterns within stands (Chaps. 6 and 12). The small-scale pattern of similar vegetation following these disturbances can be random, clumped, or regular and either isolated or in corridors, depending on specific characteristics of the disturbance. From a larger-scale perspective, each stand's vegetation is strongly determined by the initiating disturbance—and which species has the competitive advantage. The stand structure is dramatically altered whether it is subjected to a minor or a major (stand-replacing) disturbance (Henry and Swan, 1974; Leemans, 1992; Yamamoto, 1993). The landscape following a disturbance, therefore, is often a mosaic of forest conditions.

Where a disturbance affects a large proportion of similar forests in an area, it leaves a large part of the area in single cohort stands of the same age or a cohort of similar ages and species composition within many multicohort stands (Fig 16.8). Since stands change in susceptibility to disturbances as they develop (Fig. 4.6), the similar stands or cohorts are again destroyed by a single disturbance or a similar series of disturbances, once again creating many stands or cohorts within stands of a single age across a region.

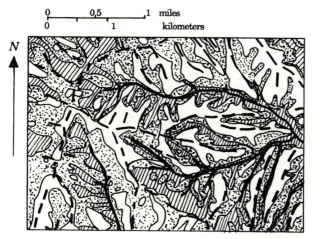

Figure 16.6 Stands of similar vegetation are often found on similar soil conditions and follow drainage patterns. Vegetation map of Little River, Virginia, area. Yellow pine forest type (blank areas) is found on ridges and south slopes, oak forest type (dotted), on midslopes, and northern hardwood type (striped), in valley bottoms, flood plains and north slopes. (Dashed lines = ridge crests; heavy solid line = streams and rivers; *Hack and Goodlett, 1960*.) (See "source notes.")

Disturbances such as fires and destructive windstorms occur at periodic intervals rather than being randomly distributed in time (Haines and Saudo, 1969; Johnson and Wowchuck, 1993), leaving clumped patterns of similar stands on an intermediate scale. Forest harvesting originally left a clumped pattern of stands, since most accessible, nearby stands were harvested. Harvesting often followed topographic features such as accessible slopes, low elevations, and well-drained areas and so was influenced by an area's geomorphology. Harvesting patterns have also been restricted by ownership patterns, which vary from large areas of single ownership in the western United States to small areas of diverse ownership in the East. The United States Forest Service had policies regarding clearcut sizes on National Forests for many years (depending on the region) but has limited the sizes and distributions of clearcuts since the National Forest Management Act of 1976. These restrictions are giving a random or regular distribution of stands of similar sizes and ages. The resulting, regular distribution can lead to fewer corridors and undisturbed habitats for animals in the landscape (Franklin and Forman,

A. FOREST TYPES

WESTERN FORESTS

⬤ Spruce - fir
◉ Larch - pine
▨ Coastal Douglas-fir
▨ Pinyon-juniper
▨ Broadleaf woodland
▤ Ponderosa pine-Douglas-fir
▦ Lodgepole pine

EASTERN FORESTS

Spruce - fir ▨
Jack-red-white pines ▨
Beech-birch-maple ▨
Oak-hickory ▨
Mixed hardwoods ▦
Oak-pine ▨
Southern pines ▥
Bottomland hardwood-cypress ⬤

Figure 16.7 On a continental scale, vegetation patterns (A) often reflect geologic provinces (B) and climatic variations (C). Vegetation patterns can be found within specific geologic provinces or can be in mountains or valleys bordering geologic provinces. (A) Forest types of the contiguous United States as they appeared before European settlement. (*Shantz and Zon, 1924; Smith, 1980.*) (See "source notes.")

B. GEOLOGIC PROVINCES

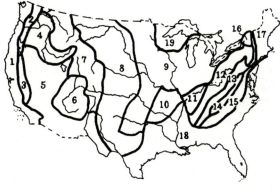

1. Pacific border	7. Rocky Mountains	13. Ridge and Valley
2. Cascades	8. Great Plains	14. Blue Ridge
3. Sierra Nevada	9. Central Lowlands	15. Piedmont
4. Columbia Plateau	10. Interior Highlands	16. Adirondacks
5. Basin and Range	11. Interior Low Plateau	17. New England
6. Colorado Plateau	12. Appalachian Plateau	18. Coastal Plain
		19. Superior Upland

C. DISTRIBUTION OF PRECIPITATION

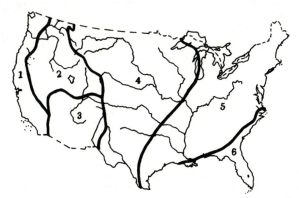

1. Pacific Type: winter maximum, summer minimum.

2. Sub-Pacific Type: wet winter moist spring, dry summer and fall.

3. Arizona Type: wet summer and winter, moist fall, dry spring.

4. Plains Type: summer maximum, winter minimum.

5. Eastern Type: evenly distributed all seasons.

6. Florida Type: rain in all seasons, heaviest in late summer and early fall.

Figure 16.7 (*Continued*) (B) Geologic provinces of the contiguous United States. (*Shimer, 1972.*) (C) Seasonal distribution of precipitation in the contiguous United States. (*Smith, 1980.*) (See "source notes.")

A. DISTURBANCE PATTERNS

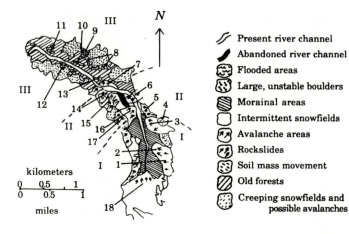

∥	Present river channel
∥	Abandoned river channel
▦	Flooded areas
▦	Large, unstable boulders
▨	Morainal areas
▢	Intermittent snowfields
▨	Avalanche areas
▨	Rockslides
▨	Soil mass movement
▨	Old forests
▨	Creeping snowfields and possible avalanches

B. STAND DEVELOPMENT STAGES   C. SPECIES DISTRIBUTIONS

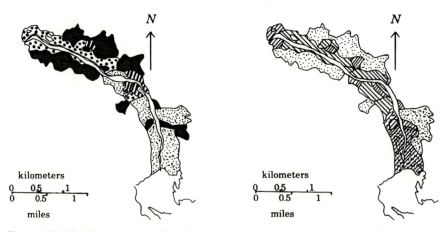

Figure 16.8 Disturbance patterns often result in patterns of similar stands by determining ages and species compositions as shown in midelevation valley (Nooksack Cirque area) in the north Cascades of Washington. (*Oliver et al., 1980; 1985.*) (A) Disturbance patterns. Primary disturbance areas: I = glacial activity within last 200 years; II = primary disturbance (possibly glacier) occurred about 600 years B.P.; III = no obvious evidence of major disturbance. Secondary disturbances: 1, 4, 16, and 17, chronic rockslides; 2, 3, 5, 6, 7, and 15, chronic snow avalanches; unstable rocks; 9, soil mass movement from ca. 1830-1880; 10, chronic soil mass movement; 11, chronic flooded area; 12, chronic beaver pond flooding; 13 and 14, avalanches from ca. 1875-1885; 18, ice calving from glacier (chronic). (B) Patterns of stands by development stage. Small dots = stand initiation stage; dark areas = stem exclusion stage; stripes = understory reinitiation; large dots = old growth. (C) Patterns of stands based on predominant species type. Small dots = herbaceous and woody angiosperms; lines sloped down to right = Pacific silver fir; lines sloped up to right = western and/or mountain hemlock; mixed dots and lines = mixed herbs, woody angiosperms, and hemlocks. (See "source notes.")

1987). On the other hand, regrowth of forests can also isolate openings and lead to fewer habitats and open corridors for migration of species requiring openings (Young, 1992).

Tornadoes and landslides generally occur over small areas and are relatively regularly distributed in time. Consequently, they do not leave large areas containing stands of primarily a single cohort and stand composition. Tornadoes (and sometimes landslides) can leave a corridor of similar vegetation in the landscape. Few disturbances in nature are regularly distributed across a landscape. Waves of destruction and regeneration caused by winds in fir stands at high elevations in the northeastern United States (Sprugel, 1976) and in Japan (Iwaki and Totsuka, 1959; Oshima et al., 1958) are a relatively unusual example of this regular distribution. Small "patchwork" harvesting by the Forest Service is another example of regular distributions of stands, and isolated woodlots in lands cleared for agriculture also have few forested corridors (Hill, 1985).

Within regions, similar stands may be clumpily distributed in time and space as well, where large disturbances occurred irregularly in time. When the 1938 hurricane in the northeastern United States blew over 243,000 ha (580,000 acres) of forests, it created a large age class which originated from that time. Large areas of northern Vancouver Island are in hemlock-spruce stands which began after a windstorm of about 1900. Fire control and grazing created many stands in the Rocky Mountains which now comprise similarly dense, single- cohort stands—often of ponderosa or lodgepole pines and other conifers (Daniel, 1980). A large cohort of ponderosa pines cover much of the Coconino Plateau area—resulting from an unusual combination of weather, available seeds, and overgrazing in 1919 (Pearson, 1950). Extensive fires in Wisconsin and Michigan in 1870 created large areas with stands of similar age, structure, and species composition (Hansen, 1980). Farm abandonment and subsidies of pine planting in the southeastern United States led to many areas of southern pine stands of similar ages. Insect outbreaks such as the spruce budworm in New England have occurred to a large enough area that many stands over the landscape are developing after them, with characteristic raspberry and hardwood species quite different from most older forests (Osawa et al., 1986).

Forest harvests have historically been in clumps, as the migratory timber industry rapidly harvested the stands in one area and moved to another. This pattern is still occurring. Corridors are often found naturally across a landscape and are related to geomorphologic features. Roads and power lines provide further corridors for some species but barriers to migration for others (Young, 1992; Fry and Money, 1994).

It is probable that each region contains and has always contained one or several major cohorts in single- and multicohort stands. During the 1600s, Indian populations in southeastern North America were dramatically reduced by diseases introduced from Europe (Lawson, 1709; Savage, 1970). Abandonment of their agricultural lands as well as reduction in areas burned for hunting probably led to a significant wave of stands of similar ages, structures, and species compositions. Agricultural clearing and abandonment in the southeastern and northeastern United States led to later waves of stands of similar ages. The shift from firewood to alternative fuels in the early twentieth century allowed a cohort of that age to grow up in single- and multicohort stands in much of the world.

Population increases have not reduced the forest land in the past few decades. During the 30 years between 1950 and 1980, the change in forest land in each region of the United States was negligible (USDA Forest Service, 1982) despite an increase in populations in each region. Much of the population increase went to urban areas (USDC Bureau of the Census, 1950-1980).

Development following disturbances

Development patterns following disturbances further affect the arrangement of stands of similar structures across a landscape. On a small scale, numbers of individuals and species decline and tree crowns spread as a stand develops. Consequently, the variation in forest structures decreases. In areas where one or a few species eventually dominate the B-stratum, the forest becomes more similar as these few species expand. Small variations in microsites often become masked by this expansion. On a larger scale, stands and regions also become more homogeneous as the many possible dominant species during stand initiation become reduced during stem exclusion. Small corridors, such as those along streams, roads, and trails, often become reduced by growing of tree branches into the openings.

Vegetation of the area

The fine-, intermediate-, and coarse-scale forest patterns are also influenced by numbers of species in the area. Where more species exist, the pattern will generally be less uniform. Each species will dominate under different conditions and accentuate any variations in soils, climates, disturbances, and stand development patterns.

Consequences of Changing Vegetation Patterns
Impacts on future stands

Changing vegetation patterns—the result of changing soils, climates, disturbance patterns, and species migrations—can cause further landscape changes. Vegetation changes can cause previously isolated species to come into contact. These contacts cause new vegetation patterns, elimination of some species through competition or—in the case of microorganisms or animals—predation, and hybridization where related species meet. Several such meetings of previously isolated populations may be present in North America (Remington, 1968, 1985).

Changing vegetation patterns can cause stands to become predisposed to disturbances—wind, fire, insects, or others—sooner or later in their development cycle. When disturbance cycles and species compositions change, development stages and stand structures (Fig. 5.6) may be extended or curtailed. Consequently, the area in each stage and structure may change over a landscape, and species favored or disfavored by each stage and structure would shift accordingly. Where many stands are clearcut while young, old growth structures are becoming less common in the landscape, and species which live primarily in old-growth-stand structures are being threatened with extinction (Gutierrez and Carey, 1985; Thompson and Baker, 1971). Changes in shapes of stands so that corridors are eliminated also threaten species (Hill, 1985). Where contact between populations is eliminated because corridors are closed, the gene pool of each isolated population may become less viable and less able to survive (Harris, 1984). All of the above changes can create conditions in which certain plant and animal species no longer can survive and so become extinct, and their protection from catastrophic fires (Sampson and Adams, 1994; Oliver and Lippke, 1995).

The changing species and development stages also dictate various socioeconomic activities, which further alter the disturbance patterns and consequently the species and development stages. Prevention of fire, overgrazing, and erosion reclamation; of degraded farmland; and other measures to ensure a future growth of forests have led to a large amount of timber volume in every region, although much of this volume is in trees of small sizes and questionable quality (USDA Forest Service, 1982). The abundance of forests has reduced the wood scarcity, and more concern has developed over maintenance of forests for other uses and protection from catastrophic fires (Sampson and Adams, 1994; Oliver and Lippke, 1995).

Impacts on species

Some plant and animal species live primarily in certain stand structures, others live in several of them, and still others live on the "edge" between structures (Budowski, 1965; Thomas, 1979; Brown, 1985; Hunter, 1990; Oliver et al., 1995; Pearson, 1993). A certain amount of each structure is needed to maintain species dependent on it. Spotted owls (*Strix occidentalis caurina*; Pacific northwestern United States; Gutierrez and Carey, 1985) and red cockaded woodpeckers (*Dendrocopos borealis*; southeastern United States; U.S.Fish and Wildlife Service, 1985) require features found in the "old growth" structures. Many butterflies (Young 1992, Fry and Money 1994) and beetles (Lenski, 1982; Bruce and Boyce, 1984), the now extinct heath hen (*Tympanuchus cupido cupido*; Robinson and Bolen, 1984), the bighorn sheep (*Ovis canadensis*; Geist, 1971), the Canadian lynx (*Felix lynx)*, and many birds (Mehlhop and Lynch, 1986; Titterington et al., 1979; Hansen and Urban, 1992) need openings to survive. Even the stem exclusion structure, which generally contains the fewest dependent species, is needed by the Kirtland's warbler (*Dendroica virens*) in jack pine stands (Walkinshaw, 1983). Some species utilize different structures in different parts of their ranges. For example, black-tail deer (*Odocoileus hemionus*) commonly browse in openings in most of their range, but often browse in old growth structures in Alaska and British Columbia (Wallmo and Schoen, 1980; Bunnell and Jones, 1984).

Many animal live on the "edge" between structures—usually between an opening and a closed stand (stem exclusion, understory reinitiation, or old growth structure; Leopold, 1933; Hunter, 1990; Robinson and Bolen, 1984). Deer and other species utilize several structures—openings for browsing, stem exclusion structures for hiding, and understory reinitiation and old growth structures for escaping winter cold and summer heat.

Other species—referred to as "interior" species—utilize parts of stands away from edges (Noss, 1983). Moose, for example, utilize the "interior" of large openings in Maine because of an incompatibility with deer, which use the edges (Hunter, 1990). Some bird species also avoid edges (Hansen and Urban, 1992).

Many species depend on specific structural features which may or may not be present in a particular stand. Red cockaded woodpeckers in the southeastern United States depend on old longleaf and other pine trees for cavity nests (Conner and O'Halloran, 1987); an old growth stand without these features would be less useful to them. Spotted owls in the Pacific Northwestern United States utilize platforms of foliage caused by the dwarf mistletoe disease, hollow trees, or large limbs for nesting and roosting (Gutierrez and Carey, 1985; Forsman et al., 1984; Barrows, 1980). Old growth stands without these features are not as favorable for their survival.

Corridors of favorable stand structures may be needed to allow migration of certain species dependent on old growth structures (Franklin and Forman, 1987) and on openings (Young, 1992). Where corridors of appropriate structures do not exist, local, isolated populations of some species may become inbred and loose viability (Hunter, 1990; Morrison et al., 1978; Young, 1992).

Animal and plant populations change as stand structures change with disturbances and regrowth. Often the changing structures cause dramatic declines of some species *("temporal bottlenecks";* Seagle et al. 1987) and dramatic increases of others. The transition from openings (stand initiation structure) to the stem exclusion structure and the transition from a closed forest (stem exclusion, understory reinitiation, or old growth structure) back to an opening are times of very distinct changes in animal and plant populations.

Regrowth of openings to stem exclusion structures can cause dramatic declines in plant and animal populations. Deer populations declined dramatically when forests regrew in the northeastern United States (Allegheny Plateau; Marquis, 1975b) and in Hokkaido, Japan (Kaji et al., 1988). Moose populations also declined as forests regrew following timber harvest and fire control in northern North America (Brossard et al., 1974; Krefting, 1974). Many songbird populations declined while others increased when stem exclusion structures replaced openings in the British Isles (Moss et al., 1979). Bighorn sheep (Geist, 1971), Canadian lynx, and several plant species (USDA Forest Service, 1990) have similarly declined with forest regrowth following catastrophic fires in parts of the Rocky Mountains. Butterfly species have declined or become extinct as open areas regrew to stem exclusion structures in the British Isles and the United States (Young, 1992; Fry and Money, 1994). Overuse and elimination of food, elimination of predators, hunting, and other factors can lead to large expansions or declines in populations (Caughley, 1970; Robinson et al., 1980; Robinson and Bolen, 1984)—often in conjunction with changing forest structures.

Many plant species survive as dormant seeds (Marks, 1974; Cook, 1978; Osawa, 1986) when stands regrow from the stand initiation to the stem exclusion structure, while seeds of others move to new openings by wind or animal vectors. Some animal species may migrate to other openings—if they are available. Other animals become concentrated ("packed") into the smaller, remaining open areas (Burel, 1992) and eventually reduce their numbers or become extirpated through starvation, disease, inbreeding, or predation. Some animals require corridors of open areas for their migrations and others do not (Young, 1992).

Browsing species can alter the habitat as they become concentrated (Leopold, 1933; Stoeckler et al., 1957; Buechner and Dawkins, 1961; Caughley, 1970; Marquis, 1975; Amaral, 1978; Hanley and Taber, 1980; Kaji and Yajima, 1987; Wolfe and Berg, 1988; Kaji et al., 1990), sometimes browsing regenerating trees in the remaining openings so severely that they keep the areas open. The can also browse plants on the forest floor in surrounding stands, creating forests with little forest floor vegetation or with only unpalatable species there. When the population of browsing species becomes high, small disturbances which create more openings allow the browsing population to expand and prevent forest regrowth in these areas, also. Predation, diseases, or starvation will reduce the population and allow open areas to regrow (Caughley, 1970; Robinson et al., 1980). A large disturbance which creates large amounts of the stand initiation stage will also allow trees to grow to the stem exclusion structure before the browsing population can browse all trees and so prevent regrowth.

A distinct transition in animal and plant populations also occurs when a closed stand changes to an opening. This transition can be quite sudden if caused by a fire, wind, or similar disturbances or can be quite slow if the trees slowly die as bogs develop in boreal conditions (Zach, 1950; Strang, 1973) or if insects and diseases slowly kill the trees.

Even though the disturbance can destroy individual animals and plants (Bendell, 1974), the greatest impact is generally to the habitat. Fires in the northwestern United States (Tillamook burn of 1933; Foothills fire of 1992; Wenatchee fires of 1994) killed deer, elk (*Cervus elaphus*), coyotes (*Canis latrans*), and young spotted owls (Isaac and Meagher, 1938; Kemp, 1967; Lucia, 1983; Mealey, 1994) and destroyed their habitats. Hurricane Hugo in the southeastern United States destroyed red cockaded woodpecker habitats.

Animals and plants which survive a disturbance adjust their population sizes to the new habitats. They may become "packed" at high concentrations in the reduced area of suitable structures, while other animals and plants survive at very low numbers in the unfavorable habitats (Lauga and Joachin, 1992), and still others immigrate from areas of previously suitable structures to the newly created suitable structures.

Transitions in plant and animal populations also occur as stands in the stem exclusion structure develop an understory (e.g, the Kirtland's warbler; Walkenshaw 1983); however, few species decline or become extirpated since relatively few species are extremely dependent on the dense stand structures.

Each species is very resilient to certain changes in structures and very sensitive to others. Openings for short-lived species such as butterflies can be created rapidly—and their populations can expand rapidly; but the species requires continuously suitable habitat for breeding since it is so short lived. On the other hand, the red cockaded woodpecker and spotted owl are relatively long lived, and so can survive several years without reproducing; however, many years may be needed to create suitable old trees for them to nest and reproduce—and reproduction is not as rapid as with butterflies.

Following a disturbance to a landscape, stands with a variety of structures are created, but there is a general increase in openings (stand initiation structure) and a general decline in closed forests ("stem exclusion," "understory reinitiation," and "old growth" structures; Oliver et al., 1995). Animal and plant populations reflect this change in structures, with declines in populations of species dependent on closed structures and increases in populations dependent on openings. With regrowth, the opening-dependent populations decline and the closed structure species again increase. The population fluctuations can be very dramatic if the shifts in structure are extreme. If the landscape contained very few openings and the disturbance is severe enough to destroy most of the trees, the shift in populations would be dramatic. It would also be dramatic if regrowth to the stem exclusion structure occurred rapidly, such as where tree seed sources are nearby or sites are very productive.

Natural disturbances occurred at a variety of severities and scales (Chapter 4). It is highly probable that species were extirpated over large areas by the extreme shifts in structures following the disturbances and again following regrowths. At the same time, the shifts in structures allowed both those species needing to migrate across openings and those needing to migrate across closed structures to have suitable landscapes, although not simultaneously. The fluctuations in populations caused by the shifting structures also allowed genetic selection during times of population reductions and genetic drift among the different, isolated populations. When the population again expanded, the genetically altered population would then interact with other, previously isolated populations.

Applications to Management
Anticipating future concerns

Forest management has described ideal management of a "regulated forests" (Davis and Johnson, 1987), with a balanced age class distribution of stands—equal numbers of hectares (acres) in each age class between regeneration and harvest—in stands managed in single-cohort patterns. Multicohort stands are regulated if they contain appropriate trees in each cohort to allow a constant flow of trees for harvest. Such regulated forests do not exist anywhere in the world because wars, natural disturbances, economic shifts, and other factors have given irregular age distributions to forests. In fact, only few places approach this regulated condition out of either accident or intention, and more opportunistic forms of management may even be preferable (Gould, 1960).

In spite of their inevitability, potential disturbances are not completely unpredictable or uncontrollable. By manipulating stand structures and edges at the correct time, foresters can control a stand's susceptibility to most disturbances. Control measures are far more costly and generally less effective if they are delayed until the disturbance is in progress. Fires, insect and disease outbreaks, and blowdown of stands require massive efforts to be stopped, especially since they generally occur during seasons and climate cycles when control is most difficult. Proactive measures such as thinning stands, reducing fuel by controlled burns, or removing the fuel are more easily taken and can mitigate the effects of other disturbances. Similarly, undesirable stand structures and species compositions which will not have a future value can be avoided by proactive measures at less expense than reactive measures of replacing an unwanted stand after it has become established. In each region, long-term problems with the stands exist and can be avoided with appropriate manipulations:

1. Dominant oaks in mixed-hardwood stands in New England (Figs. 9.1 and 9.2) are currently being harvested. If the residual stems are not treated, the stands may develop into high-graded mixed-species stands common elsewhere in the eastern and western United States (Figs. 13.1 and 13.2).

2. Very many stands in the United States' Inland West have become very crowded and contain insects and diseases because of decades of fire control and "high grade" harvesting. These stands are being attacked by insects and are extremely susceptible to fires (Everett, 1993; O'Laughlin et al., 1993; Sampson and Adams, 1994; Covington and Moore, 1994). An integration of practices including removing the small trees

through thinning and then controlled burning could reduce the fuel and insect problems of some of these stands. Others may be too unstable to be thinned, and they could be harvested or destroyed by control fires to prevent their being foci for the spread of wildfires and insects.

3. In the southeastern United States, the former pine stands regrowing to hardwoods may develop into stratified stands dominated by oaks, tulip poplars, or other species (Clatterbuck et al., 1985, 1987; Beck and Hooper, 1986; O'Hara, 1986; Clatterbuck and Hodges, 1988; Oliver et al., 1989). Their future value for timber will depend on whether the developing B-stratum trees are of desirable species and spacing. If they are desirable, their value will be greatest only if enough B-stratum trees are spaced sufficiently for rapid diameter growth to continue while relying on other species in lower strata to keep the crop trees pruned. If the B-stratum species are undesirable, they could be eliminated so more desirable species can dominate the stand.

4. Over 5.7 million ha (2.3 million acres) of young Douglas-fir plantations have been established in coastal Washington, Oregon, and British Columbia since 1965 (Oswald et al., 1986). These stands are now reaching merchantable sizes and without thinning will slow dramatically in diameter growth (Oliver, 1986a). Some stands may remain undifferentiated, become unstable, and blow over; others may differentiate slowly and develop trees with small diameters; and still others may differentiate rapidly and result in heavy mortality of the intermediate and suppressed trees. In addition, the rapid early diameter growth has led to large centers of undesirable, juvenile wood. The oldest of these stands are now at the critical time when they can be harvested immediately for low-value, small trees; thinned to allow some trees to grow to large diameters with mature (high-quality) wood surrounding the juvenile core; or left to grow, resulting in some mortality, small diameters, and possibly blowdown because of the unstable stems (large height/diameter ratios; Oliver, 1986a; 1992a,b,c).

5. Upper elevations in the Pacific silver fir zone of Washington are developing as dense stands from advance regeneration after clearcutting between about 1945 and 1985. In places these stands may be growing densely and may blow over in windstorms. They may also not develop advance regeneration for many decades, and so provide difficulties for regenerating after future harvests. Early thinning could allow remaining stems to grow more stably and avoid windthrow problems. Later thinning may also allow advance regeneration to become established before harvest.

Having large areas in the same development stage because of disturbance history has caused many problems for foresters, and various techniques are proposed to manage forests over a large scale (Oldeman, 1981; Franklin and Forman, 1987; Hunter, 1990; Oliver, 1992; Franklin, 1993; Kuehne, 1993; Boyce and McNab, 1994). Wood products that can be produced are driven by the stand's stage of development. Well-developed markets usually exist for products from stands with familiar structures from older disturbances even as these products become in short supply; good markets often do not exist for the new products from stands of new structures from new disturbances.

Animals which multiply rapidly when large areas are in the stand initiation stage are often too numerous after the stands have grown to later stages, and many animals starve to death. The animals concentrate in the decreasing areas in the stand initiation stage and browse regeneration there, often not allowing these areas to regenerate. They also heavily browse advance regeneration in areas in the understory reinitiation stage.

Where one cohort and/or species composition predominates, the unusual stands become more valuable for certain animal habitats, plant and animal survival, and species migrations. These unusual stands are the result of human or natural protection, or different human or natural disturbance patterns. Forests can also be managed to create and maintain structures and habitats favorable to these rare plants and animals and so help preserve them.

"Sustainable development," "ecosystem management," and related terms

Increasing evidence is showing that the proportion of area in each structure has always fluctuated across a large forested area. Animal and plant populations have fluctuated with the changing structures. The increasingly limited and fragmented forest areas caused by human land use changes may be reducing the ability of animals and plants to migrate to suitable structures, and human activities may cause different proportions of various structures than previously occurred.

While some fluctuation in proportions of each structure is probably beneficial to species, extreme fluctuations risk extinction of species. Species extinctions have happened periodically both before and after people became a significant influence on the earth's environment (Wilson and Peters, 1988; Pielou, 1991). Presently, many societies places a high value on keeping species from becoming extinct. There is also a general concern for keeping the earth's environment and resources sustainable. In this context, "sustainable development" has been described as development that meets the needs of the present without compromising the ability of future generations to meet their own needs" (Brundtland, 1987) Applied to forest management, it can be interpreted to mean all species (biodiversity) and soil productivity are to be maintained for present and future values while allowing people to benefit from the forests.

No human endeavor—including excluding people from certain forest areas—will guarantee that no species will become extinct and that no soils will become degraded. Also, people wish to obtain multiple, often conflicting values from forests, so interactions of people with forests will always involve conflicts and tradeoffs. The conflicts can be resolved with least rancor using systematic approaches. The approaches involve the following steps:

— identifying the various values to be provided by the forest over the long term;
— developing alternative management approaches to provide these values;
— determining the tradeoffs among approaches and values—the extent to which each alternative approach will satisfy each value;
— presenting the alternative approaches and tradeoffs to the policymaker (landowner or whomever is the deciding authority);
— implementing the chosen approach.

The most successful management will be that which can develop, convey, and implement the most creative alternatives so that fewest tradeoffs among values need to be
made. Science and technology change values, increase understanding of the relation of
various management alternatives and values, and expand the potential management alternatives.

Concern for preservation of species has led to various proposals for managing forests
under such names as "ecosystem management," "new forestry" (Franklin, 1989;
Hopwood, 1991; Seymour and Hunter 1992), "new perspectives" (Salwasser, 1990;
Salwasser et al., 1992), "high quality forestry" (Kuehne, 1993; Weigand et al., 1993),
and "landscape forestry" (Boyce 1985, 1995; Boyce and McNab, 1994; Oliver, 1992).
All proposals generally view management for biodiversity, sustainability, and commodity values. These are different from traditional "commodity-based" management, where
forests are managed for most efficient production of commodities.

The proposals utilize two approaches (Oliver 1993):

—The "landscape" approach is based on the "dynamic" theory of forest development
  (Chap. 1) and advises active silvicultural manipulations to imitate, avoid, and/or
  recover from natural disturbances to maintain the full range of stand structures, landscape patterns, processes, and species across the landscape. Maintaining forests in this
  way can involve removal of timber and other products and providing employment and
  other values (Oliver, 1992a,b,c; Boyce and McNab, 1994; Boyce, 1995). Individual
  stands could be preserved as long as they best provided a needed struture.

—The "reserves" approach establishes large areas where stands are expected to develop
  toward the *old growth development stage* over large areas by setting aside large forest
  areas where human activities are restricted or excluded (Hunter, 1990; FEMAT, 1993;
  Franklin 1989). This approach relies on the "steady state" ecological theory (Chap. 1).
  It assumes that the forests which will develop without people intervening will provide
  all stand structures needed by animals and all other values; that species requiring
  habitats not maintained by the reserves either do not exist or will exist in geographically distant areas; and/or that non-human disturbances in the reserves will occur at
  appropriate temporal and spatial scales to maintain stand structures and landscape patterns for all species. This approach also assumes the reserved area is so valuable that
  commodity values should be provided from sources outside the reserves.

Both approaches have been referred to as "ecosystem management." The term
"ecosystem" is useful for individual scientific studies; however, it is difficult to use in a
management and policy context because each scientist can define an "ecosystem" to suit
a particular study. What different scientists refer to as an "ecosystem" can be quite varied in size and temporal extent (Kimmins, 1987); a single rotting log or the entire earth
can each be considered an ecosystem. "Ecosystem management" was considered by several groups of scientists as managing ecosystems in ways compatible with both ecological processes and people's needs (e.g., Johnson and Agee, 1987; Salwasser et al., 1992;
American Forest Association, 1993; Everett, 1993; Everett et al., 1994; Jensen and

Bourgeron, 1994), similar to the "landscape" approach. Another group of scientists considered "old growth" as an ecosystem and developed large reserves (FEMAT 1993).

Mixtures of "commodity-based," "landscape," and/or "reserves" approaches are sometimes suggested (Hunter, 1990; Seymour and Hunter, 1992; Franklin, 1993); and some mixtures of these have existed in most countries (e.g., government parks, government forests, and private forests). Both the desirability of each approach and the amount of land area allocated to each approach are not agreed on. The benefits, risks, and trade-offs to all values for each approach (and with different amounts of area in each approach) can be assessed (Oliver, 1993, 1994c; Peres-Garcia, 1993; Oliver and Lippke, 1995); and creative ways to reduce adverse impacts of the balancing among values are being developed (Lippke and Oliver, 1993; Lippke, 1994; Oliver, 1994a,b). These benefits, risks, tradeoffs, and ways to reduce adverse impacts need to be understood by policymakers before informed decisions can be made.

The role of silviculturists in managing forests:
People have a variety of commodity and noncommodity values they want from the forest—production of timber and other commodities such as mushrooms; development of wildlife habitat; ensuring biodiversity; control of fires, insects, and diseases; aesthetics; and recreation, among others. These values can be obtained by managing the forest for certain *stand structures* and *landscape patterns*—distributions of stand structures within the forest. Various specialists can work with those familiar with stand dynamics to describe the desired stand structures and landscape patterns for achieving the different values. It is then the role of silviculturists to develop alternative approaches for achieving and maintaining the various structures and patterns. The silviculturists then work with other specialists to develop other components of the alternatives (e.g., incentive systems or changes in laws) and to show the degree to which each alternative will achieve each value—how the structures and patterns will change and which values will be provided under each alternative.

Rarely can a forest fully satisfy all values at the same time; consequently, there will be tradeoffs among values by alternative management approaches. The creativity of the silviculturist is challenged to provide alternatives which have the fewest tradeoffs. Once the alternatives are developed and tradeoffs analyzed, it is the role of policymakers to choose their desired alternative (although sometimes this role is delegated to a local manager). It then becomes the role of the silviculturists to implement operations to achieve and maintain the targeted structures and patterns.

There are three common, counterproductive approaches which arise when developing management alternatives:

—non-silvicultural specialists and policymakers often target the silvicultural operations to be done rather than the desired stand structures and landscape patterns.
—specialists and policymakers often assume forests will remain permanently in certain structures and patterns, rather than change through growth and disturbances.
—silviculturists and other specialists sometimes have a bias toward certain values, and inappropriately behave as policymakers by biasing their alternatives and consequences of alternatives.

REFERENCES

Abbott, H. G., and T. F. Quink. 1970. Ecology of eastern white pine seed caches made by small forest mammals, *Ecology* 51:271-278.

Abetz, P. 1970. Stand density and branch diameters in the Rhine Valley Scots pine, Allgemeine Forst und Jagdzeitung 141:233-238.

Abetz, P. 1976. Beitrage zum Baumwachstum der h/d-wert—mehr ais ein Schlankheitsgrad! *Der Forst and Holzwirt* 31:389-393.

Abetz, P., and E. Kunstle. 1982. Zur druckholzbildung bei Fichte, *Allgemeine Forst und Jagdzeitung* 153:117-127.

Abrams, M. D. 1992. Fire and the development of oak forests. *Bioscience* 42(5): 346-353.

Abrams, M. D., and J. A.Downs. 1990. Successional replacement of old-growth white oak by mixed mesophytic hardwoods in southwestern Pennsylvania. *Canadian Journal of Forest Research* 20: 1864-1870.

Abrams, M. D., and G. J. Nowacki. 1992. Historical variation in fire, oak recruitment, and post-logging accelerated succession in central pennsylvania. *Bulletin of the Botanical Club* 119(1): 19-28.

Abrams, M. D., and C. M. Ruffner. 1995. Physiographic analysis of witness-tree distribution (1765-1798) and present forest cover through north central Pennsylvania. *Canadian Journal of Forestry Research* 25(4): 659-668.

Abrams, M. D., and M. L. Scott. 1989. Disturbance-mediated accelerated succession in two Michigan forest types. *Forest Science* 35(1): 42-49.

Abrams, M. D., D.Orwig, and T.E.Demeo. 1995. Dendroecological analysis of successional dynamics for a presettlement-origin white-pine—mixed-oak forest in the southern Appalachians, U.S.A. *Journal of Ecology* 83: 123-133.

Abrams, M. D., D. G. Sprugel, and D. I. Dickman. 1985. Multiple successional pathways on recently disturbed jack pine sites in Michigan. *Forest Ecology and Management* 10: 31-48.

Adams, A. B., and V. Adams. 1987. Comparisons of vegetative succession following glacial and volcanic disturbances in the Cascade Range of Washington, U.S.A., in D. E. Bilderbeck (ed.), *Mt. St. Helens: Botanical Consequences of the Explosive Eruption*, University of California Press, Los Angeles, pp. 70-147.

Adams, D. M., and A. R. Ek. 1974. Optimizing the management of uneven-aged forest stands, *Canadian Journal of Forest Research* 4:274-287.

Adler, P. B. 1994. "New forestry" in practice: a survey of mortality in green tree retention harvest units, western Cascades, Oregon. Unpublished Bachelor of Arts thesis, Harvard College, Harvard University, Cambridge, Massachusetts. 46 pp.

Agee, J. K. 1974. Fire management in the national parks, *Western Wildlands* 1(3):27-33.

Agee, J. K. 1981. Fire effects on Pacific Northwest forests: Flora, fuels, and fauna. IN *Proceedings of the Northwest Forest Fire Council Conference*, November 23-24, 1981. Northwest Forest Fire Council, Portland, Oregon. pp. 54-66.

Agee, J. K. 1981. Fire effects on Pacific Northwest forests: Flora, fuels, and fauna in

Proceedings of the Northwest Forest Fire Council Conference, November 23-24, 1981, Northwest Forest Fire Council, Portland, Oregon, pp. 54-66.

Agee, J. K. 1987. *The Forests of San Juan Island National Historical Park* Report CPSU/UW 88-1, USDI National Park Service Cooperative Park Studies Unit, College of Forest Resources, University of Washington, Seattle, Washington, 83 pp.

Agee, J. K. 1993. *Fire ecology of Pacific Northwest forests.* Island Press, Washington, D. C. 493 pp.

Agee, J. K., and M. H. Huff. 1980. First year ecological effects of the Hoh fire, Olympic Mountains, Washington, in *Proceedings of the Sixth Meteorology Conference*, April 22-24, 1980, Society of American Foresters, Washington, D.C., pp. 175-181.

Agee, J. K., and M. H. Huff. 1986. Structure and process goals for vegetation in wilderness management, in *National Wilderness Research Conference: Current Research*, USDA Forest Service General Technical Report INT-212, pp. 17-25.

Agee, J. K., and M. H. Huff. 1987. Fuel succession in a western hemlock/Douglas-fir forest, *Canadian Journal of Forest Research* **17**:697-704.

Agee, J. K., and P. W. Dunwiddie. 1984. Recent forest development on Yellow Island, San Juan County, WA, *Canadian Journal of Forestry Research* **62**:2074-2080.

Agee, J. K., M. H. Huff, and L. Smith. 1981. Ecological effects of the Hoh fire, Olympic National Park, progress report on contract CX-9000-9-EO79, USDI, National Park Service, Pacific Northwest Region, Seattle Washington 100 pp.

Agee, J. K., R. H. Wakimoto, and H. H. Biswell. 1977. Fire and fuel dynamics of Sierra Nevada Conifers, *Forest Ecology and Management* **1**:255-265.

Agren, G. I., B. Axelsson, J. G. K. Flower-Eilis, S. Linde;, H. Persson, H. Staff, and E. Troeng. 1980. Annual carbon budget for a young Scots pine, in T. Persson (ed.), *Structure and Function of Northern Coniferous Forests—An Ecosystem Study, Ecology Bulletin* **32**:307-313.

Ahlgren, C. E. 1959. Some effects of fire on forest reproduction in northeastern Minnesota, *Journal of Forestry* **57**:194-200.

Ahlgren, C. E. 1960. Some effects of fire on reproduction and growth of vegetation in northeastern Minnesota, *Ecology* **41**:431-445.

Ahlgren, C. E. 1966. Small mammals and reforestation following prescribed burning, *Journal of Forestry* **64**:614-618.

Ahlgren, C. E. 1974. Effects of fires on temperate forests: North Central United States, in T. T. Kozlowksi and C. E. Ahlgren (eds.), *Fire and Ecosystems*, Academic, New York, pp. 195-223.

Ahlgren, C. E., and Hansen, H. L. 1957. Some effects of temporary flooding on coniferous trees, *Journal of Forestry* **55**:647-650.

Ahlgren, I. F., and C. E. Ahlgren. 1960. Ecological effects of forest fire, *Botanical Review* **26**:483-533.

Ahlstrand, G. M. 1980. Fire history of a mixed conifer forest in Guadalupe Mountains National Park, in *Proceedings of the Fire History Workshop*, October 20-24, 1980, Tucson, Arizona, USDA Forest Service General Technical Report RM-81, pp. 4-7.

Aho, P. E. 1960. Heart-rot hazard is low in *Abies amabilis* reproduction injured by logging, USDA Forest Service Research Note PNW-196, 5 pp.

Aho, P. E. 1982. Decay problems in true fir stands, in C. D. Oliver and R. M. Kenady (eds.), *Biology and Management of True Fir in the Pacific Northwest*, College of Forest Resources, University of Washington, Seattle, Institute of Forest Resources Contribution Number 45, pp. 203-208.

Aikman, D. P., and A. R. Watkinson. 1980. A model for growth and self-thinning in even-aged monocultures of plants, *Annals of Botany* **45**:419-427.

Akai, T. 1978. Studies on natural regeneration: IV. The regeneration of *Chamaecyparis obtusa* stands on several conditions in Kinki and Chugoku districts, *Bulletin of the Kyoto University Forests* 50:44-57.

Akai, T., K. Yoshimura, I. Manabe, S. Ueda, and T. Honjyo. 1983. Structure and silvicultural systems of multi-storied forest: I. On stratified mixture of Hinoki, Akamatsu, and broad-leaf trees, *Bulletin of the Kyoto University Forests* 55:1-79.

Akai, T., S. Ueda, and T. Furono. 1970. Mechanisms related to matter production in a young slash pine forest, *Bulletin of the Kyoto University Forests* 41:56-79.

Akai, T., T. Furuno, S. Ueda, and S. Sano. 1968. Mechanisms of matter production in young loblolly pine forest, *Bulletin of the Kyoto Uniuersity Forests* 40:26-49.

Akai, T., T. Sakaue, and J. Ohno. 1977. Structures and secondary succession in mixed forest of *Pinus densiflora, Chamaecyparis obtusa* and broad leaved tree, *Bulletin of the Kyoto University Forests* 49:64-80.

Alaback, P. B. 1982a. Dynamics of understory biomass in Sitka spruce-western hemlock forests of Southeast Alaska, *Ecology* 63:1932-1948.

Alaback, P. B. 1982b. Forest community structural changes during secondary succession in Southeast Alaska, in J. E. Means (ed.), *Forest Succession and Stand Development Research in the Northwest: Proceedings of the Symposium,* Oregon State University, Forest Research Laboratory, Corvallis, pp. 70-79.

Alaback, P. B. 1984a. A comparison of old-growth forest structure in the western hemlock-sitka spruce forests of southeast Alaska, in W. R. Meehan, T. R. Merrell, Jr., and T. A. Hanley (eds.), *Fish and Wildlife Relationships in Old-Growth Forests: Proceedings of a Symposium,* American Institute of Fishery Research Biologists, available from J. W. Reintjes, Morehead City, North Carolina, pp. 219-226.

Alaback, P. B. 1984b. Plant succession following logging in the sitka spruce-western hemlock forests of Southeast Alaska: Implications for management, USDA Forest Service General Technical Report PNW-173, 26 pp.

Alexander, M. F:. 1982. Calculating and interpreting forest fire intensities, *Canadian Journal of Botany* 60:349-357.

Alexander, R. R. 1975. Partial cutting in old-growth lodgepole pine, USDA Forest Service Research Paper RM-136, 17 pp.

Alexander, R. R. 1986. Silvicultural systems and cutting methods for old-growth lodgepole pine forests in the central Rocky Mountains, USDA Forest Service General Technical Report RM127, 31 pp.

Alexander, R. R., and C. B. Edminister. 1977. Uneven-aged management of old growth spruce-fir forests: Cutting methods and stand structure goals for the initial entry USDA Forest Service Research Paper RM-186, 12 pp.

Alexander, R. R., and J. H. Buell. 1955. Determining the direction of destructive winds in a Rocky Mountain timber stand, *Journal of Forestry* 53:19-23.

Allen, G. S., and J. N. Owens. 1972. *The life history of Douglas-fir,* Environment Canada Forestry Service, Information Canada, Ottawa, 139 pp.

Allen, R. H., and D. A. Marquis. 1970. Effect of thinning on height and diameter growth of oak and yellow-poplar saplings, USDA Forest Service Research Paper NE-173, 7 pp.

Allen, R. M. 1953. Release and fertilization stimulate longleaf pine cone crop, *Journal of Forestry* 51:827.

Allen, R. M. 1956. Relation of saw-palmetto to longleaf pine reproduction on a dry site, *Ecology* 37:195-196.

Allen, R. M., and N. M. Scarbrough. 1969. Development of a year's height growth in longleaf pine saplings, USDA Forest Service Research Paper S0-45, 13 pp.

Amaral, M. 1978. Black-tailed deer *(Odocoileus hemonius columbianus)* and forest reproduction: Blake Island Marine State Park, Washington, *Northwest Science* **52**:233-235.

Amaral, M. 1978. Black-tailed deer (Odocoileus hemonius columbianus) and forest reproduction: Blake Island Marine State Park, Washington. Northwest Science **52**:233-235.

American Forestry Association. 1951. American tree monarchs: Report on American big trees, Part IV, *American Forests* **57**(7):28-29.

American Forestry Association. 1993. "Building partnerships for ecosystem management on forest and range lands in mixed ownership. " A workshop convened by the Forest Policy Center, American Forestry Association, and hosted by the Yale University School of Forestry and Environmental Studies. October 22-24, 1993.

Amman, G. D. 1977. The role of the mountain pine beetle in lodgepole pine ecosystems: Impact on succession, in W. J. Mattson (ed.), *The Role of Arthropods in Forest Ecosystems,* Springer-Verlag, New York, pp. 3-18.

Amman, G. D. 1978. The biology, ecology, and causes of outbreaks of the mountain pine beetle in lodgepole pine forests, in D. L. Kibbee, A. A. Berryman, G. D. Amman, and R. W. Stark (eds.), *Proceedings of a Symposium,* Pullman, Washington, April 25-27, 1978, University of Idaho Forest Wildlife and Range Experiment Station, pp. 39-53.

Amman, G. D., and B. H. Baker. 1972. Mountain pine beetle influence on lodgepole pine stand structure, *Journal of Forestry* **70**:204-209.

Anderson, A. B. (editor). 1990. *Alternatives to Deforestation: Steps Toward Sustainable Use of the Amazon Rain Forest.* Columbia University Press, New York: 281 pp.

Anderson, G. W., and R. L. Anderson. 1963. The rate of spread of oak wilt in the Lake States, *Journal of Forestry* **61**:823-825.

Anderson, R. C., and O. L. Loucks. 1979. White-tail deer *(Odocoileus virginianus)* influence on structure and composition of Tsuga canadensis forests, *Journal of Applied Ecology* **16**:855-861.

Anderson, R. C., O. L. Loucks, and A. M. Swain. 1969. Herbaceous response to canopy cover, light intensity and throughfall precipitation in coniferous forests, *Ecology* **50**:255-263.

Andersson, M. C. 1966. Stand structure and light penetration: II. A theoretical analysis, *Journal of Applied Ecology* **3**:41-54.

Andersson, M. C. 1969. A comparison of two theories of scattering of radiation in crops, *Agricultural Meteorology* **6**:399-405.

Ando, T. 1962. Growth analysis on the natural stands of Japanese red pine *(Pinus densiflora* Sieb. et Zucc.): II. Analysis of stand density and growth, *Government Forest Experiment Station of Tokyo Bulletin* **147**:1-77.

Aplet, G. H., R. L. Laven, and F. W. Smith. 1988. Patterns of community dynamics in Colorado Engelmann spruce-subalpine fir forests, *Ecology* **69**:312-319.

Apsley, D. K., D. J. Leopold, and G. R. Parker. 1985. Tree species response to release from domestic livestock grazing, *Indiana Academy of Science* **94**:215-226.

Arend, J. L., and H. F. Scholz. 1969. Oak forests of the Lake States and their management, USDA Forest Service Research Paper NC-31, 36 pp.

Armson, K. A., and R. J. Fessenden. 1973. Forest windthrows and their influence on soil morphology, *Soil Science Society of America Proceedings* **37**:781-783.

Arno, S. F. 1980. Forest fire history in the northern Rockies, *Journal of Forestry* **78**:460-465.

Aronsson, A., and S. Elowson. 1980. Effects of irrigation and fertilization on mineral nutrients in Scots pine needles, in T. Persson (ed.), *Structure and Function of Northern Coniferous Forests—An Ecosystem Study, Ecology Bulletin* **32**:219-228.

Artsybashev, E. S. 1974. *Forest Fires and Their Control*, Lesnaya Promyshlennost, Publishers, Moscow (translated from Russian and published by the USDA Forest Service and the National Science Foundation, Washington, D.C., by Amerind Publishing Co. Pvt. Ltd., New Delhi, 160 pp.).

Ashe, W. W. 1918. "Ice storms of the Southern Appalachians" by Verne Rhoades (comment), *Monthly Weather Review* **46**:374.

Ashton, D. H. 1976. The development of even-aged stands of *Eucalyptus regnans* F. Muell. in central Victoria, *Australian Journal of Botany* **24**:397-414.

Ashton, P. M. S. 1992. Establishment and early growth of advance regeneration of canopy trees in moist mixed-species forests. *In*: (M. J. Kelty, B. C. Larson and C. D. Oliver, editors) The Ecology and Silviculture of Mixed-Species Forests: A Festschrift for David M. Smith. Kluwer Academic Publishers, Boston: 101-122.

Ashton, P. M. S., J. S. Lowe, and B. C. Larson. 1989. Thinning and spacing guidelines for blue mahoe (Hibiscus elatus SW.) *Journal of Tropical Forest Science* **2**(1):37-47.

Ashton, P. S. 1989. Species richness in tropical forests. IN L. B. Holm-Nielson, I. C. Nielson, and H. Balslev, editors. *Tropical Forests*. Academic Press, New York. pp. 239-251.

Assmann, E. 1954. Grundflachenhaltung und Zuwachsleistung Bayerischer FichtenDurchforstungsreihen, *Forstwissenschafll iches Centralblatt* **73**:257-271.

Assmann, E. 1970. *The Principles of Forest Yield Study*. Pergamon Press, New York, 606 pp.

Atkinson, W. A., and R. J. Zasoski (eds.).1976. *Western Hemlock Management*, University of Washington College of Forest Resources Institute of Forest Products Contribution Number 34, 317 pp.

Aubreville, A. 1938. La foret coloniale: Les forets de l'Afrique occidentale francaise, *Annales de l'Academie des Sciences Coloniales* **9**:1-245.

Auchmoody, L. R. 1979. Nitrogen fertilization stimulates germination of dormant pin cherry seed, *Canadian Journal of Forest Research* **9**:514-516.

Auclair, A. N., and G. Cottam. 1971. Dynamics of black cherry (*Prunus serotina* Erhr.) in southern Wisconsin oak forests, *Ecological Monographs* **41**:153-177.

Aussenac, G., A. Granier, and R. Naud. 1982. Influence d'une eclaircie sur la croissance et le bilan hydrique d'un jeune peuplement de Douglas (*Pseudotsuga menziesii* (Mirab. Franco), *Canadian Journal of Forest Research* **12**:222-231.

Awaya, H., K. Honda, T. Shiibayashi, and S. Obata. 1981. The growth and stand structure of warm-temperated broadleaved evergreen forest: II. Study of growth properties for each species and species groups, *Bulletin of the Forestry and Forest Products Research Institute* **314**:107-146.

Axelsson, B. 1981. *Site differences in yield-differenecs in biological production or in redistribution of carbon within trees.* Research Report 9, Swedish University of Agricultural Sciences, Department of Ecology and the Environment, Uppsala, Sweden, 11 pp.

Ayensu, E. S., and R. A. DeFilipps. 1978. *Endangered and Threatened Plants of the United States*, Smithsonian Institution Press, Washington D.C., 403 pp.

Aytug, B., N. Merev, and G. Edis. 1975. *Surmenet—agacbasi dolaylari ladin ormaninin tarihi ve gelecegi.* Proje No. TOAG 113; TOAG Seri No. 39, TBTAK Yayinlari No. 252, Turkiye Bilimsel ve Teknik Arastirma Kurumu, Ankara, Turkey, 64 pp.

Baker, F. S. 1925. Aspen in the central Rocky Mountain region. USDA Agricultural

Bulletin 1291, 46 pp.

Baker, F. S. 1934. *Theory and Practice of Silviculture*, McGraw-Hill, New York, 502 pp.

Baker, F. S. 1953. Stand density and growth, *Journal of Forestry* 51:95-97.

Baker, J. 1968. Effects of slash burning on soil composition and seedling growth, Canadian Forestry Service, Forestry Research Laboratory, Victoria, British Columbia, Canada, Informal Report BC-S-29, 10 pp.

Baker, J. B. 1977. Tolerance of planted hardwoods to spring flooding, *Southern Journal of Forestry* 1:23-25.

Baker, V. R. 1978. The Spokane flood controversy and the Martian outflow channels, *Science* 202:1249-1256.

Baker, W. L. 1972. Eastern forest insects, USDA Forest Service Miscellaneous Publication No. 1150, 642 pp.

Baker, W. L. 1992. Effects of settlement and fire suppression on landscape structure. *Ecology* 73:1879-1887.

Baker, W.L. 1992b. The landscape ecology of large disturbances in the design and management of nature reserves. *Landscape Ecology* 7 (3): 181-194.

Bakuzis, E. V.1969. Forest viewed in an ecosystems perspective, in G. M. van Dyne (ed.), *The Ecosystem Concept in Natural Resource Management*, Academic, New York, pp. 189-258.

Balda, R. P. 1975. The relationships of secondary cavity-nesters to snag densities in western coniferous forests, USDA Forest Service General Technical Report PNW64.

Baldwin, H. I. 1942. Natural regeneration on white pine lands following the hurricane, *Fox Forest Notes* No. 21, 2 pp.

Baldwin, J. L. 1973. *Climates of the United States*, U. S. Department of Commerce, National Oceanic and Atmospheric Administration, Environmental Data Service, Washington, D.C., 113 pp.

Balmer, W. E, and H. L. Williston. 1973. The need for precommercial thinning, USDA Forest Service Southeast Area State and Private Forestry Forest Management Bulletin, 6 pp.

Balmer, W. E., E. G. Owens, and J. R. Jorgensen. 1975. Effects of various spacings in loblolly pine growth 15 years after planting, USDA Forest Service Research Note SE-211, 7 pp.

Balslev, H., and S. S. Renner. 1989. Diversity of east Ecuadorean lowland forests. IN L. B. Holm-Nielson, I. C. Nielsen, and H. Balslev, editors. *Tropical Forests*. Academic Press, New York. pp. 287-295.

Barden, L. S. 1979. Tree replacement in small canopy gaps of a Tsuga canadensis forest in the southern Appalachians, Tennessee, *Oecologia* 44:141-142.

Barden, L. S. 1981. Forest development in canopy gaps of a diverse hardwood forest of the southern Appalachian Mountains, *Oikos* 37:205-209.

Barnes, G. H. 1962. Yield of even-aged stands of western hemlock, USDA Technical Bulletin No. 1273, 52 pp.

Barnette, R. M., and J. B. Hester. 1930. Effects of burning upon the accumulation of organic matter in forest soils, *Soils Science* 29:281-284.

Barney, R. J. 1971. Wildfires in Alaska—some historical and projected effects, in C. W. Slaughter, R. J. Barney, and G. M. Hansen (eds.), *Fire in the Northern Environment*, Alaska Section, Society of American Foresters, Fairbanks, Alaska, pp. 51-59.

Barrett, J. W. 1970. Ponderosa pine saplings respond to control of spacing and under-story vegetation, USDA Forest Service Research Paper PNW-106, 16 pp.

Barrett, J. W. 1972. Large-crowned planted ponderosa pine respond well to thinning USDA Forest Service Research Note PNW-179, 12 pp.

Barrett, J. W. 1973. Latest results from the Pringle Falls ponderosa pine spacing study, USDA Forest Service Research Note PNW-209, 22 pp.

Barrett, J. W. 1979. Silviculture of ponderosa pine in the Pacific Northwest: The state of our knowledge, USDA Forest Service General Technical Paper PNW-97, 106 pp.

Barrett, J. W. 1982. Twenty-year growth of ponderosa pine saplings thinned to five spacings in central Oregon, USDA Forest Service Research Paper PNW-301, 18 pp.

Barrett, J. W., and C. T. Youngberg. 1965. Effect of tree spacing and understory vegetation on water use in a pumice soil, *Soil Science Society of America Proceedings* 29:472-476.

Barrett, J. W., and L. F. Roth. 1985. Response of dwarf mistletoe-infested ponderosa pine to thinning: I. Sapling growth, USDA Forest Service Research Paper PNW-330, 15 pp.

Barrows, C. W. 1980. Roost selection by spotted owls: an adaptation to heat stress. *Condor* 83:302-309.

Barrows, J. S. 1951. Forest fires in the northern Rocky Mountains, Report by J. S. Barrows, Chief, Division of Fire Research, Northern Rocky Mountain Forest and Range Experiment Station, Missoula, Montana, April 1951, 253 pp.

Barry, R. G. 1981. *Mountain Weather and Climate,* Methuen, New York 313 pp.

Barton, L. V. 1965. Seed dormancy: General survey of dormancy types in seeds and dormancy imposed by external agents, *Encyclopedia of Plant Physiology* 15:699-720.

Baskerville, G. 1985. Adaptive management: Wood availability and habitat availability. *Forestry Chronicle* 61(2):171-175.

Baskerville, G. L. 1960. *Conversion to periodic selection management in a fir, spruce, and birch forest,* Department of Northern Affairs and Natural Resources of Canada, Forestry Branch, Forest Research Division Technical Note 86, 19 pp.

Baskerville, G. L. 1965. Dry-matter production in immature balsam fir stands, *Forest Science Monograph* 9:1-42.

Baskerville, G. L. 1971. *The fir-spruce-birch forest and the budworm,* Report, Forestry Service, Canadian Department of the Environment, Fredericton, New Brunswick, 111 pp.

Bates, C. G., and J. Roeser, Jr. 1928. Light intensities required for growth of coniferous seedlings, *American Journal of Botany* 15:185-244.

Baumann, R. H., and R. Adams. 1981. The creation and restoration of wetlands by natural processes in the lower Atchafalaya River system: Possible connects with navigation and flood control management, in R. H. Stovall (ed.), *Proceedings of the Eighth Annual Conference on Wetlands Restoration and Creation,* May 8-9, 1981, Volume 8, pp. 1-24.

Bazter, H. O. 1969. Forest character and vulnerability of balsam fir to spruce budworm in Minnesota, *Forest Science* 15:17-25.

Bazter, H. O., and J. L. Bean. 1962. Spruce budworm defoliation causes continued top killing and tree mortality in northeastern Minnesota, USDA Forest Service Lake States Forest Experiment Station Technical Note 621, 2 pp.

Bazzaz, F. A. 1983. Characteristics of population in relation to disturbance in natural and man-modified ecosystems, in H. A. Mooney and M. Godron (eds.), *Disturbance and Ecosystems: Components of Response,* Springer-Verlag, New York, pp. 259-275.

Bazzaz, F. A., and L. C. Bliss. 1971. Net primary production of herbs in a central Illinois deciduous forest, *Bulletin of the Torrey Botanical Club* 98:90-94.

Beals, E. W., G. Cottam, and R. J. Vogl. 1960. Influence of deer on vegetation of the Apostle Islands, Wisconsin, *Journal of Wildlife Management* 24:68-80.

Beard, J. S. 1978. The physiognomic approach, in R. H. Whittaker (ed.), *Classification of Plant Communities,* Junk, The Hague, pp. 33-64.

Beatty, S. W. 1984. Influence of microtopography and canopy species on spatial patterns of forest understory plants, *Ecology* 65:1406-1419.

Beatty, S. W., and E. L. Stone. 1986. The variety of soil microsites created by treefalls, *Canadian Journal of Forest Research* 16:539-548.

Beck, D. E., and L. Della-Bianca. 1972. Growth and yield of thinned yellow poplar, USDA Forest Service Research Paper SE-101, 20 pp.

Beck, D. E., and L. Della-Bianca. 1981. Yellow-poplar: Characteristics and management, USDA Forest Service Agricultural Handbook Number 583, 91 pp.

Beck, D. E., and R. M. Hooper. 1986. Development of a southern Appalachian hardwood stand after clearcutting, *Southern Journal of Forestry* 10:168-172.

Beck, D. E.1970. Effect of competition on survival and height growth of red oak seedlings, USDA Forest Service Research Paper SE-56, 7 pp.

Becker, R. 1995. Operational considerations of implementing uneven-aged management. IN (K. L. O'Hara, editor) *Uneven-aged management: opportunities, constraints, and methodologies.* Montana Forest and Conservation Experiment Station, School of Forestry, University of Montana, Miscellaneous Publication No. 56.

Bell, A. D. 1974. Rhizome organization in relation to vegetative spread in *Medeola virginiana, Journal of the Arnold Arboretum* 55:458-468.

Bell, D. T., and R. del Moral. 1977. Vegetation gradients in the stream-side forest of Hickory Creek, Will County, Illinois, *Bulletin of the Torrey Botanical Club* 104:127-135.

Belluschi, P. G., and N. E. Johnson. 1969. The rate of crown fade of trees killed by the Douglas-fir beetle in southwestern Oregon, *Journal of Forestry* 67:30-32.

Belsky, J. A., and C. D. Canham. 1994. Forest gaps and isolated savanna trees. *BioScience* 44(2):77-84.

Bendell, J. F. 1974. Effects of fire on birds and mammals. IN (T. T. Kozlowski and C. E. Ahlgren, editors) *Fire and ecosystems.* Academic Press, New York. 542 pp.

Bennett, F. A. 1960. Spacing and early growth of planted slash pine, *Journal of Forestry* 58:966-967.

Berlyn, G. P, 1962. Some size and shape relationships between tree stems and crowns, *Iowa State Journal of Science* 37:7-15.

Betters, D. R., and R. F. Woods. 1981. Uneven-aged stand structure and growth of Rocky Mountain aspen, *Journal of Forestry* 79:673-676.

Bey, C. F. 1964. Advance oak reproduction grows fast after clearcutting, *Journal of Forestry* 62:339-340.

Bickford, C. A., F. S. Baker, and F. G. Wilson. 1957. Stocking, normality, and measurement of stand density, *Journal of Forestry* 55:99-104.

Bickford, L. M. 1982. Silvicultural practices of east-side grand fir and associated species in the eastern Washington Cascades, in C. D. Oliver and R. M. Kenady (eds.), *Biology and Management of True Fir in the Pacific Northwest,* University of Washington College of Forest Resources, Institute of Forest Resources Contribution Number 45, pp. 251-252.

Bilderback, D. E. (ed.). 1987. *Mount St. Helens 1980: Botanical Consequences of the Explosive Eruptions,* University of California Press, Los Angeles, 360 pp.

Binkley, C. S., and B. C. Larson. 1987. Simulated effects of climate warming on the productivity of managed northern hardwood forests, in L. Kairiukstis, S. Nilsson, and

A. Straszak (eds.), *Forest Decline and Reproduction: Regional and Global Consequences*, IIASA, WP-87-75, pp. 223-230.

Binkley, D. 1984. Douglas-fir stem growth per unit of leaf area increased by interplanted Sitka alder and red alder, *Forest Science* **30**:259-263.

Binkley, D. 1986. *Forest Nutrition Management*, John Wiley, New York, 290 pp.

Binkley, D., J. D. Lousier, and K. Cromack, Jr. 1984. Ecosystem effects of Sitka alder in a Douglas-fir plantation, *Forest Science* **30**:26-35.

Biswell, H. H. 1972. Fire ecology in ponderosa pine grassland, Proceedings *Tall Timbers Fire Ecology Conference* **12**:69-97.

Black, R. A., and L. C. Bliss. 1978. Recovery sequence of Picea mariana-Vaccinium uliginosum forests after burning near Inuvik, N.W.T., Canada, *Canadian Journal of Botany* **56**:2020-2029.

Blackburn P., and J. A. Petty. 1988. Theoretical calculations of the influence of spacing on stand stability, *Forestry* **61**:236-239.

Blais, J. R. 1954. The recurrence of spruce budworm infestations in the past century in the Lao Seul area of northwestern Ontario. *Ecology* **35**:62-71.

Blais, J. R. 1958. The vulnerability of balsam fir to spruce budworm attack in northwestern Ontario, with special reference to the physiological age of a tree, *Forestry Chronicle* **34**:405-422.

Blais, J. R. 1965. Spruce budworm outbreaks in the past three centuries in the Laurentide Park, Quebec, *Forest Science* **11**:130-138.

Blais, J. R. 1968. Regional variation in susceptibility of eastern North American forests to budworm attach based on history of outbreaks. *Forestry Chronicle* **44**:17-23.

Bloomberg, W. G. 1950. Fire and spruce, *Forestry Chronicle* **26**:157-161.

Blum, B. M. 1961. Age-size relationships in all-aged northern hardwoods, USDA Forest Service Research Note NE-125, 3 pp.

Blum, B. M. 1963. Excessive exposure stimulates epicormic branching in young northern hardwoods, USDA Forest Service Research Note NE-9, 6 pp.

Blum, B. M. 1973. Some observations on age relationships in spruce-fir regeneration, USDA Forest Service Research Note NE-169.

Blum, B. M. 1975. Regeneration and uneven-aged silviculture—the state of the art, in *Uneven-Aged Silviculture and Management in the Eastern United States*, USDA Forest Service, Timber Management Research, pp. 63-83.

Blum, B. M.1966. Snow damage in young northern hardwoods and rapid recovery, *Journal of Forestry* **64**:16-18.

Boardman, N. K. 1977. Comparative photosynthesis of sun and shade plants, *Annual Review of Plant Physiology* **28**:355-377.

Bogucki, D. J. 1970. Debris slides and related flood damage with the September 1, 1951, cloudburst in the Mt. LeConte-Sugarland Mountain area, Great Smokey Mountains National Park, unpublished Ph.D. thesis, University of Tennessee, Knoxville, Tennessee, 165 pp.

Bogucki, D. J. 1977. Debris slide hazards in the Adirondack Province of New York State, *Environmental Geology* **1**:317-328.

Bollard, E. G. 1955. Trace element deficiencies of fruit crops in New Zealand, New Zealand Department of Science, Independent Research, Bulletin No. 115, 52 pp.

Bolt, C. E. 1971. Stocking and growth in even-aged forest stands: A simple, instructional model, *Proceedings of the South Dakota Academy of Sciences* **50**:79-84.

Bonnicksen, T. M., and E. C. Stone. 1981. The giant sequoia-mixed conifer forest community characterized through pattern analysis as a mosaic of aggregations, *Forest Ecology and Management* **3**:307-328.

Bonnicksen, T. M., and E. C. Stone. 1982. Reconstruction of a pre-settlement giant sequoia-mixed conifer forest community using the aggregation approach, *Ecology* 63:1134-1148.

Bormann, B. T., M. H. Brookes, E. D. Ford, A. R. Kiester, C. D. Oliver, and J. F. Weigand. 1993. *A broad, strategic framework for sustainable-ecosystem management.* Volume V of Eastside Forest Ecosystem Health Assessment. USDA Forest Service, National Forest System, Forest Service Research, Wenatchee, Washington.

Bormann, F. H. 1955. The primary leaf as an indication of physiologic condition in shortleaf pine, *Forest Science* 1:189-192.

Bormann, F. H. 1956. Ecological implications of changes in the photosynthetic response of *Pin us taeda* seedlings during ontogeny, *Ecology* 37:70-75.

Bormann, F. H. 1965. Changes in the growth pattern of white pine trees undergoing suppression, *Ecology* 46:269-277.

Bormann, F. H. 1969. A holistic approach to nutrient cycling problems in plant communities, in K. N. H. Greenidge (ed.), *Essays in Plant Geography and Ecology,* Published by the Nova Scotia Museum, Halifax, pp. 149-165.

Bormann, F. H., and G. E. Likens. 1979. Catastrophic disturbance and the steady state in northern hardwood forests, *American Scientists* 67:660-669.

Bormann, F. H., and G. E. Likens. 1981. *Pattern and Process in a Forested Ecosystem,* Springer-Verlag, New York, 253 pp.

Bormann, F. H., and M. F. Buell. 1964. Old-age stand of hemlock-northern hardwood forest in central Vermont, *Bulletin of the Torrey Botanical Club* 91:451-465.

Bormann, F. H., G. E. Likens, D. W. Fisher, and R. S. Pierce. 1968. Nutrient loss accelerated by clearcutting of a forest ecosystem, *Science* 159:882-884.

Bormann, F. H., T. G. Siccama, G. E. Likens, and R. H. Whittaker. 1970. The Hubbard Brook ecosystem study: Composition and dynamics of the tree stratum, *Ecological Monographs* 40:373-388.

Bormann, F. H.1966. The structure, function, and ecological significance of root grafts in *Pinus strobus* L., *Ecological Monographs* 36:1-26.

Botkin, D. B. 1979. A grandfather clock down the staircase: Stability and disturbance in natural ecosystems, in *Forests: Fresh Perspectives from Ecosystem Analysis,* Proceedings of the 40th Annual Biology Colloquium, Oregon State University Press, Corvallis, pp. 1-10.

Botkin, D. B. 1990. *Discordant Harmonies: a new ecology for the twenty-first century.* Oxford University Press, New York. 241 pp.

Botkin, D. B. 1994. Preface. IN (R. N. Sampson and D. L. Adams, editors). 1994. *Assessing Forest Ecosystem Health in the Inland West.* The Haworth Press, Inc., New York.

Botkin, D. B. 1995. *Our natural history: the lessons of Lewis and Clark.* G. P. Putnam's Sons, New York: 300 pp.

Botkin, D. B., and M. T. Sobel. 1975. Stability in time-varying ecosystems. *The American naturalist* 109(970):625-646.

Bourne, R. 1951. A fallacy in the theory of growing stock, *Forestry* 24:6-18.

Bowerman, B. L., and R. T. O'Connell. 1987. *Time Series Forecasting.* Second Edition. Duxbury Press, Boston, Massachusetts. 540 pp.

Bowers, F. H. 1987. Effects of windthrow on soil properties and spatial variability in southeast Alaska, unpublished Ph.D. dissertation, University of Washington, Seattle, 185 pp.

Bowersox, T. W., and W. W. Ward. 1972. Prediction of advance regeneration in mixed-oak stands of Pennsylvania, *Forest Science* 18:278-282.

Bowling, D. R., and R. C. Kellison. 1983. Bottomland hardwood stand development after clearcutting, *Southern Journal of Forestry* 7:110-116.

Boyce, J. S. 1929. Deterioration of wind-thrown timber on the Olympic Peninsula, Washington. United States Department of Agriculture, Washington, D. C. Technical Bulletin No. 104. 28 pp.

Boyce, S. G. 1961. Most oak trees in the Central States develop from seedling sprouts, *Bulletin of the Ecological Society of America* 42:96.

Boyce, S. G. 1985. Forestry decisions. USDA Forest Service GTR-SE-35: 318 pp.

Boyce, S. G. 1995. *Landscape forestry.* John Wiley & Sons, New York. 239 pp.

Boyce, S. G., and H. A. Knight. 1980. Prospective ingrowth of southern hardwoods beyond 1980, USDA Forest Service Research Paper SE-203, 33 pp.

Boyce, S. G., and W. H. McNab. 1994. Management of forested landscapes: simulations of three alternatives. *Journal of Forestry* 92:27-32.

Boyd, R., and S. Penland. 1984. Shoreface translation and the Holocene stratigraphic record in Nova Scotia, Southeastern Australia, and the Mississippi River delta plain, in B. Greenwood and R. A. Davis, Jr. (eds.), *Hydrodynamics and Sedimentation in Wave-Dominated Coastal Environments, Marine Geology* 63:391-412.

Boyd, R., and S. Penland. 1988. A geomorphologic model for Mississippi Delta evolution, *Transactions of the Gulf Coast Association of Geological Societies* 38:443-452.

Boydak, M. 1994. Personal communication with C. D. Oliver. Dr. Boydak is Professor of Silviculture, Faculty of Forestry, University of Istanbul, Buyuk Dere, Istanbul, Turkey.

Boyer, D. E., and J. D. Dell. 1980. Fire effects on Pacific Northwest forest soils, USDA Forest Service Pacific Northwest Region, 59 pp.

Boyer, W. D. 1970. Shoot growth patterns of young loblolly pine, *Forest Science* 16:472-482.

Boyer, W. D. 1958. Longleaf pine seed dispersal in south Alabama, *Journal of Forestry* 56:265-268.

Boysen-Jensen, P. 1932. Die Stoffproduktion d. Pflanzen (Jena), cited by Assmann, 1970.

Bozkus, H. F. 1988. Toros goknari *(Abies cillicica* Carr) nin Turkiye'deki dogal yayilis ve silvikulturel ozellikleri (The natural distribution and silvicultural characteristics of *Abies cilicica* Carr. in Turkey), Silvikultur Anabilim Dali, Orman Fakultesi, Istanbul Univertesi, Istanbul, Turkey, 118 pp.

Braathe, P. 1957. *Thinnings in Even-aged Stands,* University of New Brunswick, Fredericton, 92 pp.

Bradley, R. T. 1963. Thinning as an instrument of forest management, *Forestry* 36:181-194.

Brady, W. W., and T. A. Hanley. 1984. The role of disturbance in old-growth forests: Some theoretical implications for southeastern Alaska, in W. R. Meehan, T. R. Merrell, Jr. and T. A. Hanley (eds.), *Fish and Wildlife Relationships in Old-growth Forests,* proceedings of a symposium held in Juneau, Alaska, April 12-15, 1982, American Institute of Fishery Research Biologists, pp. 213-218.

Bramble, W. C., and W. M. Sharp. 1949. Rodents as a factor in direct seeding on spoil banks in central Pennsylvania, *Journal of Forestry* 47:477-478.

Brandstrom, A. J. F. 1933. *Analysis of Logging Costs and Operating Methods in the Douglas-Fir Region,* Charles Lathrop Pack Forestry Foundation, Washington, D.C., 117 pp.

Brandt, R. W. 1961. Ash dieback in the Northeast, USDA Forest Service Northeastern

Forest Experiment Station, Station Paper No. 163, 8 pp.

Brassard, J. M., E. Audy, M. Cree, and P. Grenier. 1974. Distribution and winter habitat of moose in Quebec. *Le Naturaliste Canadien* **101**:67-80.

Brassard, J. M., E. Audy, M. Crete, and P. Grenier. 1974. Distribution and winter habitat of moose in Quebec, *Le Naturaliste Canadien* **101**:67-80.

Bratton, S. P. 1975. The effect of the European wild boar *(Sus scrofa)* on gray beech forest in Great Smoky Mountains National Park, *Ecology* **56**:1356-1366.

Bratton, S. P. 1976. Resource division in an understory herb community: Responses to temporal and microtopographic gradients, *The American Naturalist* **110**:679-693.

Braun-Blanquet, J. 1932. *Plant Sociology: The Study of Plant Communities,* (trans. G. D. Fuller and H. S. Conard), translation of 1st ed. of *Pflanzensoziologie* (1928), McGraw-Hill, New York and London, 438 pp.

Braun, E. L. 1942. Forests of the Cumberland Mountains, *Ecological Monographs* **12**:413-447.

Braun, E. L. 1950. *Deciduous Forests of Eastern North America,* Macmillan, New York, 596 pp.

Bray, J. R. 1956. Gap phase replacement in a maple-basswood forest, *Ecology* **37**:598-600.

Brazier, J. D. 1977. The effect of forest practices on quality of the harvested crop, *Forestry* **50**:49-66.

Breedy, E. C. 1981. Bird communities and forest structure in the Sierra Nevada of California, *Condor* **83**:97-105.

Brender, E. V. 1960. Growth predictions for natural stands of loblolly pine in the lower Piedmont, *Georgia Forest Research Council Report No. 6,* 7 pp.

Brender, E. V., and J. C. Barber. 1966. Influence of loblolly pine overwood on advance reproduction, USDA Forest Service, Southeastern Forest Experiment Station, Station Paper No. 62, 12 pp.

Brender, E. V., and T. C. Nelson. 1954. Behavior and control of understory hardwoods after clearcutting a piedmont pine stand, USDA Forest Service Southeastern Forest Experiment Station Paper 44, 17 pp.

Brewer, R., and P. G. Merritt. 1978. Windthrow and tree replacement in a climax beech-maple forest, *Oikos* **30**:149-152.

Briegleb, P. A. 1942. Progress in estimating trend of normality percentage in second-growth Douglas-fir, *Journal of Forestry* **40**:785-793.

Brinkman, K. A. 1955. Epicormic branching in oaks in sprout stands, USDA Forest Service Central States Forest Experiment Station Technical Paper 146, 8 pp.

Britton, N. F. 1982. A model for suppression and dominance in trees, *Journal of Theoretical Biology* **97**:691-698.

Broadfoot, W. M. 1967. Shallow-water impoundment increases soil moisture and growth of hardwoods, *Soil Science Society of America Proceedings* **31**:562-564.

Broadfoot, W. M. 1970. Tree-character relations in southern hardwood stands, USDA Forest Service Research Note SO-98, 7 pp.

Broadfoot, W. M., and H. L. Williston. 1973. Flooding effects on southern forests, *Journal of Forestry* **71**:584-587.

Brokaw, N. V. L. 1985. Treefalls, regrowth, and community structure in tropical forests. IN (S. T. A. Pickett and P. S. White, editors) *The ecology of natural disturance and patch dynamics.* Academic Press, Orlando, Florida, U.S.A.: 53-69.

Bromley, S. W. 1935. The original forest types of southern New England, *Ecological Monographs* **5**:61-89.

Brooke, R. C., E. B. Peterson, and V. J. Krajina. 1970. The subalpine mountain hemlock zone, in V. J. Krajina (ed.), *Ecology of Western North America,* Volume 2, pp. 153-349.

Brookhaven National Laboratory. 1969. *Diversity and Stability in Ecological Systems,* BNL 50175 (C-56), Biology and Medicine-TID-4500, Biology Department, Brookhaven National laboratory, Upton, New York, 264 pp.

Brooks, C. F. 1939. Hurricanes into New England, *Geography Review* 29:119-127.

Brooks, C. F. 1945. The New England hurricane of September, 1944, *Geography Review* 35:132-136.

Brooks, D. J. 1985. Public policy and long-term timber supply in the south, *Forest Science* 31:342-357.

Brooks, R. R. 1987. *Serpentine and Its Vegetation,* Volume 1, Timber Press/Dioscorides Press, Portland, Oregon, 454 pp.

Brown, C. E. 1970. A cartographic representation of spruce budworm *Choristoneura fumiferana* (Clem.), infestation in eastern Canada, 1909-1966. Canada, Department of Fisheries and Forestry, Canadian Forestry Service, Publication No. 1263.

Brown, C. L., R. G. McAlpine, and P. P. Kormanik. 1967. Apical dominance and form in woody plants: A reappraisal, *American Journal of Botany* 54:153-162.

Brown, E. R. 1985. Management of wildlife and fish habitats in forests of western Oregon and Washington. USDA Forest Service R6-F&WL-192-1985.

Brown, E. R. 1985. Management of wildlife and fish habitats in forests of western Oregon and Washington, USDA Forest Service R6-F&WL-192-1985.

Brown, J. H., Jr. 1960. The role of fire in altering the species composition of forests in Rhode Island, *Ecology* 41:310-316.

Brown, J. K., and T. E. See. 1981. Downed dead woody fuel and biomass in the northern Rocky Mountains, USDA Forest Service General Technical Report INT-117, 48 pp.

Brown, K. M., and F. R. Clarke. 1980. *Forecasting forest stand dynamics:* Proceedings of a workshop, School of Forestry, Lakehead University, Thunder Bay, Ontario, 261 pp.

Brubaker, L. B. 1988. Vegetation history and anticipating future vegetation change. IN (J. K. Agee and D. K. Johnson, editors) *Ecosystem management for parks and wilderness,* University of Washington Press, Seattle, Washington: 41-61.

Brubaker, L. B. 1991. Climate change and the origin of old-growth Douglas-fir forests in the Puget Sound Lowland. IN (L. F. Ruggiero, K. B. Aubry, A. B. Carey, and M. H. Huff, editors) *Wildlife and Vegetation of Unmanaged Douglas-fir Forests.* USDA Forest Service PNW-GTR-285: 7-24.

Bruce, R. C., and S. G. Boyce. 1984. Measurements of diversity on the Nantahala National Forest. IN (J. L. Cooley and J. H. Cooley, editors) *Natural diversity in forest ecosystems.* Proceedings of the workshop. Institute of Ecology, University of Georgia, Athens, Georgia, U.S.A. pp. 71-85.

Bruenig, E. F., and Y-w Huang. 1989. Patterns of tree species diversity and canopy structure and dynamics in humid, tropical, evergreen forests on Borneo and China. IN L. B. Holm-Nielson, I. C. Nielsen, and H. Balslev, editors. *Tropical Forests.* Academic Press, New York. pp. 75-88.

Brundtland, G. H. 1987. *Our Common Future.* World Commission on Environment and Development, United Nations, Oxford University Press, New York. 400 pp.

Bruner, M. H. 1964. Epicormic sprouting on released yellow poplar, *Journal of Forestry* 62:754-755.

Buchanan, T. S., and G. H. Englerth. 1940. Decay and other volume losses in windthrown timber on the Olympic Peninsula, Washington, USDA Technical Bulletin 733, 30 pp.

Buchholz, K. 1981. Effects of minor drainages on woody species distributions in a successional floodplain forest, *Canadian Journal of Forest Research* 11:671-676.

Buckman, H. O., and N. C. Brady. 1964. *The Nature and Properties of Soils*, Macmillan, New York, 567 pp.

Buckman, R. E. 1962. Three growing-stock density experiments in Minnesota red pine: A progress report, USDA Forest Service Lake States Forest Experiment Station Paper No. 99, 9 pp.

Buckman, R. E. 1964. Effects of prescribed burning on hazel in Minnesota, *Ecology* 45:626-629.

Buckner, E., and W. McCracken. 1978. Yellow poplar: A component of climax forests? *Journal of Forestry* 76:421-423.

Budowski, G. 1965. Distribution of tropical American rain forest species in the light of successional processes. *Turrialba* 15:40-42.

Budowski, G. 1970. The distinction between old secondary and climax species in tropical central American lowland forests, *Tropical Ecology* 11:44-48.

Buechner, H. K., and H. C. Dawkins. 1961. Vegetation change induced by elephants and fire in Murchison Falls National Park, Uganda, *Ecology* 42:752-766.

Buechner, H. K., and H. C. Dawkins. 1961. Vegetation changes induced by elephants and fire in Murchison Falls National Park, Uganda. *Ecology* 42:752-766.

Buell, J. H. 1940. Effect of season of cutting on sprouting of dogwood, *Journal of Forestry* 38:649-650.

Buell, J. H. 1945. *The prediction of growth in uneven-aged timber stands on the basis of diameter distributions*, Duke University School of Forestry Bulletin No. 11, 70 pp.

Buffo, J., L. J. Fritschen, and J. L. Murphy.1972. Direct solar radiation on various slopes from 0 to 60 degrees north latitude, USDA Forest Service Research Paper PNW 142, 74 pp.

Bullock, S. H., and R. B. Primack. 1977. Comparative experimental study of seed dispersed on animals, *Ecology* 58:681-686.

Bunnell, F. L., and G. W. Jones. 1984. Black-tailed deer and old-growth forests—a synthesis. IN (W. R. Meehan, T. R. Merrell, Jr., and T. A. Hanley, editors). *Fish and wildlife relationships in old-growth forests*. American Institute of Fisheries Research Biology. Proceedings of a Symposium held April 12-15, 1982. Juneau, Alaska.

Burel, F. 1992. Effect of landscape structure and dynamics on species diversity in hedgerow networks. *Landscape Ecology* 6(3):161-174.

Burns, G. P. 1923. Measurement of solar radiant energy in plant habitats, *Ecology* 4:189-195.

Burroughs, E. R., Jr., and B. R. Thomas. 1977. Declining root strength in Douglas-fir after felling as a factor in slope stability, USDA Forest Service Research Paper INT-90, 27 pp.

Burrows, C. J., and V. L. Burrows. 1976. *Procedures for the study of snow avalanche chronology using growth layers of woody plants*, Institute of Arctic and Alpine Research Occasional Paper 23, University of Colorado, Boulder, Colorado, 54 pp.

Busgen, M., and E. Munch. 1929. *The Structure and Life of Forest Trees* (translated by T. Thompson), 3d Ed., St. Giles' Works, Norwich, Great Britain, 436 pp.

Butts, D. B. 1985. Fire policies and programs for the National Park System, in *Symposium and Workshop on Wilderness Fire*, Missoula, November 1983, USDA Forest Service Technical Report INT-182, pp. 43-48.

Cain, S. A. 1936. Synusiae as a basis for plant sociological field work, *American Midland Naturalist* 17:665-672.

Cairns, J., Jr. (ed.). 1980. *The Recovery Process in Damaged Ecosystems,* Ann Arbor Science, Ann Arbor, Michigan, 167 pp.

Cajander, A. K. 1926. The theory of forest types, *Acta-Forestalia Fennica* 29, 108 pp.

Camp, A. E. 1995. Predicting late-successional fire refugia from physiography and topography. Unpublished Ph. D. dissertation, University of Washington, College of Forest Resources. Seattle, Washington, U.S.A. 130 pp.

Camp, A. E., P. F. Hessburg, R. L. Everett, and C. D. Oliver. 1994. Spatial changes in forest landscape patterns resulting from altered disturbance regimes on the eastern slope of the Washington Cascades. In *Fire in Wilderness and Park Management,* Proceedings of a Conference held March 30-April 1, 1993. Missoula, Montana, U.S.A.

Canham, C. D., and O. L. Loucks. 1984. Catastrophic windthrow in the presettlement forests of Wisconsin. *Ecology* 65:803-809.

Canham, C. D., and P. L. Marks. 1985. The response of woody plants to disturbance: Patterns of establishment and growth, in S. T. A. Pickett and P. S. White (eds.), *The Ecology of Natural Disturbances and Patch Dynamics,* Academic, New York, pp. 197-216.

Canham, C. D., J. S. Denslow, W. J. Platt, J. R. Runkle, T. A. Spies, and P. S. White. 1990. Light regimes beneath closed canopies and tree-fall gaps in temperate and tropical forests. *Canadian Journal of Forest Research* 20:620-631.

Cannell, M. G. R., P. Rothery, and E. D. Ford. 1984. Competition within stands of *Picea sitchensis* and *Pinus contorta, Annals of Botany* 53:349-362.

Cao, K. 1995. Fagus dominance in Chinese montane forests: natural regeneration of Fagus lucida and Fagus hayatae var. pashanica. Doctoral Thesis, Wageningen Agricultural University, Wageningen, The Netherlands: 116 pp.

Carkin, R. E., J. F. Franklin, J. Booth, and C. E. Smith. 1978. Seed habits of upper-slope tree species: IV. Seed flight of noble fir and Pacific silver fir, USDA Forest Service Research Note PNW-312, 10 pp.

Carleton, T. J., and P. F. Maycock. 1981. Understory-canopy affinities in boreal forest vegetation, *Canadian Journal of Botany* 59:1709-1716.

Carmean, W. H. 1956. Suggested modifications of standard Douglas-fir site curves for certain soils in southwestern Washington, *Forest Science* 2:242-250.

Carmean, W. H. 1970a. Site quality for eastern hardwoods, in *The silviculture of oaks and associated species,* USDA Forest Service Research Paper NE-144, pp. 36-56.

Carmean, W. H. 1970b. Tree height-growth patterns in relation to soil and site, in *Tree Growth and Forest Soils, Proceedings of the Third North American Forest Soils Conference,* Oregon State University Press, Corvallis, pp. 499-511.

Carter, R. E., G. Hoyer, and C. D. Weber. 1986a. Factors affecting stem form in immature Douglas-fir stands, in C. D. Oliver, D. Hanley, and J. Johnson (eds.), *Douglas-fir: Stand Management for the Future,* Institute of Forest Resources Contribution No. 55, College of Forest Resources, University of Washington, Seattle, pp. 230-238.

Carter, R. E., I. M. Miller, and K. Klinka. 1986b. Relationships between growth form and stand density in immature Douglas-fir, *Forestry Chronicle* 62:440-445.

Carvell, K. L., and C. F. Korstian. 1955. Production and dissemination of yellow-poplar seed, *Journal of Forestry* 53:169-170.

Carvell, K. L., and E. H. Tryon. 1959. Herbaceous vegetation and shrubs characteristic of oak sites in West Virginia. *Castanea* 24:39-43.

Carvell, K. L., and E. H. Tryon. 1961. The effect of environmental factors on the abundance of oak regeneration beneath mature oak stands. *Forest Science* 7:98-105.

Carvell, K. L., E. H. Tryon, and R. P. True. 1957. Effects of glaze on the development of Appalachian hardwoods, *Journal of Forestry* **55**:130-132.

Caswell, H. 1978. Predator-mediated coexistence: A non-equilibrium model, *American Naturalist* **112**:127-154.

Cates, R. G., and G. H. Orians. 1975. Successional status and the palatability of plants to generalized herbivores, *Ecology* **56**:410-418.

Cattelino, P. J., I. R. Noble, R. O. Slatyer, and S. R. Kessell. 1979. Predicting the multiple pathways of plant succession, *Environmental Management* **3**:41-50.

Caughley, G. 1970. Eruption of ungulate populations, with emphasis on Himalayan thar in New Zealand. *Ecology* **51**:53-72.

Chabot, B. F., and H. A. Mooney (eds.). 1985. *Physiological Ecology of North American Plant Communities*, Chapman and Hall, New York, 351 pp.

Chabreck, R. H., and A. W. Palmisano. 1973. The effects of Hurricane Camille on the marshes of the Mississippi River delta, *Ecology* **54**:1118-1123.

Chandler, C., P. Cheney, P. Thomas, L. Trabaud, and D. Williams. 1983a. *Fire in Forestry* Volume I: Forest Fire Behavior and Effects, John Wiley, New York, 450 pp.

Chandler, C., P. Cheney, P. Thomas, L. Trabaud, and D. Williams. 1983b. *Fire in Forestry* Volume II: Forest Fire Management and Organization, John Wiley, New York, 298 pp.

Chang, S. J. 1984. A simple production function model for variable density growth and yield modeling. *Canadian Journal of Forest Research* **14**:783-788.

Chapman, H. H. 1926. *Factors determining natural reproduction of longleaf pine on cutover lands in LaSalle Parish, Louisiana*, Yale University School of Forestry Bulletin No. 16, 44 pp.

Chapman, H. H. 1942. *Management of loblolly pine in the pine-hardwood region in Arkansas and in Louisiana west of the Mississippi River*, Yale University School of Forestry Bulletin 49, 150 pp.

Chapman, H. H. 1944a. Natural reproduction of pines in east-central Alabama (review), *Journal of Forestry* **42**:613-614.

Chapman, H. H. 1944b. Fire and pines, *American Forests* **50**:62-64, 91-93.

Chapman, H. H., and D. B. Demeritt. 1932. *Elements of Forest Mensuration*, J. B. Lyon, Albany, New York, 452 pp.

Chen, J., and J. F. Franklin. 1992. Vegetation responses to edge environments in old-growth Douglas-fir forests. *Ecological Applications* **2**(4):387-396.

Chen, J., J. F. Franklin, and T. A. Spies. 1993. Contrasting microclimates among clearcut, edge, and interior of old-growth Douglas-fir forest. *Agricultural and Forest Meteorology* **63**:219-237.

Childs, T. W., and J. W. Edgren. 1967. Dwarf mistletoe effects on ponderosa pine growth and trunk form, *Forest Science* **13**:167-174.

Chow, Ven Te (ed. in chief). 1964. *Handbook of Applied Hydrology*, McGraw-Hill, New York.

Christensen, N. L. 1977. Changes in structure, pattern, and diversity associated with climax forest maturation in piedmont, North Carolina, *American Midland Naturalist* **97**:176-188.

Christensen, N. L., J. K. Agee, P. F. Brussard, J. Hughes, D. H. Knight, G. W. Minshall, J. M. Peek, S. J. Pyne, F. J. Swanson, J. W. Thomas, S. Wells, S. E. Williams, and H. A. Wright. 1989. Interpreting the Yellowstone fires of 1988. *BioScience* **39**:678-685.

Christy, E. J., and R. N. Mack. 1984. Variation in demography of juvenile *Tsuga hetero-*

phylla across the substratum mosaic, *Journal of Ecology* 72:75-91.

Chung, H-H., and R. L.. Barnes. 1977. Photosynthate allocation in *Pinus taeda*: I. Substrate requirements for synthesis of shoot biomass, *Canadian Journal of Forest Research* 7:106-111.

Churchill, E. D., and H. C. Hanson. 1958. The concept of climax in arctic and alpine vegetation, *Botanical Review* 24:127-191.

Clapp, R. T. 1938. The effects of the hurricane upon New England forests. *Journal of Forestry* 36:1177-1181.

Clark, F. B. 1970. Measures necessary for natural regeneration of oaks, yellow poplar, sweetgum, and black walnut, in *The Silviculture of Oaks and Associated Species*, USDA Forest Service Station Paper NE-144, pp. 1-16.

Clark, F. B., and F. G. Liming. 1953. Sprouting of blackjack oak in the Missouri Ozarks, USDA Forest Service Central States Forest Experiment Station Technical Paper 137, 22 pp.

Clark, F. B., and S. G. Boyce. 1964. Yellow-poplar seed remains viable in the forest litter, *Journal of Forestry* 62:564-567.

Clark, P. J., and F. C. Evans. 1954. Distance to nearest neighbor as a measure of spatial relationship in populations, *Ecology* 35:445-453.

Clatterbuck, W. K. and J. D. Hodges. 1988. Development of cherrybark oak and sweetgum in mixed, even-aged bottomland stands in central Mississippi, U.S.A., *Canadian Journal of Forest Research* 18:12-18.

Clatterbuck, W. K., C. D. Oliver, and E. C. Burkhardt. 1987. The silvicultural potential of mixed stands of cherrybark oak and American sycamore: Spacing is the key, *Southern Journal of Forestry* 11:158-161.

Clatterbuck, W. K., J. D. Hodges, and E. C. Burkhardt. 1985. Cherrybark oak development in natural mixed oak-sweetgum stands—preliminary results, in E. Shoulders (ed.), *Proceedings of the Third Biennial Silvicultural Research Conference*, USDA Forest Service General Technical Report SO-54, pp. 438-444.

Clausen, J. J., and T. T. Kozlowski. 1965. Heterophyllous shoots in *Betula papyrifera*, *Nature* 205:1030-1031.

Cleary, B. D., and R. H. Waring. 1969. Temperature: Collection of data and its analysis for the interpretation of plant growth and distribution, *Canadian Journal of Botany* 47:167-173.

Clements, F. E. 1905. *Research Methods in Ecology*, University Publishing Company, Lincoln, Nebraska, 334 pp.

Clements, F. E. 1916. *Plant Succession: An Analysis of the Development of Vegetation*, Carnegie Institute, Washington, D.C., 512 pp.

Clements, F. E. 1936. Nature and structure of the climax, *Journal of Ecology* 24:252-284.

Cline, A. C., and C. R. Lockard. 1925. *Mixed white pine and hardwood*, The Harvard Forest, Harvard University, Harvard Forest Bulletin No. 8, 36 pp.

Cline, A. C., and S. H. Spurr. 1942. *The virgin upland forest of Central New England: A study of old growth stands in the Pisgah Mountain section of southwestern New Hampshire*, The Harvard Forest, Harvard University, Harvard Forest Bulletin 21, 58 pp.

Cline, S., A. B. Berg, and H. M. Wright. 1980. Snag characteristics and dynamics of Douglas-fir forests, western Oregon, *Journal of Wildlife Management* 44:773-786.

Clinton, G. P., and F. A. McCormick. 1936. *Dutch elm disease, Graphium ulmi*, The Connecticut Agricultural Experiment Station Bulletin 389, New Haven, Connecticut, pp. 701-752.

Clutter, J. L. 1963. Compatible growth and yield models for loblolly pine, *Forest Science* 9:354-370.

Cobb, D. F. 1988. Development of mixed western larch, lodgepole pine, Douglas-fir, grand fir stands in eastern Washington, unpublished master's thesis, College of Forest Resources, University of Washington, Seattle, 99 pp.

Cobb, D. F., K. L. O'Hara, and C. D. Oliver. 1993. Effects of variations in stand structure on development of mixed-species stands in eastern Washington. *Canadian Journal of Forest Research* 23:545-552.

Cochran, P. H. 1969. Lodgepole pine clearcut: Size affects minimum temperatures near the soil surface, USDA Forest Service Research Paper PNW-86, 9 pp.

Cochran, P. H. 1982. Stocking levels for east-side white or grand fir, in C. D. Oliver and R. M. Kenady (eds.), *Biology and Management of True Fir in the Pacific Northwest,* University of Washington College of Forest Resources, Institute of Forest Resources Contribution Number 45, pp. 185-189.

Cochrane, L. A., and E. D. Ford. 1978. Growth of a sitka spruce plantation: Analysis and stochastic description of the development of the branching structure, *Journal of Ecology* 15:227-244.

Coile, T. S. 1937. Distribution of forest tree roots in North Carolina piedmont soils, *Journal of Forestry* 35:247-267.

Coile, T. S. 1948. *Relation of soil characteristics to site index of loblolly and shortleaf pines in the Lower Piedmont region of North Carolina,* Duke University School of Forestry Bulletin 13, 78 pp.

Coile, T. S. 1952. Soil and the growth of forests, *Advances in Agronomy* 4:329-398.

Cole, D. 1977. Ecosystem dynamics in the coniferous of the Willamette Valley, Oregon, U.S.A., *Journal of Biogeography* 4:181-192.

Cole, D. M., and C. E. Jensen. 1982. Models for describing vertical crown development of lodgepole pine stands, USDA Forest Service Research Paper INT-292, 10 pp.

Cole, D. W., S. P. Gessel, and S. F. Dice. 1967. Distribution and cycling of nitrogen, phosphorus, potassium and calcium in a second-growth Douglas-fir ecosystem, in *Symposium on Primary Productivity and Mineral Cycling in Natural Ecosystems,* University of Maine Press, Orono, pp. 197-232.

Cole, W. E., and G. D. Amman. 1969. Mountain pine beetle infestations in relation to lodgepole pine diameters, USDA Forest Service Research Note INT-95, 7 pp.

Cole, W. E., and G. D. Amman. 1980. Mountain pine beetle dynamics in lodgepole pine forests: I. Course of an infestation, USDA Forest Service General Technical Report INT-89, 56 pp.

Coleman, J. M., and L. D. Wright. 1975. Modern river deltas: variability of processes and sand bodies, in M. L. Broussard (ed.), *Deltas—Models for Exploration,* Houston Geological Society, Houston Texas, pp. 99-150.

Coleman, J. M., and S. M. Gagliano. 1964. Cyclic sedimentation in the Mississippi River deltaic plain, *Transactions of the Gulf Coast Association of Geological Societies* 14:67-82.

Coley, P. D., J. P. Bryant, and F. S. Chapin III. 1985. Resource availability and plant antiherbivore defense, *Science* 230:895-899.

Collins, B. S., K. P. Dunne, and S. T. A. Pickett. 1985. Responses of forest herbs to canopy gaps, in S. T. A. Pickett and P. S. White (eds.), *The Ecology of Natural Disturbance and Patch Dynamics,* Academic, New York, pp. 217-234.

Connell, J. H. 1978. Diversity in tropical rain forests and coral reefs, *Science* 199:1302-1310.

Connell, J. H., and R. D. Slatyer. 1977. Mechanisms of succession in natural communities and their role in community stability and organization, *American Naturalist* **111**:1119-1144.

Conner, R. N., and K. A. O'Halloran. 1987. Cavity-tree selection by red-cockaded woodpeckers as related to growth dynamics of southern pines. *Wilson Bulletin* **99**:398-412.

Cook, D. B., and S. B. Robeson. 1945. Varying hare and forest succession, *Ecology* **26**:406-410.

Cook, E. R., A. H. Johnson, and T. J. Blasing. 1987. Modelling the climate effect in tree rings for forest decline studies, *Tree Physiology* **31**:27-40.

Cook, R. E. 1978. Fragile blossoms of spring aren't shrinking violets, *Smithsonian* **8(12)**:64-70.

Cook, S. A. 1982. Stand development in the presence of a pathogen, *Phellinus weirii,* in E. Means (ed.), *Forest Succession and Stand Development Research in the Northwest,* Forest Research Laboratory, Oregon State University, Corvallis, pp. 159-163.

Cooper, C. F. 1960. Changes in vegetation, structure and growth of southwestern pine forest since white settlement, *Ecological Monographs* **30**:129-164.

Cooper, C. F. 1961. Pattern in ponderosa pine forests, *Ecology* **42**:493-499.

Cooper, R. W. 1972. Fire and sand pine, in *Sand Pine Symposium Proceedings,* USDA Forest Service Southeastern Forest Experiment Station, Asheville, North Carolina, pp. 207-212.

Cooper, W. S. 1911. Reproduction by layering among conifers, *Botanical Gazette* **52**:369-379.

Cooper, W. S. 1923. The recent ecological history of Glacier Bay, Alaska. II. The present vegetation cycle, *Ecology* **4**:223-246.

Cooper, W. S. 1931. A third expedition to Glacier Bay, Alaska, *Ecology* **12**:61-95.

Cornaby, B. W., and J. B. Waide. 1973. Nitrogen fixation in decaying chestnut logs, *Plant Soil* **39**:445 448.

Cotton, C. A. 1942. *Climatic Accidents in Landscape-Making,* Whitcombe and Tombs, London, 354 pp.

Coulson, R. N., F. P. Hain, and T. L. Payne. 1974. Radial growth characteristics and stand density of loblolly pine in relation to the occurrence of southern pine beetle, *Environmental Entomology* **3**:425-428.

Covington W. W., and S. S. Sackett. 1986. Effect of periodic burning on soil nitrogen concentrations in ponderosa pine, *Soil Science Society of America Journal* **50**:452-457.

Covington, W. W., and M. M. Moore. 1994. Postsettlement changes in natural fire regimes and forest structure: Ecological restoration of old-growth ponderosa pine forests. *Journal of Sustainable Forestry* **2**:153-182.

Cowles, H. C. 1911. The causes of vegetative cycles, *Botanical Gazette* **51**:161-183.

Craighead, F. C. 1924. General bionomics and possibility of prevention and control. IN Studies on the spruce budworm (*Cacoecia fumiferana* Clem.). Canada, Department of Agriculture, Bulletin (new series) No. 37.

Crandell, D. R. 1971. Postglacial lahars from Mount Rainier Volcano, Washington, U.S. *Geological Survey Professional Paper* 677,75 pp.

Crawford, P. D., C. D. Oliver, and J. F. Franklin. 1982. Ecology and silviculture of Pacific silver fir: 1982. Part II of Final Report for Cooperative Agreement Number 155, USDA Forest Service PNW Experiment Station and University of Washington College of Forest Resources, 260 pp.

Critchfield, W. B. 1960. Leaf dimorphism in *Populus trichocarpa, American Journal of*

Botany **47**:699-711.

Croker, T. C. 1952. Early release stimulates cone production, USDA Forest Service Southern Forest Experiment Station, Southern Forestry Notes 79.

Croker, T. C., Jr. 1958. Soil depth affects windfirmness of longleaf pine, *Journal of Forestry* **56**:432.

Croker, T. C., Jr., and W. D. Boyer. 1975. Regenerating longleaf pine naturally, USDA Forest Service Research Paper SO-105.

Cromer, D. A. N., and C. K. Pawsey. 1957. Initial spacing and growth of *Pinus radiata, Forest Timber Bureau of Australia Bulletin* 36, 46 pp.

Crouch, G. L. 1974. Interaction of deer and forest succession on clearcuttings in the Coast Range of Oregon, in H. C. Black (ed.), *Wildlife and Forest Management in the Pacific Northwest,* School of Forestry, Oregon State University, Corvallis, pp. 133-138.

Crow, T. R. 1980. A rainforest chronicle: A 30-year record of change in structure and composition at El Verde, Puerto Rico, *Biotropica* **12**:42-55.

Crow, T. R. 1988. Reproductive mode and mechanisms for self-replacement of northern red oak *(Quercus ruhra)* (a review). *Forest Science* **34**:19-40.

Curtis, J. T. 1956. The modification of mid-latitude grasslands and forests by man, in W. L. Thomas (ed.), *Man's Role in Changing the Face of the Earth,* University of Chicago Press, Chicago, pp. 721-736.

Curtis, J. T., and R. P. McIntosh. 1951. An upland forest continuum in the prairie-forest border region of Wisconsin, *Ecology* **32**:476-496.

Curtis, R. O. 1970. Stand density measures: An interpretation, *Forest Science* **16**:403-414.

Curtis, R. O. 1971. A tree area power function and related stand density measures for Douglas-fir, *Forest Science* **17**:146-159.

Curtis, R. O. 1982. A simple index of stand density for Douglas-fir, *Forest Science* **18**:92-94.

Curtis, R. O., and D. D. Marshall. 1986. Growth-growing stock relationships and recent results from the levels-of-growing-stock studies, in C. D. Oliver, D. P. Hanley, and J. A. Johnson (eds.), *Douglas-Fir: Stand Management for the Future,* University of Washington College of Forest Resources, Institute of Forest Resources Contribution No. 55, pp. 281-289.

Curtis, R. O., and D. L. Reukema. 1970. Crown development and site estimates in a Douglas-fir plantation spacing test, *Forest Science* **16**:287-301.

Cushman, M. J. 1976. Vegetation composition as a predictor of major avalanche cycles, North Cascades, Washington, unpublished master's thesis, University of Washington, Seattle, 99 pp.

Cushman, M. J. 1981. The influence of recurrent snow avalanches on vegetation patterns in the Washington Cascades, unpublished Ph.D. dissertation, University of Washington, Seattle, 175 pp.

Dale, M. E. 1975. Effects of removing understory on growth of upland oak, USDA Forest Service Research Paper NE-321, 10 pp.

Dalrymple, T. 1964. Hydrology of flow control, in V. T. Chow (ed.), *Handbook of Hydrology,* McGraw-Hill, New York, pp. 25.1-25.124.

Daniel, T. W. 1980. The middle and southern Rocky Mountain region, in J. W. Barrett (ed.), *Regional Silviculture of the United States,* 2d ed., John Wiley, New York, pp. 277-340.

Daniel, T. W., J. A. Helms, and F. S. Baker. 1979. *Principles of Silviculture,* 2d Ed. McGraw-Hill, 500 pp.

Daniels, R. F. 1976. Simple competition indices and their correlation with annual loblolly pine tree growth, *Forest Science* **22**:454-456.

Daniels, R. R. 1978. Spatial patterns and distance distributions in young seeded loblolly pine stands, *Forest Science* **24**:260-266.

Dansereau, P. M. 1946. L'erabliere Laurentienne: II Les successions et leurs indicateurs, *Canadian Journal of Research* **24**:235-291.

Dansereau, P. M. 1951. Description and recording of vegetation upon a structural basis, *Ecology* **32**:172-229.

Dansereau, P. M. 1957. *Biogeography: An Ecological Perspective*, Ronald Press, New York, 394 pp.

Daubenmire, R. 1952. Forest vegetation of northern Idaho and adjacent Washington and its bearing on concepts of vegetation classification, *Ecological Monographs* **22**:301-330.

Daubenmire, R. 1966. Vegetation: Identification of typal communities, *Science* **151**:291-298.

Daubenmire, R. 1968. *Plant Communities: A Textbook of Plant Synecology*, Harper and Row, New York, 300 pp.

Daubenmire, R., and J. B. Daubenmire. 1968. Forest vegetation of eastern Washington and Northern Idaho, Washington Agricultural Experiment Station Bulletin 60, 104 pp.

Davis, G., and A. C. Hart. 1961. Effect of seedbed preparation on natural reproduction of spruce and hemlock under dense shade. USDA Forest Service, Northeastern Forest Experiment Station, Station Paper 160, 12 pp.

Davis, J. B. 1984. Burning another empire, *Fire Management Notes* **45**(4):12-17.

Davis, J. B. 1984. Burning another empire. *Fire Management Notes* **45**(4):12-17.

Davis, J. W., G. A. Goodwin, and R. A. Ockenfels (eds.). 1983. *Snag Habitat Management: Proceedings of a Symposium*, USDA Forest Service General Technical Report RM-99, 226 pp.

Davis, K. M. 1980. Fire history of a western larch/Douglas-fir forest type in northeastern Montana, in *Proceedings of the Fire History Workshop*, October 20-24, 1980, Tucson, Arizona, USDA Forest Service General Technical Report RM-81, pp. 69-74.

Davis, K. P. 1955. Forest growing stock, *Journal of Forestry* **53**:735-739.

Davis, K. P. 1956. Determination of desirable growing stock—a central problem of forest management, *Journal of Forestry* **54**:811-815.

Davis, K. P. 1966. *Forest Management, Regulation, and Valuation*, 2d ed., McGraw-Hill, New York, 519 pp.

Davis, L. S., and K. N. Johnson. 1987. *Forest Management*, 3d ed., McGraw-Hill, New York, 790 pp.

Davis, M. B. 1976. Pleistocene biogeography of temperate deciduous forests, *Geoscience and Man* **13**:13-26.

Davis, M. B. 1981. Quaternary history and the stability of forest communities, in D. C. West, H. H. Shugart, and D. B. Botkin (eds.), *Forest Succession: Concepts and Application*, Springer-Verlag, New York, pp. 132-153.

Davis, M. B. 1986. Climatic instability, time lags, and community disequilibrium. IN (J. Diamond and T. J. Case, editors) *Community Ecology*, Harper and Row, New York: 269-284.

Davis, M. B., R. W. Spear, and L. C. K. Shane. 1980. Holocene climate of New England, *Quaternary Research* **143**:240-250.

Davis, P. H. 1965. *Flora of Turkey*, Volume 1, Edinburgh University Press, Edinburgh,

Scotland, 567 pp.

Davis, T. A. W., and P. W. Richards. 1933. The vegetation of Moraballi Creek, British Guiana: An ecological study of a limited area of tropical rain forest, part I, *Journal of Ecology* **21**:350-384.

Davis, V. B. 1935. Rate of growth in a selectively logged stand in the bottomland hardwoods, USDA Forest Service, Southern Forest Experiment Station Occasional Paper No. 41, 9 pp.

Dawson, J. W., and B. V. Sneddon. 1969. The New Zealand rain forest: A comparison with tropical rain forest, *Pacific Science* **23**:131-147.

Day, G. M. 1953. The Indian as an ecological factor in the Northeastern forest, *Ecology* **34**:329-347.

Day, M. W., and V. J. Rudolph. 1972. Thinning plantation red pine, *Michigan Agriculture Experiment Station Research Report* 151, Michigan State University, East Lansing, 10 pp.

Day, R. J. 1972. Stand structure, succession, and use of southern Alberta's Rocky Mountain forest, *Ecology* **53**:472-478.

Deal, R. L. 1987. Development of mixed western hemlock-sitka spruce stands on the Tongass National Forest, unpublished master's thesis, University of Washington College of Forest Resources, Seattle, 116 pp.

Deal, R. L., C. D. Oliver and B. T. Bormann. 1991. Reconstruction of mixed hemlock-spruce stands in coastal southeast Alaska. *Canadian Journal of Forest Research* **21**:643-654.

Deam, C. C. 1921. *Trees of Indiana,* Fort Wayne Printing Co., Department of Conservation, State of Indiana, Publication No. 13, 317 pp.

Dean, T. J., J. N. Long, and F. W. Smith. 1988. Bias in leaf area-sapwood ratios and its impact on growth analysis in *Pinus contorta, Trees* **2**:104-109.

DeBell, D. S. 1971. Phytotoxic effects of cherrybark oak, *Forest Science* **17**:180-185.

DeBell, D. S. 1989. Alternative silvicultural systems—West. IN *Proceedings of the national silviculture workshop*, USDA Forest Service, Washington, D. C.: 23-28.

DeBell, D. S., and J. F. Franklin. 1987. Old-growth Douglas-fir and western hemlock: A 36-year record of growth and mortality. *Western Journal of Forestry* **2**:111-114.

DeBell, D. S., C. D. Whitesell, and T. H. Schubert. 1989. Using N2-fixing Albizia to increase gowth of Eucalyptus plantations in Hawaii. *Forest Science* **35**:64-75.

DeBell, D. S., J. Stubbs, and D. D. Hook. 1968b. Stand development after a selection cutting in a hardwood bottom, *Southern Lumberman* **217**:126-128.

DeBell, D. S., O. G. Langdon, and J. Stubbs. 1968a. Reproducing mixed hardwoods by a seed-tree cutting in the Carolina coastal plain, *Southern Lumberman* **217**:121-123.

Decker, J. P. 1947. The effect of air supply on apparent photosynthesis, *Plant Physiology* **22**:561-571.

Deen, J. L. 1933. *Some aspects of an early expression of dominance in white pine (Pinus strobes L.),* Yale University School of Forestry Bulletin No. 36, 34 pp.

Deer, T. 1982. True fir management—Cascade silver fir stands, in C. D. Oliver and R. M. Kenady (eds.), *Biology and Management of True Fir in the Pacific Northwest,* University of Washington College of Forest Resources, Institute of Forest Resources Contribution No. 45, pp. 311-314.

DeLaune, R. D., C. J. Smith, W. H. Patrick, Jr., and H. H. Roberts. 1987. Rejuvenated marsh and bay-bottom accretion on the rapidly subsiding coastal plain of U.S. Gulf Coast: A second-order effect of the emerging Atchafalaya Delta, *Estuarine, Coastal and Shelf Science* **25**:381-389.

DeLaune, R. D., R. H. Baumann, and J. G. Gosselink. 1983. Relationships among vertical accretion, coastal submergence, and erosion in a Louisiana Gulf Coast marsh, *Journal of Sedimentary Petrology* **53**:147-157.

DeLiocourt, F. 1898. De l'amenagement des Sapinieres, *Bulletin of the Society of Foresters*, Franch-Comte Belfort, Besancon, pp. 396-409.

Della-Bianca, L., and R. E. Dils. 1960. Some effects of stand density in a red pine plantation on soil moisture, soil temperature, and radial growth, *Journal of Forestry* **58**:373-377.

Demaree, D. 1932. Submerging experiments with Taxodium, *Ecology* **13**:258-262.

Dennis, W. M., and W. T. Batson. 1974. The floating log and stump communities in the Santee Swamp of South Carolina, *Castanea* **39**:166-170.

Denny, C. S., and J. C. Goodlett. 1956. Microrelief resulting from fallen trees, in C. S. Denny (ed.), *Surficial Geology and Geomorphology of Potter County, Pennsylvania*, United States Geological Survey Professional Paper No. 228, pp. 59-66.

Derr, H. J., and T. W. Melder. 1970. Brown-spot resistance in longleaf pine, *Forest Science* **16**:204-209.

Devevan, W. M. 1992. The pristine myth: the landscape of the Americas in 1492. *Annals of the Association of American Geographers*. Vol. 82, No. 3.

Devoe, N. N. 1989. Differential seeding and regeneration in openings beneath closed canopy in subtropical wet forest. Unpublished D. For. dissertation, Yale University, School of Forestry and Environmental Studies. 316 ppp.

DeWalle, D. R., and S. G. McGuire. 1973. Albedo variations of an oak forest in Pennsylvania, *Agricultural Meteorology* **11**:107-113.

DeWiest, R. J. M. 1965. *Geohydrology*, John Wiley, New York, 366 pp.

Di Castri, F., A. J. Hansen, and M. M. Holland (eds.). 1988. A New Look at Ecotones, *Biology International* 17, (Special Issue), 162 pp.

Diamond, J., and T. J. Case. 1986. *Community Ecology*, Harper and Row, New York, 665 pp.

Dick, J. 1955. Studies of Douglas-fir seed flight in southwestern Washington, *Weyerhaeuser Timber Company Forestry Research Note* 12, Tacoma, Washington, 4 pp.

Dickerson, B. 1995. Elements of 19th-century Romanticism in contemporary forest management practices. *Journal of Forestry* 93(9): 36-40.

Dickman, A., and S. Cook 1989. Fire and fungus in a mountain hemlok forest. Canadian *Journal of Botany* **67**:2005-2016.

Dieterich, J. H. 1980. Chimney Spring forest fire history, USDA Forest Service Research Paper RM-220, 8 pp.

Dix, R. L. 1957. Sugar maple in the climax forests of Washington, D.C., *Ecology* **38**:663-665.

Dixon, A. F. G. 1971. The role of aphids in wood formation: I. The effect of the sycamore aphid, *Drepanosiphum platanoides* (Schr.)(Aphidae), on the growth of sycamore, *Acer pseudoplatanus* (L.), *Journal of Applied Ecology* **8**:165-179.

Donnelly, G. T., and P. G. Murphy. 1989. Episodic growth response to canopy gaps and relative shade-tolerance *of Fagus grandifolia* and *Acer saccharum*, Department of Botany and Plant Pathology, Michigan State University, East Lansing (submitted for publication).

Doolittle, W. T. 1958. Site index comparisons for several forest species in the southern Appalachians, *Soil Science Society of America Proceedings* **22**:455-458.

Douglass, J. E. 1960. Soil moisture distribution between trees in a thinned loblolly pine plantation, *Journal of Forestry* **58**:221-222.

Downs, A. A. 1938. Glaze damage in the birch-beech-maple-hemlock type of Pennsylvania and New York, *Journal of Forestry* 36:63-70.

Dransfield, John. 1978. Growth forms of rain forest palms. IN P. B. Tomlinson and M. H. Zimmerman, editors. *Tropical Trees as Living Systems*. Cambridge University Press, Cambridge. pp. 247-268.

Drew, T. J., and J. W. Flewelling. 1977. Some recent Japanese theories of yield-density relationships and their application to Monterey pine plantations, *Forest Science* 23:517-534.

Drew, T. J., and J. W. Flewelling. 1979. Stand density management: An alternative approach and its application to Douglas-fir plantations, *Forest Science* 25:518-532.

Driver, C . H., and R. E . Wood. 1968. Occurrence of *Fomes annosus* in intensively managed young-growth stands of western hemlock, *Plant Disease Reporter* 52:370-372.

Drolet, J. C., R. M. Newnham, and T. B. Tsay. 1972. *Branchiness of jack pine, black spruce, and balsam fir in relation to mechanized delimbing*, Forest Management Institute, Ottawa, Canada, Information Report FMR-S-34, 34 pp.

Drury, W. H., and I. C. T. Nisbet. 1973. Succession, *Journal of the Arnold Arboretum* 54:331-368.

Duerr, W. A., and W. E. Bond. 1952. Optimum stocking of a selection forest, *Journal of Forestry* 50: 12-16.

Duffield, J. W. 1972. Clearcutting, *Science* 165:1288.

Duvendeck, J. P. 1962. The value of acorns in the diet of Michigan deer, *Journal of Wildlife Management* 26:371-379.

Dyrness, C. T. 1973. Early stages of plant succession following logging and burning in the western Cascades of Oregon, *Ecology* 54:57-69.

Eagleman, J. R. 1976. *The Visualization of Climate*, Lexington Books, Lexington, Massachusetts, 227 pp.

Edwards, C. R. 1986. The human impact on the forest in Quintana Roo, Mexico. *Journal of Forest History*, July, 1986: 120-127.

Edwards, D. G. 1976. Seed physiology and germination of western hemlock, in W. A. Atkinson and R. J. Zasoski (eds.), *Western Hemlock Management Conference*, College of Forest Resources, University of Washington, Institute of Forest Products Contribution No. 34, pp. 87-102.

Eggler, W. A. 1971. Quantitative studies of vegetation on sixteen young lava flows on the Island of Hawaii, *Tropical Ecology* 12:66-100.

Egler, F. E. 1940. Berkshire plateau vegetation, Massachusetts, *Ecological Monographs* 10:146-192.

Egler, F. E. 1954. Vegetation science concepts: I. Initial floristic composition: A factor in old-field vegetation development, *Vegetatio* 4:412-417.

Ehrenfeld, J. G. 1980. Understory response to canopy gaps of varying size in a mature oak forest, *Bulletin of the Torrey Botanical Society* 107:29-41.

Eis, S. 1986. Differential growth of individual components of trees and their interrelationships, *Canadian Journal of Forest Research* 16:352-359.

Eis, S., E. H. Harman, and L. F. Ebell. 1965. Relation between cone production and diameter increment of Douglas-fir (*Pseudotsuga menziesii*[Mirb.] Franco), grand fir (*Abies grandis* [Dougl.] Lindl.), and western white pine (*Pings monticola* Dougl.), *Canadian Journal of Botany* 43:1553-1559.

Elias, T. S. 1977. Threatened and endangered species problems in North America: An overview, in G. T. Prance and T. S. Elias (eds.), *Extinction is Forever* New York Botanical Garden, New York, pp. 13-16.

Emery, K. O. 1967. *The Atlantic continental margin of the United States during the past*

70 million years, Geological Association of Canada, Special Paper Number 4, pp. 65-66.

Emmingham, W. H., and D. Hanley. 1986. Alternative thinning regimes used by private, nonindustrial landowners, in C. D. Oliver, D. Hanley, and J. Johnson (eds.), *Douglas-Fir: Stand Management for the Future,* University of Washington College of Forest Resources Institute of Forest Resources Contribution No. 56, pp. 327-334.

Emmingham, W. H., and N. M. Halverson. 1982. Community types, productivity, and reforestation: Management implications for the Pacific silver fir zone of the Cascade Mountains, in C. D. Oliver and R. M. Kenady (eds.), *Biology and Management of True Fir in the Pacific Northwest,* University of Washington College of Forest Resources, Institute of Forest Resources Contribution No. 45, pp. 291-303.

Engelhardt, N. T. 1957. Pathological deterioration of looper-killed western hemlock on southern Vancouver Island, *Forest Science* 3:125-136.

Engle, L. G. 1960. Yellow-poplar seed fall pattern, USDA Forest Service Central Forest Experiment Station, Station Note 143, 2 pp.

Englerth, G. H. 1942. *Decay of western hemlock in western Oregon and Washington,* Yale University School of Forestry Bulletin No. 50, 50 pp.

Englerth, G. H., and L. A. Isaac. 1944. Decay of western hemlock following logging injury, *Timberman* 45:34-35, 56.

Erdmann, G. G., R. M. Godman, and R. R. Oberg. 1975. Crown release accelerates diameter growth and crown development of yellow birch saplings, USDA Forest Service Research Paper NC-117, 9 pp.

Ericsson, A., S. Larsson, and O. Tenow. 1980. Effects of early and late season defoliation on growth and carbohydrate dynamics in Scots pine, *Journal of Applied Ecology* 17:747-769.

Ernst, R. L., and W. H. Knapp. 1985. Forest stand density and stocking: Concepts, terms, and the use of stocking guides, USDA Forest Service General Technical Report WO-44, 8 pp.

Esau, K. 1965. *Plant Anatomy,* 2d. Ed., John Wiley, New York, 767 pp.

Everett, R., P. Hessburg, M. Jensen, and B. Bormann. 1994. Volume I: Executive summary. USDA Forest Service PNW-GTR-317. 61 pp.

Everett, Richard. 1993. Eastside forest ecosystem health Assessment. USDA Forest Service, National Forest System, Forest Service Research. Wenatchee Forest Sciences Laboratory, Wenatchee, Washington, U.S.A. 5 Volumes.

Everitt, B. L. 1968. Use of the cottonwood in an investigation of the recent history of a floodplain, *American Journal of Science* 266:417-440.

Eversole, K. R. 1955. Spacing tests in a Douglas-fir plantation, *Forest Science* 1:14-18.

Exell, A. W. 1931. The longevity of seeds, *Gardener's Chronicle* 89:283.

Eyre, F. H. 1938. Can jack pine be regenerated without fire? *Journal of Forestry* 36:1067-1072.

Eyre, F. H., and W. M. Zillgitt. 1953. Partial cuttings in northern hardwoods of the Lake States, USDA Tecchnical Bulletin 1076, 124 pp.

Fahnestock, G. R. 1959. When will the bottom lands burn? *Forests and People,* 3d quarter, 3 pp.

Fahnestock, G. R. 1976. Fires, fuels, and flora as factors in wilderness management: The Pasayten case, *Tall Timbers Fire Ecology Conference,* 15:155-170.

Fahnestock, G. R., and J. K. Agee. 1983. Biomass consumption and smoke production by prehistoric and modern forest fires in western Washington, *Journal of Forestry* 81:653-657.

Fauss, D. L., and N. R. Pierce. 1969. Stand conditions and spruce budworm damage in a western Montana forest, *Journal of Forestry* **67**:322-325.

Felix, A. C. III, T. L. Sharik, B. S. McGinnes, and W. C. Johnson. 1983. Succession in loblolly pine plantations converted from second-growth forest in the central Piedmont of Virginia, *American Midland Naturalist* **110**:365-380.

FEMAT. 1993. Draft supplemental environmental impact statement on Management of habitat for late-successional and old-growth forest related species within the range of the northern spotted owl. (Volume and Appendix A) USDA Forest Service, USDI Bureau of Land Management.

Fenton, R. H. 1964. Production and distribution of sweetgum seed in 1962 by four New Jersey stands, USDA Forest Service Research Note NE-18, 6 pp.

Fenton, R. H., and K. E. Pfeiffer. 1965. Successful Maryland white pine-yellow poplar plantation: Its response to thinning *Journal of Forestry* **63**:765-768.

Ferguson, D. E. 1994. Advance regeneration in the Inland West: considerations for individual tree and forest health. *Journal of Sustainable Forestry* **2**(3/4):411-422.

Ferguson, D. E., A. R. Stage, and R. J. Boyd. 1986. Predicting regeneration in the grand fir-cedar-hemlock ecosystem of the northern Rocky Mountains, *Forest Science Monograph* **26**:1-41.

Ferguson, D. E., and E. L. Adams. 1980. Response of advance grand fir regeneration to overstory removal in northern Idaho, *Forest Science* **26**:537-545.

Fetherolf, J. M. 1917. Aspen as a permanent forest type, *Journal of Forestry* **15**:757-760.

Ffolliott, P. F., and W. P. Clary. 1972. *A selected and annotated bibliography of understory-overstory vegetation relationships,* University of Arizona Tucson Agricultural Experiment Station Technical Bulletin 198, 33 pp.

Fiedler, C. E. 1995. The basal area-maximum diameter-Q (BD-q) approach to regulating uneven-aged stands. IN (K. L. O'Hara, editor) *Uneven-aged management: opportunities, constraints, and methodologies.* Montana Forest and Conservation Experiment Station, School of Forestry, University of Montana, Miscellaneous Publication No. 56.

Filip, G. M., and C. L. Schmitt. 1979. Susceptibility of native conifers to laminated root rot east of the Cascade range in Oregon and Washington, *Forest Science* **25**:261-265.

Filip, S. M. 1973. Cutting and cultural methods for managing northern hardwoods in the northeastern United States, USDA Forest Service Central Technical Report NE-5, 5 pp.

Finegan, B. 1984. Forest succession. *Nature* **312**:109-114.

Finney, M. A. 1986. Effects of low intensity fire on the successional development of serial lodgepole pine forests in the North Cascades, unpublished master's thesis, University of Washington College of Forest Resources, Seattle, 140 pp.

Fisher, R. F. 1980. Allelopathy: A potential cause of regeneration failure, *Journal of Forestry* **78**:346-350.

Fisher, S. G., L. J. Gray, N. B. Grimm, and D. E. Busch. 1982. Temporal succession in a desert stream ecosystem following flash flooding, *Ecological Monographs* **52**:92-110.

Flaccus, E. 1959. Revegetation of landslides in the White Mountains of New Hampshire *Ecology* **40**:692-703.

Flint, R. F. 1971. *Glacial and Quaternary Geology,* John Wiley, New York, 892 pp.

Fogel, R., and G. Hunt. 1979. Fungal and arboreal biomass in a western Oregon Douglas-fir ecosystem: Distribution patterns and turnover, *Canadian Journal of Forest Research* **9**:245-256.

Fogel, R., and K. Cromack, Jr. 1977. Effect of habitat and substrate quality on Douglas Fir litter decomposition in western Oregon, *Canadian Journal of Botany* **55**:1632-1640.

Foiles, M. W. 1977. Stand structure, in *Uneven-aged Silviculture and Management in the Western United States*, USDA Forest Service Timber Management Research, pp. 71-80.

Forcier, L. K. 1975. Reproductive strategies and the co-occurrence of climax tree species, *Science* **189**:808-810.

Ford-Robertson, F. C. 1971. *Terminology of Forest Science, Technology, Practice, and Products*, Society of American Foresters, Washington, D.C., 349 pp.

Ford, E. D. 1975. Competition and stand structure in some even-aged plant monocultures, *Journal of Ecology* **63**:311-33.

Ford, E. D. 1984. The dynamics of plantation growth, in G. D. Bowen and E. K. S. Nambiar (eds.), *Nutrition of Plantation Forests*, Academic Press, New York, pp. 17-52.

Ford, E. D. 1985. Branching, crown structure and the control of timber production, in M. G. R. Cannell and J. E. Jackson (eds.), *Attributes of Trees as Crop Plants*, Institute of Terrestrial Ecology, Huntington, England, pp. 228-252.

Forman, R. T. T. 1974. Ecosystem size: A stress on the forest ecosystem, *Bulletin of the Ecological Society of America* **55**(2):38.

Forman, R. T. T., and M. Godron. 1981. Patches and structural components for a landscape ecology, *BioScience* **31**:733-739.

Forman, R. T. T., and M. Godron. 1986. *Landscape Ecology*, John Wiley, New York, 620 pp.

Forman, R. T. T., and R. E. Boerner. 1981. Fire frequency and the Pine Barrens of New Jersey, *Bulletin of the Torrey Botanical Club* **108**:34-50.

Forsman, E. D., E. C. Meslow, and H. M. Wight. 1984. Distribution and biology of the spotted owl in Oregon. *Wildlife Monographs* **87**:1-64.

Fowells, H. A. 1965. *Silvics of forest trees of the United States*, USDA Forest Service Agriculture Handbook No. 271, 762 pp.

Fowells, H. A., and N. B. Stark. 1971. Natural regeneration in relation to environment in the mixed conifer forest type of California, USDA Forest Service Research Paper PSW24.

Franklin J. F. 1963. Natural regeneration of Douglas-fir and associated species using modified clear-cutting systems in the Oregon Cascades. USDA Forest Service Research Paper PNW-3. 14 pp.

Franklin, J. F. 1977. Effects of uneven-aged management on species composition, in *Uneven-Aged Silviculture and Management in the Western United States*, USDA Forest Service, Timber Management Research, pp. 64-70.

Franklin, J. F. 1984. Characteristics of old-growth Douglas-fir forests, *Proceedings of the Society of American Foresters National Convention*, pp. 10-16.

Franklin, J. F. 1989. Toward a new forestry. American Forests **95**:37-44.

Franklin, J. F. 1993. Preserving biodiversity: species, ecosystems, or landscapes? *Ecological Applications* **3**(2):202-205.

Franklin, J. F., and C. E. Smith. 1974. Seeding habits of upper-slope tree species: III. Dispersal of white and shasta red fir seeds on a clearcut, USDA Forest Service Research Note PNW-215, 9 pp.

Franklin, J. F., and C. T. Dyrness. 1973. Natural vegetation of Oregon and Washington, USDA Forest Service General Technical Report PNW-8, 417 pp.

Franklin, J. F., and D. S. DeBell. 1973. Effects of various harvesting methods on forest regeneration, in R. K. Herrmann and D. P. Lavender (eds.), *Even-age Management,* Oregon State University School of Forestry Paper 848, Corvallis, pp. 29-57.

Franklin, J. F., and M. Hemstrom. 1981. Aspects of succession in the coniferous forest of the Pacific Northwest, in D. C. West, H. H. Shugart, and D. B. Botkin (eds.), *Forest Succession: Concepts and Application,* Springer-Verlag, New York, pp. 212-229.

Franklin, J. F., and R. G. Mitchell. 1967. Successional status of subalpine fir in the Cascade Range, USDA Forest Service Research Paper PNW-46, 16 pp.

Franklin, J. F., and R. H. Waring. 1979. Distinctive features of the northwestern coniferous forest: Development, structure, and function, in R. H. Waring (ed.), *Forests: Fresh Perspectives from Ecosystem Analysis,* Proceedings of the 40th Annual Biology Colloquium, Oregon State University Press, pp. 59-86.

Franklin, J. F., and R. T. T. Forman. 1987. Creating landscape patterns by forest cutting: Ecological consequences and principles, *Landscape Ecology* 1:5-18.

Franklin, J. F., F. Hall, W. Laudenslayer, C. Maser, J. Nunan, J. Poppino, C. J. Ralph, and T. Spies. 1986. Interim definitions for old-growth Douglas-fir and mixed-conifer forests in the Pacific Northwest and California, USDA Forest Service Research Note PNW-447, 7 pp.

Franklin, J. F., H. H. Shugart, and M. E. Harmon. 1987. Tree death as an ecological process, *BioScience* 37:550-556.

Franklin, J. F., J. A. MacMahon, F. J. Swanson, and J. R. Sedell. 1985. Ecosystem responses to the eruption of Mount St. Helens, *National Geographic Research* 1:198-216.

Franklin, J. F., K. Cromack, Jr., W. Denison, A. W. McKee, C. Maser, J. Sedell, F. Swanson, and G. Juday. 1981. Ecological characteristics of old-growth Douglas-fir forests, USDA Forest Service General Technical Report PNW-118, 48 pp.

Franklin, J. F., T. Spies, D. Perry, M. Harmon, and A. McKee. 1986. Modifying Douglas-fir management regimes for nontimber objectives. IN (C. D. Oliver, D. Hanley, and J. Johnson (editors), *Douglas-fir: Stand Management for the Future,* University of Washington, College of Forest Resources, Seattle, Washington, Institute of Forest Resources Contribution No. **55**:373-379.

Franklin, J. F., W. H. Moir, G. W. Douglas, and C. Wiberg. 1971. Invasion of subalpine meadows by trees in the Cascade Range, *Arctic and Alpine Research* 3:215-224.

Frederick, D. J., and M. S. Coffman. 1978. Red pine plantation biomass exceeds sugar maple on northern hardwood sites, *Journal of Forestry* 76:13-15.

Frelich, L. E., and C. G. Lorimer. 1985. Current and predicted long-term effects of deer browsing in hemlock forests in Michigan, U.S.A., *Biological Conservation* 34:99-120.

Fried, J. S., J. C. Tappeiner II, and D. E. Hibbs. 1988. Bigleaf maple seedling establishment and early growth in Douglas-fir forests, *Canadian Journal of Forest Research* 18:1226-1233.

Frissel, S. S., Jr. 1973. The importance of fire as a natural ecological factor in Itasca State Park, Minnesota, *Quaternary Research* 3:397407.

Fritschen, L. J., R. B. Walker, and J. Hsia. 1980. Energy balance of an isolated Scots pine, *International Journal of Biometeorology* 24:293-300.

Fritts, H. C., and T. W. Swetnam. 1986. *Dendroecology: A Tool for Evaluating Variations an Past and Present Forest Environments,* Utility Air Regulatory Group Acid Deposition Committee, Washington, D.C., Hunton and Williams, Printers, 61 pp.

Fritz, E. 1934. *The Story Told by a Fallen Redwood,* Save-the-Redwoods League, Berkeley, California, 7 pp.

Frothingham, E. H. 1912. Second-growth hardwoods in Connecticut, USDA Forest Service Bulletin 96, 70 pp.

Fry, M. E., and N. R. Money. 1994. Biodiversity conservation in the management of utility rights-of-way. IN *Proceedings of the 15th Annual Forest Vegetation Management Conference,* January 25-27, 1994, Redding, California, U.S.A.: 94-103.

Fryer, J. H., and F. T. Ledig. 1972. Microevolution of the photosynthetic temperature optimum in relation to the elevational complex gradient, *Canadian Journal of Botany* **50**:1231-1235.

Fujimori, T., and D. Whitehead, (editors) 1986. *Crown and canopy structure in relation to productivity.* Forestry and Forest Products Research Institute, Ibaraki, Japan.

Fuller, G. D. 1913. Reproduction by layering in the black spruce, *Botanical Gazette* **55**:452-457.

Funakoshi, S. 1985. *Shoot growth and winter bud formation in Abies sachalinensis Mast.* Research Bulletin of the College Experiment Forest Vol. 42 (40), Hokkaido University Faculty of Agriculture: 785-808.

Furniss, R. L., and W. M. Carolyn. 1977. *Western forest insects,* USDA Miscellaneous Publication No. 1339, 654 pp.

Furnival, G. M. 1987. Growth and yield prediction: Some criticisms and suggestions, in H. N. Chappell and D. A. Maguire (eds.), *Predicting Forest Growth and Yield: Current Issues, Future Prospects,* University of Washington College of Forest Resources, Institute of Forest Resources Contribution No. 58, pp. 23-30.

Gabriel, W. J. 1975. Allelopathic effects of black walnut on white birches, *Journal of Forestry* **73**:234-237.

Gadgil M., and O. T. Solbrig. 1972. The concept of r and K selection: Evidence from wild flowers and some theoretical considerations, *American Naturalist* **106**:14-31.

Gaines, E. 1952. Scrub oak as a nurse crop for longleaf pine on deep sandy soils, *Journal of the Alabama Academy of Science* **22**:107-108.

Gaiser, R. N. 1952. Root channels and roots in forest soils, *Soil Science Society Proceedings* **16**:62-65.

Galbraith, W. A., and E. W. Anderson. 1991. Grazing history of the northwest. *Rangelands* **13**:213-218.

Galloway, G. 1991. Beyond afforestation: a case study in Highland Ecuador. Unpublished Ph. D. dissertation, University of Washington, College of Forest Resources, Seattle, Washington, U.S.A. 244 pp.

Galloway, G. 1991. *Beyond afforestation: a case study in Highland Ecuador.* Unpublished Ph. D. dissertation, College of Forest Resources, University of Washington, Seattle, Washington. 245 pp.

Galloway, G., L. Ugalde, and W. Vasquez. 1995. Management of tropical plantations under stress. (In review.) Dr. Galloway, Dr. Ugalde, and Mr. Vasquez are with Madelena Project, CATIE, Turrialba, Costa Rica.

Gangloff, D. 1988. National register of big trees, *American Forests* **94**(5&6):34.

Gant, R., and E. E. C. Clebson. 1975. The allelopathic influences of *Sassafras albidum* in old-field succession in Tennessee, *Ecology* **56**:604-615.

Gara, R. I., T. Mehary, and C. D. Oliver. 1980. *Studies on the integrated pest management of the sitka spruce weevil,* 1980 Annual Report to the Sitka Spruce Committee, University of Washington College of Forest Resources, Seattle, 43 pp.

Gardiner, L. M. 1975. Insect attack and value loss in wind-damaged spruce and jack pine stands in northern Ontario, *Canadian Journal of Forest Research* 5:387-398.

Garrison, O. E. 1881. *The Upper Mississippi Region: Geologic and natural history survey of Minnesota*, 9th Annual Report (for 1880).

Gartska, W. U. 1964. Snow and snow survey, in V. T. Chow (ed.), *Handbook of Hydrology*, McGraw-Hill, New York, pp. 10.1-10.57.

Gartska, W. U. 1969. Lecture notes in forest hydrology, Yale University School of Forestry and Environmental Studies, New Haven, Connecticut.

Garver, R. D., and R. H. Miller. 1933. Selective logging in the shortleaf and loblolly pine forests of the Gulf States region, USDA Technical Bulletin No. 375, 54 pp.

Garwood, N. C., D. P. Janos, and N. Brokaw. 1979. Earthquake caused landslides: A major disturbance to tropical forests, *Science* 205:997-999.

Gates, D. J . 1982. Competition and skewness in plantations, *Journal* of *Theoretical Biology* 94:909-922.

Gates, D. J. 1978. Bimodality in even-aged plant monocultures *Journal of Theoretical Biology* 71:525-540.

Gates, D. J., R. McMurtrie, and C. J. Borough. 1983. Skewness reversal of distribution of stem diameter in plantations of *Pinus radiata, Australian Forestry Research* 13:267-270.

Gates, D. M. 1993. *Climate change and its biological consequences.* Sinauer Associates, Inc., Publishers. Sunderland, Massachusetts, U.S.A. 280 pp.

Gates, F. C. 1938. Layering in black spruce, *American Midland Naturalist* 19:589-594.

Gates, F. C., and G. E. Nichols. 1930. Relation between age and diameter in trees of the primeval northern hardwood forest, *Journal of forestry* 28:395-398.

Gauch, H. G., and E. L. Stone. 1979. Vegetation and soil pattern in a mesophytic forest at Ithaca, New York, *American Midland Naturalist* 102:332-345.

Geesink, R., and D. J. Kornet. 1989. Speciation and Malesian Luguminosae. IN L. B. Holm-Nielson, I. C. Nielsen, and H. Balslev, editors. *Tropical Forests.* Academic Press, New York. pp. 135-151.

Geiger, R. 1965. *The Climate Near the Ground,* Harvard University Press, Cambridge, Massachusetts, 611 pp.

Geist, V. 1971. *Mountain sheep—a study of behavior.* Unversity of Chicago Press, Chicago. 383 pp.

Geist, V. 1971. *Mountain Sheep—A Study of Behavior,* University of Chicago Press, Chicago, 383 pp.

Geiszler, D. R., R. I. Gara, C. H. Driver, V. F. Gallucci, and R. E. Martin. 1980. Fire, fungi, and beetle influences on a lodgepole pine ecosystem of south-central Oregon, *Oecologia* 46:239-243.

Gentry, A. H. 1982. Neotropical floristic diversity: phytogeographical connections between central and South America. Pleistocene climatic fluctuations, or an accident of the Andean orogeny. *Annals of the Missouri Botanical Garden* 69:557-593.

Gersper, G. L., and N. Holowaychuck. 1971. Some effects of stemflow from forest canopy trees on chemical properties of soil, *Ecology* 52:691-702.

Gevorkiantz, S. R., and N. W. Hosley. 1929. *Form and development of white pine stands in relation to growing space,* The Harvard Forest, Harvard University, Harvard Forest Bulletin No. 13, 83 pp.

Ghent, A. W. 1958. Studies of regeneration in forest stands devastated by the spruce budworm: II. Age, height, growth, and related studies of balsam fir seedlings, *Forest Science* 4:135-146.

Ghent, A. W., D. A. Fraser, and J. B. Thomas. 1957. Studies of regeneration of forest stands devastated by the spruce budworm: I. *Forest Science* 3:184-208.

Gholz, H. L. 1982. Environmental limits on aboveground net primary production, leaf area, and biomass in vegetation zones of the Pacific Northwest, *Ecology* **63**:469-481.

Gholz, H. L. 1986. Canopy development and dynamics in relation to primary production, in T. Fujimori and D. Whitehead (ads.), *Crown and Canopy Structure in Relation to Productivity*, IUFRO Symposium Proceedings, Forestry and Forest Products Research Institute, Ibaraki, Japan, pp. 224-242.

Gholz, H. L., and R. F. Fisher. 1982. Environmental limits on aboveground net primary production, leaf area and biomass in vegetation zones of the Pacific Northwest, *Ecology* **53**:469-481.

Gholz, H. L., C. C. Grier, A. G. Campbell, and A. T. Brown. 1979. *Equations and their use for estimating biomass and leaf area of Pacific Northwest trees, shrubs, and herbs*, Research Paper 41, Forest Research Laboratory, School of Forestry, Oregon State University, Corvallis, 37 pp.

Gibbs, C. B. 1963. Tree diameter a poor indicator of age in West Virginia hardwoods, USDA Forest Service Research Note NE 11, 4 pp.

Gilbert, A. M. 1965. Stand differentiation ability in northern hardwoods, USDA Forest Service Research Paper NE-37, 34 pp.

Gilbert, A. M., and V. S. Jensen. 1958. A management guide for northern hardwoods in New England, USDA Forest Service Northeast Forest Experiment Station Paper 112, 22 pp.

Gill, A. M. 1977. Plant traits adaptive to fires in Mediterranean land ecosystems, in *Mediterranean Climate Ecosystems*, USDA Forest Service General Technical Report WO-3, pp. 17-26.

Gilliland, R. 1982. Solar, volcanic, and CO_2 forcing of recent climatic change, *Climatic Change* **4**:111-131.

Gingrich, S. F. 1967. Measuring and evaluating stocking and stand density in upland hardwood forests in the Central States, *Forest Science* **13**:38-53.

Gingrich, S. F. 1971. Management of upland hardwoods, USDA Forest Service Research Paper NE-195, 26 pp.

Gingrich, S. F. 1975. Growth and yield, in *Uneven-aged Silviculture and Management in the Eastern United States*, USDA Forest Service, Timber Management Research, pp. 111-120.

Gleason, H. A. 1926. The individualistic concept of the plant association, *Bulletin of the Torrey Botanical Club* **46**:7-26.

Gleason, H. A. 1936. Is the synusia an association? *Ecology* **17**:444-451.

Gleason, H. A. 1939. The individualistic concept of the plant association, *American Midland Naturalist* **21**:92-110.

Glitzenstein, J. S., C. D. Canham, M. J. McDonnell, and D. R. Streng. 1990. Effects of environment and land-use history on upland forests of the Cary Arboretum, Hudson Valley, New York. *Torrey Botanical Club* **117**:106-122.

Glitzenstein, J. S., P. A. Harcombe, and D. R. Streng. 1986. Disturbances, succession, and maintenance of species diversity in an east Texas forest, *Ecological Monographs* **56**:243-258.

Glynn, P. W. 1988. El Nino-southern oscillation 1982-1983: nearshore population, community, and ecosystem responses. *Annual Review of Ecology and Systematics* **19**:309-345.

Godron, M., and R. T. T. Forman. 1983. Landscape modification and changing ecological characteristics, in H. A. Mooney and M. Godron (eds.), *Disturbance and*

Ecosystems: Components of Response, Springer-Verlag, New York, pp. 12-28.

Goff, F. G., and D. West. 1975. Canopy-understory interaction effects on forest population structure, *Forest Science* **21**:98-108.

Golomazova, G. M., E. G. Minina, and M. A. Shemberg. 1978. Rate of photosynthesis in narrow-crown and wide-crown forms of *Pinus sylvestris* L. (Translated from *Fiziologiya Rastenii* **25**:85-90), (Plenums, New York, pp. 62-66).

Good, B. J., and S. A. Whipple. 1982. Tree spatial patterns: South Carolina bottomland and swamp forests, *Bulletin of the Torrey Botanical Club* **109**:529-536.

Goodlett, J. C. 1954. *Vegetation adjacent to the border of Wisconsin drift in Potter County, Pennsylvania,* The Harvard Forest, Harvard University, Harvard Forest Bulletin 25, 93 pp.

Goodman, R. B. 1930. Conditions essential to selective cutting in northern Wisconsin hardwoods and hemlock, *Journal of Forestry* **28**:1070-1075.

Gordon, D. T. 1973. Released advance reproduction of white and red fir: Growth, damage, and mortality, USDA Forest Service Research Paper PSW-95, 12 pp.

Gordon, J. C., and P. R. Larson. 1968. Seasonal course of photosynthesis, respiration, and distribution of C^{14} in young *Pinus resinosa* trees as related to wood formation, *Plant Physiology* **43**:1617-1624.

Gorham, E. 1979. Shoot height, weight and standing crop in relation to density of monospecific plant stands, *Nature* **279**:148-150.

Gosz, J. R., G. E. Likens, and F. H. Bormann. 1976. Organic matter and nutrient dynamics of the forest and forest floor in the Hubbard Brook Forest, *Oecologia* **22**:305-320.

Gottschalk, L. C. 1964. Sedimentation, in V. T. Chow (ed.), *Handbook of Applied Hydrology,* McGraw-Hill, New York, pp. 17.1-17.67.

Goudie, J. W. 1983. *Effect of fertilization on growth characteristics of coastal Douglas-fir:* I. Final report fiscal year 1982-1983, submitted to Research Branch, British Columbia Ministry of Forests, Canada, 17 pp.

Gould, E. M., Jr. 1960. *Fifty years of management at the Harvard Forest,* The Harvard Forest, Harvard University, Harvard Forest Bulletin No. 29, 30 pp.

Gould, S. J., and R. C. Lewontin, 1979. Spandrels of San-Marco and the Panglossian paradigm—a critique of the adaptationist program, *Proceedings of the Royal Society,* **B.205**:581-598.

Graber, R. E., and D. F. Thompson. 1978. Seeds in the organic layers and soil of four beech-birch-maple stands, USDA, Forest Service Research Paper, NE-401, 8 p.

Grah, R. F. 1961. Relationship between tree spacing, knot size, and log quality in young Douglas-fir stands, *Journal of Forestry* **59**:270-272.

Graham, S. A. 1941. Climax forests of the upper peninsula of Michigan, *Ecology* **22**:355-362.

Grant, J. C. 1980. Growth comparisons between Pacific silver fir and western hemlock on the western slopes of the northern Washington Cascades, unpublished Master's thesis, University of Washington College of Forest Resources, Seattle, 160 pp.

Gratkowski, H. J. 1956. Windthrow around staggered settings in old-growth Douglas-fir, *Forest Science* **2**:60-74.

Greeley, W. B. 1907. A rough system of management for forest lands in the western Sierras, *Proceedings of the Society of American Foresters* **2**:103-114.

Greig-Smith, P. 1952. Ecological observations on degraded and secondary forest in Trinidad, British West Indies: II. Structure of the communities, *Journal of Ecology* **40**:316-330.

Grier, C. C. 1975. Wildfire effects on nutrient distribution and leaching in a coniferous ecosystem, *Canadian Journal of Forest Research* **5**:599-607.

Grier, C. C. 1978a. Biomass, productivity and nitrogen-phosphorus cycles in hemlock-spruce stands of the central Oregon coast, in W. A. Atkinson and R. J. Zasoski (eds.), *Western Hemlock Management,* University of Washington College of Forest Resources Institute of Forest Products Contribution No. 11, pp. 71-41.

Grier, C. C. 1978b. A Tsuga heterophylla-Picea sitchensis ecosystem of coastal Oregon: Decomposition and nutrient balances of fallen logs, *Canadian Journal of Forest Research* **8**:198-206.

Grier, C. C., and R. H. Waring. 1974. Conifer foliage mass related to sapwood area, *Forest Science* **20**:205-206.

Grier, C. C., and R. S. Logan. 1977. Old-growth *Pseudotsuga menziesii* (Mirb.) Franco communities of a western Oregon watershed: Biomass distribution and production budgets, *Ecological Monographs* **47**:373-400.

Grier, C. C., and S. W. Running. 1977. Leaf area of mature northwestern coniferous forests: Relation to site water balance, *Ecology* **58**:893-899.

Grier, C. C., D. W. Cole, C. T. Dryness, and R. L. Frederiksen. 1974. *Nutrient cycling in 37- and 450-year old Douglas-fir ecosystems,* in Coniferous Forest Biome Bulletin 5, University of Washington, Seattle, Washington, pp. 21-34.

Grier, C. C., K. A. Vogt, M. R. Keyes, and R. L. Edmonds. 1981. Biomass distribution and above- and below-ground production in young and mature Abies amabilis zone ecosystems of the Washington Cascades, *Canadian Journal of Forest Research* **11**:155-167.

Griffin, J. R. 1971. Oak regeneration in the upper Carmel Valley, California, *Ecology* **52**:862-868.

Griffing, C. G., and W. W. Elam. 1971. Height growth patterns of loblolly pine saplings, *Forest Science* **17**:52-54.

Griffith, B. G. 1959. The effect of thinning on a young stand of western hemlock, *Forestry Chronicle* **35**:114-134.

Griggs, R. F. 1919. The recovery of vegetation at Kodiak, *Ohio Journal of Science* **19**:1-57.

Grime, J. P. 1977. Evidence for the existence of three primary strategies in plants and its relevance to ecological and evolutionary theory, *American Naturalist* **111**:1169-1194.

Grime, J. P. 1979. *Plant Strategies and Vegetation Process,* John Wiley, New York, 222 pp.

Groome, J. S. 1988. Mutual support of trees *Scottish Forestry* **42**:12-14.

Gross, L. S. 1955. Forest growing stock, *Journal of Forestry* **53**:403-408.

Grubb, P. J. 1977. The maintenance of species-richness in plant communities: The importance of the regeneration niche, *Biological Reviews* **52**:107-145.

Grubb, P. J., J. R. Lloyd, T. D. Pennington, and T. C. Whitmore. 1963. A comparison of montane and lowland rain forest in Ecuador: I. The forest structure, physiognomy, and floristics, *Journal of Ecology* **51**:567-601.

Gruschow, G. F., and T. C. Evans. 1959. The relation of cubic-foot volume growth to stand density in young slash pine stands, *Forest Science* **5**:49-55.

Guillebaud, W. H., and F. C. Hummel. 1949. A note on the movement of tree classes, *Forestry* **23**:1-14.

Guldin, J. M. 1995. The role of uneven-aged silviculture in the context of ecosystem management. IN (K. L. O'Hara, editor) *Uneven-aged management: opportunities, constraints, and methodologies.* Montana Forest and Conservation Experiment

Station, School of Forestry, University of Montana, Miscellaneous Publication No. 56.

Guldin, J. M., and C. G. Lorimer. 1985. Crown differentiation in even-aged northern hardwood forests of the Great Lakes region, U.S.A. *Forest Ecology and Management* 10:65-86.

Guldin, J. M., and J. B. Baker. 1988. Yield comparisons from even-aged and uneven-aged loblolly-shortleaf stands. *Southern Journal of Applied Forestry* 12:107-114.

Guldin, J. M., and M. W. Fitzpatrick. 1991. Comparison of log quality from even-aged and uneven-aged loblolly pine stands in south Arkansas. *Southern Journal of Applied Forestry* 15:10-17.

Gutierrez, R. J., and A. B. Carey. 1985. Ecology and management of the spotted owl in the Pacific Northwest. USDA Forest Service General Technical Report PNW-185.

Gutierrez, R. J., and A. B. Carey. 1985. Ecology and management of the spotted owl in the Pacific Northwest. USDA Forest Service Central Technical Report PNW-185, 119 pp.

Guttenberg, S. 1952a. How far does sweetgum seed fly? USDA Forest Service Southern Forestry Notes Number 81.

Guttenberg, S. 1952b. Sweetgum seed is overrated flier, *Journal of Forestry* 50:844.

Gysel, L. W. 1951. Borders and openings of beech-maple woodlands in southern Michigan, *Journal of Forestry* 49:13-19.

Haack, R. A., and J. W. Byler. 1993. Insects and pathogens: regulators of forest ecosystems. *Journal of Forestry* 91(9):32-37.

Habeck, J. R. 1959. A vegetational study of the central Wisconsin winter deer range, *Journal of Wildlife Management* 23:273-278.

Habeck, J. R. 1972. *Fire ecology investigations in the Selway-Bitterroot Wilderness,* University of Montana Publication No. R1-72-0001.

Hack, J. T., and J. C. Goodlett. 1960. *Geomorphology and forest ecology of a mountain region in the Central Appalachians,* Professional Paper 347, U.S. Geological Survey, Washington, D.C., 66 pp.

Haight, R. G., 1987. Evaluating the efficiency of even-aged and uneven-aged stand management. *Forest Science* 33(1):116-134.

Haines, D. A., and R. W. Sando. 1969. Climatic conditions preceding historically great fires in the North Central Region, USDA Forest Service Research Paper NC-34, 19 pp.

Haines, S. G., and D. S. DeBell. 1979. Use of nitrogen-fixing plants to improve and maintain productivity of forest soils, in *Impact of Intensive Harvesting on Forest Nutrient Cycling,* College of Environmental Science and Forestry, Syracuse, New York, pp. 279-303.

Halkett, J. C. 1984. The practice of uneven-aged silviculture. *New Zealand Journal of Forestry* 2:108-118.

Hall, F. C. 1982. Ecology of grand fir, in C. D. Oliver and R. M. Kenady (eds.), *Biology and Management of True Fir in the Pacific Northwest,* University of Washington College of Forest Resources, Institute of Forest Resources Contribution Number 45, pp. 43-52.

Hall, R. C. 1933. *Post-logging decadence in northern hardwoods,* University of Michigan School of Forestry and Conservation Bulletin, 66 pp.

Hall, T. F., and W. T. Penfound. 1943. Cypress-gum communities in the Blue Girth Swamp near Selma, Alabama, *Ecology* 24:208-217.

Hall, T. G., and G. E. Smith. 1955. Effects of flooding on woody plants, *Journal of Forestry* 53:281-285.

Hall, T. H., R. V. Quenet, C. R. Layton, and R. J. Robertson. 1980. *Fertilization. and thinning effects on a Douglas-fir ecosystem at Shawnigan Lake: 6 year growth response,* Canadian Forest Service Report, BC-X-202, Victoria British Columbia, 31 pp.

Halle, F., and R. A. A. Oldeman. 1970. *Essai sur l'Architecture et la Dynamique de Crossance des Arbres Tropicaux,* Masson, Paris, 178 pp.

Halle, F., R. A. A. Oldeman, and P. B. Tomlinson. 1978. *Tropical Trees and Forests: An Architectural Analysis,* Springer-Verlag, New York, 441 pp.

Halls, L. K., and H. S. Crawford, Jr. 1960. Deer-forest habitat relationships in northern Arkansas, *Journal of Wildlife Management* 24:387-395.

Halpern, C. B. 1987. Twenty-one years of secondary succession in Pseudotsuga forests of the western Cascade Range, Oregon, unpublished Ph.D. thesis, Oregon State University, Corvallis, 239 pp.

Hamilton, G. J. 1969. The dependence of volume increment of individual trees on dominance, crown dimensions and competition, *Forestry* 42:133-144.

Hamilton, G. J. 1976. The Bowmont Norway spruce thinning experiment 1930-1974, *Forestry* 49:109-119.

Hamilton, G. J. 1981. The effects of high intensity thinning on yield, *Forestry* 54:1-15.

Hamilton, G. J., and J. M. Christie. 1974. Influence of spacing on crop characteristics and yield, British Forestry Commission Bulletin 52, 92 pp.

Hanley, D. P., W. C. Schmidt, and G. M. Blake. 1975. Stand structure and successional status of two spruce-fir forests in southern Utah, USDA Forest Service Research Paper INT-176, 16 pp.

Hanley, T. A. 1983. Black-tailed deer, elk, and forest edge in a western Cascades watershed, *Journal of Wildlife Management* 47:237-242.

Hanley, T. A., and R. D. Taber. 1980. Selective plant species inhibition by elk and deer in three conifer communities in western Washington. *Forest Science* 26:97-107.

Hanley, T. A., and R. D. Taber. 1980. Selective plant species inhibition by elk and deer in three conifer communities in western Washington, *Forest Science* 26:97-107.

Hann, D. W., and B. B. Bare. 1979. Uneven-aged forest management: State of the art (or science)? USDA Forest Service General Technical Report INT-50, 18 pp.

Hann, D. W., and M. W. Ritchie. 1988. Height growth rate of Douglas-fir: A comparison of model forms, *Forest Science* 34:165-175.

Hansbrough, T. 1968. Stand characteristics of 18-year-old loblolly pine growing at different initial spacings, Hill Farm Facts, *Forestry* 8, 4 pp.

Hansbrough, T., R. R. Foil, and R. G. Merrifield. 1964. The development of loblolly pine planted at various initial spacings, Hill Farm Facts, *Forestry* 4, 4 pp.

Hansen, A. J., and D. L. Urban. 1992. Avian response to landscape pattern: the role of species' life histories. *Landscape Ecology* 7(3):163-180.

Hansen, A. J., S. L. Garman, P. Lee, and E. Horvath. 1993. Do edge effects influence tree growth rates in Douglas-fir plantations? *Northwest Science* 67(2):113-116.

Hansen, H. L. 1980. The Lake States region, in J. W. Barrett (ed.), *Regional Silviculture of the United States,* 2d ed., John Wiley, New York,pp.67-105.

Hansen, J., D. Johnson, A. Lacis, S. Lebedeff, P. Lee, D. Rind, and G. Russell. 1981. Climate impact of increasing atmospheric carbon dioxide, *Science* 213:957-966.

Hara, T., and M. Wakahara. 1994. Variation in individual growth and the population structure of a woodland perennial herb, Paris tetraphylla. *Journal of Ecology* 82:2-12.

Hara, T., J. van der Toorn, and J. H. Mook. 1993. Growth dynamics and size structure of shoots of Phragmites australis, a clonal plant. *Journal of Ecology* 81:47-60.

Harcombe, P. A. 1986. Stand development in a 130-year-old spruce-hemlock forest based on age structure and 50 years of mortality data, *Forest Ecology and Management* **14**:41-58.

Harcombe, P. A., and P. L. Marks. 1978. Tree diameter distribution and replacement processes in southeast Texas forests, *Forest Science* **24**:153-166.

Harcombe, P. A., and P. L. Marks. 1983. Five years of tree death in a Fagus-Magnolia forest, southeast Texas USA, *Oecologia* **57**:49-54.

Harkin, D. A. 1972. Clearcut or selection: What kind of cost? *Journal of Forestry* **70**:420-421.

Harlow, W. M., E. S. Harrar, and F. M. White. 1979. *Textbook of Dendrology*, 6th ed., McGraw-Hill, New York, 510 pp.

Harmon, M. E., J. F. Franklin, F. J. Swanson, P. Sollins, S. V. Gregory, J. D. Lattin, N. H. Anderson, S. P. Cline, N. G. Aumen, J. R. Sedell, G. W. Lienkaemper, K. Cromack, Jr., and K. W. Cummins. 1986. Ecology of coarse woody debris in temperate ecosystems, *Advances in Ecological Research* **15**:133-302.

Harmon, M. E., S. P. Bratton, and P. S. White. 1983. Disturbance and vegetation response in relation to environmental gradients in the Great Smokey Mountains, *Vegetatio* **55**:129-139.

Harms, W. R., and O. G. Langdon. 1976. Development of loblolly pine in dense stands, *Forest Science* **22**:331-337.

Harper, J. L. 1977. *Population Biology of Plants*, Academic, London, 892 pp.

Harper, J. L. 1982. After description, in E. I. Newman (ed.), *The Plant Community as a Working Mechanism*, British Ecological Society, pp. 11-25.

Harper, J. L., and J. White. 1970. The dynamics of plant populations, in *Dynamics of Populations*, Proceedings of the Advanced Study Institute on Dynamics of Numbers in Populations, *Oosterbeek*, The Netherlands, pp. 41-63.

Harper, J. L., P. H. Lovell, and K. G. Moore. 1970. The shapes and sizes of seeds, *Annual Review of Ecology and Systematics* **1**:327-356.

Harper, J. L., W. T. Williams, and G. R. Sagar. 1965. The behavior of seeds in soil: I. The heterogeneity of soil surfaces and its role in determining the establishment of plants from seed, *Journal of Ecology* **53**:273-286.

Harrington, C. A., and D. L. Reukema. 1983. Initial shock and long-term stand development following thinning in a Douglas-fir plantation, *Forest Science* **29**:33-46.

Harrington, C. A., and R. L. Deal. 1982. Sitka alder, a candidate for mixed stands, *Canadian Journal of Forest Research* **12**:108-111.

Harrington, M. G., and R. G. Kelsey. 1979. Influence of some environmental factors on initial establishment and growth of ponderosa pine seedlings, USDA Forest Service Research Paper INT-230, 26 pp.

Harris, A. S. 1967. Natural reforestation on a mile-square clearcut in southeast Alaska, USDA Forest Service Research Paper PNW-52, 16 pp.

Harris, A. S. 1974. Clearcutting, reforestation and stand development on Alaska's Tongass National Forest, *Journal of Forestry* **72**:330-337.

Harris, A. S. 1989. Wind in the forests of Southeast Alaska and guides for reducing damage. USDA Forest Service PNW-GTR-244: 63 pp.

Harris, A. S., and W. A. Farr. 1974. The forest ecosystem of southeast Alaska: VII: Forest ecology and timber management, USDA Forest Service General Technical Report PNW-25, 109 pp.

Harris, L. 1984. *The Fragmented Forest: Island Biogeographic Theory and the Preservation of Biotic Diversity*, University of Chicago Press, Chicago, 211 pp.

Harris, W. F., R. S. Kinerson, and N. T. Edwards. 1978. Comparisons of belowground biomass of natural deciduous forest and loblolly pine plantations, *Pedobiologia* 17:369-381.

Harrison, W. C. and R. F. Daniels. 1988. A new biomathematical model for growth and yield of loblolly pine plantations, in A. R. Ek, S. R. Shirley, and T. E. Burk (eds.), *Forest Growth Modelling and Prediction,* Proceedings of the IUFRO conference, August 23-27, 1987, Minneapolis, Minnesota, USDA Forest Service, General Technical Report NC-120, pp. 293-304.

Hart, A. C. 1963. Spruce-fir silviculture in northern New England, *Proceedings of the Society of American Foresters* 107-110.

Hart, G., R. E. Leonard, and R. S. Pierce. 1962. Leaf fall, humus depth, and soil frost in a northern hardwood forest, USDA Forest Service, Northeastern Forest Experiment Station, Forest Research Note No. 131, 3 pp.

Hartshorn, G. 1978. Tree falls and tropical forest dynamics, in P. B. Tomlinson and M. H. Zimmerman (eds.), *Tropical Trees as Living Systems,* Cambridge University Press, Cambridge, England, pp. 617-638.

Hartshorn, G. S. 1980. Neotropical forest dynamics. *Biotropica* 12 (Supplement): 23-30.

Hartshorn, G. S. 1989a. Application of gap theory to tropical forest management: natural regeneration on strip clearcuts in the Peruvian Amazon. *Ecology* 70 (3):567-569.

Hartshorn, G. S. 1989b. Gap-phase dynamics and tropical tree species richness. IN L. B. Holm-Nielson, I. C. Nielsen, and H. Balslev, editors. *Tropical Forests.* Academic Press, New York. pp. 65-73.

Harvard Forest. 1941. *The Harvard Forest Models,* Cornwall Press, New York, 48 pp.

Harvey, A. E. 1994. Integrated roles for insects, diseases, and decomposers in fire dominated forests of the Inland Western United States: past, present, and future forest health. *Journal of Sustainable Forestry* 2: 211-220.

Hatch, A. B. 1936. The role of mycorrhizae in afforestation, *Journal of Forestry* 34:22-29.

Hatch, C. R., D. J. Gerrard, and J. C. Tappeiner II. 1975. Exposed crown surface area: a mathematical index of individual tree growth potential, *Canadian Journal of Forest Research* 5:224-228.

Hatcher, R. J. 1964. Balsam fir advance growth after cutting in Quebec, *Forestry Chronicle* 40:86-92.

Hauch, L. A. 1910. Zur variation des Wachstums bei unseren Baldvaumen mit besonderer Berucksichtigung des sogenannten Ausbreitungsvermogens, *Forstwissenschaftliches Centralblatt.* **54**:565-578.

Hawes, A. F. 1968. Connecticut: Scattering the seedy in R. R. Widner (ed.), *Forests and Forestry in the American States,* National Association of State Foresters, pp. 100-106.

Hawkes, B. C. 1980. Fire history of Kananaskis Provincial Park—mean fire return intervals, in *Proceedings of the Fire History Workshop,* October 20-24, 1980, Tucson, Arizona, USDA Forest Service General Technical Report RM-81, pp. 42-45.

Haynes, C. V. 1976. Ecology of early man in the New World, *Geoscience and Man* 13:71-76.

Haynes, R. W., and D. M. Adams. 1983. Changing perceptions of the U.S. forest sector: Implications for the RPA timber assessments *American Journal of Agricultural Economics* 65:1002-1009.

Heiberg, S. O., and P. G. Haddock. 1955. A method of thinning and forecast of yield in Douglas-fir, *Journal of Forestry* 53:10-18.

Heikkinen, O. 1982. Auringonpilkut ja niiden yhteys maapallolla havaittuihin luonnonilmoioihin (Sunspots and their relation to terrestrial phenomena) (English summary)

Terra **94**:207-214.

Heikkinen, O. 1984. Forest expansion in the subalpine zone during the past hundred years, Mount Baker, Washington, U.S.A., *Erdkunde* **38**:194-202.

Heikkinen, O., and M. Tikkanen. 1981. Ilmansaasteiden vaikutus havupuiden kasvuun esimerkkina Skoldvikin oljynjalostamon ymparisto (The effect of air pollution on growth in conifers: An example from the surroundings of the Sloldvik oil refinery) *Terra* **93**:133-144.

Heilman, P. E. 1966. Change in distribution and availability of nitrogen with forest succession on north slopes in interior Alaska, *Ecology* **47**:825-831.

Heilman, P. E. 1968. Relationship of availability of phosphorus and cations to forest succession and bog formation in interior Alaska, *Ecology* **49**:331-336.

Heinselman, M. L. 1963. Forest sites, bog processes, and peatland types in the glacial Lake Agassiz region, Minnesota, *Ecological Monographs* **33**:327-374.

Heinselman, M. L. 1970a. Restoring fire to the ecosystems of the Boundary Waters Canoe Area, Minnesota and to similar wilderness areas, *Tall Timbers Fire Ecology Conference* **10**:9-24.

Heinselman, M. L. 1970b. The natural role of fire in the northern conifer forests, *Naturalist* **21**:14-23.

Heinselman, M. L. 1973. Fire in the virgin forests of the Boundary Waters Canoe Area, Minnesota, *Journal of Quaternary Research* **3**:329-382.

Heinselman, M. L. 1981. Fire intensity and frequency as factors in the distribution and structure of northern ecosystems, in *Fire Regimes anal Ecosystem Properties,* December 11-15, 1978, Honolulu, Hawaii, USDA Forest Service General Technical Report WO-26, pp. 7-57.

Helms, J. A. 1965. Diurnal and seasonal patterns of net assimilation in Douglas-fir, *Pseudotsuga menziesii* (Mirb.) Franco, as influenced by environment, *Ecology* **46**:698-708.

Hemstrom, M. A. 1979. A recent disturbance history of forest ecosystems at Mount Rainier National Park, unpublished Ph.D. thesis, Oregon State University, Corvallis, 67 pp.

Hemstrom, M. A. 1982. Fire in the forests of Mount Rainier National Park, in J. E. Starkey, J. F. Franklin, and J. W. Matthews (eds.), *Ecological Research in the Pacific Northwest,* Oregon State University, Forest Research Laboratory, Corvallis, pp. 121-126.

Hemstrom, M. A., and J. F. Franklin. 1982. Fire and other disturbances of the forest in Mount Rainier National Park, *Quaternary Research* **18**:32-51.

Hendee, J. C., G. H. Stankey, and R. C. Lucas. 1977. Wilderness management, USDA Forest Service, Miscellaneous Publication No. 1365, 381 pp.

Henderson, J. A. 1982. Succession on two habitat types in western Washington, in J. E. Means (ed.), *Forest Succession and Stand Development Research in the Northwest,* Forest Research Laboratory, Oregon State University, Corvallis, pp. 80-96.

Henderson, J. A., and L. B. Brubaker. 1986. Response of Douglas-fir to long-term variations in precipitation and temperature in western Washington, in C. D. Oliver, D. P. Hanley, and J. A. Johnson (eds.), *Douglas-fir: Stand Management for the Future,* College of Forest Resources, University of Washington, Seattle, Institute of Forest Resources Contribution No. 55, pp. 162-167.

Henderson, J. A. 1994. The ecological consequences of long-rotation forestry. IN (J.F.Weigand, R.W. Haynes, and J. L. Mikowski, compilers).. *High Quality Forestry Workshop: the Idea of Long Rotations.* College of Forest Resources, University of Washington CINTRAFOR SP15: 43-61.

Hendricks, B. A., and J. M. Johnson. 1944. Effects of fire on steep mountain slopes in central Arizona, *Journal of Forestry* **42**:568-571.

Henry, J. D., and J. M. A. Swan. 1974. Reconstructing forest history from live and dead plant material—An approach to the study of forest succession in southwest New Hampshire, *Ecology* **55**:772-783.

Hepting, G. H. 1971. *Diseases of forest and shade trees of the United States*, USDA Handbook No. 386, 658 pp.

Herman, F. R. 1964. Epicormic branching of Sitka spruce, USDA Forest Service Research Paper PNW-18, 9 pp.

Herman, F. R. 1967. Growth comparisons of upper-slope conifers in the Cascade Range *Northwest Science* **41**:51-52.

Herman, F. R., and J. F. Franklin. 1976. Errors from application of western hemlock site curves to mountain hemlock, USDA Forest Service Research Note PNW-276, 6 pp.

Herman, R. K. 1963. Temperatures beneath various seedbeds on a clearcut forest area in the Oregon coast range, *Northwest Science* **37**:93-103.

Hermann, R. K., and D. P. Lavender (eds.). 1973. *Even-age management*. Oregon State University School of Forestry Paper 848, Corvallis, 250 pp.

Herring, L. J., and D. E. Etheridge. 1976. *Advance amabilis-fir regeneration in the Vancouver Forest District*, B.C. Forest Service/Canadian Forest Service Joint Report 5, 23 pp.

Hessburg, P. F., R. G. Mitchell, G. M. Filip. 1994. Historical and current roles of insects and pathogens in eastern Oregon and Washington forested landscapes. USDA Forest Service GTR-PNW-321: 55 pp.

Hett, J. M, and O. L. Loucks. 1968. Application of life-table analyses to tree seedlings in Quetico Provincial Park, Ontario, *Forestry Chronicle* **44**:29-32.

Hett, J. M. 1971. A dynamic analysis of age in sugar maple seedlings, *Ecology* **52**:1071-1074.

Hett, J. M., and O. L. Loucks. 1971. Sugar maple *(Acer saccharum* Marsh.) seedling mortality, *Journal of Ecology* **59**:507-520.

Hett, J. M., and O. L. Loucks. 1976. Age structure models of balsam fir and eastern hemlock, *Journal of Ecology* **64**:1029-1044.

Heusser, C. J. 1952. Pollen profiles from southeastern Alaska, *Ecological Monographs* **22**:331-352.

Heyward, F. 1933. The root system of longleaf pine on the deep sands of western Florida, *Ecology* **14**:136-148.

Hibbs, D. E. 1979. The age structure of a striped maple population, *Canadian Journal of Forest Research* **9**:504-508.

Hibbs, D. E. 1982. Gap dynamics in a hemlock-hardwood forest, *Canadian Journal of Forest Research* **12**:522-527.

Hibbs, D. E. 1983. Forty years of forest succession in central New England, *Ecology* **64**:1394-1401.

Hibbs, D. E., B. F. Wilson, and B. C. Fischer. 1980. Habitat requirements and growth of striped maple *(Acer pensylvanicum L.)*, *Ecology* **61**:490-496.

Hicks, D. J., and B. F. Chabot. 1985. Deciduous forest, in B. F. Chabot and H. A. Mooney (eds.), *Physiological Ecology of North American Plant Communities*, Chapman and Hall, New York, pp. 256-277.

Hiley, W. E. 1948. Craib's thinning prescriptions for conifers in South Africa, *Quarterly Journal of Forestry* **42**:5-19.

Hill, D. B. 1985. Forest fragmentation and its implications in central New York, *Forest Ecology and Management* **12**:113-128.

Hill, E. P., D. N. Lasher, and R.B. Roper. 1978. A review of techniques for minimizing beaver and white-tailed deer damage in southern hardwoods, in *Proceedings of the Second Symposium on Southeastern Hardwoods, pp.* 79-89.

Hinckley, T. M., and D. M. Bruckerhoff. 1975. The effects of drought on water relations and stem shrinkage of *Quercus alba, Canadian Journal of Botany* 53:62-72.

Hinckley, T. M., H. Imoto, K. Lee, S. Lacker, Y. Morikawa, K. A. Vogt, C. C. Grier, M. R. Keyes, R. O. Teskey, and V. Seymour. 1984. Impact of tephra deposition on growth in conifers: the year of the eruption. *Canadian Journal of Forest Research* 14:731-739.

Hinckley, T. M., H. Imoto, K. Lee, S. Lacker, Y. Morikawa, K. A. Vogt, C. C. Grier, M. R. Keyes, R. O. Teskey, and V. Seymour. 1984. Impact of tephra deposition on growth in conifers: The year of the eruption, *Canadian Journal of Forest Research* 14b731-739.

Hinckley, T. M., R. O. Teskey, F. Duhme, and H. Richter. 1981. Temperate hardwood forests, in T. T. Kozlowski (ed.), *Water Deficits and Plant Growth, Volume 6,* Academic, New York, pp. 153-208.

Hinds, T. E., and E. M. Wengert. 1977. Growth and decay losses in Colorado aspen, USDA Forest Service Research Paper RM-193, 10 pp.

Hintikka, V. 1972. *Wind-Induced Movements in Forest Trees,* Valtion Painatuskeskus, Helsinki, 56 pp.

Hodges, J. D. 1967. Patterns of photosynthesis under natural conditions, *Ecology* 48:234-242.

Hodges, J. D., and D. R. M. Scott. 1968. Photosynthesis in seedlings of six conifer species under natural environmental conditions, *Ecology* 49:973-981.

Hodgkins, E. J. 1970. Productivity estimation by means of plant indicators in the longleaf pine forests of Alabama, in *Tree Growth anal Forest Soils,* Third North American Forest Soils Conference, 1970, Oregon State University Press, Corvallis, pp. 461-474.

Hofmann, J. V. 1924. The natural regeneration of Douglas-fir in the Pacific Northwest, USDA Department Bulletin No. 1200, 62 pp.

Holland, P. G. 1978. Species turnover in deciduous forest vegetation, *Vegetatio* 38:113-118.

Holmsgaard, E. 1958. Comments on some German and Swedish thinning experiments in Norway spruce, *Forest Science* 4:54-60.

Holsoe, T. 1948. *Crown development and basal area growth of red oak and white ash,* The Harvard Forest, Harvard University, Harvard Forest Paper No. 3:27-33.

Honda, H., and J. B. Fisher. 1978. Tree branch angle: Maximizing effective leaf area, *Science* 199:888-890.

Hook, D. D., and J. R. Scholtens. 1978. Adaptations and flood tolerance of tree species, in D. D. Hook and R. M. M. Crawford (eds.), *Plant Life in Anaerobic Environments,* Ann Arbor Scientific Publications, Ann Arbor, Michigan, pp. 299-331.

Hook, D. D., and J. Stubbs. 1965. Selective cutting and reproduction of cherrybark and shumard oaks, *Journal of Forestry* 63:927-929.

Hook, D. D., P. P. Kormanik, and C. L. Brown. 1970. Early development of sweetgum root sprouts in coastal South Carolina, USDA Forest Service Research Paper SE-62, 6 pp.

Hook, D. D., W. P. LeGrande, and O. G. Langdon. 1967. Stump sprouts on water tupelo *Southern Lumberman* 215:111-112.

Hopkins, H. T., and R. I. Donahue. 1939. Forest tree root development as related to soil morphology, *Soil Science Society of America Proceedings* 4:353.

Hopkins, W. E. 1982. Ecology of white fir, in C. D. Oliver and R. M. Kenady (eds.), *Biology and Management of True Fir in the Pacific Northwest,* University of Washington College of Forest Resources, Institute of Forest Resources Contribution Number 45, pp. 35-41.

Hopwood, D. 1991. *Principles and practices of New Forestry.* British Columbia Ministry of Forests Land Management Report ISSN 0702; no. 71. Victoria, British Columbia, Canada. 95 pp.

Horn, H. S. 1971. *The Adaptive Geometry of Trees,* Princeton University Press, Princeton, New Jersey, 144 pp.

Horn, H. S. 1974. The ecology of secondary succession, *Annual Review of Ecology and Systematics* 5:25-37.

Horsley, S. B. 1977a. Allelopathic inhibition of black cherry by fern, grass, goldenrod, and aster, *Canadian Journal of Forest Research* 7:205-216.

Horsley, S. B. 1977b. Allelopathic inhibition of black cherry: II. Inhibition by woodland grass, ferns, and clubmoss, *Canadian Journal of Forest Research* 7:515-519.

Horton, K. W. 1956. *The ecology of lodgepole pine in Alberta and its role in forest succession,* Canadian Department of Northern Affairs National Resources, Forestry Branch, Forest Research Division Technical Note 45, 29 pp.

Hosner, J. F. 1957. Effects of water on seed germination of bottomland trees, *Forest Science* 3:67-69.

Hosner, J. F. 1960. Relative tolerance to complete inundation of fourteen bottomland tree species, *Forest Science* 6:246-251.

Hosner, J. F., and S. G. Boyce. 1962. Relative shade tolerance to water saturated soil of various bottomland hardwoods, *Forest Science* 8:180-186.

Hough, A. F. 1932. Some diameter distributions in forest stands of northwestern Pennsylvania, *Journal of Forestry* 30:933-943.

Hough, A. F. 1936. A climax forest community on East Tionesta Creek in northwestern Pennsylvania, *Ecology* 17:9-28.

Hough, A. F. 1937. A study of natural tree reproduction in the beech-birch-maple-hem-lock type, *Journal of Forestry* 35:376-378.

Hough, A. F., and R. D. Forbes. 1943. The ecology and silvics of forests in the high plateaus of Pennsylvania, *Ecological Monographs* 13:299-320.

Hough, W. A. 1973. Fuel and weather influence wildfires in sand pine forests, USDA Forest Service Research Paper SE-106, 11 pp.

Houston, D. B. 1973. Wildfires in northern Yellowstone National Park, *Ecology* 54:1111-1117.

Howe, G. E. 1995. *Genetic effects of uneven-aged management.* IN (K. L. O'Hara, editor) *Uneven-aged management: opportunities, constraints, and methodologies.* Montana Forest and Conservation Experiment Station, School of Forestry, University of Montana, Miscellaneous Publication No. 56.

Howe, H. F. 1977. Bird activity and seed dispersal of a tropical wet forest tree, *Ecology* 58:539-550.

Hoyer, G. E. 1980. Height growth of dominant western hemlock trees that had been released from understory suppression, *Washington State Department of Natural Resources Note No. 33,* 5 pp.

Hozumi, K. 1975. Studies on the frequency distribution of the weight of individual trees in a forest stand: V. The M-w diagram for various types of forest stands, *Japanese Journal of Ecology* 25:123-131.

Hozumi, K. 1977. Ecological and mathematical considerations on self-thinning in even-aged pure stands. I. Mean plant weight-density trajectory during the course of self-thinning, *Botanical Magazine of Tokyo* **90**:165-179.

Hozumi, K. 1983. Ecological and mathematical considerations on self-thinning in even-aged pure stands. III. Effect of the linear growth factor on self-thinning and its model, *Botanical Magazine of Tokyo* **96**:171-191.

Hozumi, K., H. Koyama, and T. Kira. 1955. Intraspecific competition among higher plants. IV. A preliminary account on the interaction between adjacent individuals, *Journal of the Institute of Polytechnics of Osaka City University* **6**:121-130.

Hozumi, K., T. Asahira, and T. Kira. 1956. Intraspecific competition among higher plants: VI. Effect of some growth factors on the process of competition, *Journal of the Institute of Polytechnics of Osaka City University* **7**:15-34.

Hubbell, S. P., and R. B. Foster. 1986. Commonness and rarity in a neotropical forest: implications for tropical tree conservation. IN M. E. Soule, editor. *Conservation Biology: The Science of Scarcity and Diversity.* Sinauer Associates, Sunderland, Massachusetts. pp. 205-231.

Huber, B. 1958. Recording gaseous changes under field conditions, in K. V. Thimann (ed.), *The Physiology of Forest Trees,* Ronald, New York, pp. 187-195.

Huff, M. H. 1995. Forest age structure and development following wildfires in the western Olympic Mountains, Washington. *Ecological Applications* **5**(2):471-483.

Hummel, F. C. 1951. *Increment of "free-grown" oak,* British Forestry Commission, Report on Forest Research, 1950, pp. 65-66.

Hummel, F. C. 1954. The definition of thinning treatments, in *Proceedings of the 11th IUFRO Conference,* Rome, 1953, pp. 582-588.

Hunt, R. 1982. *Plant Growth Curses,* University Park Press, Baltimore, 248 pp.

Hunter, M. L., Jr. 1990. *Wildlife, forests, and forestry.* Regents/Prentice Hall, Englewood Cliffs, New Jersey. 370 pp.

Hutchings, M. J. 1979. Weight-density relationships in ramet populations of clonal perennial herbs, with species reference to the $-3/2$ power law, *Journal of Ecology* **67**:21-33.

Hutchings, M. J., and C. S. J. Budd. 1981. Plant competition and its course through time, *BioScience* **31**:640-645.

Hutchins, R. E. 1964. *This is a Tree,* Dodd, Mead, New York, 160 pp.

Hutchinson, B. A., and D. R. Matt. 1977. The distribution of solar radiation within a deciduous forest, *Ecological Monographs* **47**:185-207.

Hutchinson, G. E. 1957. Concluding remarks, *Cold Spring Harbor Symposia on Quantitative Biology* **22**:415-427.

Hutchinson, G. E. 1959. Homage to Santa Rosalia, or why are there so many kinds of animals? *American Naturalist* **93**:145-159.

Hutchison, B. A., and D. R. Matt. 1977. The distribution of solar radiation within a deciduous forest. *Ecological Monographs* **47**:185-207.

Hutnik, R. J. 1952. Reproduction on windfalls in a northern hardwood stand, *Journal of Forestry* **50**:693-694.

Ikusima, I. K. Shinozaki, and T. Kira. 1955. Intraspecific competition among higher plants: III. Growth of duckweed, with a theoretical consideration on the C-D effect, *Journal of the Institute of Polytechnics, Osaka City University* **6**:107-119.

Irion, G. 1989. Quaternary geological history of the Amazon lowlands. IN L. B. Holm-Nielson, I. C. Nielsen, and H. Balslev, editors. *Tropical Forests.* Academic Press, New York. pp. 23-34.

Irwin, L. L., J. G. Cook, R. A. Riggs, and J. M. Skovlin. 1994. Effects of long-term grazing by big game and livestock in the Blue Mountains forest ecosystem. USDA Forest Service GTR-PNW-GTR-325: 53 pp.

Isaac, D., W. H. Marshall, and M. F. Buell. 1959. A record of reverse plant succession in a tamarack bog, *Ecology* **40**:317-320.

Isaac, L. A. 1940. Vegetative succession following logging in the Douglas-fir region with special reference to fire, *Journal of Forestry* **38**:716-721.

Isaac, L. A. 1943. *Reproductive Habits of Douglas-fir,* Charles L. Pack Foundation, Washington, D.C., 107 pp.

Isaac, L. A. 1956. Place of partial cutting in old-growth stands of the Douglas-fir region, USDA Forest Service Pacific Northwest Forest and Range Experiment Station, Research Paper No. 16, 48 pp.

Isaac, L. A., and G. S. Meagher. 1938. Natural reproduction on the Tillamook burn four years after the fire. USDA Forest Service, Pacific Northwest Forest Experiment Station, Portland, Oregon. 19 pp.

Ishikawa, K. 1982. *Guide to Quality Contol.* Asian Productivity Organization. Quality Resources, White Plains, New York. 225 pp.

Ishikawa, M., R. Nitta, T. Katsuta, and T. Fujimori. 1986. Snowbreakage-mechanisms and methods of avoiding it. *Reviews of Studies in Forestry*, No. 83. Forest Science Organization, Tokyo. 101 pp. (In Japanese)

Ishizuka, M. 1981. Development of mixed conifer-hardwood forests—analysis at Jozankei. In: *Analysis of stand structure and regeneration in natural forests (2).* Hokkaido Regional Forestry Office. 216 pp. (In Japanese; original not seen. Cited from Osawa 1992.)

Ives, R. I. 1942. The beaver-meadow complex, *Journal of Geomorphology* **5**:191-203.

Iwaki, H., and T. Totsuka. 1959. Ecological and physiological studies on the vegetation of Mt. Shimagarc: II. On the cresecnt-shaped "dead-tree-strips" in the Yatsugatke and Chichibu mountains, *Botanical Magazine of Tokyo* **72**:255-260.

Jack, W. H. 1971. The influence of tree spacing on Sitka spruce growth, *Irish Forestry* **28**:13-33.

Jackson, L. W. R. 1959. Relation of pine forest overstory opening diameter to growth of pine reproduction, *Ecology* **40**:478-480.

Jackson, M. T. 1966. Effects of microclimate on spring flowering phenology, *Ecology* **47**:407-415.

Jaeck, L. L., C. D. Oliver, and D. S. DeBell. 1984. Young stand development in coastal western hemlock as influenced by three harvesting regimes, *Forest Science* **30**:117-124.

James, R. N., and D. H. Revell. 1974. *Some effects of variation in initial stocking levels of Douglas-fir,* New Zealand Forest Service, Forest Research Institute Symposium No. **15**:138-146.

Janzen, D. H. 1969. Seed-eaters versus seed size, number, toxicity and dispersal, *Evolution* **23**:1-27.

Janzen, D. H. 1982. Neotropical anachronisms: The fruits the gomphotheres ate, *Science* **215**:19-27.

Jaynes, R. A., S. L. Anagnostakis, and N. K. Van Alfen. 1976. *Perspectives in Forest Entomology,* Academic Press, New York, pp. 61-70.

Jensen, E. C., and J. N. Long. 1983. Crown structure of a codominant Douglas-fir, *Canadian Journal of Forest Research* **13**:264-269.

Jensen, M. E., P. S. Bourgeron, editors. 1994. Ecosystem management: principles and applications. USDA Forest Service PNW-GTR-318: 376 pp.

Jensen, T. S. 1982. Seed production and outbreaks of non-cyclic rodent populations in deciduous forests, *Oecologia* **54**:184-192.

Jensen, V. S. 1943. Suggestions for the management of northern hardwood stands in the Northeast, *Journal of Forestry* **41**:180-185.

Jensen, V. S., and R. W. Wilson, Jr. 1951. Mowing of northern hardwood reproduction not profitable, USDA Forest Service Northeast Forest Experiment Station Research Note 3, 4 pp.

Jobling, J., and M. L. Pearce. 1977. *Free growth of oak,* British Forestry Commission Record 113, 17 pp.

Johnson, A. H., and T. G. Siccama. 1983. Acid deposition and forest decline, *Environmental Science and Technology* **17**:294-305.

Johnson, C. G., Jr., R. R. Clausnitzer, P. J. Mehringer, and C. D. Oliver. 1993. Biotic and abiotic processes of eastside Ecosystems: the effects of management on plant and community ecology, and on stand and landscape vegetation dynamics. IN (P. F. Hessburg) *Eastside forest health assessment. Volume III. Assessment.* (Dr. Richard Everett Team leader) USDA Forest Service, National Forest System and Forest Service Research. Wenatchee, Washington. pp. 35-100.

Johnson, D. R., and J. K. Agee. 1987. Introduction to ecosystem management. In(J. K. Agee and D. R. Johnson, editors) *Ecosystem Management for Parks and Wilderness.* University of Washington Press, Seattle. 3-14.

Johnson, E. A. 1975. Buried seed populations in the subarctic forest east of Great Slave Lake, Northwest Territories, *Canadian Journal of Botany* **53**:2933-2941.

Johnson, E. A. 1979. Fire recurrence in the subarctic and its implications for vegetation composition, *Canadian Journal of Botany* **57**:1374-1379.

Johnson, E. A. 1981a. Fire recurrence and vegetation in the lichen woodlands of the Northwest Territories, Canada, in *Proceedings of the Fire History Workshop,* USDA Forest Service General Technical Report RM-81, pp. 110-114.

Johnson, E. A. 1981b. Vegetation organization and dynamics of lichen woodland communities in the Northwest Territories, Canada *Ecology* **62**:200-215.

Johnson, E. A. 1992. *Fires and vegetation dynamics: Studies from North American boreal forest.* Cambridge University Press, Cambridge, U. K.

Johnson, E. A., and D. R. Wowchuck. 1993. Wilfires in the southern Canadian Rocky Mountains and their relationship to mid-tropospheric anomalies. *Canadian Journal of Forest Research* **23**:1213-1222.

Johnson, E. A., and J. L. Kovner. 1956. Effect on steamflow of cutting a forest understory, *Forest Science* **2**:82-91.

Johnson, E. A., and J. S. Rowe. 1977. *Fire and vegetation change in the western subarctic,* Canadian Department of Indian Affairs and Northern Development, ALUR 75-76-61 (cited in Heinselman 1981).

Johnson, J. W. 1972. Silvicultural considerations in clearcutting, in R. D. Nyland (ed.), *A Perspective on Clearcutting in a Changing World,* Proceedings of the Winter Meeting of the Society of American Foresters, pp. 19-24.

Johnson, N. E., and J. G. Zingg. 1968. The balsam woolly aphid on young Pacific silver fir in Washington, *Weyerhaeuser Forestry Paper No. 13,* 10 pp.

Johnson, P. S. 1975. Growth and structural development of red oak sprout clumps, *Forest Science* **21**:413-418.

Johnson, R. G. 1982. Brunhes-Matuyama magnetic reversal dated at 790,000 years B. P. by marine-astronomical correlation, *Quaternary Research* **17**:135-147.

Johnson, R. L. 1964. Coppice regeneration of sweetgum, *Journal of Forestry* **62**:34-35.

Johnson, R. L. 1968. Thinning improves growth in stagnated sweetgum stands, USDA Forest Service Research Note SO-82, 5 pp.

Johnson, R. L. 1970. Renewing hardwood stands on bottomlands and loess, in *Silviculture and Management of the Southern Hardwoods,* Louisiana State University 19th Annual Forestry Symposium Proceedings, pp. 113-121.

Johnson, R. L., and E. C. Burkhardt. 1976. Natural cottonwood stands—past management and implications for plantations, in *Symposium on Eastern Cottonwood and Related Species,* pp. 20-28.

Johnson, R. L., and R. C. Biesterfeldt. 1970. Forestation of hardwoods, *Forest Farmer* 30:14-15, 36-38.

Johnson, R. L., and R. M. Krinard. 1983. Regeneration in small and large sawtimber sweetgum-red oak stands following selection and seed tree harvest: 23-year results, *Southern Journal of Applied Forestry* 7:176-184.

Johnson, R. L., and R. M. Krinard. 1988. Growth and development of two sweetgum-red oak stands from origin through 29 years, *Southern Journal of Applied Forestry* 12:73-78.

Johnson, W. C., R. K. Schreiber, and R. L. Burgess. 1979. Diversity of small mammals in a powerline right-of-way and adjacent forest in east Tennessee, *American Midland Naturalist* 10:231-235.

Johnson, W. C., R. L. Burgess, and W. R. Keammerer. 1976. Forest overstory vegetation and environment on the Missouri River floodplain in North Dakota, *Ecological Monographs* 46:59-84.

Johnstone, W. D. 1976. *Variable-density yield tables for natural stands of lodgepole pine in Alberta,* Northern Forest Research Centre, Canadian Forestry Service, Forestry Technical Report 20, 110 pp.

Johnstone, W. D. 1982. *Juvenile spacing of 25-year-old lodgepole pine in Western Alberta,* Northern Forest Research Centre, Canadian Forestry Service, Information Report NOR-X-244, 19 pp.

Jones, E. P., Jr. 1977. Precommercial thinning of naturally seeded slash pine increases volume and monetary returns, USDA Forest Service Research Paper SE-164, 12 pp.

Jones, E. P., Jr. 1986. Slash pine plantation spacing study—age 30, in *Fourth Biennial Southern Silvicultural Research Conference Proceedings,* USDA Forest Service, General Technical Report, SE-42,pp.45-49.

Jones, E. W. 1945. The structure and reproduction of the virgin forest of the North Temperate zone, *New Phytologist* 44:130-148.

Jones, E. W. 1955. Ecological studies on the rain forest of southern Nigeria: IV. The plateau forest of the Okomu forest reserve: Pt I. The environment, the vegetation types of the forest, and the horizontal distribution of species, *Journal of Ecology* 43:564-594.

Jones, J. R. 1973. Rocky Mountain aspen, in *Silviculture Systems for Major Forest Tapes in the United States,* USDA Agriculture Handbook 445, pp. 49-51.

Jordan, C. F. 1982. Amazon rain forests. *American Scientist* 70:394-401.

Jordan, C. R., J. R. Kline, and D. S. Sasscer. 1972. Relative stability of mineral cycles in forest ecosystems, *American Naturalist* 106:237-254.

Jump, J. A. 1938. A study of forking in red pine, *Phytopathology* 28:798-811.

Junk, W. J. 1989. Flood tolerance and tree distribution in central Amazonian floodplains. IN L. B. Holm-Nielson, I. C. Nielsen, and H. Balslev, editors. *Tropical Forests.* Academic Press, New York. pp. 47-64.

Kairyukshtis L. A. 1964. A method of thinning mixed hardwood-spruce stands, Lithuanian Scientific Research Institute for Forestry (trans. by Israel Program for Scientific Translations, 1968), 12 pp.

Kaji, K., and T. Yajima. 1987. *Influence of shika deer on forests of Nakanoshima Island,*

Hokkaido. Trans. Congr. Int. Union Game Biol. Suppl. **18**:83-84.

Kaji, K., T. Yajima, and T. Igarashi. 1990. Forage selection by shika deer introduced on Nakanoshima Island and its effect on the forest vegetation. *Proceedings of the International Symposium on Wildlife Conservation,* in Tsukuba and Yokohama, Japan, August 21-25, 1990.

Kaji, T. Koizumi, and N. Ohtaishi. 1988. Effects of resource limitation on the physical condition of Shika deer on Nakanoshima Island, Hokkaido. Acta Theriol. **33**:187-208.

Kamitani, T., Y. Kawaguchi, H. Miguchi, and H. Takeda. 1995. *Rhus trichocarpa* patch as a contributing factor in accelerating forest dynamics. *Journal of Sustainable Forestry* (In press).

Kanzaki, N. 1987. Processes of regeneration in a subalpine mixed coniferous forest and their implications for forest management. IN (T. Fujimori and M. Kimura, editors) *Human impacts and management of mountain forests.* Forestry and Forest Products Research Institute, Ibaraki, Japan: 341-350.

Karnig, J. J., and W. H. Lyford. 1968. Oak mortality and drought in the Hudson Highlands, Harvard University, *Harvard Black Rock Forest Paper No. 29,* 13 pp.

Karsian, A. E. 1995. A 6800 year vegetation and fire history of the Bitterroot Mountain Range, Montana. Unpublished Master of Science Thesis. University of Montana, College of Forestry, Missoula, Montana, U.S.A.

Kauffert, F. H. 1933. Fire and decay injury in the southern bottomland hardwoods, *Journal of Forestry* **31**:64-67.

Kaufman, M. R., and C. A. Troendle. 1981. The relationship of leaf area and foliage biomass to sapwood conducting area in four subalpine forest tree species, *Forest Science* **27**:477-482.

Keeley, J. E. 1979. Population differentiation along a flood frequency gradient: physiological adaptation to flooding in *Nyssa sylvatica, Ecological Monographs* **49**:98-108.

Keeley, J. E., and P. H. Zedler. 1978. Reproduction of chaparral shrubs after fire: A comparison of sprouting and seeding strategies, *American Midland Naturalist* **99**:142-161.

Keen, F. P. 1936. Relative susceptibility of Ponderosa pine to bark-beetle attack, *Journal of Forestry* **34**:919-927.

Keen, F. P. 1943. Ponderosa pine tree classes redefined, *Journal of Forestry* **41**:249-253.

Kelker, G. H. 1964. Appraisal of ideas advanced by Aldo Leopold thirty years ago, *Journal of Wildlife Management* **28**:180-185.

Kellomaki, S., and M. Kanninen. 1980. Eco-physiological studies on young scots pine stands: IV. Allocation of photosynthates for crown and stem growth, *Silva Fennica* **14**:397-408.

Kelty, M. J. 1986. Development patterns in two hemlock-hardwood stands in southern New England, *Canadian Journal of Forest Research* **16**:885-891.

Kelty, M. J. 1989. Productivity of New England hemlock-hardwood stands as affected by species composition and canopy structure, *Forest Ecology and Management* **28**:237-257.

Kelty, M. J. 1992. Comparative productivity of monocultures and mixed-species stands. *In*: (M. J. Kelty, B. C. Larson and C. D. Oliver, editors) *The Ecology and Silviculture of Mixed-Species Forests: A Festschrift for David M. Smith.* Kluwer Academic Publishers, Boston:

Kelty, M. J., E. M. Gould, Jr., and M. J. Twery. 1987. Effects of understory removal in hardwood stands, *Northern Journal of Applied Forestry* **4**:162-164.

Kemp, J. L. 1967. *Epitaph for the giants: the story of the Tillamook burn*. The Touchstone Press, Portland, Oregon, U.S.A. 110 pp.

Kennedy, H. E., Jr., and R. M. Krinard. 1974. 1973 Mississippi River flood's impact on natural hardwood forests and plantations, USDA Forest Service Research Note SO-177, 6 pp.

Ker, J. W. 1953. Growth of immature Douglas-fir by tree classes *Forestry Chronicle* **29**:367-373.

Kershaw, K. A. 1963. Pattern in vegetation and its causality, *Ecology* **44**:377-388.

Kessel, S. R. 1976. Gradient modeling: a new approach to fire modeling and wilderness resource management, *Environmental Management* **1**:39-48.

Keyes, M. R., and C. C. Grier. 1981. Above- and below-ground net production in 40-year-old Douglas-fir stands on low and high productivity sites, *Canadian Journal of Forest Research* **11**:599-605.

Kilgore, B. M. 1973. The ecological role of fire in the Sierran conifer forests: Its application to national park management, *Quaternary Research* **3**:496-513.

Kilgore, B. M., and D. Taylor. 1979. Fire history of a sequoia-mixed conifer forest, *Ecology* **60**:129-142.

Kimmins, J. P. 1987. *Forest Ecology*, Macmillan, New York, 531 pp.

Kimmins, J. P. 1987. *Forest Ecology*. Macmillan Publishing Company, New York. 531 pp.

Kimney, J. W., and R. L. Furniss. 1943. Deterioration of fire-killed Douglas-firs, USDA Forest Service Technical Bulletin 851, 61 pp.

King, J. E. 1966. Site index curves for Douglas-fir in the Pacific Northwest, *Weyerhaeuser Forestry Paper 8*, Weyerhaeuser Company Research Center, Centralia, Washington, 49 pp.

King, J. E. 1970. Principles of growing stock classification for even-aged stands and an application to natural Douglas-fir forests, unpublished Ph.D. dissertation, University of Washington, Seattle, 91 pp.

Kira, T., H. Ogawa, and K. Hozumi. 1954. Intraspecific competition among higher plants: II. Further discussions on Mitscherlich's law, *Journal of the Institute of Polytechnics, Osaka City University* **5**:1-7.

Kira, T., H. Ogawa, and N. Sakazaki. 1953. Intraspecific competition among higher plants: I. Competition-density-yield interrelationship in regularly dispersed populations, *Journal of the Institute of Polytechnics, Osaka City University* **4**:1-16.

Kira, T., H. Ogawa, K. Hozumi, H. Koyama, and K. Yoda. 1956. Intraspecific competition among higher plants: V. Supplementary notes on the C-D effect, *Journal of the Institute of Polytechnics, Osaka City University* **7**:1-14.

Kira, T., N. Shinozaki, and K. Hozumi. 1969. Structure of forest canopies as related to their primary productivity, *Plant and Cell Physiology* **10**:126-142.

Kirkland, B. P., and A. J. F. Brandstrom. 1936. *Selective timber management in the Douglas-fir region*, Charles Lathrop Pack Forestry Foundation, Washington, D.C., 122 pp.

Kitamoto, T., and T. Shidei. 1972. Studies on the spatial pattern of forest trees: I. Distribution of dominant and suppressed trees in even-aged forests (with English summary), *Bulletin of the Kyoto University Forest* **43**:152-161.

Kittredge, D. B. 1988. The influence of species composition on the growth of individual red oaks in mixed stands in southern New England, *Canadian Journal of Forest Research* **18**:1550-1555.

Kittredge, J. 1952. Deterioration of site quality by erosion, *Journal of Forestry*, **50**:554-556.

Klemmedson, J. O. 1976. Effect of thinning and slash burning on nitrogen and carbon in ecosystems of young dense Ponderosa pine, *Forest Science* 22:45-53.

Klinka, K., G. J. Kayahara, R. E. Carter, and K. D. Coates. 1993. Light-growth relationships in conifers: II. Characterization of shade tolerance of seven conifers. (Unpublished manuscript)

Knight, D. H. 1975. A phytosociological analysis of species-rich tropical forest on Barro Colorado Island, Panama, *Ecological Monographs* 45:259-284.

Knight, H. A. 1987. The pine decline, *Journal of Forestry* 85:25-28.

Koevenig, J. L. 1976. Effect of climate, soil physiography, and seed germination on the distribution of river birch *(Betula nigra), Rhodora* 78:420-437.

Kohyama, H., and T. Kira. 1956. Intraspecific competition among higher plants: VIII. Frequency distribution of individual plant weight as affected by the interactions between plants, *Journal of the Institute of Polytechnics, Osaka City University* 7:73-94.

Kohyama, T. 1980. Growth pattern of *Abies mariesii* saplings under conditions of open-growth and suppression, *Botanical Magazine of Tokyo* 93:13-24.

Kohyama, T. 1983. Seedling stage of two subalpine Abies species in distinction from sapling stage: A matter-economic analysis, *Botanical Magazine of Tokyo* 96:49-65.

Kohyama, T., and N. Fujita. 1981. Studies on the Abies population of Mt. Shimagare: I. Survivorship curve, *Botanical Magazine of Tokyo* 94:55-68.

Kolstrom, T. 1992. Dynamics of uneven-aged stands of Norway spruce: a model approach. Academic dissertation presented to Faculty of Forestry, University of Joensuu. The Finnish Forest Research Institute, Department of Forest Production, Joensuu Research Station, Joensuu, Finnland. 21 pp.

Kominami, Y., H. Tanouchi, and T. Sato. 1995. Seed dispersal by birds and seedling survival of Daphniphyllum marcopodum in an old-growth evergreen broad-leaved forest. *Journal of Sustainable Forestry* (In press).

Kormanick, P. P. 1968. Suppressed buds and epicormic branching in sweetgum, unpublished Ph.D. thesis, University of Georgia, Athens, Georgia, 108 pp.

Kormanik, P. P., and C. L. Brown. 1964. Origin of secondary dormant buds in sweetgum, USDA Forest Service Research Note SE-36, 4 pp.

Kormanik, P. P., and C. L. Brown. 1967. Root buds and the development of root suckers in sweetgum, *Forest Science* 13:338-345.

Korstian, C. F. 1917. The indicator significance of native vegetation in the determination of forest site, *Plant World* 20:267-287.

Korstian, C. F., and P. W. Stickel. 1927. The natural replacement of blight-killed chestnut in the hardwood forests of the Northeast, *Journal of Agricultural Research* 34:631-648.

Korstian, C. F., and T. S. Coile. 1938. *Plant competition in forest stands*, Duke University School of Forestry Bulletin No. 3, 125 pp.

Kotar, J. 1972. Ecology of *Abies amabilis* in relation to its altitudinal distribution and in contrast to its common associate *Tsuga heterophylla*, unpublished Ph.D. dissertation, University of Washington, Seattle, 171 pp.

Kozlowski, T. T. 1943. Light and water in relation to growth and competition of Piedmont forest tree species, *Ecological Monographs* 19:208-231.

Kozlowski, T. T. 1971. *Growth and Development of Trees*, Volumes I and II, Academic, New York, Vol. I: 443 pp., Vol. II: 514 pp.

Kozlowski, T. T. 1974. *Fire and Ecosystems*, Academic, New York.

Kozlowski, T. T., and W. H. Scholtes. 1948. Growth of roots and root hairs of pine and hardwood seedlings in the Piedmont, *Journal of Forestry* 46:750-754.

Kozlowski, T. T., P. J. Kramer, and S. G. Pallardy. 1991. *The physiological ecology of woody plants.* Academic Press, New York. 657 pp.

Kraft, G. 1884. Zur Lehre von den Durch Forstungen. Schlagstellungen und Lichtungshieben, Hannover (cited by Assmann 1970 and Daniel, Helms, and Baker 1979).

Krajicek, J. E. 1955. Rodents influence red oak regeneration, USDA Forest Service Central States Forest Experiment Station, Station Note 91, 2 pp.

Krajicek, J. E. 1959. Epicormic branching in even-aged, undisturbed white oak stands, *Journal of Forestry* 57:372-373.

Krajicek, J. E., and K. A. Brinkman. 1957. Crown development: An index of stand density, USDA Forest Service, Central Forest Experiment Station, Station Note 108, 2 pp.

Krajina, V. J. 1969. Ecology of forest trees in British Columbia, *Ecology of Western North America* 2:1-147.

Kramer, H. 1966. Crown development in conifer stands in Scotland as influenced by initial spacing and subsequent thinning treatment, *Forestry* 39:40-58.

Kramer, H. 1977. Thinning in Norway spruce stands in the Federal Republic of Germany, paper presented at the University of Maine School of Forest Resources, Orono, September 27, 1977, 17 pp.

Kramer, P. J., and T. T. Kozlowski. 1960. *Physiology of Trees,* McGraw-Hill, New York, 642 pp.

Kramer, P. J., and T. T. Kozlowski. 1979. *Physiology of Woody Plants,* Academic, New York, 811 pp.

Krammes, J. S. 1960. Erosion from mountainside slopes after fire in southern California, USDA Forest Service Pacific Southwestern Experiment Station Research Note 171, 8 pp.

Krause, H. H., G. F. Weetman, and P. A. Arp. 1978. Nutrient cycling in boreal forest ecosystems of North America, in C. T. Youngberg (ed.), *Proceedings of the North American Forest Soils Conference,* Ft. Collins, Colorado, August 1978, pp. 287-319.

Krefting, L. W. 1974. Moose distribution and habitat selection in North Central North America. *Le Nauraliste Canadien* 101:81-100.

Krefting, L. W. 1974. Moose distribution and habitat selection in North Central North America, *Le Naturaliste Canadien* 101:81-100.

Krefting, L. W., and E. I. Roe. 1949. The role of some birds and mammals in seed germination, *Ecological Monographs* 19:269-286.

Kriebel, H. B. 1960. Soil moisture distribution between trees in a thinned loblolly pine plantation, *Journal of Forestry* 58:221-222.

Krinard, R. M., and R. L. Johnson. 1975. Ten-year results in a cottonwood plantation spacing study, USDA Forest Service Research Paper SO-106, 10 pp.

Kruckeberg, A. R. 1954. The ecology of serpentine soils: III. Plant species in relation to serpentine soils, *Ecology* 35:267-274.

Krueger, K. W. 1959. Diameter growth of plantation-grown Douglas-fir trees under varying degrees of release, USDA Forest Service Pacific Northwest Forest and Range Experiment Station Research Note No. 168, 5 pp.

Krugman, S. L., W. I. Stein, and D. M. Schmitt. 1974. Seed biology, in C. S. Schopmeyer (technical coordinator), *Seeds of Woody Plants in the United States,* USDA Agricultural Handbook No. 450, pp. 5-40.

Kuehne, M. J. 1993. High quality forestry: an alternative for management of national forest lands in western Washington. IN(Weigand, J. F., R. W. Haynes, and J. L.

Mikowski, compilers) *High quality forestry: the idea of long rotations.* University of Washington, College of Forest Resources, Center for International Trade in Forest Products CINTRAFOR Special Paper **15**:211-265.

Kuiper, L. C. 1988. The structure of natural Douglas-fir forests in western Washington and western Oregon, *Agricultural University Wageningen Papers* 88-5, 47 pp.

Kushima, H. 1989. Canopy stratification and its development of post-fire stands at Tokyo University Hokkaido Experiment Forest. In: *Proceedings of the 36th annual meeting of Japanese Ecological Society, Kushiro.* (In Japanese; original not seen. Cited from Osawa 1992.)

Laacke, R. J. 1982. Management experience in southern Cascades white fir stands, in C. D. Oliver and R. M. Kenady (eds.), *Biology and Management of True Fir in the Pacific Northwest,* University of Washington College of Forest Resources, Institute of Forest Resources Contribution Number 45, pp. 335-336.

Labyak, L. F., and F. X. Schumacher. 1954. The contribution of its branches to the main stem growth of loblolly pine, *Journal of Forestry* **52**:333-337.

Laessle, A. M. 1965. Spacing and competition in natural stands of sand pine, *Ecology* **46**:65-72.

Lamb, H. H. 1963. On the nature of certain climatic epochs which differ from the modern (1900-1939) norm, in *Changes of Climate,* United Nations Educational, Scientific, and Cultural Organization, Paris, France, Arid Zone Research Series 20, pp. 125-150.

Lamb, H. H. 1977. *Climatic History and the Future,* Princeton University Press, Princeton, New Jersey, 835 pp.

Lamb, H. H. 1988. *Weather, Climate, and Human Affairs,* Routledge, New York, 364 pp.

Lamb, R. C. 1973. Apparent influence of weather upon seed production of loblolly pine, USDA Forest Service Research Note SE-183, 7 pp.

Lang, A. 1963. Effects of some internal and external conditions on seed germination, *Encyclopedia of Plant Physiology* **15**:848-893.

Lang, F. C. 1952. *The Story of Trees,* Doubleday, Garden City, New York, 384 pp.

Langenheim, J. H. 1956. Plant succession on a subalpine earthflow in Colorado, *Ecology* **37**:301-317.

Langsaeter, A. 1941. Omtynning i analdret granfurvskog Maddel, *Norsite Skogfor soksvenson* **8**:131-216. (cited by Smith 1962).

Lanner, R. M. 1966. The phenology and growth habits of pines in Hawaii, USDA Forest Service Research Paper PSW-29.

Lanner, R. M. 1966. The phenology and growth habits of pines in Hawaii, USDA Forest Service Research Paper PSW-29, 25 pp.

Larcher, W. 1983. *Physiological Plant Ecology,* Springer-Verlag, New York, 303 pp.

Larsen, D. R. 1994. Adaptable stand dynamics model for integrating site-specific growth and innovative silvicultural prescriptions. *Forest Ecology and Management* **69**:245-257.

Larsen, L. T., and T. D. Woodbury. 1916. Sugar Pine, USDA Bulletin 426, 40 pp.

Larson, B. C. 1982. Development and growth of even-aged and multi-aged mixed stands of Douglas-fir and grand fir on the east slope of the Washington Cascades, unpublished Ph.D. dissertation, University of Washington, Seattle, 219 pp.

Larson, B. C. 1986. Development and growth of even-aged stands of Douglas-fir and grand fir, *Canadian Journal of Forest Research* **16**:367-372.

Larson, B. C. 1992. Pathways of development in mixed-species stands. *In*: (M. J. Kelty, B. C. Larson and C. D. Oliver, editors) *The Ecology and Silviculture of Mixed-*

Species Forests: A Festschrift for David M. Smith. Kluwer Academic Publishers, Boston:

Larson, B. C., and C. D. Oliver. 1982. Forest dynamics and fuelwood supply of the Stehekin Valley, Washington, in *Ecological Research in National Parks of the Pacific Northwest,* Oregon State University Forest Research Laboratory Publication, Corvallis, pp. 127-134.

Larson, B. C., and I. R. Cameron. 1986. Guidelines for thinning Douglas-fir: Uses and limitations, in C. D. Oliver, D. Hanley, and J. Johnson (eds.), *Douglas-fir: Stand Management for the Future,* College of Forest Resources, University of Washington, Seattle, Institute of Forest Resources Contribution No. 55, pp. 310-316.

Larson, B. C., and M. N. Zaman. 1985. Thinning guidelines for teak (Tectona grandis L.). *The Malaysian Forester* 48(4):288-297.

Larson, F. 1940. The role of bison in maintaining the short grass plain, *Ecology* 21:113-121.

Larson, P. R. 1956. Discontinuous growth rings in suppressed slash pine, *Tropical Woods* 104:80-99.

Larson, P. R. 1969. *Wood formation and the concept of wood quality,* Yale University School of Forestry and Environmental Studies Bulletin 74, 54 pp.

Larsson, S., R. Oren, R. H. Waring, and J. W. Barrett. 1983. Attacks of mountain pine beetle as related to tree vigor of ponderosa pine, *Forest Science* 29:395-402.

Lauga, J., and J. Joachim. 1992. Modelling the effects of forest fragmentation on certain spcies of forest-breeding birds. *Landscape Ecology* 6(3):183-193.

Laven, R. D., P. N. Omi, J. G. Wyant, and A. S. Pinkerton. 1980. Interpretation of fire scar data from a ponderosa pine ecosystem in the central Rocky Mountains, Colorado, in *Proceedings of the Fire History Workshop,* October 20-24, 1980, Tucson, Arizona, USDA Forest Service General Technical Report RM-81, pp. 46-49.

Lawrence, D. B. 1958. Glaciers and vegetation in southeastern Alaska, *American Scientist* 46:89-122.

Lawson, J. 1709. *A new voyage to Carolina.* London, 1709 (cited by Savage 1970).

Leak, W. B. 1961. Yellow birch grows better in mixed-wood stands than in northern hardwood old-growth stands, USDA Forest Service Northeastern Forest Experiment Station Research Note No. 122, 4 pp.

Leak, W. B. 1964. An expression of diameter distribution for unbalanced, uneven-aged stands and forests, *Forest Science* 10:39-50.

Leak, W. B. 1965. The shaped probability distribution, *Forest Science* 11:405-409.

Leak, W. B. 1973. Species and structure of a virgin northern hardwood stand in New Hampshire, USDA Forest Service Research Note NE-81, 4 pp.

Leak, W. B. 1975. Age distribution in virgin red spruce and northern hardwoods, *Ecology* 56:1451-1454.

Leak, W. B., and D. S. Solomon. 1975. Influence of residual stand density on regeneration of northern hardwoods, USDA Forest Service Research Paper NE-310, 7 pp.

Leak, W. B., and S. M. Filip. 1975. Uneven-aged management of northern hardwoods in New England, USDA Forest Service Research Paper NE-332, 15 pp.

Leak, W. B., and S. M. Filip. 1977. Thirty-eight years of group selection in New England northern hardwoods, *Journal of Forestry* 75:641-643.

Leak, W. B., D. S. Solomon, and S. M. Filip. 1969. Silvicultural guide for northern hardwoods in the Northeast, USDA Forest Service Research Paper NE-143, 34 pp.

Leary, R. A. 1987. Some factors that will affect the next generation of forest growth models, in A. R. Ek, S. R. Shirley, and T. E. Burk (eds.), *Forest Growth Modelling and Prediction,* Proceedings of the IUFRO conference, August 23-27, 1987 Minneapolis, Minnesota. USDA Forest Service. General Technical Report NC-120, pp. 22-32.

LeBarron, R. K. 1945. Adjustment of black spruce root systems to increasing depth of peat, *Ecology* 26:309-311.

LeBlanc, J. H. 1955. A mode of vegetative propagation in black spruce, *Pulp and Paper Magazine of Canada* 56:146-l 53.

Lee, Y. 1971. Predicting mortality for even-aged stands of lodgepole pine, *Forest Chronicle* 47:29-32.

Lehmkuhl, J. F., P. F. Hessburg, R. D. Ottmar, M. H. Huff, R. L. Everett, E. C. Alvarado, R. E. Vihnanek. 1994. Historical and current vegetation pattern and associated changes in insect and disease hazard, and fire and smoke conditions in eastern Oregon and Washington. USDA Forest Service GTR-PNW-328: 88 pp.

Lehmkuhl, J. F., P. F. Hessburg, R. L. Everett, M. H. Huff, and R. D. Ottmar. 1994. Historical and current forest landscapes of eastern Oregon and Washington: Part 1. Vegetation pattern and insect and disease hazards. USDA Forest Service General Technical Report PNW-GTR-328. 88 pp.

Leiberg, J. B. 1904. Forest conditions in the Absaroka Division of the Yellowstone Forest Reserve, Montana, and Livingston and Big Timber quadrangles. U. S. Geological Survey Professional Paper 29: 1-148. (Original not seen; cited from Wellner 1970.)

Lemieux, G. J. 1965. Soil-vegetation relationships in the northern hardwoods of Quebec, in C. T. Youngberg (ed.), *Forest-soil Relationships in North America* Oregon State University Press, Corvallis, pp. 163-176.

Lemon, P. C. 1945. Wood as a substratum for perennial plants in the Southeast, *American Midland Naturalist* 34:744-749.

Lemon, P. C. 1961. Forest ecology of ice storms, *Bulletin of the Torrey Botanical Club* 88:21-29.

Lemon, P. C., Jr. 1937. Tree history—six fires, USDA Forest Service Rocky Mountain Region Bulletin 20, 13 pp.

Lenski, R. E. 1982. Effects of forest cutting on two Carabus species: evidence of competition for food. *Ecology* 63:1211-1217.

Leopold, A. 1932. *Game Management,* Scribner, New York, 481 pp.

Leopold, A. C., and P. E. Kriedemann. 1975. *Plant Growth and Development,* 2d ed., McGraw-Hill, New York, 545 pp.

Leopold, A. 1924. Grass, brush, timber, and fire in southern Arizona, *Journal of Forestry* 22:1-10.

Leopold, E. B., C. S. Bortz, and J. A. Kershaw. 1989. Reconstruction of presettlement forest types from timber tax assessment cruise data, King County, Washington (submitted for publication).

Levin, S. A., and R. T. Paine. 1974. Disturbance, patch formation and community structure, *Proceedings of the National Academy of Science of the United States of America* 71:2744-2747.

Levitt, J. 1980. *Responses of Plants to Environmental Stresses: Volume II. Water, Radiation, Salt, and Other Stresses,* Academic Press, New York, 607 pp.

Libby, W. F. 1951. Radiocarbon dates, 11, *Science* 114:291-296.

Lieberman, M., D. Lieberman, and R. Peralta. 1989. Forests are not just Swiss cheese: canopy stereogeometry of non-gaps in tropical forests. *Ecology* 70.

Liegel, L. H., W. E. Balmer, and G. W. Ryan. 1985. Honduras pine spacing trial results in Puerto Rico, *Southern Journal of Applied Forestry* 9:69-75.

Lightle, P. C. 1960. Brown-spot needle blight of longleaf pine, USDA Forest Pest Leaflett 44.

Likens, G. E., F. H. Bormann, N. M. Johnson, D. W. Fisher, and R. S. Pierce. 1970. Effects of forest cutting and herbicide treatment on nutrient budgets in the Hubbard Brook watershed ecosystem, *Ecological Monographs* 40:23-47.

Liming, F. G., and J. P. Johnston. 1944. Reproduction in oak-hickory forest stands of the Missouri Ozarks, *Journal of Forestry* 42:175-180.

Lincoln, R. J., G. A. Boxshall, and P. F. Clark. 1982. *A Dictionary of Ecology, Evolution and Sytematics,* Cambridge University Press, Cambridge, England, 298 pp.

Lippke, B. 1994. Incentives for managing landscapes to meet non-timber goals: lessons from the Washington Landscape Management Project. Paper presented at the Environmental Economics Conference, Banff, Alberta, Canada. October 12-15, 1994. Professor Lippke is Director, Center for International Trade in Forest Products, College of Forest Resources, University of Washington, Seattle, Washington, U.S.A. 13 pp.

Lippke, B., and C. D. Oliver. 1993. How can management for wildlife habitat, biodiversity, and other values be most cost-effective? *Journal of Forestry.* 91:14-18.

Livingston, R. B. 1972. Influence of birds, stones, and soil on the establishment of pasture juniper, *Juniperus communis,* and red cedar, *Juniperus virginiana* in New England pastures, *Ecology* 53:1141-1147.

Livingston, R. B., and M. L. Allessio. 1968. Buried viable seed in successional field and forest stands, Harvard Forest, Massachusetts, *Bulletin of the Torrey Botanical Club* 95:58-69.

Lloyd, F. T. 1981. How many tree heights should you measure for natural, Atlantic Coastal Plain loblolly site index? *Southern Journal of Applied Forestry* 5:180-183.

Lloyd, F. T., and E. P. Jones, Jr. 1983. Density effects of height growth and its implications for site index prediction and growth projection, in *Proceedings of Second Biennial Southern Silviculture Conference,* USDA Forest Service General Technical Report SE-24, pp. 329-333.

Lloyd, F. T., and W. R. Harms. 1986. An individual stand growth model for mean plant size based on the rule of self thinning, *Annals of Botany* 67:681-688.

Loftis, D. L. 1983. Regenerating southern Appalachian mixed hardwoods with the sheltered method, *Southern Journal of Applied Forestry* 7:212-217.

Loftis, D. L. 1985. Preharvest herbicide treatment improves regeneration in southern Appalachian hardwoods, *Southern Journal of Applied Forestry* 9:177-180.

Long, E. D. 1991. Development of mixed white spruce and paper birch forest stands of western and south central Alaska. Unpublished Masters thesis, University of Washington, College of Forest Resources, Seattle, Washington. 53 pp. (appendixes additional)

Long, J. N. 1976. Forest vegetation dynamics within the *Abies amabilis* zone of a western Cascades watershed, unpublished Ph.D. thesis, University of Washington, Seattle, 174 pp.

Long, J. N. 1995. Using stand density index to regulate stocking in uneven-aged stands. IN (K. L. O'Hara, editor) *Uneven-aged management: opportunities, constraints, and methodologies.* Montana Forest and Conservation Experiment Station, School of Forestry, University of Montana, Miscellaneous Publication No. 56.

Long, J. N., and F. W. Smith. 1984. Relations between size and density in developing stands: A description and possible mechanism, *Forest Ecology and Management* 7:191-206.

Long, J. N., and J. Turner. 1975. Above ground biomass of understory and overstory in an age sequence of four Douglas-for stands, *Journal of Applied Ecology* **12**:179-188.

Long, J. N., and T. W. Daniel. 1990. Assessment of growing stock in uneven-aged stands. *Western Journal of Applied Forestry* **5**:93-96.

Long, J. N., F. W. Smith, and D. R. M. Scott. 1981. The role of Douglas-fir stem sapwood and heartwood in the mechanical and physiological support of crowns and development of stem form, *Canadian Journal of Forest Research* **11**:459-464.

Long, J. N., J. B. McCarter, and S. B. Jack. 1988. A modified density management diagram for coastal Douglas-fir, *Western Journal of Applied Forestry* **3**:88-89.

Lonsdale, W. M., and A. R. Watkinson. 1983. Plant geometry and self thinning, *Journal of Ecology* **71**:285-297.

Loope, L. L. and G. E. Gruell. 1973. The ecological role of fire in the Jackson Hole area, northwestern Wyoming, *Journal of Quaternary Research* **3**:444-464.

Lopez-Portilla, J., M. R. Keyes, A. Gonzalez, E. Cabrera C., and O. Sanchez. 1990. Los incendios de Quintana Roo: catastrofe ecologica o evento periodico ? *Ciencia Y Desarrollo* **16** 91):43-57.

Lorimer, C. G. 1977a. Stand history and dynamics of a southern Appalachian virgin forest, unpublished Ph.D. dissertation, Duke University, Durham, North Carolina, 200 pp.

Lorimer, C. G. 1977b. The presettlement forest and natural disturbance cycle of north eastern Maine, *Ecology* **58**:139-148.

Lorimer, C. G. 1980. Age structure and disturbance history of a southern Appalachian virgin forest, *Ecology* **61**:1169-1184.

Lorimer, C. G. 1981. Survival and growth of understory trees in oak forests of the Hudson Highlands, New York, *Canadian Journal of Forest Research* **11**:689-695.

Lorimer, C. G. 1983a. Eighty-year development of northern red oak after partial cutting in a mixed-species Wisconsin forest, *Forest Science* **29**:371-383.

Lorimer, C. G. 1983b. Tests of age-independent competition indices for individual trees in natural hardwood stands, *Forest Ecology and Management* **6**:343-360.

Lorimer, C. G., and A. G. Krug. 1983. Diameter distributions in even-aged stands of shade-tolerant and midtolerant tree species, *American Midland Naturalist* **109**:331-345.

Lorimer, C. G., and L. E. Frelich. 1984. A simulation of equilibrium diameter distributions of sugar maple *(Acer saccharum), Bulletin of the Torrey Botanical Club* **111**:193-199.

Lorimer, C. G., and L. E. Frelich. 1994. Natural disturbance regimes in old-growth northern hardwoods. *Journal of Forestry* **92**:33-38.

Lorimer, N. D., R. G. Haight, and R. A. Leary. 1994. The fractal forest: fractal geometry and applications in forest science. USDA Forest Service GTR-NC-170.

Lorio, P. L., Jr., and J. D. Hodges. 1977. Tree water status affects induced southern pine beetle attack and brood production, USDA Forest Service Research Paper SO-1356, 7 pp.

Lorio, P. L., Jr., V. K. Howe, and C. N. Martin. 1972. Loblolly pine rooting varies with microrelief on wet sites, *Ecology* **53**:1134-1140.

Losee, J. 1972. *A Historical Introduction to the Philosophy of Science,* Oxford University Press, Oxford, 218 pp.

Loucks, O. 1970. Evolution of diversity, efficiency, and community stability, *American Zoologist* **10**:17-25.

Lovett, G., W. Reiners, and R. Olson. 1982. Cloud droplet deposition in subalpine balsam fir forests: Hydrological and chemical inputs, *Science* **218**:1303-1304.

Lowery, R. F. 1972. Ecology of subalpine zone tree clumps in the North Cascade Mountains of Washington, unpublished Ph.D. dissertation, University of Washington, Seattle, 137 pp.

Lucia, E. 1983. *Tillamook burn country: a pictorial history.* The Caxton Printers, Ltd., Caldwell, Ohio, U.S.A. 305 pp.

Ludlum, D. M. 1963. *Early American Hurricanes 1492-1870,* American Meteorological Society, Boston, Massachusetts.

Lutz, H. J. 1928. *Trends and silvicultural significance of upland forest successions in southern New England,* Yale University School of Forestry and Environmental Studies Bulletin Number 22, 68 pp.

Lutz, H. J. 1930. The vegetation of Heart's Content, a virgin forest in northwestern Pennsylvania, *Ecology* **11**:1-29.

Lutz, H. J. 1932. Relation of forest site quality to number of plant individuals per unit area, *Journal of Forestry* **30**:34-38.

Lutz, H. J. 1940. *Disturbance of forest soil resulting from the uprooting of trees,* Yale University School of Forestry Bulletin 45, 37 pp.

Lutz, H. J. 1945. Vegetation on a trenched plot twenty-one years after establishment, *Ecology* **26**:200-202.

Lutz, H. J. 1956. *Ecological effects of forest fires in the interior of Alaska.* U. S. Department of Agriculture, Technical Bulletin No. 1133. 121 pp.

Lutz, H. J. 1959. *Aboriginal man and white man as historical causes of fires in the boreal forest, with particular reference to Alaska.* Yale University School of Forestry Bulletin 65: 49 pp.

Lutz, H. J., and F. S. Griswold. 1939. The influence of tree roots on soil morphology, *The American Journal of Science* **237**:389-400.

Lutz, H. J., and R. F. Chandler, Jr. 1946. *Forest Soils,* Wiley, London, 514 pp.

Lutz, R. J., and A. C. Cline. 1947. *Results of the first thirty years of experimentation in silviculture in the Harvard Forest, 1908-1938,* The Harvard Forest, Harvard University, Harvard Forest Bulletin 23, 182 pp.

Lyford, W. H. 1964a. Water table fluctuations in periodically wet soils of central New England, the Harvard Forest, Harvard University, Cambridge, Massachusetts, Harvard Forest Paper No. 8, 15 pp.

Lyford, W. H. 1964b. Coarse fragments in the Gloucester soils of the Harvard Forest, the Harvard Forest, Harvard University, *Harvard Forest Paper* No. 9, 16 pp.

Lyford, W. H. 1974. Narrow soils and intricate soil patterns in southern New England, *Geoderma* **11**:195-208.

Lyford, W. H. 1975. Rhizography of non-woody roots of trees in the forest floor, in J. G. Torrey and D. T. Clarkson (eds.), *The Development and Function of Roots, Third Cabot Symposium,* Academic, New York, pp. 179-196.

Lyford, W. H. 1980. Development of the root system of northern red oak *(Quercus rubra* L.), Harvard University, The Harvard Forest, Harvard University, *Harvard Forest Paper* No. 21, 30 pp.

Lyford, W. H., and B. F. Wilson. 1964. Development of the root system of *Acer rubrum* L. The Harvard Forest, Harvard University, *Harvard Forest Paper* No. 10, 17 pp.

Lyford, W. H., and B. F. Wilson. 1966. Controlled growth of forest tree roots: Technique and application, The Harvard Forest, Harvard University, *Harvard Forest Paper* No. 16, 12 pp.

Lyford, W. H., and D. W. MacLean. 1966. Mound and pit microrelief in relation to soil disturbance and tree distribution in New Brunswick, Canada, The Harvard Forest,

Harvard University, *Harvard Forest Paper* 15, 18 pp.

Lyford, W. H., J. C. Goodlett, and W. H. Coates. 1963. *Landforms, soils with fragipans, and forest on a slope in the Harvard Forest,* The Harvard Forest, Harvard University, Harvard Forest Bulletin 30, 68 pp.

Lyr, H., and G. Hoffmann. 1967. Growth rates and growth periodicity of tree roots, *International Review of Forestry Research* 2:181-236.

Macaloney, H. J. 1966. The impact of insects in the northern hardwoods type, USDA Forest Service Research Note NC-10, 3 pp.

MacArthur, R. H. 1972. Climates on a rotating earth, in *In Geographical Ecology: Patterns in the Distribution of Species,* Harper & Row, New York, pp. 5-19.

MacArthur, R. H., and J. H. Connell. 1966. *The Biology of Populations,* John Wiley, New York, 200 pp.

MacBean, A. P. 1941. *A Study of the Factors Affecting the Reproduction of Western Hemlock and Its Associates in the Quatsino Region, Vancouver Island,* British Columbia Forest Service, Victoria, British Columbia, Canada, 36 pp.

MacCleery, D. W. 1992. American forests: a history of resiliency and recovery. USDA Forest Service FS-540. 59 pp.

Mack, R. N., N. W. Rutter, V. M. Bryant, Jr., and S. Valastro. 1978a. Reexamination of postglacial vegetation history in northern Idaho: Hager Pond, Bonner County. *Quaternary Research* 10:241-255.

Mack, R. N., N. W. Rutter, V. M. Bryant, Jr., and S. Valastro. 1978b. Later Quaternary pollen record from Big Meadow, Pend Oreille County, Washington. *Ecology* 59:956-965.

MacKenzie, A. M. 1962. The Bowmont Norway spruce sample plots, *Forestry* 35:129-138.

MacLean, C. D. 1979. Relative density: The key to stocking assessment in regional analysis—a forest survey viewpoint, USDA Forest Service General Technical Report PNW-78, 5 pp.

MacLean, D. A. 1980. Vulnerability of fir-spruce stands during uncontrolled spruce budworm outbreaks: A review and discussion, *Forestry Chronicle* 56:213-221.

Magnussen, S. 1980. Wasserhaushaltuntersuchungen bei unterschiedlich Beschatteten Kustentanne, *Allgemeine Forst and Jagdzeitung* 151:227-235.

Magnussen, S., and A. Peschl. 1981. Die Einwirkung verschiedener Beschattungsgrade auf die Photosynthese und die Transpiration junger Weiss-und Kustentanne, *Allgemeine Forst und Jagdzeitung* 152:82-93.

Maguire, D. A. 1983. Suppressed crown expansion and increased bud density after precommercial thinning in California Douglas-fir, *Canadian Journal of Forest Research* 13:1246-1248.

Maguire, D. A., and M. D. Petruncio. 1995. Pruning and growth of western Cascade species: Douglas-fir, western hemlock, Sitka spruce. IN (D. P. Hanley, C. D. Oliver, D. A. Maguire, D. G. Briggs, and R. D. Fight, editors) *Forest pruning and wood quality of western North American conifers.* University of Washington, College of Forest Resources, Seattle, Washington, U.S.A. Institute of Forest Resources Contribution No. 77:179-215.

Maguire, D. A., and R. T. T. Forman. 1983. Herb cover effects on tree seedling patterns in a mature hemlock-hardwood forest, *Ecology* 64:1367-1380.

Mahoney, R. L. 1978. Lodgepole pine/mountain pine beetle risk classification methods and their application, in A. A. Berryman, G. D. Amman, and R. W. Stark (eds.), *Theory and Practice of Mountain Pine Beetle Management in Lodgepole Pine Forests,* University of Idaho Forestry, Wildlife, and Range Experiment Station,

Moscow, pp. 106-113.

Maisenhelder, L. C., and C. A. Heavrin. 1957. Silvics and silviculture of the pioneer hardwoods—cottonwood and willow, in *Proceedings of the Society of American Foresters, pp. 73-75.*

Maissurow, D. K. 1935. Fire as a necessary factor in the perpetuation of white pine, *Journal of Forestry* **33**:373-378.

Maissurow, D. K. 1941. The role of fire in the perpetuation of virgin forests in northern Wisconsin, *Journal of Forestry* **39**:201-207.

Malecki, R. A., J. R. Lassoie, E. Rieger, and T. Seamans. 1983. Effects of long-term artificial flooding on a northern bottomland hardwood forest community, *Forest Science* **29**:535-544.

Malmberg, D. G. 1965. Early thinning trials in western hemlock [*Tsuga heterophylla* (Raf.)Sarg.] related to stand structure and product development, unpublished Ph.D. thesis, University of Washington, Seattle, 138 pp.

Manion, P. D. 1981. *Tree Disease Concepts,* Prentice-Hall, Englewood Cliffs, New Jersey, 399 pp.

Manuwal, D. A., and M. H. Huff. 1987. Spring and winter bird populations in a Douglas-fir forest sere, *Journal of Wildlife Management* **51**:586-595.

Maple, W. R. 1970. Prescribed winter fire thins dense longleaf seedling stand, USDA Forest Service Research Note SO-104, 2 pp.

Mar-Moller, C. 1947. The effect of thinning, age, and site on foliage, increment, and loss of dry matter, *Journal of Forestry* **45**:393-404.

Mar-Moller, C. 1954. The influence of thinning on volume increment: I. Results of investigations, in C. Mar-Moller, J. Abell, T. Jagd, and F. Juncker (eds.), *Thinning Problems and Practices in Denmark,* State University of New York, College of Forestry, Syracuse, New York, Forestry Technical Publication 76, pp. 5-32.

Mar-Moller, C. 1960. The influence of pruning on the growth of conifers, *Forestry* **33**:37-53.

Mar-Moller, C., J. Abell, T. Jagd, and F. Juncker (eds.), 1954. *Thinning problems and practices in Denmark, State University of New York,* College of Forestry, Syracuse, Forestry Technical Publication 76, 92 pp.

Marks, P. 1974. The role of pine cherry *(Prunus pennsylvanica* L.) in the maintenance of stability in northern hardwood ecosystems, *Ecological Monographs* **44**:73-88.

Marks, P. L., and F. H. Bormann. 1972. Revegetation following forest cutting: Mechanisms for return to steady-state nutrient cycling, *Science* **176**:914-915.

Marquis, D. A. 1967. Clearcutting in northern hardwoods: Results after 30 years, USDA Forest Service Research Paper NE-85, 13 pp.

Marquis, D. A. 1973. An appraisal of clearcutting on the Monongahela National Forest, USDA Forest Service, Northeastern Forest Experiment Station, Warren, Pennsylvania, 55 pp.

Marquis, D. A. 1975a. Seed germination and storage under northern hardwood forests, *Canadian Journal of Forest Research* **5**:478-484.

Marquis, D. A. 1975b. The Allegheny hardwood forests of Pennsylvania, USDA Forest Service General Technical Report NE-15, 32 pp.

Marquis, D. A. 1977. Application of uneven-aged silviculture and management on public and private lands, in *Uneven-Aged Silviculture and Management in the Western United States,* USDA Forest Service Timber Management Research, pp. 20-54.

Marquis, D. A. 1981a. Removal or retention of unmerchantable saplings in Allegheny hardwoods: Effect of regeneration after clearcutting, *Journal of Forestry* **79**:280-283.

Marquis, D. A. 1981b. Survival, growth, and quality of residual trees following clearcutting in Allegheny hardwood forests, USDA Forest Service Research Paper NE-477, 9 pp.

Marquis, D. A. 1981c. The effect of advance seedling size and vigor on survival after clearcutting, USDA Forest Service Research Paper NE-498, 7 pp.

Marquis, D. A. 1986. Thinning Allegheny hardwood pole and small sawtimber stands, in H. C. Smith and M. C. Eye (eds.), *Guidelines for Managing Immature Appalachian Hardwood Stands,* Society of American Foresters Publication Number 86-02. West Virginia University Division of Forestry Morgantown, pp. 68-84.

Marquis, D. A. 1992. Stand development patterns in Allegheny hardwood forests, their influence on silviculture and management practices. *In:* (M. J. Kelty, B. C. Larson and C. D. Oliver, editors) *The Ecology and Silviculture of Mixed-Species Forests: A Festschrift for David M. Smith.* Kluwer Academic Publishers, Boston:

Marquis, D. A., R. L. Ernst, and S. L. Stout. 1984. Prescribing silvicultural treatments in hardwood stands of the Alleghenies, USDA Forest Service General Technical Report NE-96, 90 pp.

Marsh, E. K. 1957. *Some preliminary results from O'Conner's correlated curve trend (C.C.T.) experiments on thinnings and espacements and their practical significance* Commonwealth Forestry Conference of Australia and New Zealand, Government Printer, Pretoria, South Africa, 21 pp.

Marsh, W. M., and J. M. Koerner. 1972. Roles of moss in slope formation, *Ecology* 53:489-493.

Marshall, J. D., and R. H. Waring. 1986. Comparison of methods of estimating leaf-area index in old-growth Douglas-fir, *Ecology* 67:975-979.

Marshall, R. 1927. *The growth of hemlock before and after release from suppression,* The Harvard Forest, Harvard University, Harvard Forest Bulletin 11, 43 pp.

Marsinko, A. P. C., T. J. Straka, and J. L. Baumann. 1993. Hurricane Hugo: a South Carolina update. *Journal of Forestry* 91(9):9-17.

Martin, P. S. 1967. Prehistoric overkill, in P. S. Martin and H. E. Wright (eds.), *Pleistocene Extinctions,* Yale University Press, New Haven, Connecticut, pp. 75-120.

Martin, R. E.1982. Fire history and its role in succession, in J. E. Means (ed.), *Forest Succession and Stand Development Research in the Northwest,* Forest Research Laboratory, Oregon State University, Corvallis, pp. 92-99.

Maser, C., and J. M. Trappe, eds. 1984. The seen and unseen world of the fallen tree, USDA Forest Service General Technical Report PNW-164, 56 pp.

Maser, C., J. M. Trappe, and R. A. Nussbaum. 1978. Fungal-small mammal interrelationships with emphasis on Oregon coniferous forests, *Ecology* 59:799-809.

Matlack, G. R. 1993. Microenvironment variation within and among forest edge sites in the eastern United States. Biological Conservation 66:185-194.

Matlack, G. R. 1994. Veetation dynamics of the forest edge—trends in space and succession time. Journal of Ecology 82:113-123.

Mattoon, W. R. 1909. The origin and early development of chestnut sprouts, *Forestry Quarterly* 7:34-37.

May, R. M. 1973. *Stability and Complexity in Model Ecosystems,* Princeton University Press, Princeton, New Jersey, 265 pp.

McAndrews, J. H. 1976. Fossil history of man's impact on the Canadian flora: An example from southern Ontario, *Canadian Botanical Association Bulletin* 9:(suppl)1-6.

McArdle, R. E., W. H. Meyer, and D. Bruce. 1949. The yield of Douglas-fir in the Pacific Northwest, USDA Technical Bulletin No. 210, 74 pp.

McBride, J. R., and R. D. Laven. 1976. Scars as an indicator of fire frequency in the San Bernadino Mountains, California, *Journal of Forestry* **74**:439-442.

McCarter, J. B., and J. N. Long. 1986. A lodgepole pine density management diagram, *Western Journal of Applied Forestry* **1**:6-11.

McClay, T. A. 1955. Loblolly pine growth as affected by removal of understory hardwoods and shrubs, USDA Forest Service Southeast Forest Experiment Station Research Note 73, 2 pp.

McClurkin, D. G. 1958. Soil moisture content and shortleaf pine radial growth in north Mississippi, *Forest Science* **4**:232-238.

McClurkin, D. G. 1961. Soil moisture trends following thinning in shortleaf pine, *Soil Science Society of America Proceedings* **25**:135-138.

McColl, J. G. 1969. Soil-plant relationships in a eucalyptus forest on the south coast of New South Wales, *Ecology* **50**:364-362.

McComb, W. C., T. A. Spies, and W. H. Emmingham. 1993. Douglas-fir forests: managing for timber and mature-forest habitat. *Journal of Forestry* **91**(12):31-42.

McConnell, B. R., and J. G. Smith. 1971. Effect of ponderosa pine needle litter on grass seedling survival, USDA Forest Service Research Note PNW-155, 6 pp.

McCullough, H. A. 1948. Plant succession on decaying logs in a virgin spruce-fir forest, *Ecology* **29**:508-513.

McDermott, R. E. 1954. Effects of saturated soil on seedling growth of some bottomland hardwood species, *Ecology* **35**:36-41.

McFadden, G., and C. D. Oliver. 1988. Three-dimensional forest growth model relating tree size, tree number, and stand age: Relation to previous growth models and to self-thinning, *Forest Science* **34**:662-676.

McGaughey, R. J., and J. B. McCarter. 1995. Stand visualization system. USDA Forest Service GTR-PNW- (In review)

McGee, C. E. 1967. Regeneration in southern appalachian oak stands, USDA Forest Service Research Note SE-72, 6 pp.

McGee, C. E. 1981. Response of overtopped white oak to release, USDA Forest Service Research Note SO-273, 4 pp.

McGee, C. E. 1982. Low-quality hardwood stands: Opportunities for management in the interior uplands, USDA Forest Service General Technical Report SO-40, 22 pp.

McGee, C. E. 1984. Heavy mortality and succession in a virgin mixed mesophytic forest, USDA Forest Service Research Paper SO-209, 9 pp.

McGee, C. E. 1986a. Loss of *Quercus* spp dominance in an undisturbed old growth forest, *Journal of the Elisha Mitchell Society* **102**:10-15.

McGee, C. E. 1986b. Oaks lose dominance in an undisturbed old-growth forest, in *Proceedings of the Fourteenth Annual Hardwood Symposium*, The Hardwood Research Council, Statesville, North Carolina, pp. 115-121.

McGee, C. E., and D. L. Bivens. 1984. A billion overtopped white oak—assets or liabilities? *Southern Journal of Applied Forestry* **8**:216-220.

McGee, C. E., and D. L. Loftis. 1986. Planted oaks perform poorly in North Carolina and Tennessee, *Northern Journal of Applied Forestry* **3**:114-116.

McGinnes, B. S. 1969. How size and distribution of cutting units affect food and cover of deer, in *Whitetailed Deer in the Southern Forest Habitat*, USDA Forest Service Southern Forest Experiment Station, 130 pp.

McIntosh, R. P. 1980. The relationship between succession and the recovery process in ecosystems, in J. Cairns (ed.), *The Recovery Process in Damaged Ecosystems*, Ann Arbor Science Publishers, Inc., Ann Arbor, Michigan, pp. 11-62.

McIntosh, R. P. 1981. Succession and ecological theory, in D. C. West, H. H. Shugart, and D. B. Botkin (eds.), *Forest Succession: Concepts and Application,* Springer-Verlag, New York, pp. 10-23.

McKinnon, F. S., G. S. Hyde, and A. C. Cline. 1935. *Cut-over old field white pine lands in central New England,* The Harvard Forest, Harvard University, Harvard Forest Bulletin No. 18, 80 pp.

McKnight, J. S., and J. S. McWilliams. 1956. Improving southern hardwood stands through commercial harvest and cull-tree control, *Proceedings of the Society of American Foresters, pp.* 71-72.

McKnight, J. S., D. O. Hook, O. G. Langdon, and R. L. Johnson. 1981. Flood tolerance and related characteristics of trees of the bottomland forests of the southern United States, in J. R. Clark and J. Benforado (eds.), *Wetlands of Bottomland Hardwood Forests,* Elsevier, New York, pp. 29-69.

McLaughlin, S. B., and D. S. Shriner. 1980. Allocation of resources to defense and repair, in J. Horsfall and E. B. Cowling (eds.), *Plant Disease,* Volume 5, Academic, New York, pp. 407-437.

McLaughlin, S. B., D. J. Downing, T. J. Blasing, E. R. Cook, A. J. Johnson, and H. S. Adams. 1986. An analysis of climate and competition as contributors to the decline of red spruce in high elevation Appalachian forests (submitted for publication).

McLemore, B. F., and T. Hansbrough. 1970. The influence of light on germination of *Pinus palustris* seeds, *Physiologia Plantarum* 23:1-10.

McLintock, T. F. 1979. *Research priorities for eastern hardwoods,* Hardwood Research Council, Statesville, North Carolina, 70 pp.

McMinn, R. G. 1963. Characteristics of Douglas-fir root systems, *Canadian Journal of Botany* 41:105-122.

McMurtrie, R. 1981. Suppression and dominance of trees with overlapping crowns, *Journal of Theoretical Biology* 81:152-174.

McNicoll, C. H. 1994. Unpublished Master's thesis. University of Montana, Missoula, Montana, U.S.A.

McPhee, J. 1990. *The control of nature.* The Noonday Press, New York. 272 pp.

McQuilkin, R. A. 1989. The possible role of fire in oak regeneration (submitted for publication).

McQuillan, A. G. 1995. The potential of fractal geometry as a tool for achieving naturalism in forest aesthetics. IN (K. L. O'Hara, editor) *Uneven-aged management: opportunities, constraints, and methodologies.* Montana Forest and Conservation Experiment Station, School of Forestry, University of Montana, Miscellaneous Publication No. 56.

McWirtter, N. R. 1982. *Guiness Book of World Records,* Bantam Books, New York.

Mealey, S. P. 1994. Finding a new balance point in Boise National Forest management. Forest Supervisor's response to the Forest plan monitoring and evaluation report. USDA Forest Service, Bosie National Forest, Boise, Idaho. 14 pp.

Means, J. E. 1982. Developmental history of dry coniferous forests in the western Oregon Cascades, in J. E. Means (ed.), *Forest Succession and Stand Development Research in the Northwest,* Forest Research laboratory, Oregon State University, Corvallis, pp. 142-158.

Meentemeyer, V. 1978. Macro climate and lignin control of decomposition rates, *Ecology* 59:465-472.

Mehlhop, P. and J. F. Lynch. 1986. Bird/habitat relationships along a successional gradient in the Maryland coastal plain. The *American Midland Naturalist* 116:225-239.

Meier, M. F. 1964. Ice and glaciers, in V. T. Chow (ed.), *Handbook of Applied Hydrology,* McGraw-Hill, New York, pp. 16.1-16.32.

Melillo, J. M. 1982. Nitrogen and lignin control of hardwood leaf litter decomposition dynamics, *Ecology* **63**:621-626.

Meng, X. 1986. Crown shape of young Douglas-fir under different sunlight environment, unpublished master of science thesis, University of Washington, Seattle, 78 pp.

Mergen, F. 1954. Mechanical aspects of wind-breakage and wind-firmness, *Journal of Forestry* **52**:119-125.

Mergen, F., and H. I. Winer. 1952. Compression failures in the boles of living conifers, *Journal of Forestry* **50**:671-679.

Merkel, O. 1967. The effect of tree spacing on branch sizes in spruce. *Allgemeine Forst and Jagdzeitung* **138**:113-125.

Merrill, P. H., and R. C. Hawley. 1924. *Hemlock: Its place in the silviculture of the southern New England forest,* Yale University School of Forestry Bulletin 12, 68 pp.

Merz, R. W., and S. G. Boyce. 1956. Age of oak "seedlings," *Journal of Forestry* **54**:774775.

Metzger, F. T., and C. H. Tubbs. 1971. The influence of cutting method on regeneration of second-growth northern hardwoods, *Journal of Forestry* **69**:559-564.

Meyer, H. A. 1943. Management without rotation, *Journal of Forestry* **41**:126-132.

Meyer, H. A. 1952. Structure, growth, and drain in balanced uneven-aged forests, *Journal of Forestry* **50**:85-92.

Meyer, H. A., A. B. Recknagel, and D. D. Stevenson. 1952. *Forest Management,* Ronald, New York, 290 pp.

Meyer, H. A., and D. D. Stevenson. 1943. The structure and growth of virgin beech-birch-maple-hemlock forests in northern Pennsylvania, *Journal of Agricultural Research* **67**:465-484.

Meyer, W. H. 1930. *Diameter distribution series in even-aged forest stands,* Yale University School of Forestry Bulletin 28,108 pp.

Meyer, W. H. 1931. Thinning experiments in young Douglas-fir, *Journal of Agricultural Research* **43**:537-546.

Mielikainen, K. 1980. *Structure and development of mixed pine and birch stands, Communicationes Instituti Forestalis Fenniae* 99.3, 82 pp.

Mielikainen, K. 1985. Effect of an admixture of birch on the structure and development of Norway spruce stands, *Communicationes Instituti Forestalis Fenniae* 133,179 pp.

Mikan, C.J., D.A.Orwig, and M.D.Abrams. 1994. Age structure and successional dynamics of a presettlement-origin chestnut oak forest in the Pennsylvania Piedmont. *Bulletin of the Torrey Botanical Club* 121(1): 13-23.

Miller, D. R., J. W. Kimmey, and M. E. Fowler. 1959. White pine blister rust, USDA Forest Service Forest Pest Leaflet 36, 8 pp.

Miller, R. E., and M. D. Murray. 1978. The effects of red alder on growth of Douglas-fir, in W. A. Atkinson, D. G. Briggs, and D. S. DeBell (eds.), *Utilization and Management of Red Alder,* USDA Forest Service General Technical Report PNW-70, pp. 283-306.

Milner, K. S., and D. W. Coble. 1995. A mechanistic approach to predicting the growth and yield of stands with complex structures. IN (K. L. O'Hara, editor) *Uneven-aged management: opportunities, constraints, and methodologies.* Montana Forest and Conservation Experiment Station, School of Forestry, University of Montana, Miscellaneous Publication No. 56.

Minckler, L. S. 1961. Measuring light in uneven-aged hardwood stands, USDA Forest Service, Central States Forest Experiment Station Technical Paper No.184, 9 pp.

Minckler, L. S. 1971. A Missouri forest of the past, *Transactions of the Missouri Academy of Science* 5:48-56.

Minckler, L. S. 1973. An appraisal of clearcutting on the Monongahela National Forest, *Congressional Record,* June 8, pp. E3879-E3882.

Minckler, L. S. 1974. Prescribing silvicultural systems, *Journal of Forestry* 72:269-273.

Minckler, L. S., and J. D. Woerheide. 1965. Reproduction of hardwoods 10 years after cutting as affected by site and opening size, *Journal of Forestry* 63:103-107.

Minckler, L. S., W. T. Plass, and R. A. Ryker. 1961. Woodland management by single-tree selection: A case history, *Journal of Forestry* 59:257-261.

Mirov, N. T. 1967. *The Genus Pinus,* Ronald, New York, 602 pp.

Mitchell, A., and J. Wilkenson. 1988. *A Field Guide to the Trees of Britain and Northern Europe,* Houghton, Mifflin, Boston, 415 pp.

Mitchell, J. M., Jr. 1977. The changing climate, in Geophysics Study Committee (eds.), *Energy and Climate,* National Academy of Sciences, Studies in Geophysics, Washington, D.C., pp. 51-48.

Mitchell, K. J. 1969. *Simulation of the growth of even-aged stands of white spruce,* Yale University School of Forestry and Environmental Studies Bulletin No. 75, 48 pp.

Mitchell, K. J. 1975. Dynamics and simulated yield of Douglas-fir, *Forest Science Monograph* 17:1-39.

Mitchell, K. J., and I. R. Cameron. 1985. Managed stand yield tables for coastal Douglas-fir: Initial density and precommercial thinning, *Land Management Report* 31, Ministry of Forests, British Columbia, 69 pp.

Mitchell, K. J., and J. W. Goudie. 1980. Stagnant lodgepole pine, Progress Report E.P.850.02, British Columbia Ministry of Forestry, Victoria, British Columbia. March 31, 1980, 31 pp.

Mitchell, R. G., R. H. Waring, and G. B. Pitman. 1983. Thinning lodgepole pine increases tree vigor and resistance to mountain pine beetle, *Forest Science* 29:204-211.

Miyanishi, K., A. R. Hoy, and P. B. Cavers. 1979. A generalized law of self-thinning in plant populations (Self-thinning in plant populations), *Journal of Theoretical Biology* 78:439-442.

Mizunaga, H. 1995. Effects of disturbance on large patch formation of Magnolia obovata, a deciduous broad-leaved tree: field investigation and simulation. *Journal of Sustainable Forestry* (In press).

Moen, A. N. 1974. Turbulence and the visualization of wind flow, *Ecology* 55:1420-1424.

Mohler, C. L., P. L. Marks, and D. G. Sprugel. 1978. Stand structure and allometry of trees during self-thinning of pure stands, *Journal of Ecology* 66:599-614.

Mohn, C. A., and R. M. Krinard. 1971. Volume tables for small cottonwoods in plantations, USDA Forest Service Research Note SO-113, 4 pp.

Moir, W. H. 1966. Influence of ponderosa pine on herbaceous vegetation, *Ecology* 47:1045-1048.

Monk, C. D. 1961. Past and present influences on reproduction in the William L. Hutcheson Memorial Forest, New Jersey, *Bulletin of the Torrey Botanical Club* 88:167-175.

Monsi, M., and T. Saeki. 1953. Uber den Lichfaktor in den Pflanzengesellschaften und seine Bedeutung fur Stoff-production, *Japanese Journal of Botany* 14:22-52.

Monsi, M., Z. Uchijima, and T. Oikawa. 1973. Structure of foliage canopies and photosynthesis, *Annual Review of Ecology and Systematics* 4:301-327.

Moodie, D. W., and B. Kaye. 1969. The northern limit of Indian agriculture in North America, *Geographic Review* 59:513-529.

Mooney, H. A., and M. Godron (eds.). 1983. *Disturbance and Ecosystems: Components of Response*, Ecological Studies 44, Springer-Verlag, New York, 292 pp.

Morey, H. F. 1936. Age-size relationships of Heart's Content, a virgin forest in northwestern Pennsylvania, *Ecology* 17:251-257.

Morgan, P., G. H. Aplet, J. B. Haufler, H. C. Humphries, M. M. Moore, and W. D. Wilson. 1994. Historical range of variability: a useful tool for evaluating ecosystem change. *Journal of Sustainable Forestry* 2(1/2):87-111.

Morris, R. F. 1948. Age of balsam fir, Department of Agriculture of Canada Forest Insect Investigation bi-monthly progress report 4, (abstracted in *Forestry Abstracts* 10:136-137).

Morris, R. F. 1951. The effects of flowering on foliage production and growth of balsam fir, *Forestry Chronicle* 27:40-57.

Morris, W. G. 1935. The details of the Tillamook fire from its origin to the salvage of the killed timber. USDA Forest Service, Pacific Northwest Forest Experiment Station, Portland, Oregon. March 19, 1935.

Morrison, M. L., B. G. Marcot, and R. W. Mannan. 1978. *Wildlife-habitat relationships: concepts and applications.* The University of Wisconsin Press, Madison, Wisconsin, U.S.A. 343 pp.

Morrison, P. H. 1990. *Ancient forests on the Olympis National Forest: Analysis from a historical and landscape perspective.* The Wilderness Society, Washington, D. C.

Morrison, P. H., and F. J. Swanson. 1990. Fire history and pattern in a Cascade Range landscape. USDA Forest Service General Technical Report PNW-GTR-254.

Morrison, P. H., D. Kloepfer, D. A. Leversee, C. M. Milner, and D. L. Ferber. 1990. *Ancient forests of the Mt. Baker-Snoqualmie National Foresty: Analysis of Forest Conditions.* The Wilderness Society, Washington, D. C.

Moser, J. W., Jr. 1972. Dynamics of an uneven-aged forest stand, *Forest Science* 18:184-191.

Mosher, J. W., Jr. 1976. Specification of density for the inverse J-shaped diameter distribution, *Forest Science* 22:177-180.

Moss, D., P. N. Taylor, and N. Easterbee. 1979. The effects on song-bird populations of upland afforestation with spruce. *Forestry* 52:129-150.

Mott, D. G. 1963. The forest and the spruce budworm, in R. F. Morris (ed.), The dynamics of epidemic spruce budworm populations, *Memoirs of the Entomological Society of Canada* 13:189-202.

Mowat, E. L. 1961. Growth after partial cutting of ponderosa pine on permanent sample plots in Eastern Oregon, USDA Forest Service Pacific Northwest Forest and Range Experiment Station Research Paper 44, 23 pp.

Mueggler, W. F. 1976. Type variability and succession in Rocky Mountain aspen, in *Utilization and Marketing as Tools for Aspen Management in the Rocky Mountains,* USDA Forest Service General Technical Report RM-29, pp. 16-19.

Mueller-Dombois, D., and H. Ellenberg. 1974. *Aims and Methods of Vegetation Ecology,* John Wiley, New York, 547 pp.

Mueller-Dombois, D., J . E. Canfield, R. A. Holt, and G. P. Buelow. 1983. Tree-group death in North American and Hawaiian forests: A pathological problem or a new problem for vegetation science? *Phytocoenologia* 11:117-137.

Mueller, O. P., and M. G. Cline. 1959. Effects of mechanical soil barriers and soil wetness on rooting of trees and soil mixing by blow-down in central New York, *Soil Science* 88:107-111.

Muller, C. H. 1966. The role of chemical inhibition (allelopathy) in vegetational composition *Bulletin of the Torrey Botanical Club* 93:332-351.

Muller, R. N. 1978. The phenology, growth and ecosystem dynamics of *Erythronium Americanum* in the northern hardwood forest, *Ecological Monographs* 48:1-20.

Muller, R. N., and Bormann, F. H. 1976. Role of *Brythronium americanum* Ker. in energy flow and nutrient dynamics of a northern hardwood forest ecosystem, *Science* 193:1126-1128.

Munger, T. T. 1930. Ecological aspects of the transition from old forests to new, *Science* 72:327-332.

Munger, T. T. 1950. A look at selective cutting in Douglas-fir, *Journal of Forestry* 48:97-99.

Mustian, A. P. 1977. History and philosophy of silviculture and management systems in use today, in *Uneven-Aged Silviculture and Management in the Western United States*, USDA Forest Service Timber Management Research, pp. 3-12.

Nadelhoffer, K. J. 1983. Leaf-litter production and soil organic matter dynamic along a nitrogen-availability gradient in southern Wisconsin, *Canadian Journal of Forestry Research* 13:12-21.

Nakane, K. 1975. Dynamics of soil organic matter in different parts on a slope under evergreen oak forest, *Japanese Journal of Ecology* 25:206-216.

Nakashizuka, T. 1983. Regeneration process of climax beech *(Fagus crenata,* Blume) forests: III. Structure and development processes of sapling populations in different aged gaps, *Japanese Journal of Ecology* 33:409-418.

Nakashizuka, T. 1984a. Regeneration process of climax beech *(Fagus crenata,* Blume) forests: IV. Gap formation, *Japanese Journal of Ecology* 34:75-85.

Nakashizuka, T. 1984b. Regeneration process of climax beech *(Famous crenata,* Blume) forests: V. Population dynamics of beech in a regeneration process, *Japanese Journal of Ecology* 34:411-419.

Nakashizuka, T. 1985. Diffused light conditions in canopy gaps in a beech *(Fagus crenata,* Blume) forest, *Oecologia* 66:472-272.

Nakashizuka, T., and M. Numata. 1982a. Regeneration process of climax beech forests: I. Structure of a beech forest with the undergrowth of Sasa, *Japanese Journal of Ecology* 32:5747.

Nakashizuka, T., and M. Numata. 1982b. Regeneration process of climax beech forests: II. Structure of a forest under the influence of grazing, *Japanese Journal of Ecology* 32:473-482.

Nanson, G. C., and H. F. Beach. 1977. Forest succession and sedimentation on a meandering river flood plain, northeast British Columbia, Canada, *Journal of Biogeography* 4:229-252.

Neilson, R. P., and L. H. Wullstein. 1983. Biogeography of two southwest American oaks in relation to atmospheric dynamics, *Journal of Biogeography* 10:275-297.

Nelson, B. W., V. Kapos, J. B. Adams, W. J. Oliveira, O. P. G. Braun, and I. L. do Amaral. 1994. Forest disturbance by large blowdowns in the Brazilian Amazon. *Ecology* 75:853-858.

Newnham, R. M. 1964. The development of a stand model for Douglas-fir, unpublished Ph.D. thesis, University of British Columbia, Department of Forestry, Vancouver, Canada, 201 pp.

Newton, M., and E. C. Cole. 1987. A sustained-yield scheme for old-growth Douglas-fir, *Western Journal of Applied Forestry* 2:22-25.

Newton, M., B. A. El Hassan, and J. Zavitkovski. 1968. Role of red alder in western Oregon forest succession, in J. M. Trappe, J. F. Franklin, R. F. Tarrant, and G. M. Hansen (eds.), *Biology of Alder*, USDA Forest Service Pacific Northwest Forest and Range Experiment Station, Portland, Oregon, pp. 73-84.

Nichols, G. E. 1913. The vegetation of Connecticut, *Torreya* 13:199-215.

Nichols, G. E. 1935. The hemlock-white pine-northern hardwood region of eastern North America, *Ecology* 16:403-422.

Niering, W. A., and F. E. Egler. 1955. A shrub community of *Viburnum vintage* stable for twenty-five years, *Ecology* 36:356-360.

Niering, W. A., and R. H. Goodwin. 1974. Creation of relatively stable shrublands with herbicides: Arresting "succession" on rights-of-way and pastureland, *Ecology* 55:784-795.

Nobel, P. S. 1976. Photosynthetic rates of sun versus shade leaves of *Hyptis emoryi*, Torr. *Plant Physiology* 58:218-223.

Norberg, R. A. 1988. Theory of growth geometry of plants and self-thinning of plant populations: Geometric similarity, elastic similarity, and different growth modes of plant parts, *American Naturalist* 131:220-256.

Norby, R. J., and T. T. Kozlowski. 1980. Allelopathic potential of ground cover species on *Pinus resinosa* seedlings, *Plant and Soil* 57:363-374.

Noss, R. F. 1983. A regional landscape approach to maintain biodiversity. *Bioscience* 33:700-706.

Nowacki, G.J., and M.D.Abrams. 1992. Community, edaphic, and historical analysis of mixed oak forests of the Ridge and Valley Province of central Pennsylvania. *Canadian Journal of Forestry Research* 22: 790-800.

Nowacki, G.J., and M.D.Abrams. 1994. Forest composition, structure, and disturbance history of the Alan Seeger Natural Area, Huntington County, Pennsylvania. *Torrey Botanical Club* 12 (3): 277-291.

Nowacki, G.J., M.D.Abrams, and C.G.Lorimer. 1990. Composition, structure, and historical development of northern red oak stands along an edaphic gradient in north-central Wisconsin. *Forest Science* 36 (2): 276-292.

Nystrom, M. N. 1981. Reconstruction of pure, second-growth stands of western redcedar *(Thuja plicata* Donn.) in western Washington: The development and silvicultural implications, unpublished master's thesis, College of Forest Resources, University of Washington, Seattle, 97 pp.

Nystrom, M. N., D. S. DeBell, and C. D. Oliver. 1984. Development of young growth western redcedar stands, USDA Forest Service Research Paper PNW-324, 9 pp.

O'Conner, A. J. 1935. *Forest research with special reference to planting distances and thinning.* Union of South Africa, British Empire Forestry Conference, 30 pp.

O'Hara, K. L. (editor) 1995. *Uneven-aged management: opportunities, constraints, and methodologies.* Montana Forest and Conservation Experiment Station, School of Forestry, University of Montana, Miscellaneous Publication No. 56.

O'Hara, K. L. 1986. Development patterns of residual oaks and oak and yellow-poplar regeneration after release in upland hardwood stands, *Southern Journal of Applied Forestry* 10:244-248.

O'Hara, K. L. 1987. Thinning even-aged Douglas-fir stands: Effects of density and structure on stand volume growth, unpublished Ph.D. dissertation, College of Forest Resources, University of Washington, Seattle, 166 pp.

O'Hara, K. L. 1988. Stand structure and growing space efficiency following thinning in an even-aged Douglas-fir stand, *Canadian Journal of Forest Research* 18:859-866.

O'Hara, K. L. 1989. Stand growth efficiency in a Douglas-fir thinning trial, *Forestry* 62:409-418.

O'Hara, K. L. 1995. Early height development and species stratification across five climax series in the eastern Washington Cascade range. *New Forests* 9:53-60.

O'Hara, K. L., and C. D. Oliver. 1987. Three-dimensional representation of stand volume, age, and spacing, in A. R. Ek, S. R. Shirley, and T. E. Burk (eds.), *Forest Growth Modeling and Prediction,* Proceedings of the IUFRO conference, August 23-27, 1987, Minneapolis, Minnesota, USDA Forest Service, General Technical Report NC-120, pp. 794-801.

O'Hara, K. L., and C. D. Oliver. 1988. Three-dimensional representation of Douglas-fir volume growth: Comparison of growth and yield models with stand data, *Forest Science* **34**:724-743.

O'Hara, K. L., and K. S. Milner. 1994. Describing stand structure: some old and new approaches. IN *Mixed stands: research plots, measurement and results, models.* Proceeding from the Symposium of the IUFRO Working Groups S4. 01-03 and S4. 01-04, held in Lousa/Coimbra, Portugal, 25-29 April, **1994**:43-52.

O'Hara, K. L., and N. L. Valappil. 1995. Age class division of growing space to regulate stocking in uneven-aged stands. IN (K. L. O'Hara, editor) *Uneven-aged management: opportunities, constraints, and methodologies.* Montana Forest and Conservation Experiment Station, School of Forestry, University of Montana, Miscellaneous Publication No. 56.

O'Hara, K. L., C. D. Oliver, and D. F. Cobb. 1995. Development of western larch in mixed stands in Washington's Cascade range: implications for prioritizing thinning treatments. IN (W. C. Schmidt and K. J. McDonald, compilers) *Ecology and Management of Larix Forests: A Look Ahead.* Proceedings of an International Symposium, Whitefish, Montana, U.S.A. October 5-9, 1992. USDA Forest Service GTR-INT-**319**:259-264.

O'Hara, K. L., R. S. Seymour, S. D. Tesch, and J. M. Guldin. 1994. Silviculture and our changing profession. *Journal of Forestry* **92**(1):8-13.

O'Laughlin, J., J. G. MacCracken, D. L. Adams, S. C. Bunting, K. A. Blatner, and C. E. Keegan, III. 1993. *Forest health conditions in Idaho: Executive Summary.* University of Idaho, Idaho Forest, Wildlife, and Range Policy Analysis Group, Report No. 11: 36 pp.

O'Laughlin, J., J. M. McCracken, D. L. Adams, S. C. Bunting, K. A. Byers, and C. E. Keegan, III. 1993. *Forest health conditions in Idaho.* Report no. 11, Idaho Forest, Wildlife, and Range Policy Analysis Group, University of Idaho, Moscow.

Obeid, M., D. Machin, and J. L. Harper. 1967. Influence of density on plant to plant variation in fiber flax, *Linum usitatissimum L., Crop Science* **7**:471-473.

Odum, E. P. 1959. *Fundamentals of Ecology,* Saunders, Philadelphia, 546 pp.

Odum, E. P. 1969. The strategy of ecosystem development, *Science* **164**:262-270.

Odum, E. P. 1971. *Fundamentals of Ecology,* 3d ed., Saunders, Philadelphia, 574 pp.

Odum, S. 1965. Germination of ancient seeds: Floristic observations and experiments with archaeologically dated soil samples, *Dansk Botanish Arkiv* **24**:1-70 (cited in Harper 1977).

Ogawa, H., K. Yoda, T. Kira, K. Ogino, T. Shidei, R. Ratanawongase, and C. Apasutaya. 1965. Comparative ecological studies on three main types of forest vegetation in Thailand: I. Structure and floristic composition, *Nature and Life in Southeast Asia* **4**:13-48.

Ohkubo, T., R. Peters, H. Sawada, and M. Kaji. 1995. Fagus japonica stools in natural beech forests: age structure and growth dynamics of the sprouts. *Journal of Sustainable Forestry* (In press).

Ohmann, L. F., and R. A. Ream. 1971. Wilderness ecology: Virgin plant communities of the Boundary Waters Canoe Area, USDA Forest Service Research Paper NC-63, 55 pp.

Oinonen, E. 1967. Sporal regeneration of bracken in Finland in the light of the dimensions and age of its clones, *Acta Forestalia Fennica* **83**:3-96.

Oldeman, R. A. A. 1972. L'Architecture de la foret Guyanaise, doctor of natural sciences Thesis, Academie de Montpelier, Universite des Sciences et Techniques Du Languedoc, Montpelier, France, 248 pp.

Oldeman, R. A. A. 1981. Schaal, grootschaligheid en kleinschaligheid in de bosbouw (On large-scale and small-scale forestry), *Nederlands Bosbouw Tijdschrift* **53**:71-81.

Oliver C. D. 1986b. Silviculture—the past 30 years, the next 30 years: I. Overview, *Journal of Forestry* **84**:32-42.

Oliver, C. D. 1975. The development of northern red oak *(Quercus rubra* L.) in mixed species, even-aged stands in central New England, unpublished Ph.D. dissertation, Yale University, New Haven, Connecticut, 223 pp.

Oliver, C. D. 1976. Growth response of suppressed hemlocks after release, in W. Atkinson and R. J. Zasoski (eds.), *Western Hemlock Management,* University of Washington Institute of Forest Products Contribution No. 34, pp. 266-272.

Oliver, C. D. 1978a. *Development of northern red oak in mixed species stands in central New England,* Yale University School of Forestry and Environmental Studies Bulletin No. 91, 63 pp.

Oliver, C. D. 1978b. Subsurface geologic formations and site variation in upper Sand Hills of South Carolina, *Journal of Forestry* **76**:352-354.

Oliver, C. D. 1980. Even-aged development of mixed-species stands, *Journal of Forestry* **78**:201-203.

Oliver, C. D. 1981. Forest development in North America following major disturbances, *Forest Ecology and Management* **3**:153-168.

Oliver, C. D. 1982. Stand development—its uses and methods of study, in J. E. Means (ed.), *Forest Development and Stand Development Research in the Northwest,* Forest Research Laboratory, Oregon State University, Corvallis, pp. 100-112.

Oliver, C. D. 1986a. Silvicultural trends and influences targeting management of existing Douglas-fir plantations for specific timber quality, in *Proceedings of the 63rd Annual Washington State Forestry Conference,* The Institute of Forest Resources, College of Forest Resources, University of Washington, Seattle, pp. 69-74.

Oliver, C. D. 1992a. Enhancing biodiversity and economic productivity through a systems approach to silviculture. *The Silviculture Conference.* Forestry Canada, Ottowa, Ontario, Canada: 287-293.

Oliver, C. D. 1992b. Statement before the Subcommittee on Forests, Family Farms, and Energy; Committee on Agriculture; United States House of Representatives: A hearing to review "High Quality Forestry" and extended timber harvest rotations. March 11, 1992. Washington, D. C. 26 pp.

Oliver, C. D. 1992c. A landscape approach: achieving biodiversity and economic productivity. *Journal of Forestry* **90**:20-25.

Oliver, C. D. 1992d. Similarities of stand structure patterns based on uniformities of stand development processes throughout the world—some evidence and the application to silviculture through adaptive management. *In*: (M. J. Kelty, B. C. Larson and C. D. Oliver, editors) *The Ecology and Silviculture of Mixed-Species Forests: A Festschrift for David M. Smith.* Kluwer Academic Publishers, Boston: 11-26.

Oliver, C. D. 1993. Statement before the Subcommittee on Agricultural Research, Conservation, Forestry, and General Legislation; Committee on Agriculture, Nutrition, and Forestry; United States Senate: A hearing on the definition and implementation of ecosystem management. November 9, 1993.

Oliver, C. D. 1994. A portfolio approach to landscape management: an economically, ecologically, and socially sustainable approach to forestry. Paper presented at "Innovative Silvicultural Systems in Boreal Forests: A Changing Vision in Forest Resource Management": A conference held on October 4-8, 1994. Mayfield, Inn, Edmonton, Alberta, Canada. Canadian Forest Service, Edmonton, Alberta. Proceedings in Press.

Oliver, C. D. 1994. Ecosystem and landscape management: how do they affect timber supply? IN *Timber supply in Canada: Challenges and Choices.* Canadian Council of Forest Ministers Conference, held November 16-18, 1994. Kananaskis, Alberta, Canada. Proceedings published by Natural Resources Canada, Canadian Forest Service: 113-124.

Oliver, C. D. 1994. Ecosystem management in the United States: alternative approaches and consequences. *Proceedings of the Woodlands Section, 75th Annual Meeting,* Canadian Pulp and Paper Association. (Held April 5-8, 1994) Alberta, Canada.

Oliver, C. D. 1994. Silvicultural opportunities for creating high quality wood and other values. IN (J. F. Weigand, R. W. Haynes, and J. L. Mikowski, compilers). High *Quality Forestry Workshop: the Idea of Long Rotations.* College of Forest Resources, University of Washington CINTRAFOR SP15: 43-61.

Oliver, C. D. and B. C. Lippke. 1995. Wood supply and other values and ecosystem management in western interior forests. IN *Ecosystem management in western interior forests.* Proceedings of a symposium held May 3-5, 1994, Spokane, Washington, U.S.A. Washington State University.

Oliver, C. D., A. B. Adams, A. R. Weisbrod, J. Dragovon, R. J. Zasoski, and K. Bardo. 1980. Nooksack cirque natural history final report, University of Washington College of Forest Resources, Seattle, 127 pp.

Oliver, C. D., A. B. Adams, and R. J. Zasoski. 1985. Disturbance patterns and forest development in a recently deglaciated valley in the northwestern Cascade Range of Washington, U.S.A., *Canadian Journal of Forest Research* 15:221-232.

Oliver, C. D., A. Osawa, and A. E. Camp. 1995. Forest dynamics and resulting plant and animal population changes. *Journal of Sustainable Forestry* (In press).

Oliver, C. D., and B. C. Larson. 1990. *Forest Stand Dynamics.* McGraw-Hill, New York. 467 pp.

Oliver, C. D., and E. P. Stephens. 1977. Reconstruction of a mixed species forest in central New England, *Ecology* 58:562-572.

Oliver, C. D., and J. B. McCarter. 1995. Developments in decision support for landscape management. IN *Proceedings of the Ninth Annual Symposium on Geographic Information Systems,* held March 27-30, 1995. Vancouver, British Columbia, Canada .

Oliver, C. D., and L. N. Sherpa. 1990. *The effects of browsing and other disturbances on the forest and shrub vegetation of the Hongu, Inkhu, and Dudh Koshi Valleys.* The Makalu-Barun Conservation Project Working Paper Publication Series, Report 9. Department of National Parks and Wildlife Conservation, HMG of Nepal, Kathmandu; and Woodlands Mountain Institute Mount Everest Ecosystem Conservation Program, Kathmandu, Nepal and Franklin, West Virginia, U.S.A. 40 pp.

Oliver, C. D., and M. D. Murray. 1983. Stand structure, thinning prescriptions, and density indexes in a Douglas-fir thinning study, western Washington, U.S.A., *Canadian Journal of Forest Research* 13:126-136.

Oliver, C. D., and R. M. Kenady (eds.). 1982. *Biology and Management of True Fir in the Pacific Northwest,* University of Washington College of Forest Resources, Institute of Forest Resources Contribution Number 45, 344 pp.

Oliver, C. D., C. Harrington, M. Bickford, R. Gara, W. Knapp, G. Lightner, and L. Hicks. 1994. Maintaining and creating old growth structural features in previously disturbed stands typical of the eastern Washington Cascades. *Journal of Sustainable Forestry* 2(3/4):353-387.

Oliver, C. D., D. R. Berg, D. R. Larsen and K. L. O'Hara. 1992. Integrating management tools, ecological knowledge, and silviculture. *In*: (R. Naiman and editor) *New Perspectives for Watershed Management.* Springer-Verlag, New York: 361-382.

Oliver, C. D., E. C. Burkhardt, and W. K. Clatterbuck. 1989. Spacing and stratification patterns of cherrybark oak and American sycamore in mixed, even-aged stands in the southeastern United States, *Forest Ecology and Management* 29:(in press).

Oliver, C. D., K. L. O'Hara, G. McFadden, and I. Nagame. 1986b. Concepts of thinning regimes, in C. D. Oliver, D. Hanley, and J. Johnson (eds.), *Douglas-fir: Stand Management for the Future,* College of Forest resources, University of Washington, Seattle, Institute of Forest Resources Contribution No. 55, pp. 246-257.

Oliver, C. D., L. L. Irwin, and W. H. Knapp. 1994. Eastside forest management practices: hsitorical overview, extent of their applications, and their effects on sustainability of ecosystems. USDA Forest Service PNW-GTR-324: 73 pp.

Oliver, C. D., W. Michalec, L. DuVall, C. A. Wierman, and H. Oswald. 1986a. Silvicultural costs of growing Douglas-fir of various spacings, sites, and wood qualities, in C. D. Oliver, D. Hanley, and J. Johnson (eds.), *Douglas-fir: Stand Management for the Future,* College of Forest Resources, University of Washington, Seattle, Washington, Institute of Forest Resources Contribution No. 55, pp. 132-142.

Oliver, W. W. 1967. Ponderosa pine can stagnate on a good site, *Journal of Forestry* 56:814-816.

Oliver, W. W. 1984. Brush reduces growth of thinned ponderosa pine in northern California, USDA Forest Service Research Paper PSW-172, 6 pp.

Olmsted, N. W., and J. D. Curtis. 1947. Seeds of the forest floor, *Ecology* 28:49-52.

Olson, A. R. 1965. *Natural changes in some Connecticut woodlands during 30 years,* the Connecticut Agricultural Experiment Station Bulletin 669, 52 pp.

Olson, J. S. 1958. Rates of succession and soil changes on southern Lake Michigan sand dunes, *Botanical Gazette* 119:125-170.

Olson, J. S. 1963. Energy storage and the balance of producers and decomposers in ecological systems, *Ecology* 44:322-331.

Omule, S. A. Y. 1984. Results from a correlated curve trend experiment on spacing and thinning of coastal Douglas-fir, Province of British Columbia, Ministry of Forests, Research Note No. 93, 22 pp.

Omule, S. A. Y. 1985. Response of coastal Douglas-fir to precommercial thinning on a medium site in British Columbia, Province of British Columbia, Ministry of Forests, Research Note No. 100, 56 pp.

Oohata, S., and K. Shinozaki. 1979. A statical model of plant form—further analysis of the pipe model theory, *Japanese Journal of Ecology* 29:323-335.

Oosting, H. J. 1956. *The Study of Plant Communities,* W. H. Freeman, San Francisco, 440 pp.

Oregon State University. 1974. *Annual report of the Forest Research Laboratory,* Oregon State University, Corvallis, 46 pp.

Orians, G. H. 1982. The influence of treefalls in tropical forests on tree species richness, *Tropical Ecology* 23:255-279.

Orr, T. J. 1965. Some uses, abuses, and potential uses of the soil-vegetation concept, in C. T. Youngberg (ed.), *Forest-soil Relationships in North America,* Oregon State

University Press, Corvallis, pp. 495-501.

Orwig, D.A., and M. D. Abrams. 1994a. Contrasting radial growth and canopy recruitment patterns of Liriodendron tulipifera and Nyssa sylvatica: gap-obligate versus gap-facultative tree species. *Canadian Journal of Forestry Research* 24: 2141-2149.

Orwig, D.A., and M.D.Abrams. 1994b. Land-use history (1720-1992), composition, and dynamics of oak-pine forests within the Piedmont and Coastal Plain of northern Virginia. *Canadian Journal of Forestry Research* 24: 1216-1225.

Osawa, A. 1986. Patch dynamics of spruce-fir forests during a spruce budworm outbreak in Maine, unpublished Ph.D. dissertation, Yale University, New Haven, Connecticut, 128 pp.

Osawa, A. 1989. Causality in mortality patterns of spruce trees during a spruce budworm outbreak. *Canadian Journal of Forest Research* 19:632-638.

Osawa, A. 1992. Development of a mixed-conifer forest in Hokkaido, northern Japan, following a catastrophic windstorm: A "parallel" model of plant succession. IN (M. J. Kelty, B. C. Larson, and C. D. Oliver, editors) *The Ecology and Silviculture of Mixed-Species Forests*, Kluwer Academic Publishers, Dordrecht, The Netherlands: 29-52.

Osawa, A. 1994. Seedling responses to forest canopy disturbance following a spruce budworm outbreak in Maine. *Canadian Journal of Forest Research* 24 (in press).

Osawa, A., and S. Sugita. 1989. The self-thinning rule: Another interpretation of Weller's results, *Ecology* 70:279-283.

Osawa, A., C. J. Spies III, and J. B. Diamond. 1986. *Patterns of tree mortality during an uncontrolled spruce budworm outbreak in Baxter State Park, 1983*, Maine Agricultural Experiment Station Technical Bulletin 121, 69 pp.

Osawa, A., T. C. Maximov, and B. I. Ivanov. 1994. Forest fire history and tree growth patterns in east Siberia. IN (G. Inoue, editor) *Proceedings of the Second Symposium on Joint Siberian Permafrost Studies between Japan and Russia in 1993.* Held January 12-11, 1994, at National Institute for Environmental Studies, Tsukuba, Japan.

Oshima, Y., M. Kimura, H. Iwake, and S. Kuroiwa. 1958. Ecological and physiological studies on the vegetation of Mt. Shimagare: I. Preliminary survey of the vegetation of Mt. Shimagare, *Botanical Magazine of Tokyo* 71:289-300.

Oswald, D. D., F. Hegyi, and A. Becker. 1986. The current status of coast Douglas-fir timber resources in Oregon, Washington, and British Columbia, in C. D. Oliver, D. Hanley, and J . Johnson (eds.), *Douglas-fir: Stand Management for the Future*, College of Forest resources, University of Washington, Seattle, Institute of Forest Resources Contribution No. 55, pp. 26-31.

Oswald, H. 1984a. Le Douglas en France: Sylviculture et production, *Revue Forestiere Francais* 36:56-68.

Oswald, H. 1984b. Production et sylviculture du Douglas en plantations, *Revue Forestiere Francais* 36:268-278.

Oswald, H., and J. Parde. 1976. Une experience d'espacement de Douglas en foret domaniale d'Amance, *Revue Forestiere Francais* 28:185-192.

Ovington, J. D. 1957. Dry matter production by *Pinus sylvestris* L., *Annals of Botany* 21:287-314.

Owens, J. N., and M. Molder. 1977. Vegetative bud development and cone differentiation in *Abies amabilis, Canadian Journal of Botany* 55:992-1009.

Packee, E. D., C. D. Oliver, and P. D. Crawford. 1982. Ecology of Pacific silver fir, in C. D. Oliver and R. M. Kenady (eds.), *Biology and Management of True Fir in the*

Pacific Northwest, College of Forest Resources, University of Washington, Seattle, Institute of Forest Resources Contribution Number 45, pp. 19-34.

Parker, G, R., and P. T. Sherwood. 1986. Gap phase dynamics of a mature Indiana forest, *Indiana Academy of Science* **95**:217-223.

Parker, G. R. 1989. Old-growth forests of the central hardwood region, *Natural Area Journal* **9**:5-11.

Parker, G. R., and D. J. Leopold. 1983. Replacement of *Ulmus americana* L. in a mature east-central Indiana woods, *Bulletin of the Torrey Botanical Club* 110:482-488.

Parker, G. R., D. J. Leopold, and J. K. Eichenberger. 1985. Tree dynamics in an old-growth, deciduous forest, *Forest Ecology and Management* **11**:31-57.

Parsons, D. J., and S. H. DeBenedetti. 1979. Impact of fire suppression on a mixed-conifer forest, *Forest Ecology and Management* **2**:21-33.

Parsons, D. J., D. M. Graber, J. K. Agee, and J. W. van Wagtendonk. 1986. Natural fire management in national parks, *Environmental Management* **10**:21-24.

Pase, C. P. 1971. Effect of a February burn on Lehmann lovegrass, *Journal of Range Management* **24**:454-456.

Patton, D. R. 1975. A diversity index for quantifying habitat edge, *Wildlife Society Bulletin* 394:171-173.

Payandeh, B. 1974. Spatial patterns of trees in the major forest types of northern Ontario, *Canadian Journal of Forest Research* 4:8-14.

Pearson, G. A. 1931. Forest types in the Southwest as determined by climate and soil, USDA Technical Bulletin 247, 144 pp.

Pearson, G. A. 1950. Management of ponderosa pine in the Southwest, USDA Forest Service Agricultural Monograph 6, 218 pp.

Pearson, S. M. 1993. The spatial extent and relative influence of landscape-level factors on wintering bird populations. *Landscape Ecology* **8**(1):3-18.

Peet, R. K. 1981. Forest dynamics of the Colorado Front Range: Composition and dynamics, *Vegetatio* **45**:3-75.

Peet, R. K., and N. L. Christensen. 1580. Succession: A population process, *Vegetatio* **43**:131-140.

Peet, R. K., and N. L. Christensen. 1987. Competition and tree death, *BioScience* **37**:586-595.

Peres-Garcia, J. M. 1993. Global forestry impacts of reducing softwood supplies from North America. *CINTRAFOR Working Paper 43.* University of Washington, College of Forest Resources, Center for International Trade in Forest Products: 35 pp.

Perley, S. 1891. *Historic Storms of New England,* Salem Press, Salem, Massachusetts, 341 pp.

Perry, D. A. 1984. A model of physiological and allometric factors in the self-thinning curve, *Journal of Theoretical Biology* **106**:383-401.

Perry, D. A. 1985. The competition process in forest stands, in M. G. R. Cannell and J. E. Jackson (eds.), *Attributes of Trees as Crop Plants,* Institute of Terrestrial Ecology, Abbots Ripton, Hunts, England, pp. 481-506.

Pessin, L. J. 1944. Stimulating the early height growth of longleaf pine seedlings, *Journal of Forestry* **42**:95-98.

Petersen, J. D. 1992. Grey ghosts in the Blue Mountains. *Evergreen Magazine* (January/February). pp. 3-8.

Peterson, D. L., and G. L. Rolfe. 1985. Temporal variation in nutrient status of a floodplain forest soil. *Forest Ecology and Management* **12**:73-82.

Peterson, G. B. 1971. Determination of the presence, location and allelopathic effect of substances produced by *Juniperus scopulorum* Sarg., *Dissertation Abstracts International* **B32**(7):3811-3812.

Petruncio, M. D. 1994. Effects of pruning on growth of western hemlock (Tsuga hetero-phylla (Raf.) Sarg.) and Sitka spruce (Picea sitchensis (Bong.) Carr.) in Southeast Alaska. Unpublished Ph. D. dissertation, University of Washington, College of Forest Resources, Seattle, Washington, U.S.A. 160 pp.

Pezeshki, S. R., and J. L. Chambers. 1985. Responses of cherrybark oak seedlings to short-term flooding, *Forest Science* 31:760-771.

Pezeshki, S. R., and T. M. Hinckley. 1982. The stomatal response of red alder and black cottonwood to changing water status, *Canadian Journal of Forest Research* 12:761-771.

Phares, R. E., and R. D. Williams. 1971. Crown release promotes faster diameter growth of pole-size black walnut, USDA Forest Service Research Note NC-124, 4 pp.

Phillips, E. A. 1959. *Methods of vegetation study,* Holt, Rinehart, & Winston, New York, 197 pp.

Phillips, F. J. 1910. The dissemination of junipers by birds, *Forestry Quarterly* 8:60-73.

Phillips, J. J. 1963. Advance reproduction under mature oak stands of the New Jersey Coastal Plain, USDA Forest Service Research Note NE-4, 5 pp.

Pickett, S. T. A. 1976. Succession: An evolutionary interpretation, *American Naturalist* 110:107-119.

Pickett, S. T. A. 1983. Differential adaptation of tropical species to gaps and its role in community dynamics, *Tropical Ecology* 24:68-84.

Pickett, S. T. A., and J. N. Thompson. 1978. Patch dynamics and the design of nature reserves, *Biological Conservation* 13:27-37.

Pickett, S. T. A., and P. S. White (eds.). 1985. *The Ecology of Natural Disturbance and Patch Dynamics,* Academic, New York, 472 pp.

Pielou, E. C. 1959. The use of point-to-plant distances in the study of the pattern of plant populations, *Journal of Ecology* 47:607-713.

Pielou, E. C. 1974. Spatial pattern of trees in the major forest types of northern Ontario, *Canadian Journal of Forest Research* 4:8-14.

Pielou, E. C. 1991. *After the ice age: the return of life to glaciated North America.* The University of Chicago Press, Chicago, Illinois, U.S.A. 366 pp.

Pienaar, L. V. 1965. Quantitative theory of forest growth, unpublished Ph.D. thesis, University of Washington, Seattle, 176 pp.

Pienaar, L. V. 1979. An approximation of basal area growth after thinning based on growth in unthinned plantations, *Forest Science* 25:223-232.

Pienaar, L. V., and B. D. Shiver. 1984a. An analysis and models of basal area growth in 45-year-old unthinned and thinned slash pine plantation plots, *Forest Science* 30:933-942.

Pienaar, L. V., and B. D. Shiver. 1984b. The effect of planting density on dominant height in unthinned slash pine plantations, *Forest Science* 30:1059-1066.

Pienaar, L. V., and K. J. Turnbull. 1973. The Chapman-Richards generalization of Von Bertalanffy's growth model for basal area growth and yield in even-aged stands, *Forest Science* 19:1-22.

Pienaar, L. V., B. D. Shiver, and G. E. Grider. 1985. Predicting basal area growth in thinned slash pine plantations, *Forest Science* 31:731-741.

Pierce, F. R. 1921. Tornado destroys great forest. *Popular Mechanics,* May, 1921. pp. 659-662.

Piper, C. V. 1906. *Flora of the State of Washington,* Vol. 11, Contribution to U.S. National Herbarium, 637 pp.

Piussi, P. 1966. Some characteristics of a second-growth northern hardwood stand, *Ecology* 47:860-864.

Platt, W. J. 1975. The colonization and formation of equilibrium plant species associations on badger disturbances in a tall-grass prairie, *Ecological Monographs* 45:285-305.

Plochman, R., F. R. von Paar, D. Kraft, and C. D. Oliver. 1983. Pollution is killing German forests, *Journal of Forestry* 81(9):back page.

Plummer, F. G. 1912. Forest fires: Their causes, extent, and effects, with a summary of recorded destruction and loss, USDA Forest Service Bulletin 117, 39 pp.

Pollock, B. M., and V. K. Toole. 1961. After ripening, rest period, and dormancy, in A. Stefferud (ed.), *Seeds; the Yearbook of Agriculture, 1961*, U.S. Department of Agriculture, pp. 106-113.

Pomerleau, R. 1953. History of hardwood species dying in Quebec, in *Canada Department of Agriculture Report on the Symposium on Birch Dieback*, Part 1, Canada Department of Agriculture, Ottawa, Canada, pp. 10-11.

Pons, L. J., N. Van Breemen, and P. M. Driessen. 1982. Physiography of coastal sediments and development of potential soil acidity, in *Acid Sulfate Weathering*, Soil Science Society of America, Madison, Wisconsin, pp. 1-18.

Priester, D. S., and M. T. Pennington. 1978. Inhibitory effects of broomsedge extracts on the growth of young loblolly pine seedlings, USDA Forest Service Research Paper SE-182.

Pritchett, W. L. 1979. *Properties and Management of Forest Soils*, John Wiley, New York, 500 pp.

Pukkala, T., and T. Kolstrom. 1992. Stochastic spatial regeneration model for Scots pine. *Scandanavian Journal of Forest Research* 7.

Purdie, R. W., and R. O. Slatyer. 1976. Vegetation succession after fire in sclerophyll woodland communities in southeastern Australia, *Australian Journal of Ecology* 1:223-236.

Putnam, J. A. 1951. Management of bottomland hardwoods, USDA Forest Service Occasional Paper No. 116, Southern Forest Experiment Station, 60 pp.

Putnam, J. A., G. M. Furnival, and J. S. McKnight. 1960. Management and inventory of Southern hardwoods, U. S. Department of Agriculture, Agriculture Handbook Number 181, 102 pp.

Putz, F. E., P. D. Coley, K. Lu, A. Montaluo, and A. Aiello. 1983. Uprooting and snapping of trees: Structural determinants and ecological consequences, *Canadian Journal of Forest Research* 13:1011-1020.

Pyne, S. J. 1982. *Fire in America: A Cultural History of Wildland and Rural Fire*, Princeton University Press. Princeton, New Jersey, 654 pp.

Pyne, S. J. 1984. *Introduction to Wildland Fire*, John Wiley, New York, 455 pp.

Quick C. R. 1961. How long can a seed remain alive? in A. Stefferud (ed.), *Seeds: The Yearbook of Agriculture*, 1961, U.S. Department of Agriculture, pp. 94-99.

Ralston, R. A. 1953. Some effects of spacing on jack pine development in Lower Michigan after 25 years, *Michigan Academy of Science, Arts, and Letters, Paper* 37:137-143.

Rangnekar, P. V., and D. F. Forward. 1973. Foliar nutrition and wood growth in red pine: Effects of darkening and defoliation on the distribution of C^{14} in young trees, *Canadian Journal of Botany* 51:103-108.

Rangnekar, P. V., D. F. Forward, and N. J. Nolan. 1969. Foliar nutrition and wood growth in red pine: The distribution of radiocarbon photoassimilated by individual branches of young trees, *Canadian Journal of Botany* 47:1701-1711.

Ranney, J. W., M. C. Bruner, and J. B. Levenson. 1981. The importance of edge in the structure and dynamics of forest islands, in R. L. Burgess and D. M. Sharpe (eds.), *Forest Island Dynamics in Man-Dominated Landscapes,* Ecological Studies, Volume 41, Springer-Verlag, New York, pp. 67-95.

Raunkiaer, C. 1928. Dominansareal, Artstaethed og Formation sdominanter. kg. *Danske Vidensk Selsk. Biol. Med.,* 7,1 (cited by Shimwell, 1971).

Raup, H. M. 1941. An old forest in Stonington, Connecticut, *Rhodora* 43.67-71.

Raup, H. M. 1942. Trends in the development of geographic botany, *Annals of the Association of American Geographers* 32:319-354.

Raup, H. M. 1946. Phytogeographic studies of the Athabaska-Great Slave Lake Region II, *Journal of the Arnold Arboretum* 27:1-85.

Raup, H. M. 1957. Vegetation adjustment to the instability of site, in *Proceedings and Papers, Sixth Technical Meeting of the International Union,* Conservation of Natural and National Resources, Edinburgh, pp. 36-48.

Raup, H. M. 1964. Some problems with ecological theory and their relation to conservation, *Journal of Ecology* 52:(Suppl): 19-28.

Raup, H. M., and G. W. Argus. 1982. *The Lake Athabasca sand dunes of Northern Saskatchewan and Alberta, Canada: I. The land and vegetation,* National Museums of Canada Publications in Botany No. 12, 96 pp.

Reifsnyder, W. E., and H. W. Lull. 1965. Radiant energy in relation to forests, USDA Forest Service Technical Bulletin No. 1344, 111 pp.

Reineke, L. H. 1933. Perfecting a stand-density index for even-aged forests, *Journal of Agricultural Research* 46:627-638.

Reiners, N. M, and W. A. Reiners. 1965. *Natural Harvesting of Trees,* William L. Hutcheson Memorial Forest Bulletin 2:9-17.

Reiners, W. A. 1983. Disturbance and basic properties of ecosystem energetics, in H. A. Mooney and M. Godron (eds.), *Disturbance and Ecosystems: Components of Response,* Springer-Verlag, New York, pp. 83-96.

Reiners, W. A., I. A. Worley, and D. B. Lawrence. 1971. Plant diversity in a chronosequence at Glacier Bay, Alaska, *Ecology* 52:55-69.

Remington, C. L. 1968. Suture-zones of hybrid interaction between recently joined biotas, *Evolutionary Biology* 2:321-428.

Remington, C. L. 1985. Genetical differences in solutions to the crises of hybridization and competition in early sympatry, *Bulletin of Zoology* 52:21-43.

Reukema, D. L. 1961. Response of individual Douglas-fir trees to release, USDA Forest Service Pacific Northwest Forest and Range Experiment Station Research Note No. 208, 4 pp.

Reukema, D. L. 1970. Forty-year development of Douglas-fir stands planted at various spacings, USDA Forest Service Research Paper PNW-100, 21 pp.

Reukema, D. L. 1972. Twenty-one-year development of Douglas-fir stands repeatedly thinned at varying intervals, USDA Forest Service Research Paper PNW-141, 23 pp.

Reukema, D. L. 1975. Guidelines for precommercial thinning of Douglas-fir, USDA Forest Service General Technical Report PNW-30, 10 pp.

Reukema, D. L. 1979. Fifty-year development of Douglas-fir stands planted at various spacings, USDA Forest Service Research Paper PNW-253, 21 pp.

Reukema, D. L., and D. Bruce. 1977. Effect of thinning on yield of Douglas-fir: Concepts and some estimates obtained by simulation, USDA Forest Service General Technical Report PNW-58, 36 pp.

Reynolds, K. M., and W. J. Bloomberg. 1982. Estimating probability of intertree root contact in second-growth Douglas-fir, *Canadian Journal of Forest Research* **12**:493-498.

Reynolds, R. R. 1954. Growing stock in the all-aged forest, *Journal of Forestry* **52**:744-747.

Reynolds, R. R. 1969. Twenty-nine years of selection timber management on the Crossett Experimental Forest. USDA Forest Service Research Paper SO-40.

Rice, E. L. 1974. *Allelopathy*, Academic, New York, 353 pp.

Rich, L. R. 1962. Erosion and sediment movement following a wildfire in a ponderosa pine forest of central Arizona, USDA Forest Service, Rocky Mountain Experiment Station Research Note 76, 12 pp.

Richards, B. N., and G. K. Voigt. 1963. Nitrogen accretion in coniferous forest ecosystems in C. T. Youngberg (ed.), *Forest-soil Relationships in North America*, Oregon State University Press, Corvallis, pp. 105-116.

Richards, P. W. 1952. *The Tropical Rain Forest: An Ecological Study*, Cambridge University Press, Cambridge, 450 pp.

Richardson, J. L. 1980. The organismic concept of vegetation: Resilience of an embattled concept, *Bioscience* **30**:465- 471.

Rickerd, J., and D. Hanley. 1982. *Diameter limit cutting—a questionable practice*, University of Idaho College of Agriculture Cooperative Extension Service Current Information Series No. 630, 2 pp.

Ricklefs, R. E. 1977. Environmental heterogeneity and plant species diversity: A hypothesis, *American Naturalist* **111**:376-381.

Ridley, H. N. 1930. *The Dispersal of Plants Throughout the World*, L. Reeve, Ashford, England, 744 pp.

Rietveld, W. J. 1975. Phytotoxic grass residues reduce germination and initial root growth of ponderosa pine, USDA Forest Service Research Paper RM-153, 15 pp.

Riley, D., and A. Young. 1966. *World Vegetation*, Cambridge University Press, Cambridge, 96 pp.

Rinehart, M. 1979. Spatial patterns in natural second-growth stands of coastal western hemlock, master of forest science thesis. College of Forest Resources, University of Washington, Seattle, 69 pp.

Roach, B. A. 1977. A stocking guide for Allegheny hardwoods and its use in controlling intermediate cuttings, USDA Forest Service Research Paper NE-373, 30 pp.

Roach, B. A., and S. F. Gingrich. 1968. Even-aged silviculture for upland central hardwoods, USDA Forest Service Agriculture Handbook 355, 39 pp.

Robinson, W. L., and E. G. Bolen. 1984. *Wildlife ecology and management*. Macmillan Publishing Company, New York. 478 pp.

Robinson, W. L., L. H. Fanter, A. G. Spalding, and S. L. Jones. 1980. Biological impacts of political mismanagement of whitetailed deer in Pictured Rocks National Lakeshore. *Proceedings of the Second Conference on Scientific Research in National Parks* **7**:283-292.

Roe, A. L. 1967. Seed dispersal in a bumper spruce seed year, USDA Forest Service Research Paper INT-39, 10 pp.

Rogers, W. E. 1922. Ice storms and trees, *Torreya* **22**:61-63.

Rogers, W. E. 1923. Resistance of trees to ice storm injury, *Torreya* **23**:95-99.

Rollins, G. 1982. Management experience with grand fir in the northern Blue Mountains in C. D. Oliver and R. M. Kenady (eds.), *Biology and Management of True Fir in the Pacific Northwest*, College of Forest Resources, University of Washington, Seattle, Institute of Forest Resources Contribution Number 45, pp. 253-255.

Romberger, J. A. 1963. Meristems, growth, and development in woody plants, USDA Forest Service Technical Bulletin No. 1293, 214 pp.

Romme, W. H. 1980. Fire frequency in subalpine forests of Yellowstone National Park, in *Proceedings of the Fire History Workshop,* October 20-24, 1980, Tucson, Arizona, USDA Forest Service General Technical Report RM-81, pp. 27-30.

Romme, W. H. 1982. Fire and landscape diversity in subalpine forests of Yellowstone National Park, *Ecological Monographs* 52:199-221.

Romme, W. H., and D. H. Knight. 1981. Fire frequency and subalpine forest succession along a topographic gradient in Wyoming, *Ecology* 62:319-326.

Rook, D. A., and M. J. Carlson. 1978. Temperature and irradiance and the total daily photosynthetic production of the crown of a *Pinus radiata* tree, *Oecologia* 36:371-382.

Ross, M. A., and J. L. Harper. 1972. Occupation of biological space during seedling establishment, *Journal of Ecology* 60:77-88.

Roth, E. R., and B. Sleeth. 1939. Butt rot in unburned sprout oak stands, U.S. Department of Agriculture Technical Bulletin 684, 43 pp.

Roth, E. R., and G. H. Hepting. 1943. Origin and development of oak stump sprouts as affecting their likelihood to decay, *Journal of Forestry* 41:27-36.

Roth, L. F., and J. W. Barrett. 1985. Response of dwarf mistletoe-infested ponderosa pine to thinning: II. Dwarf mistletoe propagation, USDA Forest Service Research Paper PNW-331, 20 pp.

Rowe, J. S., and G. W. Scotter. 1973. Fire in the boreal forest, *Journal of Quaternary Research* 3:444-464.

Roy, D. F. 1955. Hardwood sprout measurements in northwestern California, USDA Forest Service Research Note PSW-95, 6 pp.

Royama, T. 1984. Population dynamics of the spruce budworm *Choristoneura fumiferana. Ecological Monographs* 54: 429-462.

Ruark, G. A., D. L. Mader, and T. A. Tattar. 1982. The influence of soil compaction and aeration on the root growth and vigour of trees—a literature review. Part I, *Arboriculture Journal* 6:251-265.

Rudolph, T. D. 1964. Lammas growth and prolepsis in jack pine in the lake state, *Forest Science Monograph* 6:1-70.

Runkle, J. R. 1981. Gap regeneration in some old-growth forests of the eastern United States, *Ecology* 62:1041-1051.

Runkle, J. R. 1982. Patterns of disturbance in some old-growth mesic forests in eastern North America, *Ecology* 63:1533-1546.

Runkle, J. R. 1984. Development of woody vegetation in treefall gaps in a beech-sugar maple forest, *Holarctic Ecology* 7:157-164.

Running, S. W. 1980. Relating plant capacitance to the water relations of *Pinus contorta, Forest Ecology and Management* 2:237-252.

Ruth, R. H. 1970. Effect of shade on establishment and growth of salmonberry, USDA Forest Service Research Paper PNW-96, 10 pp.

Ruth, R. H., and R. A. Yoder. 1953. Reducing wind damage in the forests of the Oregon coast range, USDA Forest Service Pacific Northwest Forest and Range Experiment Station Research Paper Number 7, 30 pp.

Safranyik, L., and R. Jahren. 1970. Emergence patterns of the mountain pine beetle from lodgepole pine, Canadian Department of Fishery and Forestry, *Bimonthly Research Notes* 26(2):11-12.

Salinas, L. M., R. D. DeLaune, and W. H. Patrick, Jr. 1986. Changes occurring along a rapidly submerging coastal area, *Journal of Coastal Research* 2:269-282.

Salisbury, F. B., and C. Ross. 1969. *Plant Physiology,* Wadsworth, Belmont, California, 747 pp.

Salisbury, H. E. 1989. *The Great Black Dragon Fire: a Chinese inferno.* Little, Brown, & Co., Boston.

Salo, J., and M. Rasanen. 1989. Heirarchy of landscape patterns in western Amazon. IN L. B. Holm-Nielson, I. C. Nielsen, and H. Balslev, editors. *Tropical Forests.* Academic Press, New York. pp. 35-45.

Salwasser, H. 1990. Gaining perspective: forestry for the future. *Journal of Forestry* **88**(11):32-38.

Salwasser, H., D. MacCleery, and T. Snellgrove. 1992. *New perspectives for managing the U. S. National Forest system.* Report to the North American Forestry Commission, 16th Session, Cancun, Mexico. February, 1992.

Sampson, R. N, and D. L. Adams (editors). 1994. *Assessing Forest Ecosystem Health in the Inland West.* The Haworth Press, Inc., New York. 461 pp.

Sampson, R. N., D. L. Adams, S. Hamilton, S. P. Mealey, R. Steele, and D. Van DeGraff. 1994. Assessing forest ecosystem health in the Inland West. *American Forests Magazine* 100, March/April. pp. 13-16.

Sander, I. L. 1966. Composition and distribution of hardwood reproduction after harvest cutting, in: *Proceedings of Symposium on Hardwoods of the Piedmont and Coastal Plain,* Georgia Forest Research Council, pp. 30-33.

Sander, I. L. 1971. Height growth of new oak sprouts depends on size of advance reproduction, *Journal of Forestry* 69:809-811.

Sander, I. L. 1972. Size of oak advance reproduction: Key to growth following harvest cutting, USDA Forest Service North Central Forest Experiment Station Research Paper NC-79, 6 pp.

Sander, I. L., and F. B. Clark. 1971. Reproduction of upland hardwood forests in the Central States, U.S. Department of Agriculture Handbook 405, 25 pp.

Sanford, R. L., Jr. 1987. Apogeotropic roots in an Amazon rain forest, *Science* **235**:1062-1064.

Sanford, R. L., Jr., J. Saldarriaga, K. Clark, C. Uhl, and R. Herrara. 1985. Amazon rainforest fires. *Science* **227**:53-55.

Santantonio, D., R. K. Hermann, and W. S. Overton. 1977. Root biomass studies in forest ecosystems, *Pedobiologia* **17**:1-9.

Sargent, C. S. 1884. *Report on the forests of North America,* U.S. Department of the Interior, Census Office, 612 pp.

Sarigumba, T. I., and G. A. Anderson. 1979. Response of slash pine to different spacings and site-preparation treatments, *Southern Journal of Applied Forestry* 3:91-94.

Sartwell, C. 1971. Thinning ponderosa pine to prevent outbreaks of mountain pine beetle, in *Precommercial Thinning of Coastal and Intermountain Forests in the Pacific Northwest, Proceedings of a Short Course,* Washington State University, Pullman, pp. 41-52.

Sato, A. 1983. Idea of new density control chart involving time axis, *Hoppo Ringyo* **35**:311-315.

Satoo, T., K. Negis, and M. Senda. 1959. Materials for the studies of growth in stands: V. Amount of leaves and growth in plantations of *Zelkowa serratapp.* with crown thinning, *Bulletin of the Tokyo University Forest* **55**:101-125.

Savage, H., Jr. 1970. *Lost Heritage,* Morrows, New York, pp. 329.

Scagel, R. F., R. J. Bandoni, G. E. Rouse, W. B. Schofield, J. R. Stein, and T. M. C. Taylor. 1969. *Plant Diversity: An Evolutionary Approach,* Wadsworth, Belmont, California, 460 pp.

Schenck, C. A. 1924. Der Waldbau des Urwalds, *Allgemeine Forst und Jagdzeitung* **100**:377-388.

Schlaegel, B. E. 1978. Growth and yield of natural hardwood stands: Concepts, practices, and problems, in *Proceedings of the Second Symposium on Southeastern Hardwoods,* Dothan, Alabama, pp. 120-129.

Schlaegel, B. E. 1981. Sycamore plantation growth: 10-year results of spacing trials, in J. P. Barrett (ed.), *Proceedings of the First Biennial Southern Silviculture Research Conference,* USDA Forest Service General Technical Report SO-34, pp. 149-155.

Schmidt, R. L. 1957. *The silvics and plant geography of the genus Abies in the coastal forests of British Columbia,* British Columbia Forest Service Technical Publication T-46, 31 pp.

Schmidt-Vogt, H. 1986. *Die Fichte.* Band II/1. Verlag Paul Parey, Hambur. 563 pp. (In German)

Schnur, G. L. 1934. Diameter distributions for old-field loblolly pine stands in Maryland, *Journal of Agricultural Research* **49**:731-743.

Schoeneweiss, D. F. 1975. Predisposition, stress, and plant disease, *Annual Review of Phytopathology* **13**:193-211.

Scholz, H. F. 1931. *Physical properties of the cover soils on the Black Rock Forest,* Harvard University, Harvard Black Rock Forest Bulletin 2, 59 pp.

Scholz, H. F. 1948. Diameter-growth studies of northern red oak and their possible silvicultural implications, *Iowa State College Journal of Science* **22**:421-429.

Schopmeyer, C. S. 1974. Seeds of woody plants in the United States, USDA Forest Service Agriculture Handbook No. 450, 883 pp.

Schroeder, M. J., and C. C. Buck. 1970. Fire weather, USDA Forest Service, Agriculture Handbook 360, 288 pp.

Schumacher, F. X., and T. S. Coile. 1960. *Growth and Yields of Natural Stands of the Southern Pines,* T. S. Coile, Durham, North Carolina, 115 pp.

Scott, D. R. M. 1962. The Pacific Northwest region, in J. W. Barrett (ed.), *Regional Silviculture of the United States,* Ronald, New York, pp. 503-570.

Scott, D. R. M. 1980. The Pacific Northwest region, in J. W. Barrett (ed.), *Regional Silviculture of the United States,* 2d ed., Ronald, New York, pp. 447-494.

Scott, D. R. M., J. N. Long, and J. Kotar. 1976. Comparative ecological behavior of western hemlock in the Washington Cascades, in W. A. Atkinson and R. J. Zasoski (eds.), *Proceedings of the Western Hemlock Management Conference,* College of Forest Resources, University of Washington, Seattle, Institute of Forest Products Contribution No. 34, pp. 26-33.

Scully, N. J. 1942. Root distribution and environment in a maple-oak forest, *Botanical Gazette* **103**:492-517.

Seagle, S. W., R. A. Lancia, D. A. Adams, M. R. Lennartz, and H. A. Devine. 1987. Integrating timber and red-cockaded woodpecker habitat management. *Transactions of the North American Wildlife Natural Resources Conference* **52**: 41-52. (Original not seen; cited from Hunter, 1990.)

Segura, G. 1991. Dynamics of old-growth forests of Abies amabilis impacted by tephra from the 1980 eruption of Mount St. Helens, Washington. Unpublished Ph. D. dissertation, University of Washington, College of Forest Resources, Seattle, Washington: 170 pp.

Segura-Warnholtz, G., and L. Snook. 1992. Population dynamics and disturbance patterns of a pinon pine forest in Central East Mexico. *Forest ecology and management* **47**:175-194.

Segura, G., L. B. Brubaker, J. F. Franklin, T. M. Hinckley, D. A. Maguire, and G.

Wright. 1994. Recent mortality and decline in mature Abies amabilis: the interaction between site factors and tephra deposition from Mount St. Helens. *Canadian Journal of Forest Research* 24:1112-1122.

Seidel, K. W. 1977. Suppressed grand fir and shasta red fir respond well to release, USDA Forest Service Research Note PNW-288, 7 pp.

Seidel, K. W. 1980. Diameter and height growth of suppressed grand fir saplings after overstory removal, USDA Forest Service Research Paper PNW-275, 9 pp.

Selm, H. R. 1952. Carbon dioxide gradients in a beech forest in central Ohio, *Ohio Journal of Science* 52:187-198.

Senda, M., and T. Satoo. 1956. Materials for the study of growth in stands: II. White pine *(Pinus strobus)* stands of various densities in Hokkaido, *Bulletin of the Tokyo University Forest* 52:15-31.

Sessions, J., and J. B. Sessions. 1992. Scheduling and network analysis program (SNAPII) User's guide, Version 2. 02. Corvallis, Oregon, U.S.A.

Seton, E. T. 1912. *The Forester's Manual,* Doubleday, New York.

Seymour, R. S. 1992. The red spruce-balsam fir forest of Maine: evolution of silvicultural practice in response to stand development patterns and disturbances. *In*: (M. J. Kelty, B. C. Larson and C. D. Oliver, editors) *The Ecology and Silviculture of Mixed-Species Forests: A Festschrift for David M. Smith.* Kluwer Academic Publishers, Boston:

Seymour, R. S., and D. M. Smith. 1987. A new stocking guide formulation applied to eastern white pine, *Forest Science* 33:469-484.

Seymour, R. S., and M. L. Hunter, Jr. 1992. New forestry in eastern spruce-fir forests: principles and applications to Maine. *Maine Agricultural Experiment Station Miscellaneous Publication 716*: 36 pp.

Seymour, R. S., D. G. Mott, S. M. Kleinschmidt, P. H. Triandifillou, and R. Keane. 1983. Green woods model: a forecasting tool for planning timber harvesting and protection of spruce-fir forests attacked by the spruce budworm, USDA Forest Service. General Technical Report NE-91, 38 pp.

Seymour, V. A., T. M. Hinckley, Y. Morikawa, and J. F. Franklin. 1983. Foliage damage in coniferous trees following volcanic ashfall from Mt. St. Helens. *Oecologia* 59:339-343.

Seymour, V. A., T. M. Hinckley, Y. Morikawa, and J. F. Franklin. 1983. Foliage damage in coniferous trees following volcanic ashfall from Mt. St. Helens, *Oecologia* 59:339-343.

Shamida, A. 1978. Kermes oaks in the land of Israel, *Israel Land and Nature* 6:9-16.

Shantz, H. L., and R. Zon. 1924. Natural vegetation, in *Atlas of American Agriculture,* U.S. Government Printing Office, Washington, D.C.

Sharp, R. F. 1975. Nitrogen fixation in deteriorating wood: The incorporation of $^{15}N_2$ and the effects of environmental conditions on acetylene reduction, *Soil Biology Biochemistry* 7:9-14.

Sharp, R. P. 1960. *Glaciers,* Condon Lectures, Oregon State System of Higher Education, Eugene, 78 pp.

Shea, K. R. 1960a. Decay in logging scars in western hemlock and sitka spruce, *Weyerhaeuser Company Forestry Research Note 25*, Forestry Research Division, Centralia, Washington, 13 pp.

Shea, K. R. 1960b. Deterioration—a pathological aspect of second-growth management in the Pacific Northwest, *Weyerhaeuser Company Forestry Research Note 28*, Forestry Research Division, Centralia, Washington, 17 pp.

Shea, K. R. 1967. *Effect of artificial root and bole injuries on diameter increment of Douglas-fir*, Weyerhaeuser Company Forestry Paper 11, Forestry Research Center, Centralia, Washington, 11 pp.

Shea, K. R., and N. E. Johnson. 1962. *Deterioration of wind-thrown conifers three years after blowdown in southwestern Washington*, Weyerhaeuser Company Forestry Research Note 44, Forestry Research Division, Centralia, Washington, 17 pp.

Shearer, R. C. 1974. Early establishment of conifers following prescribed broadcast burning in western larch/Douglas-fir forests, *Tall Timbers Fire Ecology Conference* 14:481-500.

Sheffield, R. M., and N. D. Cost. 1987. Behind the decline. Why are natural pine stands in the Southeast growing slower? *Journal of Forestry* 85:29-33.

Shelford, V. E. 1954. Some lower Mississippi valley flood plain biotic communities: Their age and elevation, *Ecology* 35:126-142.

Shepard, F. P. 1963. Thirty-five thousand years of sea level, in *Essays in Marine Geology in Honor of K. O. Emery*, University of Southern California Press, Los Angeles, pp. 1-10.

Shigo, A. L. 1985. Wounded forests, starving trees, *Journal of Forestry* 83:668-673.

Shigo, A. L., and E. H. Larson. 1969. A photo guide to the patterns of discoloration and decay in living northern hardwood trees, USDA Forest Service Research Paper NE-127, 100 pp.

Shimer, J. A. 1972. *Field Guide to Landforms in the United States*, Macmillan, New York, 272 pp.

Shimwell, D. W. 1971. *The Description and Classification of Vegetation*, University of Washington Press, Seattle, 322 pp.

Shinozaki, K., and T. Kira. 1956. Intraspecific competition among higher plants: VII. Logistic theory of the C-D effect. *Journal of the Institute of Polytechnics, Osaka City University* 7:35-72.

Shinozaki, K., K. Yoda, K. Hozumi, and T. Kira. 1964a. A quantitative analysis of plant form—the pipe model theory: I. Basic analyses, *Japanese Journal of Ecology* 14:97-105.

Shinozaki, K., K. Yoda, K. Hozumi, and T. Kira. 1964b. A quantitative analysis of plant form—the pipe model theory: II. Further evidence of the theory and its application in forest ecology, *Japanese Journal of Ecology* 14:133-139.

Shipman, R. D. 1955. Furrows improve longleaf survival in scrub oak, *Southern Lumberman* (June 15).

Shirley, H. L. 1945. Reproduction of upland conifers in the Lake States as affected by root competition and light, *American Midland Naturalist* 33:537-612.

Show, S. B., and E. I. Kotok. 1924. The role of fire in California pine forests, USDA Bulletin No. 1294, 80 pp.

Shugart, H. H., D. C. West, and W. R. Emanuel. 1981. Patterns and dynamics of forests: An application of simulation models, in D. C. West, H. H. Shugart, and D. B. Botkin (eds.), *Forest Succession: Concepts and Application*, Springer-Verlag, New York, pp. 74-94.

Shugart, H. H., Jr., and D. C. West. 1980. Forest succession models, *BioScience* 30:308-313.

Siccama, T. G., F. H. Bormann, and G. E. Likens. 1970. The Hubbard Brook ecosystem study: Productivity, nutrients, and phytosociology of the herbaceous layer, *Ecological Monographs* 40:389-402.

Siccama, T. G., G. Weir, and K. Wallace. 1976. Ice damage in a mixed hardwood forest in Connecticut in relation to Vitis infestation, *Bulletin of the Torrey Botanical Club*

103:180-188.

Sidle, R. C., A. J. Pearce, and C. L. O'Loughlin. 1985. *Hillslope Stability and Land Use,* American Geophysical Union, Washington, D.C., 140 pp.

Siemon, G. R, G. B. Wood, W. G. Forrest. 1976. Effects of thinning on crown structure in radiata pine, *New Zealand Journal of Forest Science* 6:57-66.

Sigafoos, R. S., and E. L. Hendricks. 1969. *The time interval between stabilization of alpine glacial deposits and establishment of tree seedlings,* U.S. Geological Survey Professional Paper 650-B.

Sjolte-Jorgensen, J. 1967. The influence of spacing on the growth and development of coniferous plantations, in J. A. Romberger and P. Mikola (eds.), *International Review of Forestry Research,* Volume 2, Academic, New York, pp. 43-94.

Skeen, J. N. 1976. Regeneration and survival of woody species in a naturally-created forest opening, *Bulletin of the Torrey Botanical Club* 103:259-268.

Skeen, J. N. 1981. The pattern of natural regeneration and height stratification within a naturally-created opening in an all-aged Piedmont deciduous forest, in *First Biennial Southern Silvicultural Research Conference Proceedings, USDA* Forest Service General Technical Report SO-34, pp. 259-268.

Skinner, B. J., and S. C. Porter. 1987. *Physical Geology,* John Wiley, New York, 750 pp.

Slobotkin, L. B. 1961. *Growth and Regulation of Animal Populations,* Holt, Rinehart, & Winston, New York, 184 pp.

Smith F. W., and J. N. Long. 1987. Elk hiding and thermal cover guidelines in the context of lodgepole pine stand density, *Western Journal of Applied Forestry* 2:6-10.

Smith, A. P. 1973. Stratification of temperate and tropical forests, *American Naturalist* 107:671-683.

Smith, B. E., and G. Cottam. 1967. Spatial relationships of mesic forest herbs in southern Wisconsin, *Ecology* 48:546-558.

Smith, C. C. 1970. The coevolution of pine squirrels *(Tamiasiurus)* and conifers, *Ecological Monographs* 40:349-371.

Smith, D. M. 1946. Storm damage in New England forests, master of forestry thesis, Yale University School of Forestry and Environmental Studies, New Haven, Connecticut, 173 pp.

Smith, D. M. 1951. *The influence of seedbed conditions on the regeneration of eastern white pine,* the Connecticut Agricultural Experiment Station Bulletin 545, New Haven, Connecticut, 61 pp.

Smith, D. M. 1962. *The Practice of Silviculture,* 7th ed., Wiley, New York, 578 pp.

Smith, D. M. 1973. Silvicultural practice and stand development in eastern deciduous forests, paper presented at Symposium on *Ecological Impact of Management Practices in Eastern Deciduous Forests,* Ecological Society of America, Amherst, Massachusetts, July 18, 1973.

Smith, D. M. 1976. Changes in eastern forests since 1600 and possible effects, in J. F. Anderson and H. K. Kaya (eds.), *Perspectives in Forest Entomology,* Academic, New York, pp. 3-20.

Smith, D. M. 1980. The forests of the United States, in J. W. Barrett (ed.), *Regional Silviculture of the United States,* 2d ed., John Wiley, New York, pp. 1-23.

Smith, D. M. 1982. Patterns of development of forest stands, in J. E. Means (ed.), *Forest Succession and Stand Development Research in the Northwest,* Forest Research Laboratory, Oregon State University, Corvallis, pp. 1-4.

Smith, D. M. 1984. Terminology of stand dynamics, paper presented at Stand Dynamics Interest Group Meeting, Cottonwood, Mississippi, October 15, 1984, 5 pp.

Smith, D. M. 1986. *The Practice of Silviculture,* 8th ed., Wiley, New York, 527 pp.

Smith, D. W., and N. E. Linnartz. 1980. The southern hardwood region, in J. W. Barrett (ed.), *Regional Silviculture of the United States*, Wiley, New York, pp. 145-230.

Smith, F. E. 1975. Ecosystems and evolutions *Bulletin of the Ecological Society of America* 56:2-6.

Smith, H. C. 1966. Epicormic branching on eight species of Appalachian hardwoods, USDA Forest Service Research Note NE-53, 4 pp.

Smith, H. C. 1983. Growth of Appalachian hardwoods kept free to grow from 2 to 12 years after clearcutting, USDA Forest Service Research Paper NE-528, 6 pp.

Smith, H. W., Jr. 1963. Establishment of yellow-poplar (*Liriodendron tulipifera* L.) in canopy openings, unpublished doctor of forestry dissertation, School of Forestry and Environmental Studies, Yale University, New Haven, Connecticut, 176 pp.

Smith, J. H. G. 1961. Comments on "Relationship between tree spacing, knot size, and log quality in young Douglas-fir stands," *Journal of Forestry* 59:682-683.

Smith, J. H. G. 1986. Projections of stand yields and values to age 100 for Douglas-fir, western redcedar, and western hemlock and implications for management from the UBC forest spacing trials, unpublished paper, Faculty of Forestry, University of British Columbia, Vancouver, British Columbia, 16 pp.

Smith, J. H. G., J. W. Ker, and J. Csizmazia. 1965. *Economics of reforestation of Douglas-fir, western hemlock, and western redcedar in the Vancouver Forest District*, Faculty of Forestry, University of British Columbia, Vancouver, British Columbia, Forestry Bulletin No. 3, 144 pp.

Smith, L. 1973. Indication of snow avalanche periodicity through interpretation of vegetation patterns in the North Cascades, Washington, in *Avalanche Studies (1972-1973)* State Highway Department Research Progress Report 8.4.

Smith, N. J., and D. W. Hann. 1984. A new analytical model based on the $-3/2$ power rule of self-thinning, *Canadian Journal of Forest Research* 14:605-609.

Smith, S. T. 1982. Management experience in northern California true fir stands: experience of a private land owner, in C. D. Oliver and R. M. Kenady (ads.), *Biology and Management of True Fir in the Pacific Northwest*, College of Forest Resources, University of Washington, Seattle, Institute of Forest Resources Contribution Number 45, pp. 325-329.

Smith, T. M., and K. Grant. 1986. The role of competition in the spacing of trees in a *Burkea africana-Terminalia sericea* Savanna, *Biotropica* 18:219-223.

Society of American Foresters. 1950. *Forestry Terminology*, 2d ed., Society of American Foresters, Baltimore, Maryland.

Sollins, P., C. C. Grier, F. M. McCorison, K. Cromack, Jr., R. Fogel, and R. L. Fredriksen. 1980. The internal element cycles of an old-growth Douglas-fir ecosystem in western Oregon, *Ecological Monographs* 50:261-285.

Solomon, D. S. 1977. The influence of stand density and structure on growth of northern hardwoods in New England, USDA Forest Service Research Paper NE-362.

Solomon, D. S., and B. M. Blum. 1967. Stump sprouting of four northern hardwoods, USDA Forest Service Research Paper NE-59, 13 pp.

Sonderman, D. L. 1985. Stand density—a factor affecting stem quality of young hardwoods, USDA Forest Service Research Paper NE-561, 8 pp.

Sorensen, R. W., and B. F. Wilson. 1964. The position of eccentric stem growth and tension wood in leaning red oak trees,The Harvard Forest, Harvard University, *Harvard Forest Paper* 12, 10 pp.

Sousa, W. P. 1984. The role of disturbance in natural communities. *Annual Review of Ecology and Systematics* 15:353-391.

South, D. B., and J. P. Barnett. 1986. Herbicides and planting date affect early performance of container-grown and bare-root loblolly pine seedlings in Alabama, *New Forests* 1:17-28.

Sparhawk, W. N. 1918. Effect of grazing upon western yellow pine reproduction in central Idaho., U.S. Department of Agriculture Bulletin No. 738, 31 pp.

Spaulding, V. M., and B. E. Fernow. 1899. *The white pine,* USDA Division of Forestry Bulletin 22, 185 pp.

Spaulding, W. G., and P. S. Martin. 1979. Ground sloth dung of the Guadalupe Mountains, in H. H. Genoways and R. J. Baker (eds.), *Biological Investigations in the Guadalupe Mountains National Park, Texas,* Proceedings and Transactions Series No. 4, USDI National Park Service, pp. 259-269.

Spring, P. E., M. L. Brewer, J. R. Brown, and M. E. Fanning. 1974. Population ecology of loblolly pine *(Pinus taeda)* in an old field community, *Oikos* 25:1-6.

Spring, S. N. 1922. Studies in reproduction—the Adirondack hardwood type, *Journal of Forestry* 20:571-580.

Sprugel, D. G. 1974. Natural disturbance and ecosystem responses in wave-generated Abies balsamea forests, unpublished Ph.D. thesis, Yale University, New Haven, Connecticut, 287 pp.

Sprugel, D. G. 1976. Dynamic structure of wave-generated Abies balsamea forests in the northeastern United States, *Journal of Ecology* 64:889-912.

Sprugel, D. G. 1985. Natural disturbances and ecosystem energetics, in S. T. A. Pickett and P. S. White (eds.), *The Ecology of Natural Disturbance and Patch Dynamics,* Academic, New York, pp. 335-352.

Sprugel, D. G. 1989. Woody-tissue respiration and photosynthesis, in J. P. Lassoie and T. M. Hinckley (eds.), *Methods and Approaches in Tree Ecophysiology,* Chemical Rubber Company Press, pp. 1-19.

Sprugel, D. G. 1991. Disturbance, equilibrium, and Environmental Variability: What is "Natural" Vegetation in a Changing Environment? *Biological Conservation* 58:1-18.

Sprugel, D. G., and T. M. Hinckley. 1988. The branch autonomy theory in W. E. Winner and L. B. Phelds (eds.), *Response of Trees to Air Pollution: the Role of Branch Studies,* Proceedings of an Environmental Protection Agency Workshop held November 5-6 1987, in Boulder, Colorado, pp. 1-19.

Spurr, S. H. 1952. *Forest Inventory,* Ronald, New York, 476 pp.

Spurr, S. H. 1954. The forests of Itasca in the nineteenth century as related to fire, *Ecology* 35:21-25.

Spurr, S. H. 1956. Natural restocking of forests following the 1938 hurricane in central New England, *Ecology* 37:443-451.

Spurr, S. H. 1964. *Forest Ecology,* Ronald, New York, 352 pp.

Spurr, S. H., and B. V. Barnes. 1980. *Forest Ecology,* 3d ed., John Wiley, New York, 687 pp.

Staebler, G. R. 1955. Extending the Douglas-fir yield tables to include mortality, *Proceedings of the Society of American Foresters, pp.* 54-59.

Staebler, G. R. 1956a. Effect of controlled release on growth of individual Douglas-fir trees, *Journal of Forestry* 54:567-568.

Staebler, G. R. 1956b. Evidence of shock following thinning of young Douglas-fir *Journal of Forestry* 54:339.

Staebler, G. R. 1958. Some mensurational aspects of the level-of-growing-stock problem in even-aged stands, *Journal of Forestry* 56:112-115.

Staebler, G. R. 1959. Optimum levels of growing stock for managed stands, *Proceedings of the Society of American Foresters, pp.* 110-113.

Staebler, G. R. 1960. Theoretical derivation of numerical thinning schedules for Douglas-fir, *Forest Science* 6:98-109.

Staebler, G. R. 1964. Height and diameter growth for four years following pruning of Douglas-fir, *Journal of Forestry* 62:406.

Staff, H., and B. Berg. 1982. Accumulation and release of plant nutrients in decomposing Scots pine needle litter. Long-term decomposition in a Scots Pine forest, II, *Canadian Journal of Botany* 60:1561-1568.

Staley, J. M. 1965. Decline and mortality of red and scarlet oaks, *Forest Science* 11:2-17.

Stanley, R. G., and W. L. Butler. 1961. Life processes of the living seed, in A. Stefferud (ed.), *Seeds, the Yearbook of Agriculture, 1961,* U.S. Department of Agriculture, pp. 88-94.

States, J. B. 1976. Local adaptations in chipmunk *(Eutamias amoenus)* populations and evolutionary potential at species' borders, *Ecological Monographs* 46:221-256.

Stauffer, H. B. 1977. Application of the Indices of nonrandomness of Pielou, Hopkins and Skellam, and Clark and Evans, Canadian Forestry Service Pacific Forest Research Centre, Victoria, British Columbia, Canada, Report BC-X-166, 32 pp.

Stearns, F. W. 1951. The composition of the sugar maple-hemlock-yellow birch association in northern Wisconsin, *Ecology* 32:245-265.

Steele, R. W. 1955a. Growth after precommercial thinning in two stands of Douglas-fir, USDA Forest Service Pacific Northwest Forest and Range Experiment Station Research Note No. 117, 6 pp.

Steele, R. W. 1955b. Thinning nine-year-old Douglas-fir by spacing and dominance methods, *Northwest Science* 29:84-89.

Steele, R. W. and N. P. Worthington. 1955. Increment and mortality in a virgin Douglas-fir forest, USDA Forest Service Pacific Northwest Forest and Range Experiment Station Research Note No. 110, 6 pp.

Steen, H. K. 1976. *The U. S. Forest Service: a history.* University of Washington Press, Seattle. 356 pp.

Steinbrenner, E. C. 1951. Effect of grazing on floristic composition and soil properties of farm woodlands in southern Wisconsin, *Journal of Forestry* 49:906-910.

Steinbrenner, E. C., and S. P. Gessel. 1956. Windthrow along cutlines in relation to physiography on the McDonald Tree Farm, *Weyerhaeuser Timber Company Forestry Research Note* 15, Forestry Research Division, Centralia, Washington, 11 pp.

Stephens, E. P. 1955a. The historical-developmental trend of determining forest trends, unpublished Ph.D. thesis, Harvard University, Cambridge, Massachusetts, 228 pp.

Stephens, E. P. 1955b. Research in the biological aspects of forest production, *Journal of Forestry* 53:183-186.

Stephens, E. P. 1956. The uprooting of trees: A forest process, *Proceedings of the Soil Science Society of America* 20:113-116.

Stephens, G. R. 1976. The Connecticut forest, in J. F. Anderson and H. K. Kaya (eds.), *Perspectives in Forest Entomology,* Academic, New York, pp. 21-30.

Stephens, G. R., and P. E. Waggoner. 1970. *The forests anticipated from 40 years of natural transitions in mixed hardwoods,* the Connecticut Agricultural Experiment Station Bulletin No. 707, 68 pp.

Stephens, G. R., and P. E. Waggoner. 1980. *A half century of natural transitions in mixed hardwood forests,* the Connecticut Agricultural Experiment Station Bulletin No. 783, 43 pp.

Sterba, H., and R. A. Monserud. 1993. The maximum density concept applied to uneven-aged mixed-species stands. *Forest Science* 39:432-452.

Stevens, W. K. 1990. New eye on nature: the real constant is eternal turmoil. *New York Times*. Science article. Tuesday, July 31, 1990. B5-B6.

Stewart, R. E. 1975. Allelopathic potential of western bracken, *Journal of Chemical Ecology* 1:161-169.

Stickney, J. S., and P. R. Hoy. 1881. Toxic action of black walnut, *Transactions of the Wisconsin State Horticultural Society* 11:166-167.

Stiell, W. M. 1964. *Twenty-year growth of red pine planted at three spacings*, Canadian Department of Forestry Publication No. 1045, 24 pp.

Stiell, W. M. 1966. *Red pine crown development in relation to spacing*, Canadian Department of Forestry Publication No. 1145.

Stiell, W. M. 1976. White spruce: Artificial regeneration in Canada, Canadian Forestry Service, Forest Management Institute, Ottawa, Ontario, Information Report FMR-X-85, 275 pp.

Stiell, W. M. 1982. Growth of clumped vs. equally spaced trees, *Forestry Chronicle* 58:23-25.

Stoeckler, J. H., R. O. Strothmann, and L. W. Krefting. 1957. Effects of deer browsing in northern hardwood-hemlock type in northeastern Wisconsin. *Journal of Wildlife Management* 21:75-80.

Stoeckler, J. H., R. O. Strothmann, and L. W. Krefting. 1957. Effects of deer browsing in northern hardwood-hemlock type in northeastern Wisconsin, *Journal of Wildlife Management* 21:75-80.

Stone, E. C., and R. B. Vasey. 1962. Redwood physiology: Key to recreational management in redwood state parks, *California Agriculture* 16:2-3.

Stone, E. C., and R. B. Vasey. 1968. Preservation of coast redwood on alluvial flats, *Science* 159:157-161.

Stone, E. L., and S. M. Cornwell. 1968. Basal bud burls in *Betula populifolia*, *Forest Science* 14:64-65.

Stone, E. L., Jr. 1952. Evaporation in nature, *Journal of Forestry* 50:746-747.

Stoneburner, D. L. 1978. Evidence of hurricane influence on Barrier Island slash pine forests in the northern Gulf of Mexico, *American Midland Naturalist* 99:234-238.

Stout, B. B. 1952. *Species distribution and soils in the Harvard Forest*, The Harvard Forest, Harvard University, Harvard Forest Bulletin No. 24, 29 pp.

Stout, B. B. 1966. *Studies of the root systems of deciduous trees*, The Harvard Forest, Harvard University, Harvard Black Rock Forest Bulletin No. 15, 45 pp.

Strahler, A. N., and A. H. Strahler. 1978. *Modern Physical Geography*, John Wiley, New York, 544 pp.

Strain, B. R. 1978. Report on the Workshop on Anticipated Plant Responses to Global Carbon Dioxide Enrichment, Duke Environmental Center, Duke University, Durham, North Carolina.

Strang, R. M, and A. H. Johnson. 1981. Fire and climax spruce forests in central Yukon, *Arctic* 34:60-61.

Strang, R. M. 1973. Succession in unburned subarctic woodlands, *Canadian Journal of Forest Research* 3:140-143.

Struick, G. J., and J. T. Curtis. 1962. Herb distribution in an *Acer saccharum* forest, *American Midland Naturalist* 68:285-296.

Stuart, J. D., D. R. Geiszler, R. I. Gara, and J. K. Agee. 1983. Mountain pine beetle scarring of lodgepole pine in south-central Oregon, *Forest Ecology and Management*5:207-214.

Stubblefield, G. W. 1978. Reconstruction of a red alder/Douglas-fir/western hemlock/western redcedar mixed stand and its biological and silvicultural implications, unpublished master of science thesis, College of Forest Resources, University of Washington, Seattle, 141 pp.

Stubblefield, G. W. and C. D. Oliver. 1978. Silvicultural implications of the reconstruction of mixed alder/conifer stands, in W. A. Atkinson, D. Briggs, and D. S. DeBell (eds.), *Utilization and Management of Red Alder,* USDA Forest Service General Technical Report PNW-70, pp. 307-320.

Stubbs, J. 1964. Many-aged management compared with even-aged management in Coastal Plain bottomland hardwoods, in *Proceedings of the Forty-third Annual Meeting of the Appalachian Section of the Society of American Foresters,* pp. 7-9.

Sugita, S. 1990. Palynological records of forest disturbance and development in the Mountain Meadows Watershed, Mt. Rainier, Washington. Unpublished Ph. D. dissertation, University of Washington, Seattle, Washington, U.S.A. 213 pp.

Sumida, A. 1993. Growth of tree species in a broadleaved secondary forest as related to the light environments of crowns. *Journal of the Japanese Forestry Society* 75:278-286.

Sundstrom, F. J., and S. R. Pezeshki. 1988. Reduction of *Capsicum annuum* L. growth and seed quality by soil wooding, *HortScience* 23:574-576.

Sutton, R. F. 1969. *Form and development of conifer root systems,* Commonwealth Forest Bureau Technical Communication No. 7, Oxford, England, 131 pp.

Sutton, W. R. J. 1968. Initial spacing and financial return of *Pinus radiata* on coastal sands, *New Zealand Journal of forest Science* 13:203-219.

Swaine, J. M. 1933. The relation of insect activities to forest development as exemplified in the forests of eastern North America, *Science in Agriculture* 14:8-31.

Swank, W. T., and H. T. Schreuder. 1974. Comparison of three methods of estimating surface area and biomass for a forest of young eastern white pine, *Forest Science* 20:91-100.

Swanson, F. J. 1979. Geomorphology and ecosystems, in: *Forests: Fresh Perspectives from Ecosystem Analysis,* Proceedings of the 40th Annual Biology Colloquium, Oregon State University Press, Corvallis, pp. 159-170.

Swanson, F. J., G. W. Lienkaemper, and J. R. Sedell. 1976. History, physical effects, and management implications of large organic debris in western Oregon streams, USDA Forest Service General Technical Report PNW-56, 15 pp.

Swift, L. W., Jr., and K. R. Knoerr. 1973. Estimating solar radiation on mountain slopes, *Agricultural Meteorology* 12:329-336.

Switzer, G. L., and L. E. Nelson. 1972. Nutrient accumulation and cycling in loblolly pine plantation ecosystems: The first twenty years, *Soil Science Society of America Proceedings* 36:143-147.

Switzer, G. L., L. E. Nelsen, and W. H. Smith. 1966. The characterization of dry matter and nitrogen accumulation by loblolly pine *(Pinus taeda* L.), *Soil Science Society of America Proceedings* 30:114-119.

Tadaki, Y. 1964. Effect of thinning of stem volume yield studied with competition-density effect, *Government Forest Experiment Station of Tokyo Bulletin* 154:1-19.

Tadaki, Y. 1966. Some discussions on the leaf biomass of forest stands and trees, *Tokyo Forest Experiment Station Bulletin* 184:135-161.

Tadaki, Y. 1977. Leaf biomass, in T. Shidei and T. Kira (eds.), *Primary Productivity of Japanese Forests,* JIBP Synthesis 16, University of Tokyo Press, Tokyo, pp. 39-44.

Tallis, J. H. 1991. *Plant Community History: Long-term Changes in Plant Distribution and Diversity.* Chapman and Hall, New York. 398 pp.

Tanaka, H., M. Shibata, and T. Nakashizuka. 1995. A mechanistic approach to wind dispersal and its role in tree population dynamics. *Journal of Sustainable Forestry* (In press).

Tanaka, N. 1988. Tree invasion into patchy dwarf-bamboo thickets within a climax beech-fir forst in Japan. IN (M. J. A. Werger and J. H. Willems, editors) *Diversity and pattern in plant communities.* SPB Academic Publishing, The Hague, The Netherlands: 253-261.

Tande, G. F. 1977. Forest fire history around Jasper Townsite, Jasper National Park, Alberta, unpublished master of science thesis, University of Alberta, Edmonton, Alberta, Canada.

Tannehill, I. R. 1944. *Hurricanes: Their Nature and History,* 5th ed., Princeton University Press, Princeton, New Jersey.

Tansley, A. G. 1935. The use and abuse of vegetational concepts and terms, *Ecology* 16:284-307.

Tappeiner, J. C. II, and J. A. Helms. 1971. Natural regeneration of Douglas-fir and white fir on exposed sites in the Sierra Nevada of California, *American Midland Naturalist* 86:358-370.

Tappeiner, J. C. II, and P. M. McDonald. 1984. Development of tanoak understories in conifer stands, *Canadian Journals of Forest Research* 14:271-277.

Tappeiner, J. C. II, J. F. Bell, and J. D. Brodie. 1982. *Response of young Douglas-fir to 16 years of intensive thinning,* Forestry Research Laboratory of Oregon State University, Corvallis, Research Bulletin 38, 17 pp.

Tappeiner, J. C. II, T. B. Harrington, and J. D. Walstad. 1984. Predicting recovery of tanoak *(Lithocarpus densiflorus)* and Pacific Madrone *(Arbutus menziesii)* after cutting or burning, *Weed Science* 32:413-417.

Tappeiner, J. C. II. 1971. Invasion and development of beaked hazel in red pine stands in northern Minnesota, *Ecology* 52:514-519.

Tarbox, E. E., and P. M. Reed. 1924. *Quality and growth of white pine as influenced by density, site, and associated species,* The Harvard Forest, Harvard University, Harvard Forest Bulletin No. 7, 30 pp.

Tarrant R. F., and J. M. Trappe. 1971. The role of *Alnus* in improving the forest environment *Plant Soil* (special volume):335-348.

Taylor, A. R. 1969. Tree-bole ignition in superimposed lightning scars, USDA Forest Service Research Note INT-90, 4 pp.

Taylor, W. P. 1934. Significance of extreme or intermittent conditions in distribution of species and management of natural resources, with a restatement of Liebig's law of the minimum, *Ecology* 15:274-379.

Teensma, P. D. A., J. T. Rienstra, and M. A. Yeiter. 1991. Preliminary reconstruction and analysis of change in forest stand age classes of the Oregon Coast Range from 1850 to 1940. U. S. Department of the Interior, Bureau of Land Management Technical Note T/N OR-9 (Filing Code: 9217). 9 pp.

Tesch, S. D. 1981. Comparative stand development in an old-growth Douglas-fir *(Pseudotsuga menziesii* var. *glauca)* forest in western Montana, *Canadian Journal of Forest Research* 11:82-89.

Teskey, R. O., and T. M. Hinckley. 1977. *Impact of water level changes on woody riparian and wetland communities. Volume I: Plant and soil responses,* U.S. Government Printing Office, Washington, D.C. FWS/OBS-77/58, 30 pp.

Teskey, R. O., and T. M. Hinckley. 1978. *Impact of water level changes on woody riparian and wetland communities. Volume V. Northern forest region,* U.S. Government Printing Office, Washington, D.C. FWS/OBS-78/88, 54 pp.

The Timberman. 1921. Storm inflicts heavy damage to Olympic Peninsula timber. *The*

Timberman, February, 1921. pp. 33-35.

Thibalut, J. R., J. A. Forten, and V. A. Smirnoff. 1982. In vitro allelopathic inhibition of nitrification by balsam poplar and balsam fir, *American Journal of Botany* **69**:676-679.

Thomas, J. W. (ed.). 1979. *Wildlife habitats in managed forests: The blue mountains of Oregon and Washington,* USDA Forest Service. Agricultural Handbook No. 553, 512 pp.

Thomas, T. L., and J. K. Agee. 1986. Prescribed fire effects on mixed conifer forest structure at Crater Lake, Oregon, *Canadian Journal of Forest Research* **16**:1082-1087.

Thompson, D. Q., and R. H. Smith. 1970. The forest primeval in the Northeast—A great myth? *Proceedings of the Tall Timbers Fire Ecology Conference* **10**:255-265.

Thompson, J. N. 1980. Treefalls and colonization patterns of temperate forest herbs, *American Midland Naturalist* **104**:176-183.

Thompson, J. N., and M. F. Wilson. 1978. Disturbance and the dispersal of fleshy fruits, *Science* **200**:1161-1163.

Thompson, L. 1916. *To the American Indian,* Cummins Print Shop, Eureka, California, 214 pp.

Thompson, R. L., and W. W. Baker. 1971. A survey of red-cockaded woodpecker nesting habitat requirement, in *The Ecology and Management of the Red-Cockaded Woodpecker,* USDI Department of the Interior, Bureau of Sports Fisheries and Wildlife, pp. 170-188.

Thornburgh, D. A. 1969. Dynamics of the true fir-hemlock forests of the west slope of the Washington Cascade Range, unpublished Ph.D. thesis, University of Washington, Seattle, 210 pp.

Thornthwaite, C. W. 1941. Atlas of climatic types in the United States: 1900-1939, USDA Soil Conservation Service, Miscellaneous Publication No. 421, 55 pp.

Thornthwaite, C. W., and J. R. Mather. 1955. The water balance, *Publications in Climatology,* No. 8, 104 pp.

Tinnin, R. O. 1981. Interrelationships between *Arceuthobium* and its hosts, *The American Midland Naturalist* **106**:126-132.

Tinnin, R. O., F. G. Hawksworth, and D. M. Knutson. 1982. Witches' broom formation in conifers infected by *Arceuthobium* spp.: An example of parasitic impact upon community dynamics, *American Midland Naturalist* **107**:351-359.

Titterington, R. W., H. S. Crawford, and B. N. Burgason. 1979. Songbird responses to commercial clear-cutting in Maine spruce-fir forets. *Journal of Wildlife Management* **43**:602-609.

Tomaselli, R. 1977. The degradation of the Mediterranean maquis, *Ambio* **6**:356-362.

Toole, E. H., and V. K. Toole. 1961. Until time and place are suitable, in A. Stefferud (ed.) *Seeds; the Yearbook of Agriculture, 1961,* U.S. Department of Agriculture, pp. 99-106.

Toole, E. R. 1959. Decay after fire injury to southern bottom-land hardwoods, U.S. Department of Agriculture Technical Bulletin 1189, 25 pp.

Toole, E. R. 1961. Fire scar development, *Southern Lumberman* **203**:111-112.

Toole, E. R. 1965a. Deterioration of hardwood logging slash in the South, U.S. Department of Agriculture Technical Bulletin 1328, 27 pp.

Toole, E. R. 1965b. Fire damage to commercial hardwoods in southern bottom lands, *Tall Timbers Fire Ecology Conference* **4**:144-151.

Toole, E. R. 1966. Decay of hardwood logging slash, Mississippi State University, Mississippi, Agricultural Experiment Station Information Sheet 932, 2 pp.

Toole, E. R., and W. M. Broadfoot. 1959. Sweetgum blight as related to alluvial soils of the Mississippi River floodplain, *Forest Science* **5**:1-9.

Toole, V. K., E. H. Toole, S. B. Hendricks, H. A. Borthwick, and A. G. Snow, Jr. 1961. Responses of seeds of *Pinus virginiana* to light, *Plant Physiology* **36**:285-290.

Toumey, J. W. 1929a. Initial root habit in American trees and its bearing on regeneration, *Proceedings of the International Congress of Plant Sciences* **1**:713-728.

Toumey, J. W. 1929b. The vegetation of the forest floor: Light versus soil moisture, *Proceedings of the International Congress of Plant Sciences* **1**:575-590.

Toumey, J. W., and R. Kienholz. 1931. *Trenched plots under forest canopies*, Yale University School of Forestry and Environmental Studies Bulletin No. 30, 31 pp.

Tranquillini, W. 1955. Die Bedeutung des Lichtes und der Temperatur fur die Kohlensaureassimilation von *Pinus cembra* Jungwuchs an einem hochalpinen Standort, *Planta* **46**:154-178.

Trenbath, B. R. 1974. Biomass productivity of mixtures, *Advances in Agronomy* **26**:177-210.

Trimble, G. R., Jr. 1963. Cull development under all-aged management of hardwood stands, USDA Forest Service Research Paper NE-10, 10 pp.

Trimble, G. R., Jr. 1965. Species composition changes under individual tree selection cutting in cove hardwoods, USDA Forest Service Research Note NE-30, 6 pp.

Trimble, G. R., Jr. 1968a. Form recovery by understory sugar maple under uneven-aged management, USDA Forest Service Research Note NE-89, 8 pp.

Trimble, G. R., Jr. 1968b. Multiple stems and single stems of red oak give same site index, *Journal of Forestry* **66**:198.

Trimble, G. R., Jr. 1970. Twenty years of intensive uneven-aged management: Effect on growth, yield, and species composition in two hardwood stands in West Virginia, USDA Forest Service Research Paper NE-154, 12 pp.

Trimble, G. R., Jr. 1973. The regeneration of Central Appalachian hardwoods with emphasis on the effect of site quality and harvesting practice, USDA Forest Service Research Paper NE-282.

Trimble, G. R., Jr., and E. H. Tryon. 1966. Crown encroachment into openings cut in Appalachian hardwood stands, *Journal of Forestry* **64**:104-108.

Trimble, G. R., Jr., and G. Hart. 1961. Appraisal of early reproduction after cutting in Northern Appalachian hardwood stands, USDA Forest Service Northeastern Forest Experiment Station Paper No. 162, 21 pp.

Trimble, G. R., Jr., and H. C. Smith. 1976. Stand structure and stocking control in Appalachian mixed hardwoods, USDA Forest Service Research Paper NE-340, 10 pp.

Troedsson, T., and W. H. Lyford. 1973. Biological disturbance and small-scale spatial variations in a forested soil near Garpenberg, Sweden. *Studia Forestalia Suecica* 109, 23 pp.

Troup, R. S. 1928. *Silvicultural systems*. Clarendon Press, Oxford.

Troup, R. S. 1952. *Silvicultural systems*. Second edition. Oxford University Press, London.

Trousdell, K. B. 1955. Loblolly pine seed tree mortality, USDA Forest Service Southeastern Forest Experiment Station Paper 61, 11 pp.

Tryon, E. H., and G. R. Trimble, Jr. 1969. Effect of distance from stand border on height of hardwood reproduction in openings, *Proceedings of the West Virginia Academy of Sciences* **41**:125-133.

Tryon, E. H., and K. L. Carvell. 1958. Regeneration under oak stands, West Virginia University, Morgantown, West Virginia, Agricultural Experiment Station Bulletin

424T (November), 22 pp.

Tubbs, C. H. 1968. Natural regeneration, in *Sugar Maple Conference*, sponsored by USDA Forest Service, School of Forestry and Wood Products, Michigan Technical University, Houghton, Michigan, pp. 75-81.

Tubbs, C. H. 1973. Allelopathic relationship between yellow birch and sugar maple seedlings, *Forest Science* 19:139-145.

Tubbs, C. H. 1975. Stand composition in relation to uneven-aged silviculture, in *Uneven-Aged Silviculture in the Eastern United States*, USDA Forest Service Timber Management Research, pp. 84-99.

Tubbs, C. H. 1977. Age and structure of a northern hardwood selection forest, 1929-1976, *Journal of Forestry* 75:22-24.

Tucker, G. F., and W. H. Emmingham. 1977. Morphological changes in leaves of residual western hemlock after clear and shelterwood cutting, *Forest Science* 23:195-203.

Tucker, G. F., T. M. Hinckley, J. Leverenz, and S. Jiang. 1987. Adjustments of foliar morphology in the acclimation of understory Pacific silver fir following clearcutting, *Forest Ecology and Management* 21:249-268.

Turner, J. 1977. Effect of nitrogen availability on nitrogen cycling in a Douglas-fir stand, *Forest Science* 23:307-316.

Turner, K. B. 1952. The relation of mortality of balsam fir *Abies balsamea* (L.) Mill., caused by the spruce budworm, *Choristoneura fumiferana* Clem., to forest composition in the Algoma forest of Ontario, Canada. Canadian Department of Agriculture, Publication No. 875. 107 pp.

Twery, M. J. 1987. Changes in the vertical distribution of xylem production in hardwoods defoliated by gypsy moth. Unpublished Ph. D. dissertation. Yale University, New Haven, Connecticut, U.S.A. 96 pp.

Uemura, S., S. Tsuda, and S. Hasegawa. 1990. Effects of fire on the vegetation of Siberian taiga predominated by *Larix dahurica*. *Canadian Journal of Forest Research* 20:547-553.

Ugolini, F. C. 1982. Soil development in the Abies amabilis zone of the central Cascades, Washington, in C. D. Oliver and R. M. Kenady (eds.), *Biology and Management of True Fir in the Pacific Northwest*, College of Forest Resources, University of Washington, Seattle, Institute of Forest Resources Contribution Number 45, pp. 165-176.

Ugolini, F. C., and D. H. Mann. 1979. Biopedological origin of peatlands in South East Alaska, *Nature* 281(5730):366-368.

Ugolini, F. C., B. Bormann, and F. H. Bowers. 1989. The role of tree windthrow on forest soil development in southeast Alaska (submitted for publication).

Uhl, C., and I. C. Guimaraes Vieira. 1989. Ecological impacts of selective logging in the Brazilian Amazon: a case study from the Paragominas Region of the state of Para. *Biotropica* 21(2):98-106.

Uhl, C., and P. G. Murphy. 1982. Composition, structure, and regeneration of a tierra firme forest in the Amazonian Basin of Venezuela, *Tropical Ecology* 22:219-237.

Uhl, C., D. Nepstad, R. Buschbacher, K. Clark, B. Kauffman, and S. Subler. 1990. Studies of ecosystem response to natural and anthropogenic disturbances provide guidelines for designing sustainable land-use systems in Amazonia. IN (A. B. Anderson, editor) *Alternatives to deforestation: steps toward sustainable use of the Amazon rain forest*. Columbia University Press, New York. pp. 24-42.

Uhl, Christopher, Daniel Nepstad, Robert Buschbacher, Kathleen Clark, Boone Kauffman, and Scott Subler. 1990. Studies of ecosystem response to natural and anthropogenic disturbances provide guidelines for designing sustainable land-use systems in Amazonia. IN A. B. Anderson, editor. *Alternatives to Deforestation: Steps Toward Sustainable Use of the Amazon Rain Forest.* Columbia University Press, New York. pp. 24-42.

United States Congress 1994. Hearing on Bill H. R. 1164:"Forest biodiversity and clearcutting prohibition act of 1993. " Hearing before the Subcommittee on specialty crops and natural resources of the Committee on Agriculture, House of Representatives. 103rd Congress. October 28, 1993; Serial No. 103-43: United States Government Printing Office, Washington, D. C., U.S.A.

USDA Forest Service. 1929. *Volume, yield, and stand tables for second-growth southern pines,* Miscellaneous Publication 50, 202 pp.

USDA Forest Service. 1964. *Winds over wild lands—a guide for forest management,* USDA Forest Service, Agricultural Handbook No. 272, 33 pp.

USDA Forest Service. 1981. Fire *regimes and ecosystem properties,* USDA Forest Service General Technical Report WO-26, 594 pp.

USDA Forest Service. 1982. An analysis of the timber situation in the United States: 1952-2030, Forest Resource Report 23, 499 pp.

USDA Forest Service. 1988. The South's fourth forest: Alternatives for the future, USDA Forest Service Report No. 24, 512 pp.

USDC *Bureau of the Census. 1950-1980.* United States Department of Commerce, U.S. Census of the Population, U.S. Government Printing Office, Washington, D.C.

Utzig, G., and L. Herring. 1974. Factors significant to high elevation forest management: A report emphasizing soil stability and stand regeneration, British Columbia Forestry Service Research Division, Vancouver, British Columbia, Canada.

Vaartaja, O. 1959. Evidence of photoperiodic ecotypes in trees, *Ecological Monographs* 29:91-111.

Van Cleve, K., and C. T. Dyrness. 1983. Introduction and overview of a multidisciplinary research project: the structure and function of a black spruce (*Picea mariana*) forest in relation to other fire-affected taiga ecosystems. *Canadian Journal of Forest Research* 13:675-702.

Van der Wal, D. W. 1985. A proposed concept of branch autonomy and non-ring production in branches of Douglas-fir and grand firs unpublished master's thesis, University of Washington, Seattle, 96 pp.

Van Dersal, W. R. 1938. Utilization of woody plants as food by wildlife, in *Third North American Wildlife Conference Transactions,* pp. 768-775.

Van Devender, T. R., and W. G. Spaulding. 1979. Development of vegetation and climate in the southwestern United States, *Science* 204:701-710.

Van Devender, T. R., W. G. Spaulding, and A. M. Phillips. 1979. Late pleistocene plant communities in the Guadalupe Mountains, Culberson County, Texas, in H. H. Genoways and R. J. Baker (eds.), *Biological Investigations in the Guadalupe Mountains National Park, Texas,* Proceedings and Transactions Series No. 4, USDI National Park Service, pp. 13-30.

Van Laar, A. 1973. *Needle biomass, growth, and growth distribution of Pinus radiata in South Africa in relation to pruning and thinning,* Forschungsberichte, Forstliche Forschungsanstalt, Munchen, 258 pp.

Van Wagner, C. E. 1978. Age-class distribution and the forest fire cycle, *Canadian Journal of Forest Research* 8:220-227.

Vander Wall, S. B. V., and R. P. Balda. 1977. Coadaptation of the Clark's nutcracker and the Pinon pine for efficient seed harvest and dispersal, *Ecological Monographs* 47:89-111.

Vandermeer, John. 1982. The evolution of mutualism, in B. Shorrocks (ed.), *Evolutionary Ecology: The 23rd Symposium of the British Ecological Society, Leeds,* Blackwell, London, pp. 221-232.

VanderPijl, L. 1972. *Principles of Dispersal in Higher Plants,* Springer-Verlag, New York, 162 pp.

Vankat, J. L., W. H. Blackwell, Jr., and W. E. Hopkins. 1975. The dynamics of Hueston Woods and a review of the question of the successional status of the southern beech-maple forest, *Castanea* 40:290-308.

Veblen T. T., and D. H. Ashton. 1978. Catastrophic influences on the vegetation of the valdivian Andes, Chile, *Vegatatio* 36:149-167.

Veblen, T. T. 1982. Growth patterns of chusquea bamboos in the understory of Chilean nothofagus forests and their influences in forest dynamics, *Bulletin of the Torrey Botanical Club* 109:474-487.

Veblen, T. T. 1985. Forest development in tree-fall gaps in the temperate rain forests of Chile. NGR/Spring, pp. 162-183.

Veblen, T. T. 1986. Treefalls and the coexistence of conifers in subalpine forests of the central rockies, *Ecology* 67:644-649.

Veblen, T. T. and D. C. Lorenz. 1988. Recent vegetation changes along the forest-steppe ecotone of northern Patagonia, *Annals of the Association of American Geographers* 78:93-111.

Veblen, T. T., A. T. Veblin, and F. M. Schlegel. 1979. Understory patterns in mixed evergreen-deciduous nothofagus forests in Chile, *Journal of Ecology* 67:809-823.

Veblen, T. T., and D. C. Lorenz. 1986. Anthropogenic disturbance and recovery patterns in montane forests, Colorado front range, *Physical Geography* 7:1-24.

Veblen, T. T., and G. H. Stewart. 1980. Comparison of forest structure and regeneration on Bench and Stewart Islands, New Zealand, *New Zealand Journal of Ecology* 3:50-68.

Veblen, T. T., and G. H. Stewart. 1982a. On the conifer regeneration gap in New Zealand: The dynamics of *Libocedrus bidwillii* stands on South Island, *Journal of Ecology* 70:413-436.

Veblen, T. T., and G. H. Stewart. 1982b. The effects of introduced wild animals on New Zealand forests, *Annals of the Association of American Geographers* 72:372-397.

Veblen, T. T., D. Claudio Z., F. M. Schlegel, and B. Escobar R. 1981. Forest dynamics in south-central Chile, *Journal of Biogeography* 8:211-247.

Veblen, T. T., F. M. Schlegel, and B. Escobar R. 1980. Structure and dynamics of old-growth nothofagus forests in the Valdivian Andes, Chile, *Journal of Ecology* 68:1-31.

Vezina, P. E., and M. M. Grandtner. 1965. Phenological observation of spring geophytes in Quebec, *Ecology* 46:869-872.

Viereck, L. A. 1973. Wildfire in the taiga of Alaska, *Journal of Quaternary Research* 3:465-495.

Viers, S. D. 1980a. The influence of fire in coastal redwood forests, in *Proceedings of the Fire History Workshop,* October 20-24, 1980, Tucson, Arizona, USDA Forest Service General Technical Report RM-81, pp. 93-95.

Viers, S. D. 1980b. The role of fire in northern coast redwood forest dynamics, in *Proceedings of the Conference on Scientific Research in the National Parks,* San Francisco, California, November 26-30, 1979, Volume 10: *Fire Ecology,* USDI National Park Service, pp. 190-209.

Viers, S. D. 1982. Coast redwood forest: Stand dynamics, successional status, and the role of fire, in J. E. Means (ed.), *Forest Succession and Stand Development*

Research in the Northwest, Oregon State University Forest Research Laboratory, Corvallis, pp. 119-141.

Vines, R. A. 1977. *The Trees of East Texas,* University of Texas Press, Austin, Texas.

Vitousek, P. M., and W. A. Reiners. 1975. Ecosystem succession and nutrient retention: A hypothesis, *BioScience* **25**:376-381.

Vitousek, P. M., J. R. Gosz, C. C. Grier, J. M. Melillo, W. A. Reiners, and R. L. Todd. 1979. Nitrate losses from disturbed ecosystems, *Science* **204**:469-474.

Vogt, K. A., R. L. Edmonds, and C. C. Crier. 1981a. Dynamics of ectomycorrhizae in Abies amabilis stands: The role of *Cenococcum graniforme (*Sow.*)* Fred and Winge, *Holarctic Ecology* **4**:167-173.

Vogt, K. A., R. L. Edmonds, and C. C. Crier. 1981b. Seasonal changes in biomass and vertical distribution of mycorrhizal and fibrous-textured conifer fine roots in 23- and 180-year-old subalpine Abies amabilis stands, *Canadian Journal of Forest Research* **11**:223-229.

Voigt, G. K. 1968. Variation in nutrient uptake by trees, in *Forest Fertilization: Theory and Practice,* Tennessee Valley Authority, Muscle Shoals, Alabama, pp. 20-27.

Von Bertalanffy, L. 1972. The history and status of general systems theory, in G. J. Klir (ed.), *Trends in General Systems Theory,* Wiley, London, pp. 21-41.

Wagener, W. W. 1961. Past fire incidence in Sierra Nevada forests, *Journal of Forestry* **59**:739-748.

Wagner, R. G. 1980. Natural regeneration at the edge of an Abies amabilis zone clearcut on the west slope of the Central Washington Cascades, unpublished master of science thesis, University of Washington, Seattle, 148 pp.

Wahlenberg, W. G. 1930. Effect of ceanothus brush on western yellow pine plantations in the northern Rocky Mountains, *Journal of Agricultural Research* **41**:601-612.

Wahlenberg, W. G. 1934. Dense stands of reproduction and stunted individual seedlings of longleaf pine, USDA Forest Service Southern Forest Experiment Station Occasional Paper 39, 16 pp.

Wahlenberg, W. G. 1946. *Longleaf Pine,* Charles Lanthrop Pack Forestry Foundation and USDA Forest Service, Washington, D.C., 429 pp.

Wahlenberg, W. G. 1948. Effect of forest shade and openings on loblolly pine seedlings, *Journal of Forestry* **46**:832-834.

Wahlenberg, W. G. 1950. Epicormic branching of young yellow poplar, *Journal of Forestry* **48**:417-419.

Wales, B. A. 1972. Vegetation analysis of north and south edges of a mature oak-hickory forest, *Ecological Monographs* **42**:451-471.

Walkinshaw, L. H. 1983. *Kirtland's Warbler,* Cranbrook Institute of Science, Bloomfield Hills, Michigan, 207 pp.

Wallerstein, I. 1983. An agenda for world-systems analysis, in W. R. Thompson (ed.), *Contending Approaches to World System Analysis,* Sage, Beverly Hills, California, pp. 299-301.

Wallis, G. W., and D. J. Morrison. 1975. Root rot and stem decay following commercial thinning in western hemlock and guidelines for reducing losses, *Forestry Chronicle* **51**:203-207.

Wallmo, O. C., and J. W. Schoen. 1980. Response of deer to secondary forest succession in Southeast Alaska, *Forest Science* **26**:448-462.

Walter, H. 1971. *Ecology of Tropical and Subtropical Vegetation* (D. Mueller-Dombois, translator, and J. H. Burnett, ed.), Oliver and Boyd, Edinburgh, 539 pp.

Walter, H. 1973a. *Vegetation of the Earth and Ecological Systems of the Geobiosphere,* 2d ed., (trans. in 1979 by Joy Wieser), Springer-Verlag, New York, 274 pp.

Walter, H. 1973b. *Vegetation of the Earth: In Relation to Climate and the Eco-Physiological Conditions*, Springer-Verlag, New York, 237 pp.

Walter, Heinrich. 1973. *Vegetation of the Earth and Ecological Systems of the Geo-biosphere*. Second Edition. (translated in 1979 by Joy Wieser), Springer-Verlag, New York. 274 pp.

Walters, Carl. 1986. *Adaptive Management of Renewable Resources*. Macmillan Publishing Company. 374 pp.

Walters, D. T., and A. R. Gilmore. 1976. Allelopathic effects of fescue on the growth of sweetgum, *Journal of Chemical Ecology* 2:469-479.

Walters, M. A., R. O. Teskey, and T. M. Hinckley. 1980. Impact of water level changes on woody riparian and wetland communities: VIII. Pacific Northwest and Rocky Mountain regions. FWS/OBS-78/94, 47 pp.

Wampler, M. 1993. Growth of Douglas-fir under partial overstory retention. Unpublished Master of Science thesis, University of Washington, College of Forest Resources, Seattle, Washington, U.S.A. 97 pp.

Ward, H. A., and L. H. McCormick. 1982. Eastern hemlock allelopathy, *Forest Science* 28:681-686.

Ward, R. T. 1961. Some aspects of the regeneration habits of American beech, *Ecology* 42:828-832.

Wardle, R. 1959. The regeneration of *Fraxinus excelsior* in woods with a field layer of *Mercurialis perennis*, *Journal of Ecology* 47:483-497.

Waring, R. H. 1983. Estimating forest growth and efficiency in relation to canopy leaf area, in A. MacFadyen and E. D. Ford (eds.), *Advances in Ecological Research*, Academic, New York, pp. 327-354.

Waring, R. H. and W. H. Schlesinger. 1985. *Forest Ecosystems: Concepts and Management*, Academic, New York, 340 pp.

Waring, R. H., and G. B. Pitman. 1980. A simple model of host resistance to bark beetles, Forest Research Laboratory Research Note 65, School of Forestry, Oregon State University, Corvallis, 2 pp.

Waring, R. H., and G. B. Pitman. 1983. Physiological stress in lodgepole pine as a precursor for mountain pine beetle attack, *Zeitschrift fuer Angewandte Ichthyologie* 96:265-270.

Waring, R. H., and J. F. Franklin. 1979. Evergreen coniferous forests of the Pacific Northwest, *Science* 204:1380-1386.

Waring, R. H., H. L. Golz, C. C. Grier, and M. L. Plummer. 1977. Evaluating stem conducting tissue as an estimator of leaf area in four woody angiosperms, *Canadian Jounal of Botany* 55:1474-1477.

Waring, R. H., K. Newman, and J . Bell. 1981. Efficiency of tree crowns and stemwood production at different canopy leaf densities, *Forestry* 54:129-137.

Waring, R. H., W. G. Thies, and D. Muscato. 1980. Stem growth per unit of leaf area: A measure of tree vigor, *Forest Science* 26:112-117.

Warrack G. C. 1959. Forecast of yield in relation to thinning regimes in Douglas-fir, British Columbia Forest Service, Technical Publication T.51, 56 pp.

Warrack, G. C. 1949. Treatment of red alder in the coastal region of British Columbia, British Columbia Forestry Service Research Note 12, 7 pp.

Warrack, G. C. 1952. Comparative observations of the changes in classes in a thinned and natural stand of immature Douglas-fir, *Forestry Chronicle* 28:46-56.

Washington State Library. 1975. Natural Disasters in the State of Washington, 2d ed., compiled for the Washington State Department of Emergency Services by the Washington State Library, Olympia, 251 pp.

Watanabe, S. 1985. Dendrosociological studiesof the natural forest in Hokkaido, Japan. Hokkaido Regional Forestyr Office. 196 pp. (In Japanese; original not seen. Cited from Osawa 1992.)

Watson, D. J. 1947. Comparative physiological studies on the growth of field crops: I. Variation in net assimilation rate and leaf area between species and varieties, and within and between years, *Annals of Botany* 11:41-76.

Watt, A. S. 1947. Pattern and process in the plant community, *Journal of Ecology* 35:1-22.

Watt, A. S. 1970. Contributions to the ecology of bracken *(Pteridium aquilinum):* VII. Bracken and litter: The cycle of change, *New Phytologist* 69:431-449.

Weatherhead, D. J. 1974. Fuel treatment systems for partially cut stands, Progress Report, ED & T 2423, USDA Forest Service Equipment Development Center, Missoula, Montana, 45 pp.

Weaver, H. 1951. Fire as an ecological factor in the southwestern ponderosa pine forests, *Journal of Forestry* 49:93-98.

Weaver, H. 1959. Ecological changes in the ponderosa pine forest of the Warm Springs Indian Reservation in Oregon, *Journal of Forestry* 57:15-20.

Weaver, H. 1967. Fire and its relationship to ponderosa pine, *Tall Timbers Fire Ecology Conference Proceedings* 7:127-149.

Webb, D. P. 1977. Root regeneration and bud dormancy of sugar maple, silver maple, and white ash: Effects of chilling, *Forest Science* 23:474-483.

Webb, L. J. 1958. Cyclones as an ecological factor in tropical lowland rain forest, North Queensland, *Australian Journal of Botany* 6:220-228.

Webb, L. J. 1978. A structural comparison of New Zealand and south-east Australian rain forests and their tropical affinities, *Australian Journal of Ecology* 3:7-21.

Webb, L. J., J. G. Tracey, and W. T. Williams. 1972. Regeneration and pattern in the subtropical rain forest, *Journal of Ecology* 60:675-695.

Weber, C. D., Jr. 1983. Height growth patterns in a juvenile Douglas-fir stand, effects of planting site, microtopography and lammas occurrence. unpublished master of science thesis, College of Forest Resources, University of Washington, Seattle, 65 pp.

Weigand, J. F., R. W. Haynes, and J. L. Mikowski (compilers). 1993. *High quality forestry: the idea of long rotations.* University of Washington, College of Forest Resources, Center for International Trade in Forest Products CINTRAFOR Special Paper 15. 265 pp.

Wein, R. W., and D. A. MacLean. 1983. *The role of fire in northern circumpolar ecosystems.* SCOPE 18, John Wiley and Sons, Chichester, England. 322 pp.

Wein, R. W., and J. M. Moore. 1977. Fire history and rotations in the New Brunswick Acadian Forest, *Canadian Journal of Forest Research* 7:285-294.

Weiner, J. 1984. Neighborhood interference amongst *Pinus rigida* individuals, *Journal of Ecology* 72:183-195.

Weiner, J., and S. C. Thomas. 1986. Size variability and competition in plant monocultures, *Oikos* 47:211-222.

Weller, D. E. 1987. A reevaluation of the $-3/2$ power rule of plant self-thinning, *Ecological Monographs* 57:23-43.

Wellner, C. A. 1970. IN *The role of fire in the intermountain West.* Proceedings of a symposium sponsored by the Intermountain Fire Research Council, Missoula, Montana. pp. 42-64.

Wells, B. W., and I. V. Shunk. 1931. The vegetation and habitat factors of the coarser sands of the North Carolina Coastal Plain: An ecological study, *Ecological*

Monographs 1:465-520.

Wells, C. G., R. E. Campbell, L. F. DeBano, C. E. Lewis, R. L. Fredrickson, E. C. Franklin, R. C. Froehlich, and P. H. Dunn. 1979. Effects of fire on soils, USDA Forest Service General Technical Report WO-7, 34 pp.

Wells, J. R., P. W. Thompson, and G. P. Fons. 1983. Some Michigan upper peninsular big trees and their age estimates, *Michigan Botanist* 23:5-10.

Wendell, G. W. 1977. Longevity of black cherry, wild grape, and sassafrass seed in the forest floor, USDA Forest Service Research Paper NE-375.

Wenger, K. F. 1954. The stimulation of loblolly pine seed trees by preharvest release, *Journal of Forestry* 52:115-118.

Wenger, K. F., T. C. Evans, T. Lotti, R. W. Cooper, and E. V. Brender. 1958. The relation of growth to stand density in natural loblolly pine stands, USDA Forest Service Southeastern Forest Experiment Station Paper 97, 10 pp.

Went, F. W., G. Juaren, and M. C. Juaren. 1952. Fire and biotic factors affecting germination, *Ecology* 33:351-359.

West, D. C., H. H. Shugart, Jr., and D. B. Botkin (eds.). 1981a. *Forest Succession: Concepts and Applications*, Springer-Verlag, New York, 517 pp.

West, D. C., H. H. Shugart, Jr., and J. W. Ranney. 1981b. Population structure of forests over a large area, *Forest Science* 27:701-710.

West, P. W., and C. J. Borough. 1983. Tree suppression and the self-thinning rule in a monoculture of *Pin us Radiata* D. Don. *Annals of Botany* 52:149-158.

Westoby, M. 1977. Self-thinning driven by leaf area not by weight, *Nature* 265:330-331.

Westoby, M. 1981. The place of the self-thinning rule in population dynamics, *American Naturalist* 118:581-587.

Westoby, M. 1984. The self-thinning rule, *Advances in Ecological Research* 14:167-225.

Westoby, M., and J. Howell. 1981. Sclf-thinning: The effect of shading on glass-house populations of silver beet *(Beta vulgaris)*, *Journal of Ecology* 69:359-365.

Westveld, M. 1954. A budworm vigor-resistance classification for spruce and balsam fir, *Journal of Forestry* 52:11-24.

Whipple, S. A. 1980. Population dispersion patterns of trees in a southern Louisiana hardwood forest, *Bulletin of the Torrey Botanical Club* 107:71-76.

White, A. S. 1985. Presettlement regeneration patterns in a southwestern ponderosa pine stand, *Ecology* 66:589-594.

White, J. 1980. Demographic factors in populations of plants, in O. T. Solbrig (ed.), *Demography and Evolution in Plant Populations*, University of California Press, Berkeley, pp. 2148.

White, J. 1981. The allometric interpretation of the self-thinning rule, *Journal of Theoretical Biology* 89:475-500.

White, J. 1985. The thinning rule and its application to mixtures of plant populations, in J. White (ed.), *Studies on Plant Demography: A Festschrift for John Harper*, Academic, New York, pp. 291-309.

White, J., and J. L. Harper. 1970. Correlated changes in plant size and number in plant populations, *Journal of Ecology* 58:467-485.

White, P. S. 1979. Pattern, process, and natural disturbance in vegetation, *Botanical Review* 45:229-299.

White, P. S., and S. T. A. Pickett. 1985. Natural disturbance and patch dynamics: An introduction, in S. T. A. Pickett and P. S. White (eds.), *The Ecology of Natural Disturbance and Patch Dynamics*, Academic, New York, pp. 3-13.

Whitehead, A. N. 1925. *Science and the Modern World: Lowell Lectures, 1925*, Free Press, Macmillan, New York, 121 pp.

Whitehead, D. 1978. The estimation of foliage area from sapwood basal area in Scots pine, *Forestry* 51:137-149.

Whitehead, D. R. 1972. Developmental and environmental history of the Dismal Swamp, *Ecological Monographs* 42:301-315.

Whitford, P. B. 1949. Distribution of woodland plants in relation to succession and clonal growth, *Ecology* 30:199-208.

Whitmore, T. C. 1974. Change with time and the role of cyclones in tropical rain forest of Kolombangara, Solomon Islands, Oxford University Commonwealth Forestry Institute, Institute Paper No. 46 (cited by White 1979).

Whitmore, T. C. 1975. *Tropical Rain Forests of the Far East.* Clarendon Press, Oxford. 282 pp.

Whitmore, T. C. 1978. Gaps in the forest canopy, in P. B. Tomlinson and M. H. Zimmerman (eds.), *Tropical Trees as Living Systems,* Cambridge University Press, New York, pp. 639-655.

Whitmore, T. C. 1984. *Tropical Rain Forests of the Far East,* Clarendon, New York.

Whitmore, T. C. 1989. Canopy gaps and the two major groups of forest trees. *Ecology* 70(3):536-538.

Whitney, G. G., and J. R. Runkle. 1981. Edge versus age effects in the development of a beech-maple forest, *Oikos* 37:377-381.

Whittaker, R. H. 1953. A consideration of climax theory: the climax as a population and pattern. *Ecological Monographs* 23:41-78.

Whittaker, R. H. 1965. Dominance and diversity in land plant communities, *Science* 147:250-260.

Whittaker, R. H. 1975. *Communities and Ecosystems,* 2d ed., Macmillan, New York, 385 pp.

Whittaker, R. H., and S. A. Levin. 1977. The role of mosaic phenomena in natural communities, *Theoretical Population Biology* 12:117-139.

Wierman, C. A., and C. D. Oliver. 1979. Crown stratification by species in even-aged mixed stands of Douglas-fir/western hemlock, *Canadian Journal of Forest Research* 9:1-9.

Wilde, S. A., and H. F. Scholz. 1934. Subdivision of the upper peninsula experimental forest on the basis of soils and vegetation, *Soil Science* 38:383-399.

Wiley, K. N. 1965. Effects of the October 12,1962, windstorm on permanent growth plots in southwest Washington. *Weyerhaeuser Forestry Paper* No. 7, Weyerhaeuser Forestry Research Division, Centralia, Washington, 11 pp.

Wiley, K. N. 1968. Thinning of western hemlock: A literature review, *Weyerhaeuser Forestry Paper* No. 12. Weyerhaeuser Forestry Research Center, Centralia, Washington, 12 pp.

Wiley, K. N., and M. D. Murray. 1974. Ten-year growth and yield of Douglas-fir following stocking control, *Weyerhaeuser Forestry Paper* No. 14, Weyerhaeuser Forestry Research Center, Centralia, Washington, 88 pp.

Williams-Linera, G. 1990. Vegetation structure and environmental conditionis of forest edges in Panama. *Journal of Ecology* 78:356-373.

Williamson, A. W. 1913. Cottonwood in the Mississippi Valley, U.S. Department of Agriculture Bulletin No. 24, 62 pp.

Williamson, G. B. 1975. Pattern and seral composition in an old-growth beech-maple forest, *Ecology* 56:727-731.

Williamson, R. L. 1973. Results of shelterwood harvesting of Douglas-fir in the Cascades of western Oregon. USDA Forest Service Research Paper PNW-161. 13 pp.

Williamson, R. L. 1978. Natural regeneration of western hemlock, in W. A. Atkinson and R. J. Zasoski (eds.), *Proceedings of the Western Hemlock Management Conference,* College of Forest Resources, University of Washington, Seattle, Institute of Forest Products Contribution No. 34, pp. 166-169.

Williamson, R. L., and G. R. Staebler. 1965. A cooperative level-of-growing-stock study in Douglas-fir. Report No. 1: Description of study and existing study areas, USDA Forest Service Research Paper PNW-111, 12 pp.

Williamson, R. L., and R. O. Curtis. 1984. I,evels-of-growing-stock cooperative study in Douglas-fir: Report No. 7: Preliminary results Stampede Creek, and some comparisons with Iron Creek and Hoskins, USDA Forest Service Research Paper PNW-323, 42 pp.

Willis, G. L., and J. A. Johnson. 1978. Regeneration of yellow birch following selective cutting of old-growth northern hardwoods, Michigan Technological University, *Ford Forestry Center Research Note* 26, 13 pp.

Wilson, B. F. 1964. Structure and growth of woody roots of *Acer rubrum* L., The Harvard Forest, Harvard University, *Harvard Forest Paper* 11, 14 pp.

Wilson, B. F. 1966. Development of the shoot system of *Acer rubrum* L., The Harvard Forest, Harvard University, *Harvard Forest Paper* 14, 21 pp.

Wilson, B. F. 1968. Red maple stump sprouts: Development the first year, The Harvard Forest, Harvard University, *Harvard Forest Paper* 18, 10 pp.

Wilson, B. F. 1970a. Evidence for injury as a cause of tree root branching, *Canadian Journal of Botany* 48:1497-1498.

Wilson, B. F. 1970b. *The Growing Tree,* University of Massachusetts Press, Amherst, 152 pp.

Wilson, C. C. 1948. Fog and atmospheric carbon dioxide as related to apparent photosynthetic rate of some broadleaf evergreens, *Ecology* 29:507-508.

Wilson, D. S. 1976. Evolution on the level of communities, *Science* 192:1358-1360.

Wilson, D. S. 1980. *The Natural Selection of Populations and Communities,* Benjamin Cummings, Menlo Park, California, 186 pp.

Wilson, E. O., and F. M. Peters (editors). 1988. *Biodiversity.* National Academy Press, Washington, D. C. 521 pp.

Wilson, F. G. 1946. Numerical expression of stocking in terms of height, *Journal of Forestry* 44:758-761.

Wilson, F. G. 1951. Control of growing stock in even-aged stands of conifers, *Journal of Forestry* 49:692-695.

Wilson, R. W. J. 1953. How second-growth northern hardwoods develop after thinning, USDA Forest Service Northeastern Forest Experiment Station, Station Paper 62, 12 pp.

Wisler, C. O., and E. F. Brater. 1959. *Hydrology,* John Wiley, London, 408 pp.

Wolfe, M. L., and F-C. Berg. 1988. Deer and forestry in Germany: half a century after Leopold. *Journal of Forestry* 86(5):25-31.

Wood, O. M. 1939. Persistence of stems per sprout clump in oak coppice stands of southern New Jersey, *Journal of Forestry* 37:269-270.

Woodman, J. N. 1971. Variation of net photosynthesis within the crown of a large forest-grown conifer, *Photosynthetica* 5:50-54.

Woods, F. W., and R. K. Shanks. 1959. Natural replacement of chestnut by other species in the Great Smokey Mountains National Park, *Ecology* 40:349-361.

Woodward, F. I. 1987. *Climate and Plant Distribution,* Cambridge University Press, New York, 174 pp.

Woodwell, G. M. 1970. Effects of pollution on the structure and physiology of ecosystems, *Science* **168**:429-433.

Woodwell, G. M., and D. B. Botkin. 1970. Metabolism of terrestrial ecosystems by gas exchange techniques: The Brookhaven approach, in D. E. Reichle (ed.), *Analysis of Temperate Forest Ecosystems,* Springer-Verlag, New York, pp. 73-95.

Worley, D., and H. A. Meyer. 1951. The structure of an uneven-aged white pine-hemlock-hardwood forest in Luzerne County, Pennsylvania, *Pennsylvania State Forest School Research Paper* 15, Pennsylvania State University, 4 pp.

Worthington, N. P. 1953. Reproduction following small group cuttings in virgin Douglas-fir. USDA Forest Service PNW Research Note 84. 5 pp.

Worthington, N. P. 1953. Reproduction following small group cuttings in virgin Douglas-fir, USDA Forest Service Pacific Northwest Experiment Station Research Note 84, 5 pp.

Worthington, N. P. 1961. Some observations on yield and early thinning in a Douglas-fir plantation, *Journal of Forestry* **59**:331-334.

Worthington, N. P., D. L. Reukema, and G. R. Staebler. 1962. Some effects of thinning on increment in Douglas-fir in western Washington, *Journal of Forestry* **60**:115-119.

Wright, E., and L. A. Isaac. 1956. Decay following logging injury to western hemlock, Sitka spruce, and true fir, USDA Technical Bulletin 1148.

Wright, H. A., and A. W. Bailey. 1982. *Fire Ecology: United States and Southern Canada,* John Wiley, New York, 501 pp.

Wright, H. E., Jr. 1968. The roles of pine and spruce in the forest history of Minnesota and adjacent areas, *Ecology* **49**:937-955.

Wright, H. E., Jr. 1974. Landscape development, forest fires and wilderness management, *Science* **186**:487-495.

Wright, H. E., Jr., and M. L. Heinselman (eds.). 1973. The ecological role of fire in natural conifer forests of western and northern America, *Journal of Quaternary Research* **3**:317-513.

Wright, H. E., Jr., J. E. Kutzbach, T. Webb III, W. F. Ruddiman, F. A. Street-Perrott, and P. J. Bartlein, editors. 1993. *Global climates since the last glacial maximum.* University of Minnesota Press, Minneapolis, Minnesota, U.S.A. 569 pp.

Wright, K. H., and G. M. Harvey. 1967. The deterioration of beetle-killed Douglas-fir in western Oregon and Washington, USDA Forest Service Research Paper PNW-50, 20 pp.

Yaltirik, F. 1984. *Turkiye Meseleri.* Tarim Orman ve Koyisleri Bakanligi, Genel Mudurlugi Yayini, 64 pp.

Yamaguchi, D. K. 1983. New tree-ring dates for recent eruptions of Mt. St. Helens, *Journal of Quaternary Research* **20**:246-250.

Yamaguchi, D. K. 1986. The development of old-growth Douglas-fir forests northeast of Mount St. Helens, Washington, following an A.D. 1480 eruption, unpublished Ph.D. dissertation, University of Washington, Seattle, 100 pp.

Yamaguchi, D. K. 1993. Forest history, Mount St. Helens. *National Geographic Research and Exploration* **9**:294-325.

Yamakura, T. 1987. An empirical approach to the analysis of forest stratification: I. Proposed graphical method derived by using an empirical distribution function, *Botanical Magazine of Tokyo* **100**:109-128.

Yamamoto, S. 1988. Seedling recruitment of *Chamaecyparis obtusa* and *Sciadopitys verticillata* in different microenvironments in an old-growth *Sciadopitys verticillata* forest. *The Botanical Magazine of Tokyo* **101**:61-71.

Yamamoto, S. 1992. Preliminary studies on the species composition, stand structure and regeneration characteristics of an old-growth *Pseudotsuga japonica* forest at the Sannoko on the Kii Peninsula, southwestern Japan. *Japanese Journal of Forest Environment* **34**:50-58.

Yamamoto, S. 1993. Structure and dynamics of an old-growth Chamaecyparis forest in the Akasawa Forest Reserve, Kiso district, central Japan. *Japanese Journal of Forest Environment* **35**:32-41.

Yamamoto, S., and T. Tsutsumi. 1979. The population dynamics of naturally regenerated hinoki seedlings in artificial hinoki stands (I) The mortality process of firest year seedlings. *Journal of the Japanese Forestry Society* **61**: 287-293. (Abstract and figures in English)

Yamamoto, S., and T. Tsutsumi. 1984. The population dynamics of naturally regenerated hinoki seedlings in artificial hinoki stands (III) Dynamics of the seed population on the forest floor. *Journal of the Japanese Forestry Society* **66**:483-490. (Abstract and figures in English)

Yamamoto, S., and T. Tsutsumi. 1985. The population dynamics of naturally regenerated hinoki seedlings in artificial hinoki stands (V) The development and survivorship of seedlingson the various kinds of forest floors. *Journal of the Japanese Forestry Society* **67**:427-433. (Abstract and figures in English)

Yamamoto, S., and T. Tsutsumi. 1985. The population dynamics of naturally regenerated hinoki seedlings in artificial hinoki stands (VI) Mortality factors of seedlings and their working mechanisms. *Journal of the Japanese Forestry Society* **67**:486-494. (Abstract and figures in English)

Yarranton, M., and G. A. Yarranton. 1975. Demography of a jackpine stand, *Canadian Journal of Botany* **53**:310-314.

Yeaton, R. I. 1978. Competition and spacing in plant communities: Differential mortality of white pine *(Pinus strobus* L.) in a New England woodlot, *American Midland Naturalist* **100**:285-293.

Yocum, H. A. 1971. Releasing shortleaf pines increases cone and seed production, USDA Forest Service Research Note SO-125, 2 pp.

Yoda, K. 1974. Three-dimensional distribution of light intensity in a tropical rain forest of West Malaysia, *Japanese Journal of Ecology* **24**:247-254.

Yoda, K., T. Kira, and K. Hozumi. 1957. Intraspecific competition among higher plants: IX. Further analysis of the competitive interactions between adjacent individuals, *Journal of the Institute of Polytechnics, Osaka City University* **8**:161-178.

Yoda, K., T. Kira, H. Ogawa, and K. Hozumi. 1963. Intraspecific competition among higher plants: II. Self-thinning in overcrowded pure stands under cultivated and natural conditions, *Journal of Biology of Osaka City University* **14**:107-129.

Yoneda, T. 1975a. Studies on the rate of decay of wood litter on the forest floor: I. Some physical properties of decaying wood, *Japanese Journal of Ecology* **25**:40-46.

Yoneda, T. 1975b. Studies on the rate of decay of wood litter on the forest floor: II. Dry weight loss and CO_2 evolution of decaying wood, *Japanese Journal of Ecology* **25**:132-140.

Young, M. R. 1992. Conserving insect communities in mixed woodlands. IN (M. G. R. Cannell, D. C. Malcolm, and P. A. Robertson, editors) *The ecology of mixed-species stands of trees.* Blackwood Scientific Publications, London. 277-296.

Young, T. P., and S. P. Hubbell. 1991. Crown asymmetry, treefalls, and repeat disturbances of broad-leaved forest gaps. *Ecology* **72**(4):1464-1471.

Young, T. P., and V. Perkocha. 1994. Treefalls, crown asymmetry, and buttresses. *Journal of Ecology* **82**:319-324.

SCIENTIFIC NAMES OF PLANT SPECIES DISCUSSED

acacias	*Acacia* spp.
alder, red	*Alnus rubra* Bong.
ash, European	*Fraxinus excelsior* L.
ash, green	*Fraxinus pennsylvanica* Marsh.
ash, white	*Fraxinus americana* L.
aspen, bigtooth	*Populus grandidentata* Michx
aspen, quaking or trembling	*Populus tremuloides* Michx
baldcypress	*Taxodium distichum* (L.) Rich.
beech, American	*Fagus grandifolia* Ehrh
beech, European	*Fagus sylvatica* L.
beech, Japanese	*Fagus crenata* Bl.
beech, oriental	*Fagus orientalis* Lipsky
birch, black	*Betula lenta* L.
birch, gray (white)	*Betula populifolia* Marsh.
birch, paper (white)	*Betula papyrifera* Marsh.
birch, river	*Betula nigra* L.
birch, yellow	*Betula alleghaniensis* Britton
blackberry	*Rubus* spp.
boxelder	*Acer negundo* L.
cascara	*Rhamnus purshiana* DC.
cedar, Alaskan yellow-	*Chamaecyparis nootkatensis* (D.Don) Spach
cedar, eastern red-	*Juniperus virginiana* L.
cedar, of Lebanon	*Cedrus libani* Loud.
cedar, western red-	*Thuja plicata* Donn
cherry, bitter	*Prunus emarginata* Dougl. ex Eaton
cherry, black	*Prunus serotina* Ehrh.
cherry, choke	*Prunus uirginiana* L.
cherry, pin	*Prunus pensyvuanica* L. f.
chestnut, American	*Castanea dentata (Marsh.)* Borkh.
chestnut, Chinese	*Castanea mollissima* Bl.
chestnut, Japanese	*Castanea crenata* (Sieb. and Zucc.)
cottonwood, black	*Populus balsamifera* ssp. *trichocarpa* (Torr. & Gray) Brayshaw
cottonwood, eastern	*Populus deltoides* Bartr.

cypress (listed under baldcypress)

dogwood, flowering — *Cornus florida* L.

Douglas-fir — *Pseudotsuga menziesii* (Mirb.) Franco

elm, American — *Ulmus americana* L.

elm, winged — *Ulmus alata* Michx.

eucalyptus — *Eucalyptus* spp.

firs, true — *Abies* spp.

fir balsam — *Abies balsamea* (L.) Mill.

fir, European silver — *Abies alba* A.

fir, grand — *Abies grandis* (Dougl.) Lindl.

fir, noble — *Abies procera* Rehd.

fir, Pacific silver — *Abies amabilis* (Dougl.) Forbes

fir (Turkey) — *Abies bornmuellerana* Mattf.

Franklinia bush (or tree) — *Franklinia alatamaha* Marsh.

gum, black — *Nyssa sylvatica* Marsh

hackberry — *Celtis occidentalis* L.

hemlock, eastern — *Tsuga canadensis* (L.) Carr.

hemlock, mountain — *Tsuga mertensiana* (Bong.) Carr.

hemlock, western — *Tsuga heterophylla* (Raf.) Sarg.

hickories — *Carya* spp.

hickory, water — *Carya aquatica* (Michx. f.) Nutt.

hophornbeam or ironwood — *Ostrya virginiana* (Mill.) K. Koch

honeysuckle, Japanese — *Lonicera japonica* Thunb.

Juniper, common — *Juniperus communis* L.

larch, western — *Larix occidentalis* Nutt.

locust, honey — *Gleditsia triacanthos* L.

magnolia, evergreen — *Magnolia grandiflora* L.

maple, red — *Acer rubrum* L.

maple, striped — *Acer pensylvanicum* L.

maple, sugar — *Acer saccharum* Marsh.

maple, vine — *Acer circinatum* Pursh.

oaks — *Quercus* spp.

oak, black — *Quercus velutina* Lam.

oak, cherrybark — *Quercus falcata* var. *pagodaefolia* Ell.

oak, chestnut — *Quercus prinus* L.

oak, Nuttall — *Quercus nuttallii* Palmer

oak, overcup — *Quercus lyrata* Walt.

oak, (northern) red — *Quercus rubra* L.

oak, southern red — *Quercus falcata* Michx.

oak, swamp white — *Quercus bicolor* Willd.

oak, water — *Quercus nigra* L.

oak, white — *Quercus alba* L.

oak, willow	*Quercus phellos* L.
pecan, sweet	*Carya illinoensis* (Wangenh.) K. Koch
pecan, swamp	*Carya aquatica* (Michx.f.) Nutt.
pines	*Pinus* spp.
pine, caribbean	*Pinus caribaea* Morelet
pine, eastern white	*Pinus strobus* L.
pine, jack	*Pinus banksiana* Ten.
pine, loblolly*	*Pinus taeda* L.
pine, lodgepole	*Pinus contorta* Dougl.
pine, longleaf*	*Pinus palustris* Mill.
pine, radiata	*Pinus radiata* D. Don
pine, pitch	*Pinus rigida* Mill.
pine, ponderosa	*Pinus ponderosa* Laws.
pine, red	*Pinus resinosa* Ait.
pine, sand*	*Pinus clausa* (Chapm.) Vasey
pine, Scots	*Pinus sylvestris* L.
pine, shortleaf*	*Pinus echinata* Mill.
pine, slash*	*Pinus elliottii* Engelm.
pine, spruce*	*Pinus glabra* Walt.
pine, Swiss	*Pinus cembra* L.
pine, Virginia*	*Pinus virginiana* Mill.
pine, western white	*Pinus monticola* Dougl.
poplar, yellow or tulip	*Lirodendron tulipifera* L.
raspberry	*Rubus* spp.
redbud	*Cercis canadensis* L.
redwood (coastal)	*Sequoia sempervirens* (D. Don) Endl.
salal	*Gaultheria shallon* Pursh
sassafras	*Sassafras albidum* (Nutt.) Nees
sequoia	*Sequoiadendron giganteum* (Lindl.) Bucholz
sourwood	*Oxydendrum arboreum* (L.) DC
spruces	*Picea* spp.
spruce, black	*Picea mariana* (Mill.) B.S.P.
spruce, Norway	*Picea abies* (L.) Karst.
spruce, red	*Picea rubens* Sarg.
spruce, sitka	*Picea sitchensis* (Bong.) Carr.
spruce, white	*Picea glauca* (Moench) Voss
sugarberry	*Celtis laevigata* Willd.
sumac	*Rhus glabra* L.
sweetgum	*Liquidambar styraciflua* L.
sycamore, American	*Plantanus occidentalis* L.
swamp tupelo	*Nyssa sylvatica* var. *biflora* (Walt.) Sarg.
walnut, black	*Juglans nigra* L.
willows	*Salix* spp.
willow, black	*Salix nigra* Marsh.

SOURCE NOTES

Page 3: Figure 1.1
From "Stand development—its uses and methods of study," C. D. Oliver in *Forest Development and Stand Development Research in the Northwest*, J.E. Means (ed.), Oregon State University, Corvallis, OR, pp. 100-112.

Page 13: Figure 2.1
Reprinted from the *Journal of Forestry* 78 published by the Society of American Foresters, 5400 Grosvenor Lane, Bethesda, MD 20814-2198. Not for further reproduction.

Page 31: Figure 2.6
From *The Visualization of Climate*, J.R.Eagleman, copyright © 1976, by J.R.Eagleman, Lexington Books, Inc., Lexington, MA, pp. 37-66. Reprinted by permission.

Page 53: Figure 3.4
"Figure based on information from Dr. James W. Goudie, British Columbia Ministry of Forestry, Victoria, British Columbia, Canada. Used with permission."

Page 61: Figure 3.8A
"Reprinted from *Forest Science* 2, published by the Society of American Foresters, 5400 Grosvenor Lane, Bethesda, MD 20814-2198. Not for further reproduction."

Page 61: Figure 3.8B
From *The Development of Norther Red Oak (Quercus rubra L.) in Mixed Species, Even-Aged Stands in Central New England*, C. D. Oliver, copyright © 1975 by C. D. Oliver. Ph.D.Dissertation, Yale University, New Haven CT, pg. 84. Reprinted by permission.

Page 62: Figure 3.9
From *Site Index for Douglas-fir in the Pacific Northwest*, J.E. King, copyright © 1966 by Weyerhaeuser Company, Weyerhaeuser Forestry Paper 8, Centralia, WA, pg. 23. Reprinted by permission. From *Volume, Stand, and Yield Tables for Second-Growth Southern Pines*, USDA Forest Service, Miscellaneous Publication 50. 1929.

Page 63: Figure 3.10
Reprinted from *Forest Science Monograph* 17, published by the Society of American Foresters, 5400 Grosvenor Lane, Bethesda, MD 20814-2198. Not for further reproduction.

Page 64: Figure 3.11
From "Growth Response of Western Hemlocks after Release," C. D. Oliver, in W. Atkinson and R. J. Zasoski (eds.) *Western Hemlock Management*, 1976, University of Washington Institute of Forest Products Contribution 34, pg. 269. Reprinted by permission.

Page 66: Figure 3.12
Figure 3.12A From *Development of the Root System of Northern Red Oak (Quercus rubra L.)*, by W. H. Lyford, copyright © 1980 by The Harvard Forest, Harvard University, Petersham, MA. Harvard Forest Paper 21, pg. 15. Reprinted by permission. Figure 3.12B from *Studies of the Root Systems of Deciduous Trees)*, by B. B. Stout, copyright © 1956 by The Harvard Forest, Harvard University, Petersham, MA. Black Rock Forest Bulletin 15, pg. 33. Reprinted by permission.

Page 81: **Figure 3.17**
From "Silvicultural Costs of Growing Douglas-fir at various spacings, sites, and wood qualities," C. D. Oliver, W. Michalec, L. DuVall, C. A. Wierman, and H. Oswald, in C. D. Oliver, D. Hanley, and J. Johnson (eds.) *Douglas-fir: Stand Management for the Future*, copyright © 1986 by University of Washington, College of Forest Resources, Seattle, WA, pg. 135. Reprinted by permission.

Page 96: **Figure 4.2A**
From *Studies of Douglas-fir Seed Flight in Southern Western Washington*, James Dick, copyright © 1955 by Weyerhaeuser Company, Weyerhaeuser Timber Company Forestry Research Note 12, Centralia, WA, pg. 2. Reprinted by permission.

Page 105: **Figure 4.3A&B**
Figure 4.3A from *Climates of the United States*, J. J. Baldwin, 1973, U. S. Department of Commerce, Washington, D. C. Figure 4.3B from *Fire weather*, M. J. Schroeder and C. C. Buck, 1970, U. S. D. A. Forest Service, Agriculture Handbook 360.

Page 106: **Figure 4.4**
From *Introduction to Wildland Fire*, S. J. Pyne, copyright © 1984 by John Wiley & Sons, New York. Reprinted by permission of John Wiley & Sons, Inc.

Page 112: **Figure 4.5**
From *Climates of the United States*, J. J. Baldwin, 1973, U. S. Department of Commerce, Washington, D. C.

Page 115: **Figure 4.6**
From *Hydrology*, C. O. Wisler and E. F. Brater, copyright © 1959 by John Wiley & Sons, New York. Reprinted by permission of John Wiley & Sons, Inc.

Page 121: **Figure 4.7**
From *Field Guide to Landforms in the United States*, J. A. Shimer, copyright © 1972 by Macmillan Publishing Co., New York. pg. 169. Reprinted by permission.

Page 146: **Figure 5.1**
From "Vegetation Science Concepts. I. Initial floristic composition", by F. E. Egler, copyright by Kluwer Academic Publishers, Dordrecht, The Netherlands. *Vegetatio* 4, pp. 414-415. Reprinted by permission of Kluwer Academic Publishers.

Page 149: **Figure 5.2**
From "Forest development in North America following major disturbances", by C. D. Oliver, copyright © 1981 by Elsevier Science Publishers B.V., Amsterdam, The Netherlands. *Forest Ecology and Management* 3, pp. 156. Reprinted by permission.

Page 168: **Figure 5.6**
From "Forest development in North America following disturbances", by C. D. Oliver, copyright © 1981 by Elsevier Science Publishers B.V., Amsterdam, The Netherlands. *Forest Ecology and Management* 3, pp. 164. Reprinted by permission.

Page 178: **Figure 6.2A&B**
From "The development of northern red oak in mixed stands," by C. D. Oliver, copyright © 1978 by Yale University School of Forestry and Environmental Studies, New Haven, Connecticut. *Yale University School of Forestry and Environmental Studies Bulletin* 91, pg. 19. Reprinted by permission.

Page 179: **Figure 6.2C**
From "Distinctive features of the northwestern coniferous forest: development, structure, and function," by J. F. Franklin and R. H. Waring (ed.) Forests: *Fresh Perspectives from Ecosystem Analysis*, Proceedings of the 40th Annual Biology Colloquium, copyright © 1979 by Oregon State University Press, Corvallis. pg. 75. Reprinted by permission.

Page 181: **Figure 6.3A**
From "Reconstruction of a mixed-species forest in central New England," by C. D. Oliver and E. P. Stephens, copyright © 1977 by The Ecological Society of America, Washington, D. C. *Ecology* 58, pp. 562-572. Reprinted by permission.

Page 187: **Figure 6.4A&B**
From "Narrow soils and intricate soil patterns in southern New England," by W. H. Lyford, copyright © 1974 by Elsevier Science Publishers B. V., Amsterdam, The Netherlands. *Geoderma* 11, pg. 205. Reprinted by permission. Figure 6.4B unpublished figures from Professor J. D. Hodges, School of Forest Resources, Mississippi State University, Mississippi. Printed by permission.

Page 188: **Figure 6.4C**
From *Effects of windthrow on soil properties and spatial variability in southeast Alaska*, F. H. Bowers, copyright © 1987 by F. H. Bowers. Ph.D. Dissertation, University of Washington, Seattle, WA, pg. 129. Reprinted by permission.

Page 189: **Figure 6.5A**
From "Disturbance of forest soil resulting from the uprooting of trees," by H. J. Lutz, copyright © 1940 by Yale University School of Forestry and Environmental Studies, New Haven, Connecticut. *Yale Forestry School Bulletin* 45, pg. 43. Reprinted by permission. Figure 6.5A also from "The influence of tree roots on soil morphology," by H. .J. Lutz and F. S. Griswold, copyright © 1939 by *Americal Journal of Science*, Yale University, New Haven, Connecticut. American Journal of Science 237, pg. 396. Reprinted by permission of *American Journal of Science*.

Page 215: **Figure 8.1**
From "Concepts of thinning regimes," C. D. Oliver, K. L. O'Hara, G. McFadden, and I. Nagame, in C. D. Oliver, D. Hanley, and J. Johnson (eds.) *Douglas-fir: Stand Management for the Future*, and copyright © 1986 by University of Washington, College of Forest Resources, Seattle, WA, pp. 246-257. Reprinted by permission.

Page 219: **Figure 8.3**
Photo from Dr. K. J. Mitchell, British Columbia Ministry of Forestry, Victoria, British Columbia, Canada. Used with permission.

Page 231: **Figure 8.9**
From "Concepts of thinning regimes," C. D. Oliver, K. L. O'Hara, G. McFadden, and I. Nagame, in C. D. Oliver, D. Hanley, and J. Johnson (eds.) *Douglas-fir: Stand Management for the Future*, and copyright © 1986 by University of Washington, College of Forest Resources, Seattle, WA, pp. 246-257. Reprinted by permission.

Page 239: **Figure 9.1**
From "The development of northern red oak in mixed stands," by C. D. Oliver, copyright © 1978 by Yale University School of Forestry and Environmental Studies, New Haven, Connecticut. *Yale University School of Forestry and Environmental Studies Bulletin* 91, pg. 39. Reprinted by permission.

Page 303: **Figure 13.4**
From *Development and growth of even-aged and multi-aged mixed stands of Douglas-fir and grand fir on the east slopes of the Washington Cascades*, copyright © 1982 by B. C. Larson. Ph.D. Dissertation, University of Washington, Seattle, WA, pp. 219. Reprinted by permission.

Page 304: **Figure 13.5**
From "Reconstruction of mixed hemlock-spruce stands in coastal southeast Alaska," by R. L. Deal, C. D. Oliver, and B. T. Bormann, copyright © 1991 by National Research Council of Canada. *Canadian Journal of Forest Research* 21, pg. 650. Reprinted by permission.

Page 306: **Figure 13.7**
From "Reconstruction of mixed hemlock-spruce stands in coastal southeast Alaska," by R. L. Deal, C. D. Oliver, and B. T. Bormann, copyright © 1991 by National Research Council of Canada. *Canadian Journal of Forest Research* 21, pg. 651. Reprinted by permission.

Page 307: **Figure 13.8**
From "Reconstruction of a mixed-species forest in central New England," by C. D. Oliver and E. P. Stephens, copyright © 1977 by The Ecological Society of America, Washington, D. C. *Ecology* 58, pp. 562-572. Reprinted by permission.

Page 309: **Figure 13.9**
From "Reconstruction of a mixed-species forest in central New England," by C. D. Oliver and E. P. Stephens, copyright © 1977 by The Ecological Society of America, Washington, D. C. *Ecology* 58, pp. 562-572. Reprinted by permission.

Page 338: **Figure 15.1C**
From "Concepts of thinning regimes," C. D. Oliver, K. L. O'Hara, G. McFadden, and I. Nagame, in C. D. Oliver, D. Hanley, and J. Johnson (eds.) *Douglas-fir: Stand Management for the Future*, and copyright © 1986 by University of Washington, College of Forest Resources, Seattle, WA, pp. 246-257. Reprinted by permission.

Page 348: **Figure 15.5**
Reprinted from *Forest Science* 25 published by the Society of American Foresters, 5400 Grosvenor Lane, Bethesda, MD   20814-2198. Not for further reproduction.

Page 348: **Figure 15.6**
Reprinted from *Forest Science* 34 published by the Society of American Foresters, 5400 Grosvenor Lane, Bethesda, MD   20814-2198. Not for further reproduction.

Page 357: **Figure 15.11**
From "Some factors that will affect the next generation of forest growth models, by R. A. Leary, in A. R. Ek, S. R. Shirley, and T. E. Burk (eds.), *Forest Growth Modelling and Prediction*. U. S. D. A. Forest Service General Technical Report NE-143, pp. 22-32.

Page 359: **Figure 15.12**
From *Development and growth of even-aged and multi-aged mixed stands of Douglas-fir and grand fir on the east slopes of the Washington Cascades*, copyright © 1982 by B. C. Larson. Ph.D. Dissertation, University of Washington, Seattle, WA, pp. 219. Reprinted by permission.

Page 361: **Figure 15.13**
Reprinted from *Western Journal of Applied Forestry* 1 published by the Society of American Foresters, 5400 Grosvenor Lane, Bethesda, MD   20814-2198. Not for further reproduction.

Paage 362: **Figure 15.14**
Reprinted from *Forest Science* 25(3) published by the Society of American Foresters, 5400
Grosvenor Lane, Bethesda, MD 20814-2198. Not for further reproduction.

Page 362: **Figure 15.15**
Reprinted from *Forest Science* 13(1) published by the Society of American Foresters, 5400
Grosvenor Lane, Bethesda, MD 20814-2198. Not for further reproduction.

Page 370 & 371: **Figure 16.1**
From *The Harvard Forest Models*, copyright © 1941 by The Harvard Forest, Harvard University,
Petersham, Massachusetts. Reprinted by permission.

Page 375: **Figure 16.2**
From *Nooksack Cirque Natural History final report*, by C. D. Oliver, A. B. Adams, A.R. Weisbrod,
J. Dragovon, R. J. Zasoski, and K. Bardo. 1980. Report to the USDI Park Service, Cooperative
Park Service Unit, University of Washington, College of Forest Resources, Seattle, Washington.
pp. 127.

Page 384: **Figure 16.5**
From *Field Guide to Landforms in the United States*, J. A. Shimer, copyright © 1972 by Macmillan
Publishing Co., New York. pp. 171-174. Reprinted by permission.

Page 385: **Figure 16.6**
From *"Geomorhology and forest ecology of a mountain region in the Central Appalachians,"*
1960. U.S. Geological Survey, Professional paper 347, pp. 66. Washington, D. C.

Page 386: **Figure 16.7A**
From "Natural Vegetation," by H. L. Shantz and R. Zon. 1924. in *Atlas of American Agriculture*,
U. S. Government Printing Office, Washington, D.C.; and from "The forests of the United States,"
by D. M. Smith, in J. W. Barrett (ed.), *Regional Silviculture of the United States*, copyright © 1980
by John Wiley & Sons, New York. Reprinted by permission of John Wiley & Sons, Inc.

Page 387: **Figure 16.7B & C**
Figure 16.7B from *Field Guide to Landforms in the United States*, J. A. Shimer, copyright © 1972
by Macmillan Publishing Co., New York. pg. 158. Reprinted by permission. Figure 16.7C from
"The forests of the United States," by D. M. Smith, in J. W. Barrett (ed.), *Regional Silviculture of
the United States*, copyright © 1980 by John Wiley & Sons, New York. Reprinted by permission of
John Wiley & Sons, Inc.

Page 388: **Figure 16.8**
From *Nooksack Cirque Natural History final report*, by C. D. Oliver, A. B. Adams, A.R. Weisbrod,
J. Dragovon, R. J. Zasoski, and K. Bardo. 1980. Report to the USDI Park Service, Cooperative
Park Service Unit, University of Washington, College of Forest Resources, Seattle, Washington.
pp. 127; and from "Disturbance patterns and forest development in a recently deglaciated valley in
the northwestern Cascade Range of Washington, U.S.A.," by C. D. Oliver, A. B. Adams, and R. J.
Zasoski, copyright © 1985 by National Research Council of Canada. *Canadian Journal of Forest
Research* 15, pp. 224-225. Reprinted by permission.

INDEX

Page numbers in *italic* refer to figures; page numbers in **boldface** refer to tables.

Flower and seed production, for photosynthates, 75
Fluorescence growth, terminal, *47*, 51
Forests. See also *Plants, Species, Stands, Trees, Vegetation.*
  age distribution in, 12
  fire, *106*
  logging practices in, 11
  long-term patterns in. See *Long-term forest patterns.*
  migrations in, 11
  overview, developmental, 164
Forest floor
  disturbances to, 262
  environmental factors in, 261
  stratum, 156
  vegetation, 262-266
Forest zones, 220
Free-growing microorganisms, nutrient extraction by, 24-26
Free growth, 48, 72
  competition during, *149*, **174**, 205-206
  defined, 48, 72
  in yearly stand development, *63*, 204
  of crown, 229
Freezing, damage because of, 126
Fuel buildup
  following windstorms, 114
  in fires, 101
Functional crown, 73
Fungi, new roots and, 24

G
Gaps
  in multicohort stands, 182
  light in, 327
  prevalence of, 327-329, *329*
  within stand, 316
Gap phase development, 159, 296
Genetic variation in stand development, *215, 220, 223, 224*, 225
Geologic engineering, vs. silviculture, 6
Geomorphology in forest patterns, *182, 382-383, 384, 385, 386*
Genets in trees, 128

Glaciation, in long-term forest patterns, 377
Glaciers
  secondary disturbances because of, 121
  soil patterns and, 184
  valley, 120, 122
Grand fir, development of, 246, *247*
Grasses, growth patterns of, 198
Gravimorphism, in crown shape, 54
Grazing, wildlife. See also *Browsing, wildlife.*
  changes because of, 373
  damage because of, 125
Gross growth, formula, 332
Gross tree yield, 333
Growing space
  availability of, 36-38, *37*
  climatic fluctuations in, 32-34, 38
  daily fluctuations in, 29
  forest development and, 35
  limitations of, 28
  seasonal changes in, 29-30, *31*
  species variation and, 28
  variations in, 34-35
  weather and, 38
Growth
  disturbances to, 34-35
  efficiency of, *82*, 335
    following thinning, 352
    photosynthesis and, 80-81
  gross, 332
  height. See *Height growth.*
  limitations, 21-27
    carbon dioxide in, 27
    management in, 38-40
    nutrients in, 23-24
    oxygen in, 26
    pH in, 23
    stability and diversity in, 38
    sunlight in, 22
    temperature in, 26
    water in, 23
  measurement of, 331-333
  mortality and, in single-species single-cohort stand, 340
  old. See *Old growth.*
  preformed, 46-48

About the Authors

CHAD OLIVER is Professor of Silviculture at the University of Washington, Seattle

BRUCE LARSON is Associate Professor of Forestry at Yale University

CPSIA information can be obtained at www.ICGtesting.com
Printed in the USA
BVOW07s0449220713

326445BV00003B/10/A